2nd Annual International Conference on Mechanical Engineering and Materials (ICMEM 2019)

IOP Conference Series: Materials Science and Engineering
Volume 746

Guangzhou, China
20 - 21 December 2019

ISBN: 978-1-7138-3499-1
ISSN: 1757-8981

Printed from e-media with permission by:

Curran Associates, Inc.
57 Morehouse Lane
Red Hook, NY 12571

Some format issues inherent in the e-media version may also appear in this print version.

Institute of Physics
Dirac House, Temple Back
Bristol BS1 6BE UK

Phone: 44 1 17 929 7481
Fax: 44 1 17 920 0979

techtracking@iop.org

Additional copies of this publication are available from:

Curran Associates, Inc.
57 Morehouse Lane
Red Hook, NY 12571 USA
Phone: 845-758-0400
Fax: 845-758-2633
Email: curran@proceedings.com
Web: www.proceedings.com

TABLE OF CONTENTS

MATERIAL SCIENCE AND APPLICATION

MECHANICAL DESIGN AND TECHNOLOGY

IOP Conf. Series: Materials Science and Engineering **746** (2020) 011001 doi:10.1088/1757-899X/746/1/011001

Preface of ICMEM2019

2nd Annual International Conference on Mechanical Engineering and Materials [ICMEM2019] was held on December 20-21, 2019, Guangzhou, China. ICMEM2019 aims to bring researchers, engineers, and students to the areas of mechanical engineering and materials.

We received over 152 submissions from various parts of the world. The Technical Program Committee worked very hard to have all manuscripts reviewed before the review deadline. All the accepted papers have been submitted to strict peer-review by 2-4 referees, and selected based on originality, significance and clarity for the purpose of the conference. The conference program is extremely rich, profound and featuring high-impact presentations of selected papers and additional late-breaking contributions. We sincerely hope that the conference would not only show the participants a broad overview of the latest research results on related fields, but also provide them with a significant platform for academic connection and exchange. There are three keynote speakers: Prof. Chen-Kuei Chung, Prof. Zhiyong Hao and Dr. Weimin Huang. The keynote speakers are internationally recognized leading experts in their research fields, who have demonstrated outstanding proficiency and have achieved distinction in their profession. The proceedings would be published by IOP Conference Series: Materials Science and Engineering (MSE).

We would like to express our sincere gratitude to all the members of Technical Program Committee and organizers for their enthusiasm, time, and expertise. Our deep thanks also go to many volunteers and staffs for the long hours and hard work they have generously given to ICMEM2019. The last but not least, we would like to thank all the authors, speaker and participants for their great contributions to the success of ICMEM2019.

ICMEM2019 Organizing Committee

Peer review statement

All papers published in this volume of *IOP Conference Series: Materials Science and Engineering* have been peer reviewed through processes administered by the proceedings Editors. Reviews were conducted by expert referees to the professional and scientific standards expected of a proceedings journal published by IOP Publishing.

Thermal Effect of Short-term Overload on High-voltage Cross-linked Polyethylene (XLPE) Cable Insulation

Wenwei Zhu[1], Xiangbing Wang[1], Aiwu Liang[1], Yifeng Zhao[2], Xinghui Fan[2] and Gang Liu[2]

[1]Guangdong Power Grid Co. Ltd Grid Planning Research Centre, GuangZhou, 510000, China
[2]School of Electric Power, South China University of Technology, GuangZhou, 510000, China
m18202729711@163.com

Abstract. This paper describes changes on micro-structure and aggregation structure of cross-linked polyethylene (XLPE) with different thermal histories after different times of overheating treatment to analogize the short-term fault or emergency overload in actual operating cables. 5, 10 and 20 times overheating treatment were conducted on a spare cable and a retired cable with service years of 32, whose insulation were analyzed by the diagnostic measurements of Fourier transform infrared spectroscopy (FTIR), differential scanning calorimetry (DSC) and thermogravimetry (TG). The results show that the spare cable has presented a trend of improvement on the insulation properties after overheating treatment below 20 times, but slight deterioration occurs after 20 times overheating treatment; the retired cable with service years of 32 has presented a trend of degradation on the insulation properties at the first 5 times overheating treatment, and a severe deterioration occurs after 20 times overheating treatment.

1. Introduction

Cross-linked polyethylene (XLPE) has been widely used as an insulating material for high-voltage cables because of its excellent electrical and thermal properties [1~3]. Under its working condition, the cables are subjected to the thermal, electrical, mechanical stresses and environment factors, which eventually leads to the aging phenomenon [4]. It is well evidenced that long-term electric field and thermal field can change the morphology of XLPE insulation deeply and lead to the degradation of the insulation [5]. In the actual operation of the cable, the insulation is influenced by many factors. Therefore, the aging process of the insulation is complicated and the insulation properties can be enhanced in some certain situations.

It is clearly pointed out that the XLPE cable can be superimposed with a large current for a long-term operating temperature of 90°C in AEIC standard [6]. Generally speaking, the threshold temperatures are set as 130°C or 250°C with a limited overheating time for the cross-linked polyethylene cables. The evaluation of this standard ignore the thermal histories on the cables with different service year, or precisely aging status of the cables. On the other hand, most of the research rarely considered the situation of accelerating heat aging in the process of instantaneous changing temperature caused by short-term faults or emergency overload in actual operating cables. Therefore, it is meaningful to conduct a thermal aging to analogize the short-term fault or emergency overload in actual operating cables with different service year to analyze the resistance to overheating situation of the cables with different statuses.

This paper has focused on two cables with different aging statuses to explore the ability to resist the multiple overheating situation. 5, 10 and 20 times overheating treatment in vacuum condition were conducted on a spare cable and a retired cable with service years of 32, whose insulation were analyzed by the diagnostic measurements to observe changes on micro-structure and aggregation structure.

2. Experimental

2.1. Preparation of XLPE Samples

Two XLPE high-voltage cables produced by Japan Showa Cable Factory were selected as experimental objects, including a retired cable which have operated for 32 years and a spare cable at the same period. It can be observed that cables sheath and the insulating layer were not damaged, and there was no water permeation before the thermal aging test. The critical parameters of the cables are listed in Table 1.

Each cable insulation was peeled parallel to the conductor surface, and the tape-like XLPE peels were obtained. Peels near the inner semi-conductive layer were taken as the samples, because these positions of the XLPE endured the most severe electrical and thermal stresses.

Table 1. Critical parameters of the cables

Sample	Voltage Level (kV)	Conductor Area (mm^2)	Conductor Diameter (mm)	Insulation Diameter (mm)	Operation Years
XLPE-0	110/63.5	700	32	68	-
XLPE-30	110/63.5	700	32	68	1985-2017

2.2. Thermal Aging

The samples were divided into 4 groups and thermal aged by three different overloading times. The first three groups were treated by overheated recycling for 5 times, 10 times and 20 times. The last group was untreated group which played a role of reference. The detailed testing conditions are listed in Table 2. The testing samples were set in the aging oven under vacuum condition with the program regulating the temperature curve as shown in Figure 1 measured by the thermo-couple sensors.

The thermal aging includes three phases: transient heating phase of 80 minutes, isothermal phase of 2 hours and cooling phase of 6 hours. These phases characterize the process of superposition of additional current under short-term fault or emergency overload in actual operating condition, the process of insulations' ability to withstand high temperature and the process of withdrawing superposition current after eliminating short-term fault or emergency overload, respectively.

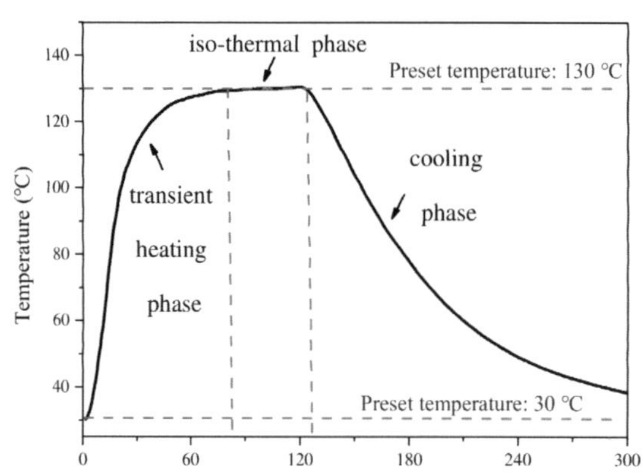

Figure 1. Temperature curve of overheating thermal aging test changed with time (one cycle)

Table 2. Introduction about the testing conditions of cable samples

Sample	Aging temperature /°C	Overloading number	Overloading time/h
A-0	130	0	0
A-5	130	5	10
A-10	130	10	20
A-20	130	20	40
B-0	130	0	0
B-5	130	5	10
B-10	130	10	20
B-20	130	20	40

2.3. Diagnostic Measurements

The Fourier transform infrared spectroscopy (FTIR), the differential scanning calorimetry (DSC) measurement, X-ray diffracttometry and the breakdown field strength measurement were adopted to analyze the properties changes on the specimens after the thermal effects.

Micro-structure change of the samples was analysed by the VERTEX 70 infrared spectrometer manufactured by German. Each sample was tested at 32 scans in the range of 400-4000 cm^{-1}.

Aggregation structure changes on the specimens were analyzed by the DSC NETZSCH-DSC 214 instrument manufactured by German. 5 mg specimens were prepared to the test with the program of two heating phases and a cooling phase under nitrogen atmosphere to avoid thermal degradation. The temperature was increased from 25 to 140°C at a rate of 10°C /min and maintained at 140°C for 5 min, and then cooled to 25°C. This scanning was repeated twice per measurement, and the first cooling and second heating phases were analyzed in this paper.

Thermo-stability on the specimens were analyzed by thermos-gravimetric analyzer (type: NETZSCH TG 209 F1 Libra) to measure the TG weight loss curve of XLPE insulation. During the experiment, the temperature was set from 30°C to 800°C at a heating rate of 10°C/min, and the weight loss of each sample with increasing temperature was observed. In a single experiment, the sample weighed 5 mg.

3. Results and Discussions

We define the sample of spare cable and the retired cable which have operated for 32 years as A and B with corresponding overheating number. That means A-0, A-5, A-10 and A-20 for the spare cable with different overheating numbers; B-0, B-5, B-10 and B-20 for the retired cable which have operated for 32 years with different overheating numbers.

3.1. Result of FTIR Spectroscopy Measurement

XLPE is a high molecular polymer whose methylene groups (-CH_2) are the most characteristic group of XLPE. This absorption peak appears at a wavelength of 720 cm^{-1}, and it is confirmed by the peaks that locate at 1471 cm^{-1}, 2856 cm^{-1} and 2937 [7]. Dicumyl peroxide (DCP) is commonly used as cross-linking agent and its by-product mainly consists of acetophenone, cumyl alcohol and α-methylstyrene. The corresponding characteristic groups of these compounds are phenylenyl, carbonyl 3371 cm^{-1}, respectively [8]. In addition, absorption peaks ranging from 1700 cm^{-1} to 1800 cm^{-1} can be considered as the thermo-oxidative products [9]. In this range, carboxylic acid absorption appears at 1701 cm^{-1}, ketone absorption locates at 1718 cm^{-1} and aldehyde absorption situates at 1741 cm^{-1}. The peak at 1635 cm^{-1} is assigned to the unsaturated groups absorption, which can indicate the molecular decomposition process. In order to judge the oxidation and decomposition situation of each samples under the different equivalent accelerated aging test condition, carbonyl index and double band index were chosen for research. The indexes are shown in Table 3.

Figure 2 (a) shows the FTIR spectrum of the spare cable after 5, 10 and 20 overheating treatment. It can be seen that obvious absorption peaks appear in the range of 1600 cm^{-1}~1760 cm^{-1} in the untreated sample XLPE-0. The main reason for this phenomenon may be attributed to the introduction of a large amount of impurities during the production of the cable. As the number of overheating treatment increases, the intensity of the absorption peaks in the range of 1500~1680 cm^{-1} have increased gradually. On the contrary, the intensity of absorption peaks in the range of 1680~1760 cm^{-1} have decreased comparing with XLPE-0. This situation is ascribed to a reduction of volatile impurities and a conversion of groups under the thermal effects.

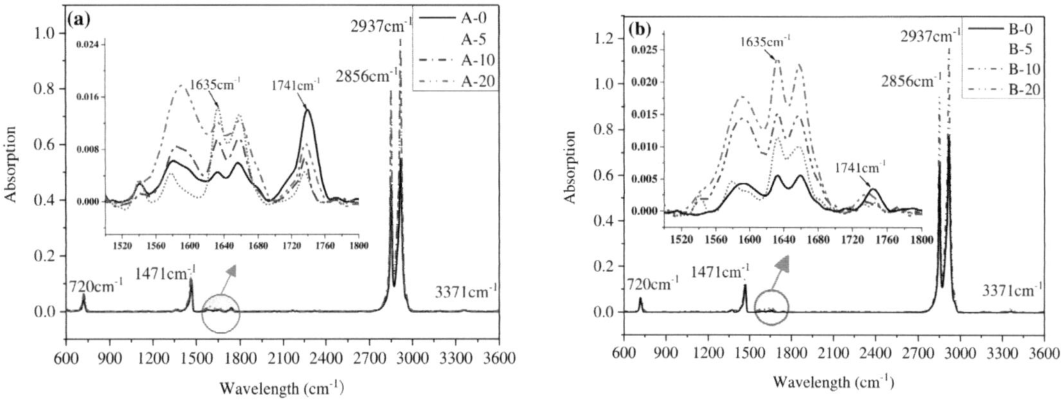

Figure 2. FTIR spectrum of each sample before and after the overheating thermal aging test. (a): the spare cable; (b): the retired cable with service years of 32

Table 3. Carbonyl and double band indexes

Sample	Carbonyl index ($\mathbf{I}_{1741/1471}$)	Double band index ($\mathbf{I}_{1635/1471}$)
A-0	0.146	0.048
A-5	0.041	0.111
A-10	0.041	0.072
A-20	0.065	0.091
B-0	0.028	0.047
B-5	0.009	0.079
B-10	0.009	0.098
B-20	0.023	0.198

From the changes on the carbonyl index and the double band index in Table 3, the following regularities can be found: the carbonyl index has decreased comprehensively after overheating treatment, but the values have ascended monotonously with the overheating times increases. It indicates the overheating treatment have promoted the process of impurities volatilization in the preliminary phase of thermal aging. Subsequently, the thermal oxidative degradation has played a dominant part gradually in the XLPE insulation with the numbers of overheating treatment increases. In the inspection on double band index, we can also find that the decomposition process of the samples to generate a certain quantity of broken molecular chains occurred distinctly after overheating treatment.

Figure 3 (b) displays the FTIR spectrum of the cable with service years of 32 after 5, 10 and 20 times of overheating treatment. We can observe that absorption peaks intensity of the untreated sample in the range of 1720~1800 cm^{-1} is weaken than the spare cable's. it may be explained by the fact that the cable with long-term operation was basically free of the volatile impurities and thermal oxidation was not distinct in a relatively low-temperature operating condition. As the number of overheating treatment increases, the intensity of the absorption peak in the range of 1500~1680 cm^{-1} increases monotonously, while the intensity of the absorption peaks in the range of 1720~1800 cm^{-1} decreases properly.

We can also figure the regularities from the changes on carbonyl index and double band index in Table 3. The double band index among the samples shows a linear ascending trend with the increasing times of overheating treatment, which indicates the crystalline morphology of the insulation has been basically stable in the long-term actual operation process, and thermal degradation dominates in the case of overheating treatment, aggravating the destruction of the molecular chains. In the aspect of carbonyl index, it declines preliminarily after 5 and 10 times of overheating treatment. Subsequently, abrupt increase occurs after 20 times of overheating treatment. The drop on carbonyl index is probably attributed to the transformation of the groups under the thermal effects. The elevation on carbonyl index indicates the dominant role of thermo-oxygen degradation.

3.2. Result of DSC Measurement

With the results of DSC measurement, two phases were analyzed, including the heating and cooling phase, which can be used to analyze crystalline structure and the ability of recrystallization among the samples. Figure 3 and Figure 4 show the heating and cooling phase thermosgrams of each sample.

Figure 3 (a) shows the DSC heat flow curve of heating phase on the spare cable samples. It can be observed that the endothermic peak of all the samples drifts to higher temperatures after the overheating treatment. But the melting range of each sample is enlarged at the same time. This phenomenon is mainly attributed to the nucleation and development of crystalline structure have not reached steady state. With the thermal effect activating the adequate motion of macromolecular chains, the scission of tie molecules traversing the amorphous regions have rearranged into the lattice to increase crystallinity [10, 11]. From the results of FTIR, it can be deduced that the generation of broken chains and small molecular chains would also promote the formation of secondary crystal to increase crystallinity.

Figure 3 (b) shows the DSC heat flow curve of heating phase on the retired cable samples. It can be seen that endothermic peaks gradually drift to a lower temperature with the numbers of overheating treatment increases. During the long-term operation of the retired cable, the development of the internal crystalline nucleus and crystals basically have reached the steady state. That means the potential of crystalline structure to recrystallize to better one has been declined. Therefore, the deterioration of the crystallite structure becomes more distinct as the overheating treatment increases with the phenomenon of the melting range expanding.

IOP Conf. Series: Materials Science and Engineering **746** (2020) 012001 doi:10.1088/1757-899X/746/1/012001

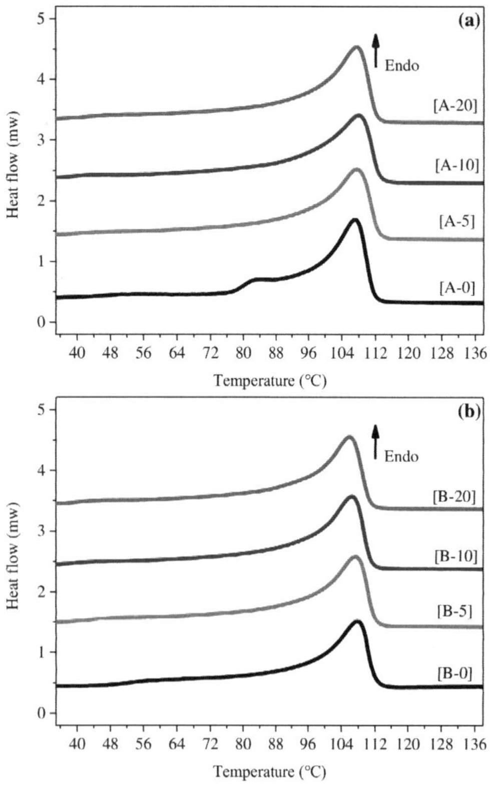

Figure 3. DSC spectrum of heating phase before and after the accelerated thermal aging test. (a): the spare cable; (b): the retired cable with service years of 32

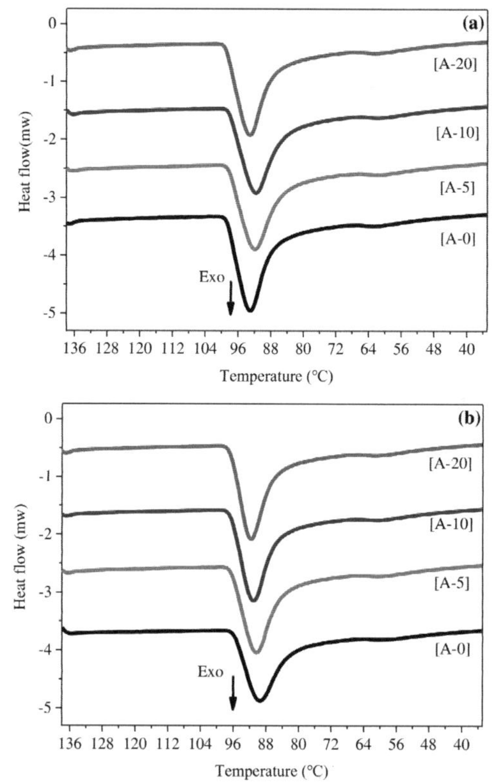

Figure 4. DSC spectrum of cooling phase before and after the accelerated thermal aging test. (a): the spare cable; (b): the retired cable with service years of 32

Figure 4 (a) shows the DSC heat flow curve of cooling phase on the spare cable samples. It can be seen that the heat flow curves drift to lower temperature after 5 times and 10 times of overheating treatment, but slightly drifts to high temperature after 20 times of overheating treatment comparing to untreated one (A-0). This phenomenon indicates the deterioration on re-crystalline structure occurs in the beginning of the overheating treatment but has been improved after 20 times of overheating treatment, which reveals the cable insulation has not reached steady state with high potential to recrystallize for better forms.

Figure 4 (b) shows the DSC heat flow curve of cooling phase on the retired cable samples. Abrupt elevation in enthalpy of crystallization and diminution in degree of super-cooling can be associated with the increase of broken chains and small molecular chains after overheating treatment, which is prone to generate large quantities of secondary crystalline structure.

3.3. Result of TG Measurement

Table 4 shows the critical parameters obtained from TG measurement, where T_0 is the initial weight loss temperature, which characterizes the first melting parts of the secondary or imperfect crystalline structure; $T_{50\%}$ is the temperature at 50% weight loss of the samples; and T_P is the peak value of DTG, which characterizes the temperature corresponding to the most violent reaction of weight loss.

It can be seen from Table 4 that the thermal stability of the insulation is enhanced after 5 times of overheating treatment of the spare cable insulation, which is manifested by an increase in T_0 and T_P. However, the thermal stability of the samples has been slightly decreased with the increasing of the overheating treatment of 10 and 20 times. We can also find that the T_P is increased distinctly after 5 and 10 times overheating treatment, which demonstrates the main crystalline structure becomes more compact.

From Table 4, there is a distinct regularity can be found. The crystalline structure of the cable with service years of 32 has basically reached the steady state. The parameters of T_0, $T_{50\%}$ and T_P exhibit a linear decrease with the overheating times increases. This phenomenon can be ascribed to the aggravation of destruction on the macromolecular chains of the samples to generate more broken chains and small molecular chains, which form a large quantity of incompact crystals with thin lamellar thickness under the overheating process. Therefore, the thermal stability of the samples has been declined with the overheating times increases.

Table 4. Critical parameters obtained from TG measurement

Sample	T_0 (°C)	$T_{50\%}$ (°C)	T_P (°C)
A-0	445.9	464.4	468.7
A-5	446.9	464.0	470.1
A-10	445.2	463.7	470.9
A-20	443.6	462.2	469.3
B-0	450.1	467.6	473.5
B-5	449.1	466.1	472.8
B-10	448.3	465.8	471.0
B-20	447.0	464.2	470.3

4. Conclusion

This paper describes changes on micro-structure and aggregation structure of a spare cable and a retired cable with service years of 32 after different times of overheating treatment to analogize the short-term fault or emergency overload in actual operating cables. Distinct differences on the changing trend between these two cables indicate the thermal history plays a decisive role on the resistance to overheating treatment of the cable. Following conclusions can be made:

1) The spare cable has presented a trend of improvement on the insulation properties after overheating treatment below 20 times, but slight deterioration occurs after 20 times overheating treatment;

2) The retired cable with service years of 32 has presented a trend of degradation on the insulation properties at the first 5 times overheating treatment, and a severe deterioration occurs after 20 times overheating treatment.

Therefore, it can be deduced that the spare cable has a high potential to resist multiple overheating situation below 20 times due to the integrate structure of XLPE macromolecular chains and the unstable status of the crystal nucleus and crystalline structure; On the contrary, the retired cable with service years of 32 can hardly resist multiple overheating over 10 times. This kind of cables should avoid multiple overheating situation in the actual operating condition.

5. Acknowledgments

This research is truly supported by the foundation item: The Technical Projects of China South-ern Power Grid (No. GDKJXM20172797).

6. Reference

[1] Pan Luo, Yang Xu, Xiao Gu, Yanqun Liao, Jiangjing Cui, Zhihua Lu, and Zhigang Ren, “Thermal and mechanical properties analysis for EHV XLPE cables with different operating years.” Annual Report Conf. Electrical Insulation and Dielectric Phenomena, 2013.

[2] Antonios Tzimas, Simon Rowland, Leonard A. Dissado, Mingli Fu, and Ulf H. Nilsson, “Effect of Long-time Electrical and Thermal Stresses upon the Endurance Capability of Cable Insulation Material.” IEEE Transactions on Dielectrics and Electrical Insulation, vol. 16, No.5, October 2009.

[3] Larbi Boukezzi, Ahmed Boubakeur, Christian Laurent, and Mouhamed Lallouani, "Observations on Structural Changes under Thermal Ageing of Cross-linked Polyethylene Used As Power Cables Insulation," in Iranian Polymer Journal, vol. 17 (8), 2008, pp. 611-624.
[4] A. Campus, B. Lafin and T. Sterzynski, "Structure morphology modification of cable insulation polymers" IEEE 6th Int'l. Conf. Solid Dielectr. (ICSD), Vasteras, Sweden, pp. 357-360, 1998.
[5] L. Boukezzi1, A. Boubakeur, and M. Lallouani, "Effect of Artificial Thermal Aging on the Crystallinity of XLPE Insulation Cables: X-ray Study." Annual Report Conf. Electrical Insulation and Dielectric Phenomena, 2007.
[6] Association of Edison Illuminating Companies. AEIC CS7 Specifications for Crosslinked Polythene In-sulated Shieled Power Cables Rated 69 Through 138kV,1983.
[7] L. Boukezzi1, A. Boubakeur, and M. Lallouani, "Effect of Artificial Thermal Aging on the Crystallinity of XLPE Insulation Cables: X-ray Study." Annual Report Conf. Electrical Insulation and Dielectric Phenomena, 2007.
[8] Gamez-Garcia M, Bartnikas R, Wertheimer MR, "Modification of XLPE exposed to partial discharges at elevated temperature." IEEE Trans Elec Insul, 25, 688-692, 1990.
[9] Nedjar M, Beroual A, and Boubakeur A, "Influence of thermal aging on the electrical properties of poly (vinylchloride)," in Applied Polymer Science, vol. 102 (5), 2006, pp. 4728-4733.
[10] Fu Qiang, Liu JiePing, He TianBai, "Non-isothermal Crystallization Behavior and Kinetics of Metallocene Short Chain Branched Polyethylen." In Chemical Research In Chinese Universities Journal, 2002 (06):1183-1188.
[11] Larbi Boukezzi, Ahmed Boubakeur, Christian Laurent, and Mouhamed Lallouani, "Observations on Structural Changes under Thermal Ageing of Cross-linked Polyethylene Used As Power Cables Insulation." in Iranian Polymer Journal, vol. 17 (8), 2008, pp. 611-624.

IOP Conf. Series: Materials Science and Engineering **746** (2020) 012002 doi:10.1088/1757-899X/746/1/012002

Effect of Polyethylene Glycol and Chloride Ions on Local Electrochemical Deposition of Micro Copper Columns

Qixin Qing [1], Fuliang Wang [1,*], Yufei Xiao [1] and Lin Wang[2]

[1]School of Mech. and Elect. Engineering, Central South University, No.932, South Lushan Road, Changsha City, 410083, Hunan Province, P.R. China
[2]Shenzhen Coship Electronics Co., Ltd. Rainbow Technology Mansion, Hi-tech Park North, Nanshan District, Shenzhen, 518000, P.R. China
[*]Corresponding author: E-mail: wangfuliang@csu.edu.cn

Abstract. Local electrochemical deposition (LECD) is a very promising three-dimensional (3D) metal microstructure fabrication method. To fabricate smaller and better micro structures, a 20 μm micro anode and suppressor were used to reduce the diameter of a micro copper column and improve its surface morphology. Micro copper columns were deposited with a copper sulfate electrolyte, with or without polyethylene glycol (PEG) and chloride (Cl^-) ions. The morphology and size of all deposited columns were analyzed at the same voltage, and the deposition mechanisms with PEG and Cl^- were discussed. Results indicated that the 20 μm micro anode formed a micro copper column of diameter 30-40 μm, and the use of PEG and Cl^- further decreased the diameter to 28.1 μm.

1. Introduction

Local electrochemical deposition (LECD) is a very promising three-dimensional (3D) metal microstructure fabrication method [1], as it only required simple and inexpensive equipment, a general environment, simple procedures, and widely available chemical materials. Ever since LECD was proposed by Madden in 1995 [2], it has been used to machine microstructures in the lab. Currently, microstructures fabricated by LECD include high-aspect-ratio copper columns, spring, nickel columns, and copper walls [2-5].

These studies have promoted the development of LECD. As the size of the anode is directly related to the deposited diameter of the micro column, different anodes (diameters ranging from 20 μm to 200 μm) were used [6-8]. Movement control methods for the anode were also studied to improve the deposited surface morphology, including shape of the bump and the deposited root [7, 9]. Additionally, the initial deposition gap has proven a key factor influencing the deposited threshold voltage and morphology of the microstructure [3, 10]. In order to fabricate different microstructures by LECD, complex programs designed by computer software were used to control the 3D movement of the anode and manufacture microstructures of different shapes [2, 4, 8, 11]. A hollow copper cylinder deposited by rotating the anode and the copper wall are the most typical examples of all microstructure shapes [5, 8].

In addition to these physical parameters and models, the deposition morphology is also affected by the additives. Additives are crucial to further improve the morphology and reduce the size. According to previous work, additives can cause grain refinement, inhibit or accelerate deposition, and burnish surfaces [5, 12-15]. In particular, after combining different additive species in a base solution, the effect of inhibition or acceleration is more distinct. For example, with bis-(3-sulfopropyl) disulfide, which is an accelerator used in electrochemistry, the deposition rate is higher, as it acts synergistically with chloride (Cl^-) ions [16]. However, Polyethylene glycol (PEG), which is a suppression additive, is used in the electrochemical deposition of copper [17, 18]. Although Cl^- shows an inhibition effect, the interaction of PEG and Cl^- can enhance the inhibition effect [19-22]. However, the effect of PEG and/or Cl^- on LECD has not been studied yet. Theoretically, the suppressor can lower the deposition rate of copper and may reduce the diameter of the micro copper column.

In this study, the effect of PEG and/or Cl^- on the deposition of a micro copper column via LECD was analyzed. The micro copper columns were deposited in three forms – without the additive, with only PEG as the suppressor (molecular weight = 10000), and with a combination of PEG and Cl^-. The morphology and size of all deposited columns were analyzed at the same voltage, and the deposition mechanism with PEG and Cl^- was discussed.

2. Experimental

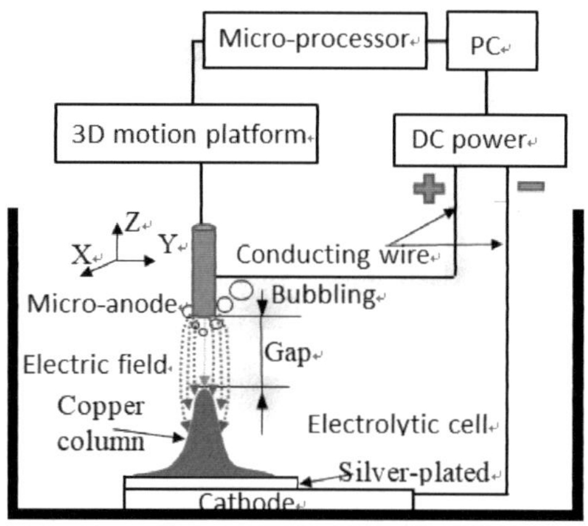

Figure 1. Schematic of the experimental set-up for LECD

Figure 1 shows the experimental setup. The anode used in this study was a 20 μm diameter platinum wire sealed in a glass tube (inner diameter = 100 μm), and the gap between the tube and wire was filled by an epoxy resin adhesive. One end of the Pt wire was welded with the conducting wire and connected to a DC power supplier, whereas the other end was ground by sandpaper, and polished with 0.5 μm Al_2O_3 with a grinding machine, to expose the anode to the electroplating solution. Subsequently, the anode was installed in a 3D motion platform. The circuit contains the anode as well as the cathode. The cathode was a copper plate deposited with silver on a surface of dimensions 5 mm × 5 mm × 0.2 mm. It was then affixed to a metal substrate by a conductive adhesive. The substrate cured by the epoxy was placed at the bottom of the container. Finally, the cathode was connected to the ground of the DC power supplier by a conducting wire.

The experiment was conducted in two steps. In the first step, the anode was in contact with the cathode. After switching on the power, the anode was moved toward the cathode with a high resolution of 1 μm per step, until the current exceeded the threshold (100 mA), at which value the anode and cathode were in contact. The power was then turned off. The second step was the deposition of the column. The initial gap and deposition voltage were set by a oftware, such that the anode would ascend a certain distance (for example, 1 μm) toward the cathode. After setting the initial parameters, the electrolyte was poured into the container and the power was turned on again to commence electroplating.

LECD was carried out as follows: copper ions were reduced and deposited on the cathode and grew toward the anode under the action of electric field. When the growing column almost approached the anode, a sudden rise in current was detected, which exceeded the threshold and triggered the next step. The power was turned off immediately and electroplating was paused. After the anode ascended a certain distance to maintain the predefined gap, the power was turned on and electroplating continued again. This cycle was repeated until the copper columns reached an expected height. In this study, all Cu columns had the same height of 1000 μm.

3. Results and Discussion

3.1. Copper Column Deposited with Only Copper Sulfate Solution

Figure 2 shows the SEM morphologies of the copper micro columns deposited at different voltages (2.7 V, 2.8 V, 3.0 V, 3.2 V, and 3.3 V) without any additive. The concentration of $CuSO_4{\cdot}5H_2O$ was 200 g/L.

At 2.7 V, the copper column does not possess branches, as shown in Figure 2a; however, the surface is covered with some small particles, and the top shows cracks. On increasing the voltage to 2.8 V, the surface morphology does not exhibit particles or branches, and the top is flatter than at 2.7 V; meanwhile, the diameter decreases from 40.8 μm to 30.1 μm, as shown in Figure 2b. On further increasing the voltage to 3.0 V, branches and bumps appear on the trunk and top, respectively, as shown in Figure 2c. With increasing voltage to 3.2 V and 3.3 V, the branches and bump structure became increasingly pronounced, and the shape is no longer a pillar, as shown in Figures 2d and 2e.

Figure 2 indicates that the voltage has a significant effect on the deposition of the column, and a voltage of 2.8 V is optimum, as it provides better surface morphology and smaller diameter.

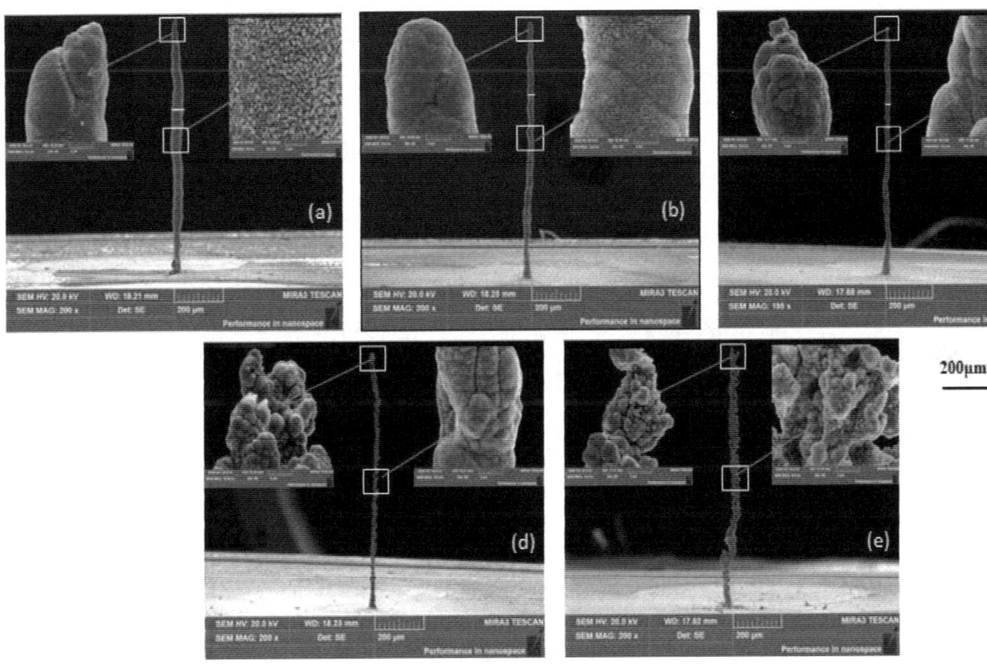

Figure 2. SEM images of the copper columns deposited at different voltages. (a) 2.7 V, (b) 2.8 V, (c) 3.0 V, (d) 3.2 V, and (e) 3.3 V with only copper sulfate solution

3.2. Copper Column Deposited with PEG Additive

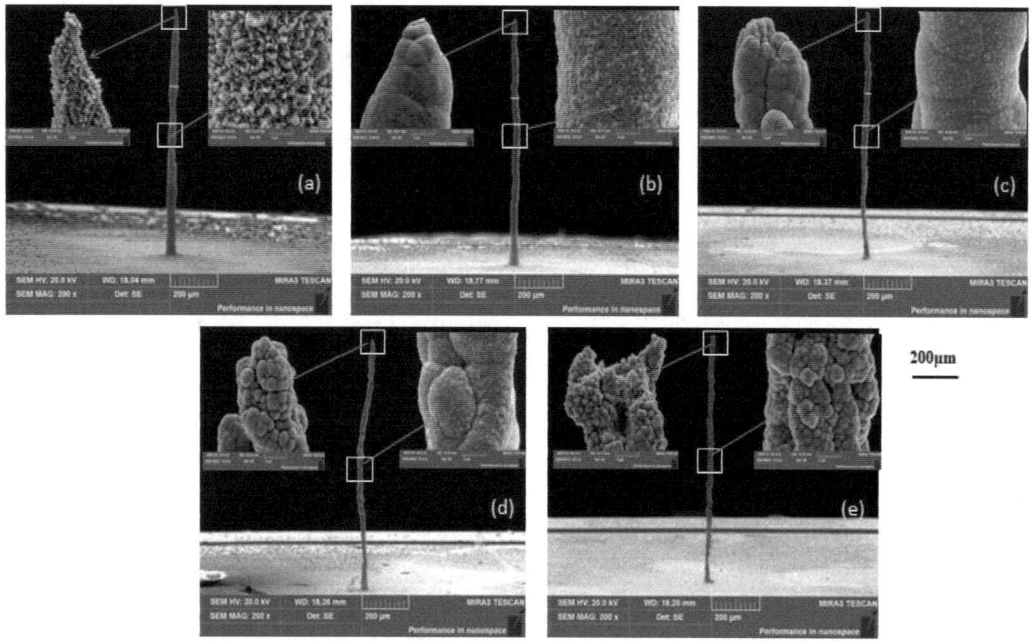

Figure 3. SEM images of copper columns deposited at different voltages, with only PEG added to the sulfate electroplating solution (a) 2.7 V, (b) 2.8 V, (c) 3.0 V, (d) 3.2 V, and (e) 3.3 V

Figure 3 shows the effect of deposition voltage on the morphology of the copper column deposited in the sulfate solution with the PEG additive. PEG is a common suppressor and can block the copper surface and inhibit deposition [18]. In the experiment, the concentrations of $CuSO_4{\cdot}5H_2O$ and PEG were 200 g/L and 6 mg/L, respectively.

Figure 3a shows that the column deposited at a low voltage (2.7 V) has a larger size, with a diameter of 36.3 μm; further, the surface is rough and without branches. However, the surface morphology appears better at the 2.8 V condition, for which the diameter decreases to 29.6 μm (Figure 3b). When the voltage is increased to 3.0 V, branches and bump structures appear (Figure 3c). With a further increase in voltage to 3.2-3.3 V, the columns transform into vines with many small branches, as shown in Figures 3d and 3e.

Based on Figure 3, the morphology at 2.8 V was deduced to be the best in different deposition voltages, with only PEG as the additive. Hence, the higher the voltage, the poorer the morphology.

3.3. Copper Column Deposited with PEG Additive and Cl^-

Comparing Figures 2 and 3, the PEG additive alone does not appear to cause improvement in the deposition quality of the copper columns, as the suppression effect of PEG requires Cl^- [21]. Therefore, PEG and Cl^- were added simultaneously, at concentrations of 6 mg/L and 0.5mg/L, respectively. The concentration of $CuSO_4{\cdot}5H_2O$ was 200 g/L. Figure 4 shows the morphologies of the copper micro columns deposited in the copper sulfate solution containing the PEG additive and Cl^- ions.

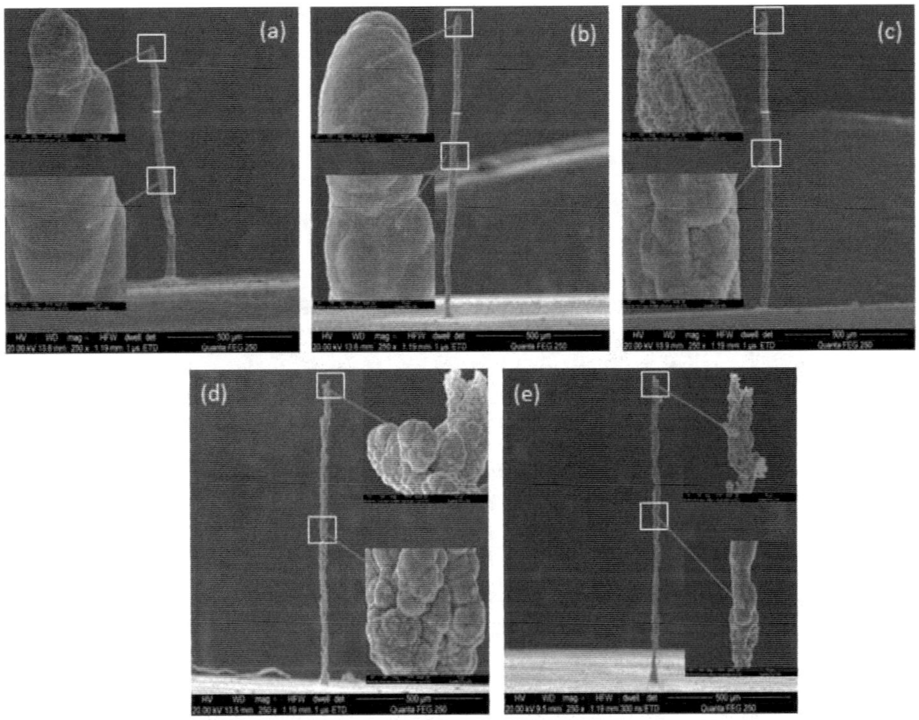

Figure 4. SEM images of copper columns deposited at different voltages, with PEG and Cl^- in the sulfate electroplating solution. (a) 2.7 V, (b) 2.8 V, (c) 3.0 V, (d) 3.2 V, and (e) 3.3 V

Figure 4a and 4b show the columns deposited at 2.7 V and 2.8 V. PEG and Cl^- together significantly improve the surface morphology, which does not exhibit grains, bumps, or branches in the middle or top of the column, and decrease the diameter to 30.5 μm and 28.1 μm, respectively. However, the column fabricated at 3.0 V shows many bumps, which make the surface rougher, as shown in Figure 4c. With increasing deposition voltage, the columns transform into vines with many small branches, as shown in Figure 4d and 4e.

Therefore, the combined addition of PEG and Cl^- can dramatically improve the surface morphology and decrease the diameter of the columns at deposition voltages of 2.7 V and 2.8 V. The suppression effect of PEG and Cl^- can decrease the grain size [17], thereby enhancing the smoothness and decreasing the diameter of the column.

3.4. Mechanism of Suppression Effect on Deposition of Copper Column

Copper ions driven by electric field are reduced to copper and grow into copper column at the cathode. Thus, the distribution of electric field on the surface of the copper column is not uniform as shown in figure1. No uniformity lead to some problems of the surface grain size and cross-sectional diameter. Therefore, we hope to use PEG additive and Cl^- to obtain good surface morphology.

The schematic diagram of mechanism of PEG is shown in figure 5a. It was adsorbed on the surface of the column with potential-dependence, thereby impeding the deposition of copper ions on the surface, and eventually resulting in a column with a smaller diameter than that without the PEG additive. However, the electric field become weak with copper column grown to make the absorption of PEG weak to desorb from the surface. In the meantime, owing to the chain molecular structure of PEG, PEG molecules are not very close together in the weak electric field region. Based on some reasons, there are some gaps between PEG molecules or PEG and copper column. And the gaps lead to weaker suppression effect than elsewhere. Consequently, these gaps enable the rapid deposition of copper ions driven by the electric field, resulting in a rougher surface. Therefore, the morphology in this case is poorer compared with the others. Although these gaps cause the surface to be uneven and unsmooth, the average diameter decreases by 4.5 μm, compared with the diameter without using the additive at 2.7 V. With increasing voltage, the change in the diameter of the micro column becomes smaller.

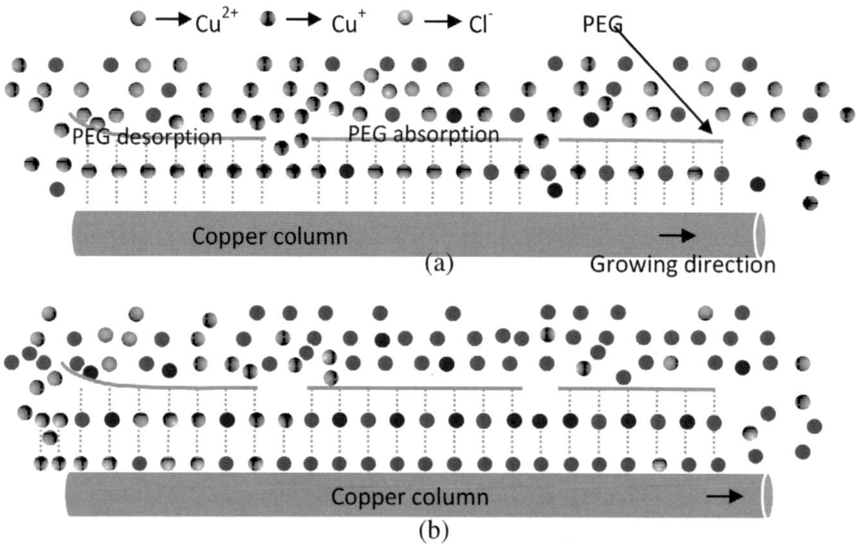

Figure 5. Illustration of the suppression effect of the PEG additive and Cl^- ions on column deposition: (a) with only PEG and (b) with PEG and Cl^-

However, Cl^- ions can overcome this disadvantage. The mechanism diagram is shown in figure 5b. PEG interacts synergistically with Cl^- and Cu^{2+} / Cu^+ to inhibit the metal deposition [21, 22]. On the surface of the copper column, PEG and Cu^{2+} / Cu^+ and Cl^- combine to absorb the surface of copper column to form the clathrate that impede copper ions deposition. Cl^- absorbed is closest to the copper column. At the same time, Cl^- and Cu^+ can impede the deposition at these gaps region. Under the action of PEG and chloride ions, copper ions are firstly reduced to cuprous ions and then to copper on the surface of copper column. At the same voltage, the deposition rate is lower than that without Cl^- ions. The suppression effect of PEG-Cu^{2+}/ Cu^+-Cl^- can lead to smaller-diameter columns and the more suitable the surface morphology. Because of the absorbed effect, the diameter is also reduced by about 10 μm at 2.7 V and 2 μm at 2.8 V, compared to that without the additive. The mechanism explains that PEG and Cl^- not only improve the quality of the morphology, but also reduce the cross-sectional diameter in the experiment. Of course, some unknown mechanisms of PEG need to be explored further.

4. Conclusions

The effect of adding PEG and Cl^- ions on LECD was studied. Micro copper columns were deposited with a micro anode of diameter 20 μm in three forms: (1) with PEG, (2) without PEG, and (3) with both PEG and Cl^-. The morphology and diameter of the micro copper columns were compared. Some conclusions can be drawn, as follows:

1) The 20 μm micro anode formed a micro copper column of diameter 30-40 μm. The voltage had a significant effect on the deposition of the column, and 2.8 V was the optimum voltage.

2) The addition of PEG slightly decreased the diameter, but resulted in poor morphology due to gaps between the adsorbed PEG molecules.

3) The interaction of PEG and Cl^- decreased the diameter of the micro copper column 28.1 μm and avoided poor morphology.

IOP Conf. Series: Materials Science and Engineering **746** (2020) 012002 doi:10.1088/1757-899X/746/1/012002

5. Acknowledgments

This work was supported by the China Department of Science & Technology Program 973 (Contract No. 2015CB057202).

6. References

[1] J. D. Madden, S. R. Lafontaine, and I. W. Hunter 1995 *Proc. Int. Symp. Micro Mach. Hum. Sci. pp. 77-81.*
[2] J. D. Madden, and I. W. Hunter 1996 *J. Microelectromech. Syst.* **5**, 24.
[3] F. Wang, et al. 2017 *J. Electrochem. Soc.* **164**, D297.
[4] F. Wang, et al. 2017 *J. Electrochem. Soc.* **164**, D758.
[5] S. Pané, et al.2011 *Electrochem. Commun.* **13**, 973.
[6] L. Yong, et al. 2003 *Sens. Actuators, A,* **108,** 144.
[7] J. C. Lin, et al.2005 *J. Micromech. Microeng.* **15,** 2405.
[8] S. H. Yeo and J. H. Choo, 2001 *J. Micromech. Microeng.* **11,** 435.
[9] C. S. Lin, et al.2005 *Electrochem. Solid-State Lett.* **8,** c125.
[10] H. Xiao, et al. 2016 *Int. Conf. Electron. Packag. Technol.* 69-72.
[11] M. A. Habib, S. W. Gan, and M. Rahman, 2009 *J. Mater. Process. Technol.* **209**, 4453.
[12] J. H. Yang, et al. 2009 *J. Micromech. Microeng.* **19,** 025015.
[13] Y. Li, J. Li, and F. Wang. 2016 *Int. Conf. Electron. Packag. Technol.* 175-179.
[14] F. Wang, et al. 2017 *J. Electrochem. Soc.* **164,** D419.
[15] W. -P. Dow, et al.2005 *J. Electrochem. Soc.* **152**, C425.
[16] M. Tan, et al. 2007 *J. Electrochem. Soc.* **154**, D78.
[17] W. -P. Dow, et al.2005 *J. Electrochem. Soc.* **152**, C769.
[18] S. -L. Ko, et al.2008 *Thin Solid Films*, **516,** 5046.
[19] Z. Nagy, 1995 *J. Electrochem. Soc.* **142**, L87.
[20] T. P. Moffat, D. Wheeler, and D. Josell, 2004 *J. Electrochem. Soc.* **151,** C262.
[21] Z. V. Feng, X. Li, and A. A. Gewirth, 2003 *J. Phys. Chem. B* **107,** 9415.
[22] J. J. Kelly, 1998 *J. Electrochem. Soc.* **145,** 3477.

IOP Conf. Series: Materials Science and Engineering **746** (2020) 012003 doi:10.1088/1757-899X/746/1/012003

Experimental Research on the Heat Transfer Characteristics on Train Body of Bullet Train

Weibing Shi[1], Li Li[1], Mengmeng Li[2, 3] and Xiao Zeng[2]

[1]R&D Center OF CRRC QINGDAO SIFANG CO, LTD, Qingdao 266111, Shandong, China

[2]School of Energy science and engineering, Central South University, Changsha 410083, Hunan, China

Email: 2960944083@qq.com

Abstract. In order to know the characteristic of heat transfer performance on train body, the DRH-300 instrument was used to investigate equivalent thermal conductivity coefficient of hollow-structure aluminum of bullet train "Fuxing". It was analyzed that the same hollow-structure aluminium was under different temperature difference; while the equivalent thermal conductivity coefficient with different position of train body was under same temperature difference. The experimental results showed that the equivalent thermal conductivity coefficient of the same hollow-structure aluminum was the same under different temperature difference; the impact of the equivalent thermal conductivity coefficient on different hollow-structure aluminum was different under the same temperature difference. It showed that heat convection in the hollow was tiny, compared with heat conduction. On the practical application, the reasonable structure was selected in accordance with practical needs, so as to reduce heat transfer and energy consumption of train.

1. Research Background and Significance

With the higher and higher demand of speed for rail train body, it is obviously necessary to further realize the lightness of train body structure and obtain better heat insulation under the condition of making sure the train body strength and stiffness. After 1990s, hollow-structure aluminum became main materials of high-speed train body in the world [1]. Also, the first section of aluminum train body on the railway is completed to manufacture at Changchun bus factory [2], until now, the train body material of high-speed railway, motor car and subway have been mostly realized the change from carbon steel to stainless, aluminum alloy, and the body structure replaces the solid material with hollow-structure.

With the requirement of passengers for the comforts of trip, the new demand on the air conditioner of train is asked, while the impact on train body material's heat transfer characteristics of air conditioner's loading is very large, but the study of heat transfer of the hollow-structure aluminum for train body is still weak. Although there are some studies on this aspect [3-5], it will take a long time to get the results. In order to provide reference for optimized design of train body structure, some hollow-structure aluminum of Bullet train "Fuxing" are tested and analyzed in this essay.

2. Test Principle and Equipment

2.1. Test Principle

The experimental instrument used is the DRH-300 steady state thermal conductivity tester, which is measured on the thermal conductivity rate for the samples based on heating plate method. Based on

Fourier's one-dimensional steady-state heat transfer principle, the thermal conductivity is calculated through testing temperature difference, heating flow and thickness between cold and heat plate:

$$\lambda=\frac{Q\cdot d}{S\cdot\Delta Z\cdot\Delta T}=\frac{P\cdot d}{S\cdot\Delta T}=\frac{I\cdot U\cdot d}{S\cdot\Delta T}[W/(m\cdot K)] \quad (1)$$

Where: ΔT—the average temperature difference between cold and heat plate, K;
ΔZ—test interval, h;
I —hot pole current, A;
U—electric heating voltage, V;
d—sample thickness, m;
S—the effective heating area of sample, m^2;

2.2. Test Equipment
The experimental instrument used is the DRH-300 steady state thermal conductivity tester, which instrument can measure the thermal conductivity of plastics, glass, fiber, foam, insulation materials, etc. And the instrument has good stability, convenient operation, and can quickly and easily export experimental data. The physical map of the equipment is shown in figure 1.

Figure 1. The steady state thermal conductivity tester (DRH300)

The main technical indication of instrument:
1. The requirement of sampling size: 300 × 300 × (10-90) (mm) single sample;
2. The instrument can be controlled experimental temperature, the heating range: room temperature ~ 100°C, temperature resolution 0.01°C, testing accuracy 0.05°C, cooling range: 0 ~ 60°C, temperature resolution 0.01°C (constant water sink with high precision of -5°C to 99°C);
3. Heating power 35W ± 1%;
4. Automatic test of computing control with automatic calibration function;
5. Working condition: 1) environmental temperature 10 ~ 35°C; 2) relative humidity ≤ 80%RH.

2.3. Test Objects
The main structure of the train body is shown in Figure 2. It is mainly constituted by top plate, side wall plate and floor, and it is composed by longitudinal distribution of multiple cavity structure, and welded by extruded hollow-structure aluminum of hollow section with triangular hollow cavity.

IOP Conf. Series: Materials Science and Engineering **746** (2020) 012003 doi:10.1088/1757-899X/746/1/012003

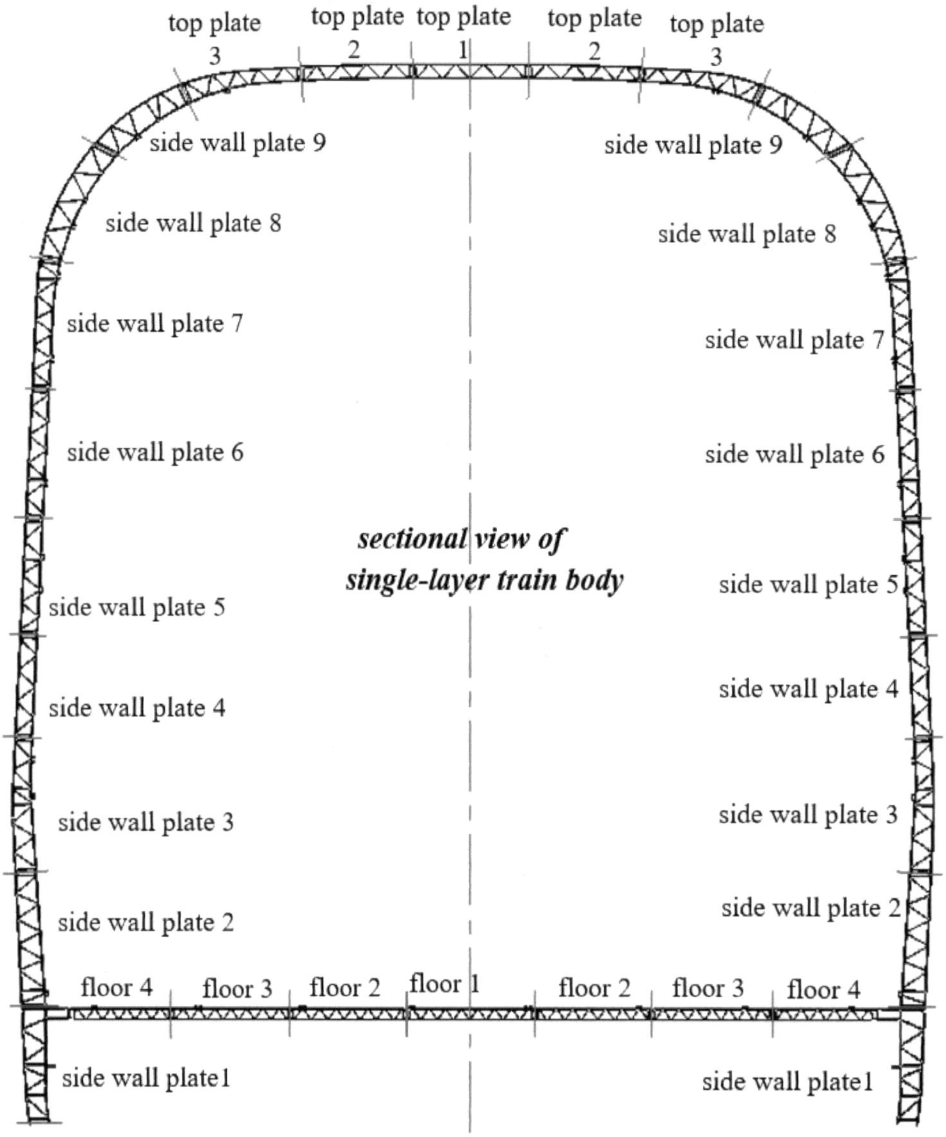

Figure 2. The main structure of train body

In accordance with the structural characteristics and train body contour distribution, the experimental sample, which including top plate 1, side wall plate 9, side wall plate 3, side wall plate 2 and floor 1, was selected, and the area of sectional samples is 300x300.See figure 3 for details.

Due to the existence of chute and material surfaced, the thermal conductive silicon wafers are laid on the top and bottom of the hollow-structure aluminum in order to satisfy with the requirement of measuring instrument sampling.

IOP Conf. Series: Materials Science and Engineering **746** (2020) 012003 doi:10.1088/1757-899X/746/1/012003

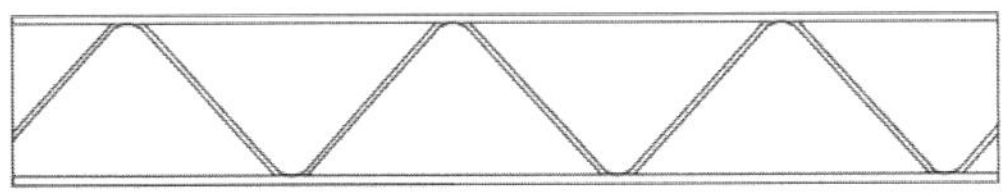

(a) top plate 1.

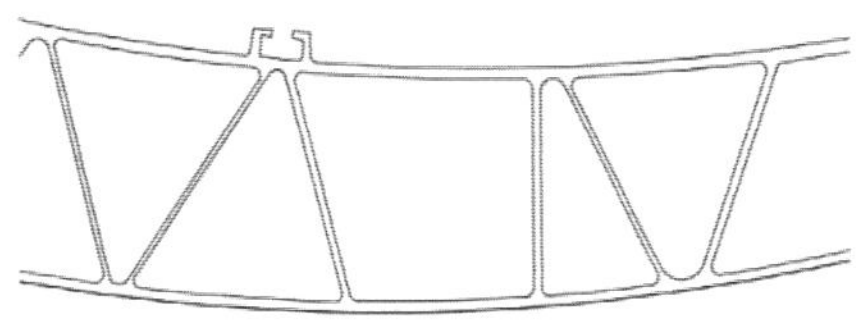

(b) side wall plate 9.

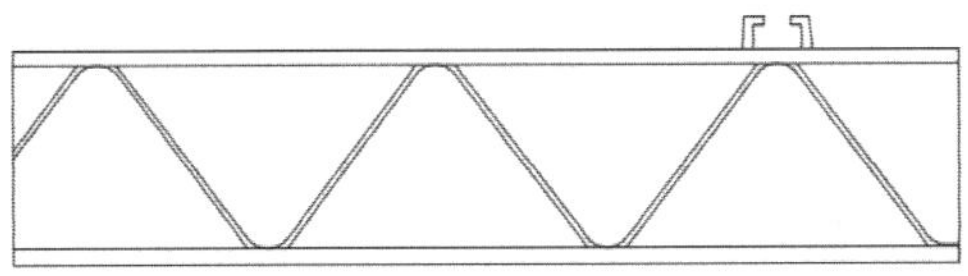

(c) side wall plate 3.

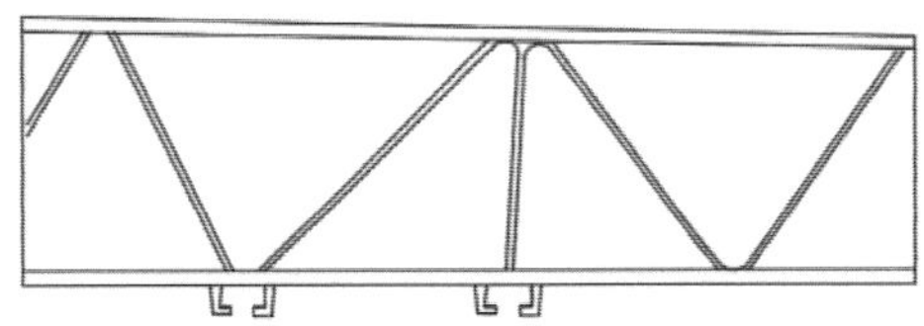

(d) side wall plate 2.

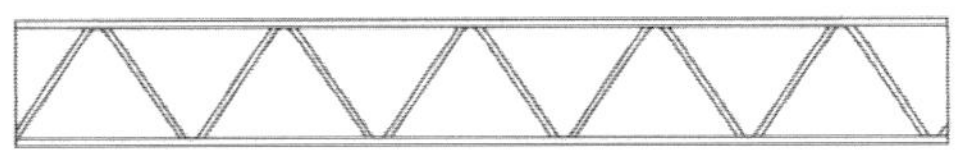

(e) floor 1.

Figure 3. Test objects

3. Experimental Results and Analysis

3.1. Analysis of Equivalent Thermal Conductivity of Hollow-structure Aluminum Under Different Temperature Differences

According to different seasons in actual operation, the temperature difference between the inside and outside of the train is constantly changing. The change of equivalent thermal conductivity coefficient of the floor 1 and side wall plate 2 at 5 different temperature of 2°C, 5°C, 10°C, 15°C, 20°C, 30°C is listed at table 1.

Table 1. Equivalent thermal conductivity of hollow-structure aluminum under different temperature differences W/ (m·K)

material	temperature difference					
	2°C	5°C	10°C	15°C	20°C	30°C
floor 1	12.3664	12.3658	12.3642	12.3689	12.3652	12.3641
side wall plate 2	10.0135	10.0141	10.0108	10.0123	10.0120	10.0146

According to the data in Table 1, it can be clearly seen that there is no obvious change in the equivalent thermal conductivity of the hollow-structure aluminum with temperature changes. Therefore, the impact of the temperature difference on equivalent thermal conductivity coefficient of hollow-structure aluminum can be ignored basically, and the natural convection heat transfer in the cavity is limited.

3.2. The Analysis of Parts on Different Position

According to the experimental result of temperature difference, the rest hollow-structure aluminum material is taken measure under the condition of 20°C for temperature difference, count the concrete parameter of experimental sample as well as organize experimental data, which is listed at table 2.

Table 2. Hollow-structure aluminum parameters at different position

material	number of cavities	sample thickness mm	Wall thickness mm	Rib thickness mm	effective thermal conductivity W/ (m·K)
floor 1	11	40	2.2	2	12.3652
top plate 1	7	50	2.2	2	8.2200
side wall plate 9	7	84.5	3.5	2	11.9810
side wall plate 3	6	65	4	2	10.4340
side wall plate 2	6	83.95	5	2	10.0120

It can be concluded as below in accordance with the data analysis of table 2:

1. For the hollow-structure aluminum represented by floor 1, its density of hollow cavity (the number of hollow cavities with unit length) is biggest and the equivalent thermal conductivity coefficient is also biggest.

2. For the hollow-structure aluminum material represented by top plate 1 and side wall plate 3, its density of hollow cavity is basically same, but the whole thickness of top plate 1 is larger and its equivalent thermal conductivity coefficient is also larger.

3. From the analysis of heat transfer characteristics, the heat transfer of hollow-structure aluminum is the coupling of heat conduction of aluminum material and natural convective heat transfer of cavity, however, for the temperature difference of changing hollow-structure aluminum's two ends, its equivalent thermal conductivity coefficient is basically stable, therefore, the main heat transfer process that determines the equivalent thermal conductivity of hollow-structure aluminum is the thermal conductivity of aluminum, among which, the natural convective heat transfer doesn't play an important role in the cavity.

4. Conclusion and Prospect

From the consequence of experimental research, conclusions can be drawn.

1. The thermal conductivity of aluminium is the main factor affecting the heat transfer characteristics and thermal conductivity of hollow-structure aluminum. And the natural convection heat transfer in the cavity is limited.

2. Among the structural parameter of hollow-structure aluminum, main impact on equivalent thermal conductivity coefficient is the density (the number of hollow cavities with unit length) of hollow cavity and thickness of hollow-structure aluminum. The bigger the hollow cavity's density is, the bigger thickness of hollow-structure aluminum is, and the bigger equivalent thermal conductivity coefficient is also.

3. On the subsequent study works, the means of experimental study and numeric simulation computing will be combined, the other structure's parameter will be carried out to study for hollow-structure aluminum in addition to hollow density and thickness, and explore its effect on the heat transfer of hollow-structure aluminum.

5. References

[1] Jingmin Dai 1995 Large extruded aluminum for high speed trains Light Alloy Processing Technology. 23 (5) 2-7.

[2] Zhixuan Li 1999 Application of aluminum alloy in locomotive and vehicle Electric Locomotive Technology. 3 17-18.

[3] Sian Yu 1988 Study on Rapid Measurement Method of K Value of Refrigerated Vehicle (Chengdu: Southwest Jiaotong University).

[4] Runquan Wang 1988 A new method for calculating heat transfer coefficient of car body

insulation Railway transaction. 12 (2) 11-15.
[5] Yufeng Du 2009 Study on Thermal Performance of Passenger Car Body Insulation Wall (Chengdu: Southwest Jiaotong University).

IOP Conf. Series: Materials Science and Engineering **746** (2020) 012004 doi:10.1088/1757-899X/746/1/012004

Research on the Durability of Tools and the Burrs of Workpiece When Milling T700 Carbon Fibre-reinforced Polymer Composites

Dazhen Wang and Peifeng Feng

College of Mechanical and Energy Engineering, Jimei University, No. 9 Shigu Road, Jimei District, Xiamen, China

Email: dazhenwang@163.com

Abstract. This research aimed to analyse the life and the wear of the coated tools with a diamond thin film, the uncoated cemented carbide tools, and the quality of machined surface, respectively, when milling T700 carbon fibre-reinforced polymer (CFRP) composite using the optimal cutting parameters. The results indicate that (1) in the same cutting condition, the wear resistance of the coated tools with CVD diamond thin film is better, whose life is 7.5 times that of the uncoated cemented carbide tools. (2) The life of tools directly affects the quality of the machined surface. The size and density of burrs on the machined surface are proportional to the width of blade wear of tool when carbon fibre-reinforced polymer (CFRP) composite was machined. The coated tools with diamond thin film by Chemical vapour deposition (CVD) can effectively prevent the chemical wear and diffusion wear of tools, which will make the blade wear of tools slow and uniform. (3) In practice, based on the quality, efficiency and processing economy, the tools need to be regularly checked when the workpiece is cut to 5.4m, 40.5m length for uncoated tools and coated tools, respectively.

1. Introduction

The carbon fibre-reinforced polymer composite (CFRP) is an attractive class of engineering materials characterised by a series of superior performance, such as light quality, high specific strength, good wear resistance, high fatigue resistance and corrosion resistance, etc. [1, 2]. They have been widely utilised in many fields such as aerospace industry, aircraft manufacturing, atomic energy industry, automobile parts, sports equipment, agricultural machinery and medical equipment [3].

T700 carbon fibre-reinforced polymer composite is categorised to a kind of difficult-to-process materials [4-6]. The main challenges can be found in several aspects, specifically, it is quite likely to produce the stratification during further processing because of inherent layered graphite structure and relatively vulnerable shear stress and bending strength. [7-9].

The composites are anisotropically easy to split, thus, it is suggested that the direction of the reinforced fibres must be considered in processing. The high hardness and abrasion resistance of carbon fibres make the cutting tools worn rapidly. Moreover, it is necessary to employ cold compressed air during the process due to the poor thermal conductivity of composites.

At present, the studies on the carbon fibre-reinforced polymer composites (CFRP) remains at the beginning stage domestically as well as internationally. [8, 10]. Therefore, it is of great significance to find qualified tools suitable for machining the high strength carbon fibre-reinforced polymer composite (CFRP), determine the service life and wear mechanism and performance of tools, analyse the relationship between the blade wear of the tool and the quality of the machined parts. That is the main reasons the researches emphasis on speeding up the practical engineering application of high strength carbon fibre-reinforced polymer composite (CFRP).

2. Experimental Materials and Procedure

2.1. Experimental Materials

The test materials are made of carbon fibre-reinforced polymer composites produced by Toray Company of Japan. The performance of the materials was shown in Table 1. The size of the sample is 330 mm × 333 mm × 1.2 mm (length × width × height).

Table1. Physical and mechanical properties of the samples

Reinforce -ment Material	Matrix Material	Laying Method /(°)	Volume Fraction of Fibre /%	Density /g.cm^{-3}	Linear Density /mg/m	Elongation percentage /%	Longitudinal Tensile Strength /GPa	Transverse Tensile Strength /GPa
T700	Epoxy Resin	90	60 ± 3	1.80	800	2.1	4.9	230

2.2. Experimental Procedure and Detection Instruments

The test has experimented on the TAKAM TJ-700 CNC milling machine. The processing range of the machine is 800 mm × 550 mm (length × width). The rated power of the main motor is 7.5 kW, and the maximum speed of the spindle is 8000 rpm/min. The maximum speed of cutting feed (X, Y, Z axis) is 6000 mm/min.

The experimental result indicates two kinds of tools, which were uncoated cemented carbide tools and CVD diamond thin film coated carbide tools, were selected for the test, as shown in Figure 1. The uncoated cemented carbide tools were adopted the UBRS4 series 4-blade flat end mills produced by Xiamen Jinlu Special Alloy Co., Ltd, and the diamond film coated carbide tools were adopted the 12-edge rhombic tools of the company EDW series. The parameters of the tools are shown in Table 2.

Table2. Parameters of the test tools

Materials	Diameter /mm	Number of teeth	Front angle/(°)	Flank angle / (°)	Helix angle/(°)
Cemented carbide	8	4	7	15	30
Diamond thin film	8	4	5	20	40

In order to ensure the success of the milling test, the special fixture designed by us was used for clamping the test material. The burr length of the machined surface was measured with a vernier calliper, and the macroscopic appearance of the burrs was filmed by a Fuji F505 EXR digital camera.

In order to reduce the number of tests, the cutting distance of this experiment is limited to 1.8 m, 3.6 m, 5.4 m, respectively. The test is expected to be finalised when the readings of uncoated tools reach the critical value of 4.5 m, 9 m, 13.5 m, 18 m, 22.5 m, 27 m, 31.5 m, 40.5 m, respectively. Followed by the measurement of the amount of flank wear of the tool and the size of the burrs of machined surface; and filming the macroscopic appearance of the burrs after processing. In order to ensure the accuracy of the test results, the method of averaging multi-blade measurement and multiple measurements are adopted.

Cooling method: dry cutting and using 301 high-power industrial grade vacuum cleaners produced by Zhejiang Yongkang Jeno Industry & Trade Co., Ltd. to cool the vacuum, as shown in Figure 2.

In order to obtain the readings of tool life value under the optimal cutting parameters, according to the proposed orthogonal test method before the subject, it would be necessary to combine with actual production and the cutting test results obtained. The optimised cutting parameters are determined after the comparison: cutting speed v = 5000 rpm, feed rate f = 250 mm / min, depth of cut a_p = 1 mm.

The average wear amount of the flank of the tool VB = 0.1mm is set as the blunt standard of the tool, and the test is finished when the flank wear value is close to or equal to the blunt standard.

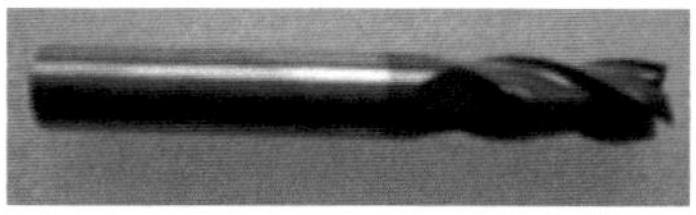

(a) Uncoated tool

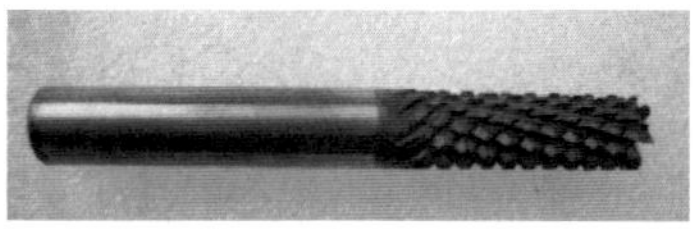

(b) Coated tool with diamond

Figure 1. Test tools

Figure 2. Industrial vacuum cleaner

3. Analysis and Discussion of Test Results

3.1. Variation of Flank Wear Width of Uncoated Tools

As shown in Figure 3, it is the flank wear width curve of the uncoated tool.

It is clear from Figure 3 when the cutting distance is approaching to 5.4 m, the tool flank wear amount is close to the blunt standard value.

3.2. Variation of Flank Wear Width of Coated Tools

As shown in Figure 4, it is the flank wear width curve of the coated diamond film tool.

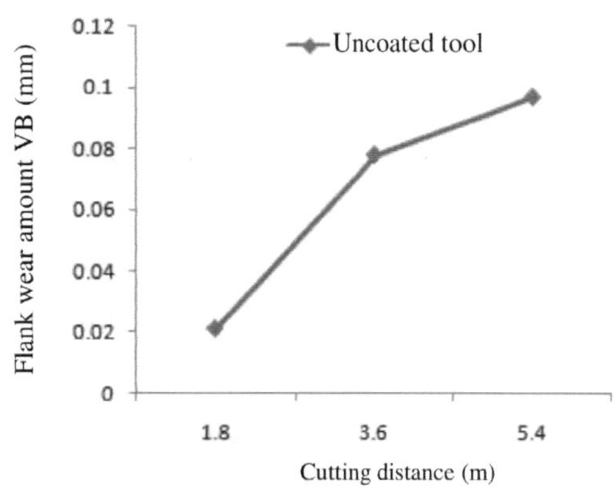

Figure 3. Flank wear curve of uncoated tool

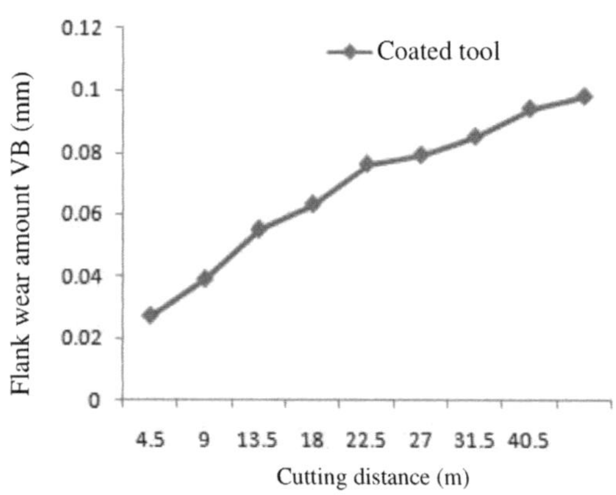

Figure 4. Flank wear curve of coated tool

It is clear from Figure 4 when the cutting distance is approaching to 40.5 m; the tool flank wear amount is close to the blunt standard value.

3.3. Tool Life Analysis

When milling T700 carbon fibre-reinforced polymer composites, flank wear is the main failure mode of the tool. The standard of the tool life is defined as the cutting distance of milling when the width of the flank wear band of the tool reaches the blunt threshold readings.

It can be seen from Figure 3 and Figure 4 that the uncoated tool is quickly worn, while the coated tool is slowly worn. Based on this fact it is believed that the uncoated tool experiences quick wear in the beginning stage (the curve is steep), however, the coated tool exhibits good wear resistance during the beginning and normal stage (the curve is gentle). It is also clear that the milling length of the uncoated tool and the coated tool reached blunt standard at 5.4m and 40.5m respectively which can be calculated that the useful life of the coated tool is approximate 7.5 times longer than that of the uncoated tool.

From the test data above, when T700 carbon fibre-reinforced polymer composite was milled by the CVD diamond film coated tool, the curve is relatively flat, the change of wear amount is more stable, and the milling length is much larger than that of the uncoated cemented carbide tool. The reason is that the diamond coating of the tool can reduce the coefficient of friction between the tool and the workpiece, and it has a high thermal hardness to withstand a large amount of cutting heat generated by the machining. Since the surface of the uncoated cemented carbide tool has no thermal barrier, the huge cutting pressure between the tool and the surface of the workpiece will generate severe friction, and the temperature of the body of the tool will be instantaneously increased, which softens the material of the tool base, makes that the tool rapidly worn so that the life of the tool is greatly reduced.

The diamond coating material of the coated tool and the carbon fibre reinforcement in the composite are all carbon materials. Therefore, graphite acts as a solid lubricant when machining, which reduces the thermal stress of the friction, the adhesion of chips to the surface of the tool, prevents the formation of a built-up edge at the same time. The conclusion is that the tool life of CVD diamond film coated tools is much greater than that of uncoated cemented carbide milling tools when milling T700 carbon fibre-reinforce polymer composite.

3.4. Blade Wear Analysis

When T700 carbon fibre-reinforced polymer composites are milled with the uncoated and diamond coated carbide tools respectively, the blade wear profiles are shown in Figure 5.

It is obvious to witness from the figures that the wear resistance of the coated tool is better than that of the uncoated tool. Due to the periodic cutting in and out, the intermittent cutting is formed when machining. The cutting edge of the uncoated cemented carbide tool has a wide groove only after the milling length of 5.4 m, and the flank wear value of the tool has been close to blunt standard.

Compared with the uncoated cemented carbide tool, diamond coated tools show good wear resistance. When the milling length reaches 4.5 m, the flank wear amount is only 0.027 mm, and the cutting edge remains sharp. Simultaneously, when the milling length reaches 40.5 m, the flank wear value is just close to the blunt standard. At this time, the blunt region becomes larger, and most of the coating material is worn away, and greyish white matrix of the tool is exposed.

Based on the analysis of the wear mechanism, abrasive wear occurs in the uncoated and coated tools during cutting because the carbon fibres in the composite are hard points composed of graphite crystallites, With respect to uncoated cemented carbide tools, the high cutting temperatures will be produced due to friction between high hardness carbon fibres and tougher epoxy matrix materials and the tools during high-speed milling. The tool is accompanied by oxidative wear, diffusion wear and bonding, the number of wear increases sharply, and the blade tips or blades are easily broken to chipping. However, the CVD diamond coating material of tools is coated with a diamond film on the surface of the cemented carbide substrate of tools by chemical vapour deposition (CVD), which avoids the occurrence of chemical wear and diffusion wear in the cutting process. At the beginning stage of the cutting, the hard particles in the workpiece material continuously produce “micro-cut” to the surface of the tool coating, and the cutting edge of the tool will be manufactured to a similar “mechanical furrow” cutting mark. In the middle and late period of the cutting, as the enhancement of the reaction from abrasive particles and the rapid interaction between the workpiece and the tool, the powder of the diamond peeled off from the coating of the tool. Carbon fibre graphite microcrystalline powder from the composite plays a significant role in lubricating and polishing for the flank of the tool, so the wear of the flank face of the tool is relatively uniform and slow.

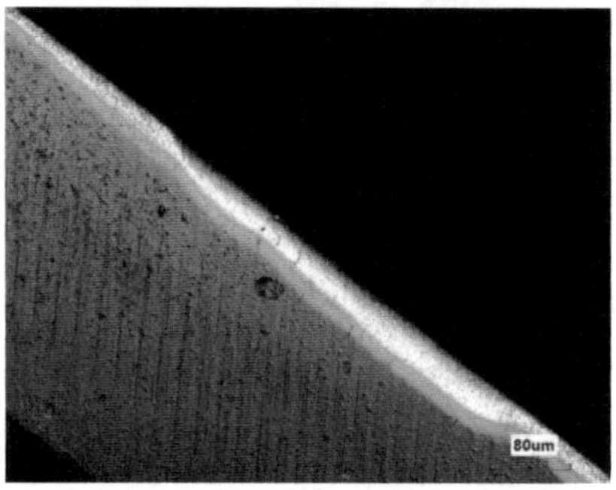

(a) uncoated tool (L = 1.8 m) × 400

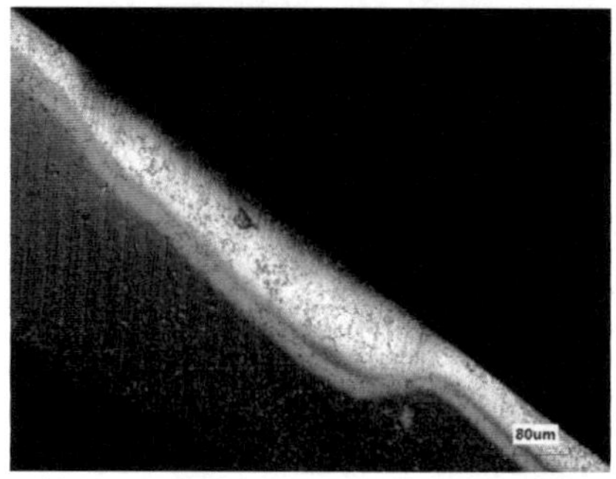

(b) uncoated tool (L = 5.4 m) × 400

(c) coated tool (L = 4.5 m) x 400

(d) coated tool (L = 40.5 m) × 400

Figure 5. Profiles of worn tools

3.5. Influence of Tool Life on Surface Quality

The burr size of the machined surface by cutting the carbon fibre composites is shown in Figure 6 and Figure 7.

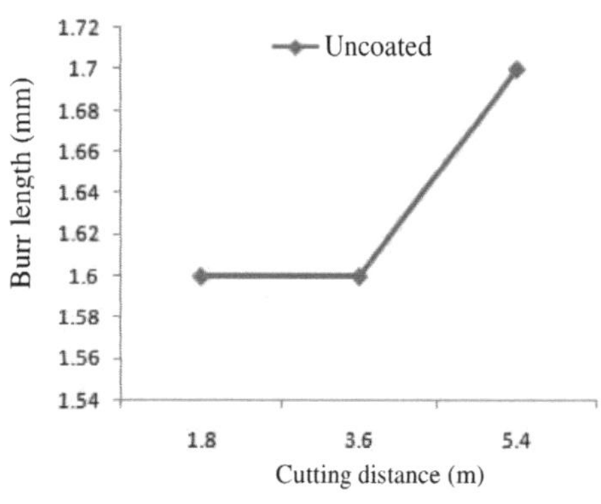

Figure 6. Burr length curve with uncoated tool

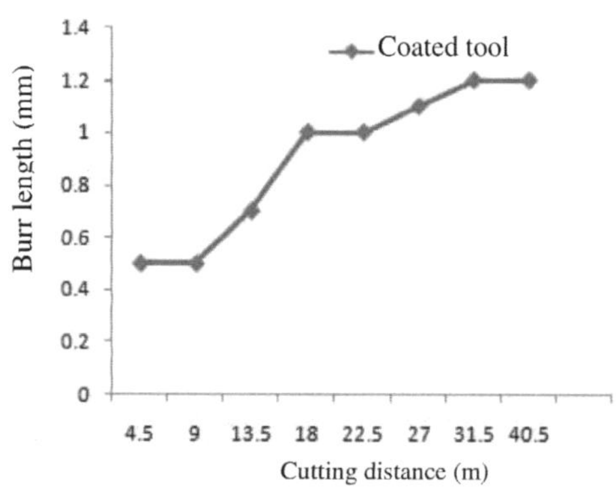

Figure 7. Burr length curve with coated tool

It can be seen from Figure 7, when the milling length of the diamond coated tool reaches 40.5 m, at this time, the flank wear amount of the tool is close to the blunt standard value, the generated burrs on the machined surface are only 1.2 mm length, which is much shorter than that of the machined surface produced by cutting to the length of 1.8 m with the uncoated tools. Therefore, the use of the coated tool for processing can achieve a better surface quality of the workpiece.

The quality of the machined surface after cutting T700 carbon fibre composite is not only related to the length of the burrs generated, but also closely affect the density of the burrs. The macroscopic morphologies of the burr density produced by processing T700 carbon fibre composite are shown in Figure 8 and Figure 9.

It can be seen from Figure 8 and Figure 9, when machining with the diamond coated cemented carbide tool, the length of the burrs is tiny and the density of the burrs is sparse, and the quality of the machined surface processed in the whole life cycle of the tool is qualified and stable.

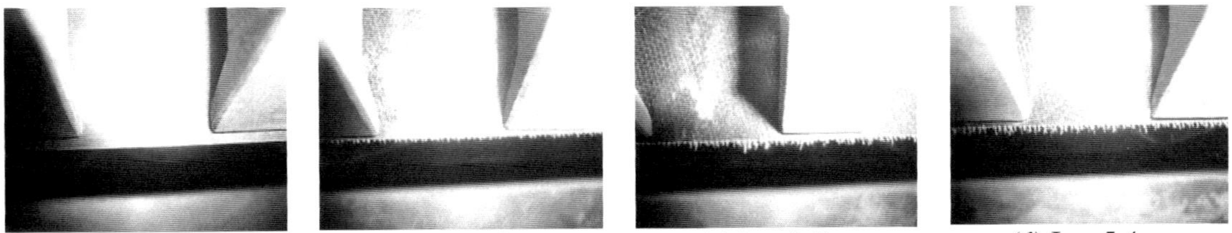

(a) L = 0 m (b) L = 1.8 m (c) L = 3.6 m (d) L = 5.4 m

Figure 8. Macroscopic appearance of burrs with uncoated tools at different milling lengths

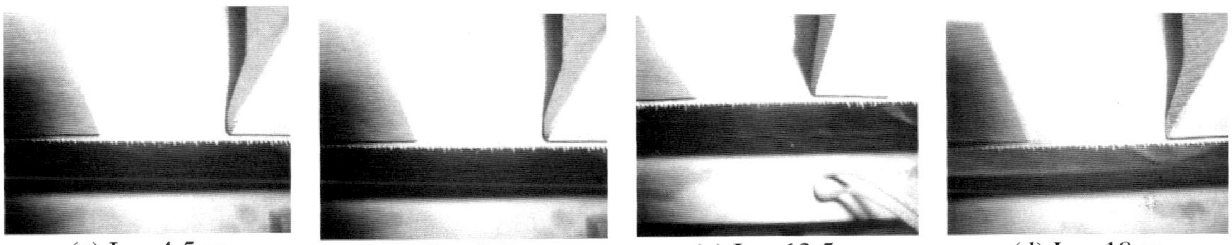

(a) L = 4.5 m (b) L = 9.0 m (c) L = 13.5 m (d) L = 18 m

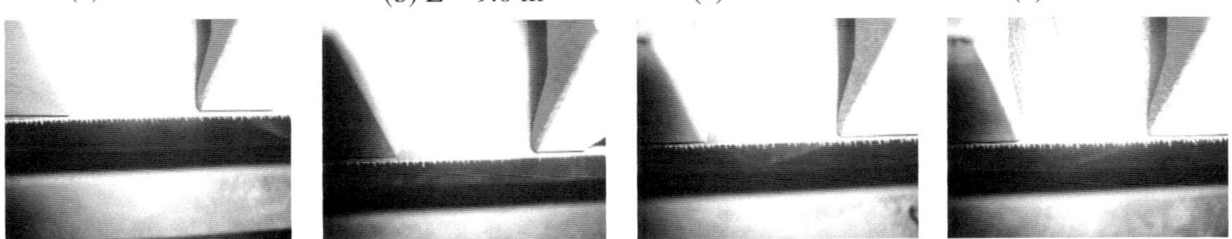

(e) L = 22.5 m (f) L = 27 m (g) L = 31.5 m (h) L = 40.5 m

Figure 9. Macroscopic appearance of burrs with coated tools at different milling lengths

According to the variation curves of the burr length (Figure 6 and Figure 7) and the microstructures of the workpiece burrs (Figure 8 and Figure 9), it is obvious to conclude that the variation law of the burr length of the workpiece is positively correlated with the change curve of the tool life (Figure 3 and Figure 4). A gentle trend in the curve can be seen in the wear curves of the diamond coated tool and the curves of the burr length of the workpiece compared to the uncoated tool. With the milling length increases, the increases of width in the tool's flank wear can be expected and the burrs tend to be coarse and dense, which is more obvious when machining with the uncoated tools.

The wear of coated tools is evenly and slowly consumed in processing because the diamond coating on the tool surface has a higher hardness than the T700 carbon fibre. Therefore, the burrs of the workpiece after processing are small, and the distance between the burrs is loose.

4. Conclusion

The T700 carbon fibre-reinforced polymer composite can be machined by milling, and the quality of the product is closely related to the type of machining tool selected.

When milling with CVD diamond film coated carbide tools, the burrs of the machined surface obtained are less than that of workpieces with the uncoated cemented carbide tools in the whole tool life cycle. At the same time, the burr density is lower, and the surface quality of the workpiece is superior.

CVD diamond film coated carbide tools can maintain long-term wear resistance during milling. Meanwhile, they have low wear and stability, and the tool's durability can be as 7.5 times as that of uncoated carbide tools.

The main wear mechanism of diamond coated tools is abrasive wear caused by broken carbon fibre particles and resin matrix material for continuous wear and impact of the tool. The main damage forms of the tools are micro-cracking of the cutting edge and tip of the tools, flank coating wear and coating peeling caused by coating crack propagation.

5. Acknowledgments

This work was financially supported by the Natural Science Foundation of Fujian of China (No. 2019J01327).

6. References

[1] Jiang Z B, Yuan S M and Li Z Research on Ultrasonic Assisted Cutting Carbon Fiber Reinforced Mg-Matrix Composite *Aviation Manufacturing Technology* 2019,vol. 62 (11), pp55-62.

[2] Shan C W and Lv X B Research Progress on Milling and Drilling of Carbon Fiber Reinforced Composites *Aviation Manufacturing Technology* 2016, (15), pp34-41.

[3] Zhou J W, Qi J F and H P H Research on Tool Wear During Trimming of CFRP *Mechanical Manufacturing* 2018, (5), pp53-57.

[4] Wang C Y, Wen L and Ming W W Research progress in milling technology of carbon fiber reinforced composites *Aviation Manufacturing Technology* 2015, vol.14, pp 76-80.

[5] Ye X Z, Wang D Z, Liu J D Experimental Study on Wear of Milled Carbon Fiber Composite Tools *Electromechanical Technology* 2014, vol. 6, pp 75-77.

[6] Duan C Z, Li P X and Zhang F Y Wear mechanism of tools made of different materials in milling carbon fiber reinforced plastics *Journal of Harbin Engine*ering University 2018, vol. 39 (10), pp1655-1660.

[7] Zhang L Y, Liu X D and Sun J J Ultrasonic Assisted Milling Carbon Fiber Recombination Materials Technology Research *Modular Machine Tool & Automatic Manufacturing Technique* 2018, (6), pp1471502.

[8] Su F and Sun F J Research on Milling Tool Wear of Carbon Fiber Reinforced Composite *Equipment Manufacturing Technology* 2018, (7), pp219-222.

[9] Lin Y X, Yu J and Lin H Tool Wear Mechanism of High Speed Milling of Carbon Fiber *China Surface Engineering* 2016, vol. 29 (5), pp 138-145.

[10] Wu H, Chen Y and Han S H Study on Milling Force of Carbide Tool Milling Carbon Fiber Composites *Mechanical Science and Technology* 2014, vol. 33 (8), pp 1254-1257.

Quality Evaluation of Tourmaline Green Appearance Based on CIECAM16 Color Appearance Model

Bin Yuan[1], Ying Guo[1, 2, *], Yinuo Li[1] and Haoyuan Jia[1]

[1] School of Gemology, China University of Geoscience, Beijing 100083, China
[2, *]Email: guoying@cugb.edu.cn

Abstract: Tourmaline minerals are mainly produced in pegmatite, which is mainly produced in pegmatite. At present, the color evaluation of green tourmaline in China needs to be supplemented. In this paper, the CIECAM16 color appearance model is used to calculate the color appearance attribute parameters of green tourmaline samples, and the tristimulus XYZ of green tourmaline color measured by Colori5 colorimeter is used as the input data of CIECAM16 color appearance model to predict the color appearance attribute of green tourmaline under different conditions. The CIECAM16 color appearance model written by computer C++ language is used to calculate the color appearance attribute parameters under different illuminance levels under D65 light source, different environment (Average, Dim, Dark) conditions and different illumination light source color temperatures (D65, D50, A). Through analysis and calculation of the color appearance attribute parameters, it is found that that the increase of illuminance has a more intuitive effect on the visual brightness Q, while the change of ambient light and shade has a greater effect on the brightness than the change of light source. The A light source has a more obvious effect on the tourmaline of color than D65 and D50 light sources.

1. Introduction

Tourmaline is a common gem mineral [1]. The chemical composition of tourmaline is extremely complex. Its chemical formula is (Na, Ca) RA_{16} $[Si_6O_{18}BO_3]_3$ (O, OH, $F)_4$, which is a cyclic borosilicate mineral. The boron element and hydroxyl contained in tourmaline are closely related to its pegmatite origin. Changes in tourmaline's chemical composition directly affect changes in its physical properties.

According to the different cations in the chemical formula, they are mainly divided into iron tourmaline, magnesium tourmaline, lithium tourmaline, etc. Their colors and transparency are different. For example, iron tourmaline is black, while gemstone tourmaline is mainly lithium tourmaline, with dark green, green, light green, blue, yellow, rosered and other colors, as well as colorless ones [2, 3].

Tourmaline is widely distributed, not only in various acidic rocks bodies and dikes, but also in various hydrothermal deposits and volcanic massive sulfide deposits. Therefore tourmaline has potential application as ore guide [4]. In addition, tourmaline has piezoelectricity and pyroelectricity, so tourmaline has environmental functions such as emitting far infrared rays, releasing anions, shielding electromagnetic interference, and water treatment [5]. Gem-grade tourmaline is mainly produced in granite pegmatite, and a small amount can also be found in boron-rich hydrothermal or metasomatic, rocks. At present, the main mining areas of tourmaline which are discovered include Sri Lanka, Madagascar, Brazil, Mozambique, Angola, etc. China's mining areas mainly include Altai Mountain and Tianshan Mountain in Xinjiang, Luobei in Heilongjiang, and the southern tip of Ailao Mountain in Yunnan, etc. Almost all of them are produced in pegmatite.

At present, the scientific research on non-gem tourmaline mainly focuses on its performance as a mineral material. Mariano Mercurio et al [6] used scanning electron microscope, energy dispersion spectrum, SEM-EDS, electron microprobe analysis, wavelength dispersion spectrum, EMPA-WDS, laser ablation, inductively coupled plasma mass spectrometry LA-ICP-MS and other instruments to study the preliminary chemical characterization of tourmaline and classify the tourmaline studied. S.O. Olabanji et al. [7] used complementary accelerator-based pixel electron probe analysis (EMPA) for the first time to characterize the natural tourmaline minerals in Nigeria, in order to understand the local occurrence characteristics and quality of tourmaline. Nilo F. Cano et al. [8] studied green tourmaline samples by using TL and EPR techniques to investigate the center responsible for their TL characteristics and their possible application in gamma radiation dosimetry, and to conduct optical absorption measurements. K. Krambrock et al [9] studied the yellow color center produced by γ radiation in colorless natural tourmaline by using electron paramagnetic resonance (EPR) and electron nuclear double resonance (ENDOR) techniques.

At present, researches on the color of tourmalines in China are mainly focused on yellow-green and pink tourmalines and red tourmalines. Lack of color evaluation of green tourmalines. However, there are many researches on color in other green jewelry and jade. For example, there are many sets of color classification systems for green jadeite, Zhang Li Li et al. [10] obtained color threshold intervals of different green jadeite in HSL color space model and established their color grading. Guo Ying et al. [11] evaluated the color of jadeite and studied the influence of different light sources on the color of jadeite in CIE 1976 L*A*B* uniform color space. Yin Ke et al. [12] studied the origin of the color of "Youqingzhong" jadeite jade by X-ray diffraction and infrared spectroscopy. Wang Yongya et al. [13] obtained the green genesis mechanism of Serpentine Jade in China by using electron paramagnetic resonance spectroscopy and other instruments. Guo Ying et al. [14] studied the influence of brightness on emerald green brightness in different achromatic backgrounds. Luo Hongyu et al. [15] used X-ray fluorescence spectrometer to study the color-forming mechanism of chrysoberyl. Therefore, it is very urgent and necessary to evaluate the color of green tourmaline.

Although progress has been made in the research on the influence of components on color, the evaluation and comparison of color mostly rely on subjective observation of human eyes, lacking objective and unified evaluation standards. This subjective evaluation is easily influenced by environmental factors (such as lighting source, background environment, etc.), so that observers will have different color senses when observing the same sample under different conditions. In order to quantitatively describe color, these influencing factors must be taken into account, and color evaluation can be carried out in color space at this time. Florian Schiller et al. [16] used different color spaces to give examples to study the correspondence between saturation of color space and perceived saturation of human beings. Wu Jiahui et al. [17] studied the evaluation of manganese almandine in CIE1976L*a*b homogeneous color space. Guo Ying et al. [18] evaluated the quality of jadeite lightness in CIE1976L*a*b uniform color space. However, there are still many problems that may be encountered when evaluating the color: for example, when describing and evaluating the color of a gem, the evaluation is made under specific light source irradiation, specific observer and environmental conditions. when any one of the light source, external conditions and the observer changes, it is impossible to ensure that the same tourmaline has the same color perception, thus obtaining a consistent color evaluation result. Therefore, many factors are needed to comprehensively evaluate the color of gemstones. The CIE1976L*a*b uniform color space, which is often used, is not sufficient to fully describe the color of gemstones, and a complex color appearance model is needed to describe the color of gemstones [19].

Since CIE recommended CIECAM02 color appearance model in 2002 [20], CIECAM02 color appearance model has been widely used in scientific research and industry [21], Snjezana Soltic et al. [22] have applied CIECAM02 color appearance model to predict the effect of gamma on CRT display color. Andrew Thwaites et al. [23] used CIECAM02 model and CIELAB model to study human cortex. Yang Xiaoli et al. [24] compared CIECAM02 model color space with CIELAB color space in predicting hue and lightness. Dong Shuwen et al. [25] used CIECAM02 model to predict and evaluate the color appearance of printing chromatography with different glossiness. At present, the latest color appearance model is CIECAM16 color appearance model. CIECAM16 color appearance model is

IOP Conf. Series: Materials Science and Engineering **746** (2020) 012005 doi:10.1088/1757-899X/746/1/012005

improved [26-28] on the basis of CIECAM02 color appearance model, which solves some disadvantages of CIECAM02 color appearance model, and also makes the structure simpler. CIECAM 16 color appearance model is better than CIECAM02 color appearance model in predicting visual effects, and the evaluation of gem color is more accurate [29-31].

2. Tourmaline Sample Color Test

2.1. Selection of Experimental Samples

(1) The size is moderate, mainly 6 mm x 8 mm, which meets the requirements of aperture of X-Rite SP62 colorimeter.

(2) The visual color is uniform, with strong, light, dark and light green under the naked eye. The samples are mainly in green hues, including green (yellow green) and green (slight blue) hues. There are many kinds of green, and the data are representative and convincing.

(3) Moderate transparency. Using reflection method to test, the sample with higher transparency will make the test light completely penetrate and increase the test error.

(4) Fine texture and no inclusion.

2.2. Test Instruments

The color measuring instrument used in this paper is the U.S. X-Rite SP62 integrating sphere spectrophotometer. It is convenient to carry and can be used for measuring various colors. Its fixed aperture is 8mm measurement area and 13mm illumination area, with 9 kinds of light sources including C, D50, D65, D75, A, F2, F7, F11 and F12. SCI (Specular Component Included Including Specular Reflectance) and SCE (Specular Component Excluded Excluding Specular Reflectance) data can be measured at the same time.

The test conditions of this instrument are: D65 standard light source, excluding specular reflection mode (SCE mode),spectral measurement range 400 nm-700 nm, measurement time less than 2.5 s, wavelength interval 10 nm, voltage 220 V, current 50 Hz-60 Hz, and operating temperature controlled between 10° and 40°.

Uniform color space CIE1976L*a*b* is used as a color system for quantitative representation and analysis. CIECAM16 color appearance model is used to calculate the color appearance attribute of tourmaline.

2.3. Color Appearance Model

The CIECAM16 color appearance model is used in this paper. CIECAM16 color appearance model changes the structure of CIECAM02 model and adapts to the color and brightness of light source in the same space instead of two different spaces. In addition, the new CAM16 model is simpler than the original CIECAM02 model. Considering only color adaptation, a new transform cat16 can be used to replace the previous cat02 transform. Finally, a new cam16-ucs uniform color space is proposed to replace the previous cam02-ucs space.CIECAM16 color appearance model has two models: forward model and reverse model. In this paper, the forward model is used to calculate the color appearance attribute of tourmaline.

2.4. Calculation Steps of CIECAM 16 Color Appearance Model

2.4.1. Enter parameters

1) Sample relative tristimulus values X, Y, Z.

2) Reference white point tristimulus values Xw, Yw, Zw under test conditions.

3) Adaptive field brightness $LA = E/5\pi$.

4) Background relative brightness Yb.

5) Lightness contrast factor F, surrounding environment factor c, color induction factor Nc.

6) Reference white dot isoenergetic white Xwr = Ywr = Zwr = 100. It is applied in CAT16 color adaptive transformation.

7) Ambient environment selection: Average, Dim, Dark.

8) Standard light source selection: D65 light source, D50 light source, A light source
9) Three illumination levels under D65 standard light source.

The relative tristimulus values of the samples can be quickly and accurately measured by various color measuring instruments. In this paper, tristimulus values X, Y and Z of 53 tourmaline samples have been measured by Color i5 colorimeter. The reference white point tristimulus values X_W, Y_W, Z_W can be obtained by looking up the table below. The background relative brightness is the brightness of the background relative to the reference white. The background brightness is set to 20. After the environmental conditions are selected, the parameters F, c and Nc can be obtained by looking up the table below.

2.4.2. Color adaptation transformation

1) Transform from the tristimulus value $X_1Y_1Z_1$ of the sample to the normalized cone response space RGB.
2) Transform the adaptive front cone response to the adaptive rear cone response.
3) Calculation of dynamic response function.

2.4.3. Nonlinear Compression before calculating the color appearance, there is a nonlinear response compression process to simulate the dynamic non-response of the cone response of the vision system Linear characteristic.

2.4.4. Color appearance attribute calculation the value of the perceived color appearance attribute is calculated, and the output parameters are: lightness J, apparent lightness Q, chroma C, apparent chroma M, saturation s, and hue angle h.

2.5. Experimental Conditions

2.5.1. Under different illuminance

The standard light source is D65 light source. The determination of theoretical illuminance value mainly refers to the International Lighting Committee Lighting Standard and the national standard GB/T26189-2010 Lighting for Indoor Workplaces. Set to 3 groups, namely 750lx, 1000lx and 1500lx. Through calculation, the corresponding adaptive field brightness is LA1=47.75cd/m^2, LA2=63.66cd/m^2, and LA3=95.49 cd/m^2, respectively. The tristimulus values X_W, Y_W and Z_W of the adaptive field white point are the coordinates X_W=95.04, Y_W=100 and Z_W=108.88 of the adaptive field white point under D65 light source. Select the general environment and look up the table to wait for relevant parameters. The surrounding environment factor c=0.69, the color induction factor Nc=1.0, and the Lightness contrast factor F=1.0. The remaining parameters remain unchanged.

2.5.2. Under different environments

The standard light source is D65 light source. LA2=63.66cd/m^2 is selected for adaptive field brightness. The tristimulus values XW, YW and ZW of the adaptive field white point are the coordinates of the adaptive field white point under D65 light source. Three kinds of ambient environment, general environment, Dim environment and Dark environment, were selected respectively. By looking up the table, the surrounding environment factor c, color induction factor Nc and Lightness contrast factor F can be obtained. The remaining parameters remain unchanged.

Table 1-1. Environmental Factors in Three Ambient Environments in CIECAM 16 Color Appearance Model

Surrounding environment	c	Nc	F
General environment	0.69	1.0	1.0
Dim environment	0.59	0.9	0.9
Dark room environment	0.525	0.8	0.8

2.5.3. Under different light sources

D65 light source, D50 light source and A light source are respectively selected as standard light sources. Adaptive field brightness LA2=63.66cd/m^2.Three stimulation values X_W, Y_W and Z_W for adaptive field white point are selected according to the light source, as shown in the table. Select the general environment and look up the table to wait for relevant parameters. The surrounding environment factor c=0.69, the color induction factor Nc=1.0, and the Lightness contrast factor F=1.0 . The remaining parameters remain unchanged.

Table 1-2. Parameters of Three Standard Light Sources

Light Sources	X_W	Y_W	Z_W	Color temperature (K)	Remarks
D65	95.04	100	108.88	6504	Northern day natural sunlight
D50	96.42	100	82.51	5003	Horizon light
A	109.85	100	35.58	2856	Incandescent lamp/tungsten lamp

3. CIECAM16 Color Appearance Model Predicts Color Appearance of Green Tourmaline

3.1. Color Appearance Prediction of Tourmaline Color under Different Illuminations

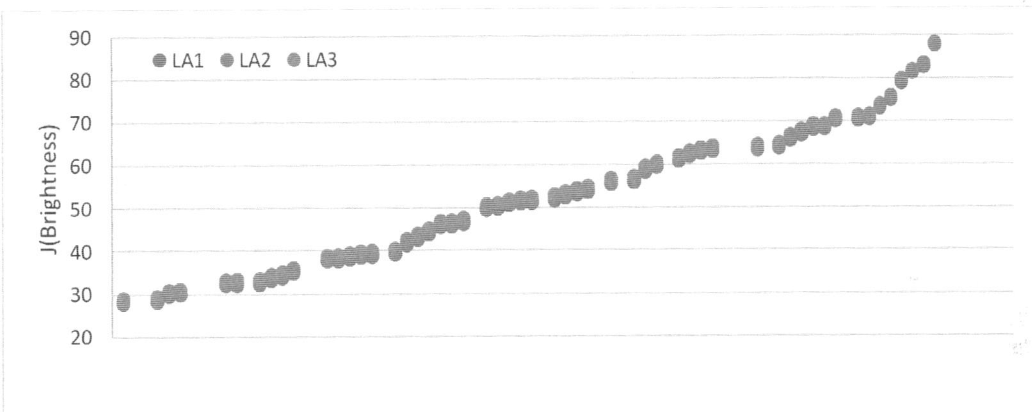

Figure 1-1. Distribution of Sample Brightness under Different Illuminations

Figure 1-2. Distribution of Sample Apparent Value under Different Illuminations

From LA1 to LA3, the light intensity increases gradually. As shown in figures 1-1, the samples are arranged in ascending order according to the lightness value under LA1 adaptive field brightness. The abscissa is the sample number and has no specific numerical meaning (the scatter chart below is the same). The ordinate is the sample lightness value. The lightness J of green tourmaline has little difference under the three illuminances. The average values of lightness J under the three illuminances are 52.37, 52.86 and 53.63 respectively. The lightness J is a relative brightness, and the change of illuminance intensity does not make the relative brightness ratio between green tourmaline in human vision and the brightest area in the corresponding environment change obviously. However, the apparent brightness Q of tourmaline gradually increases with the increase of illuminance, and the change is obvious compared with the brightness. As shown in figures 1-2, the average value of tourmaline apparent brightness q under illuminance LA1 is 721.87, the average value of q under illuminance LA2 is 820.24, and the average value of q under illuminance LA3 is 979.66. This indicates that the overall amount of light perceived by human eyes on tourmaline has increased. With the increase of illuminance, the increase of brightness Q becomes larger. Comparing the difference between the maximum value and the minimum value of the apparent brightness Q under the three illuminances, the increase rate of illuminance LA2 is 12.19% under the relative illuminance LA1, and the increase rate of illuminance 3 is 31.3% under the relative illuminance 1. This shows that under the condition of increasing brightness illumination,the light-dark difference between tourmaline specimens becomes larger, and the difference of apparent brightness between the brightest specimen and the darkest specimen increases with the increase of illuminance. The change result of brightness Q confirms Stevens effect of color appearance phenomenon. Stevens effect shows that the contrast of apparent brightness increases with the increase of brightness.

Table 2-1. Average Values of Luminance J, Apparent Luminance Q, Apparent Color M and Saturation S under Different Illuminations

	LA1=47.75	LA2=63.66	LA3=95.49	Illumination LA2 is based on the increase of illumination LA1	Illumination LA3 is based on the increase of illumination LA1
J	52.37	52.86	53.62	0.93%	2.40%
Q	721.87	820.24	979.66	13.62%	35.71%
M	19.22	19.53	19.94	1.6%	3.73%
s	16.08	15.20	14.06	-5.44%	-12.583%

Table 2-2. Table of Changes of Tourmaline Color Appearance Attribute Apparent Value Q under Different Illuminations

	Maximum value	Minimum value	The difference between the maximum and minimum values	Difference increase (based on illuminance 1)
LA1=47.7464	945.05	531.51	413.54	——
LA2=63.6619	1069.93	605.96	463.97	12.19%
LA3=95.4930	1270.56	727.55	543.01	31.3%

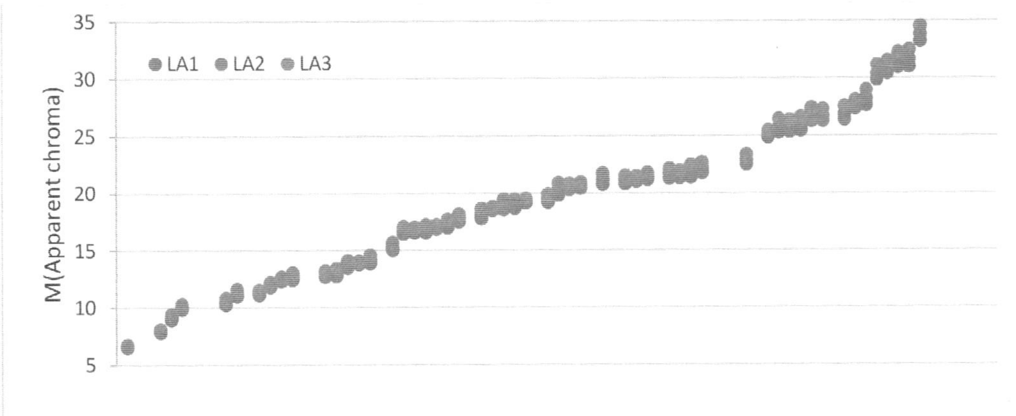

Figure 1-3. Distribution of Sample Apparent Colorimetry under Different Illuminations

Stenves effect is closely related to Hunt effect. As shown in figures 1-3, the apparent chroma M of tourmaline specimen increases with the increase of illuminance. The CIECAM16 color appearance model predicts that the apparent chromaticity change of tourmaline color under different illuminance conforms to Hunt model, that is, the apparent chromaticity increases with the increase of brightness. The results show that when the lighting conditions increase, the illuminance increases and the brightness increases, the tourmaline color in human eyes will increase and the tourmaline color will be more vivid. On the other hand, if the lighting condition is weakened, the illuminance becomes lower, and the brightness decreases, a higher visual chroma is required to match the same color perception before.

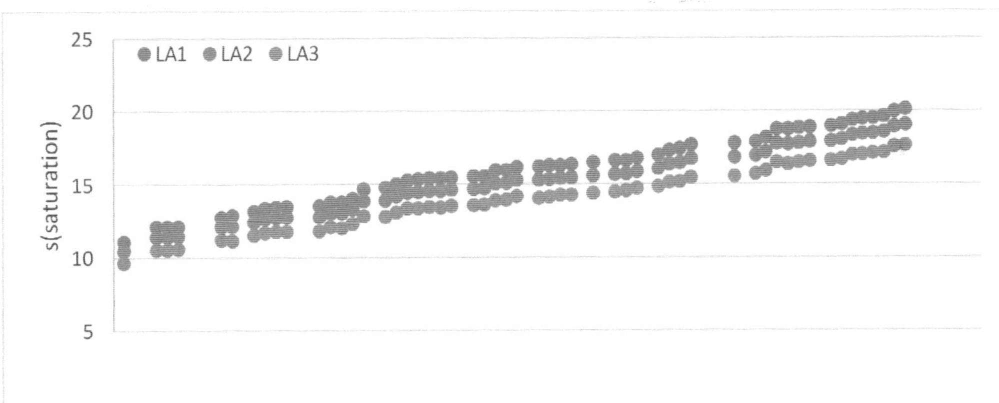

Figure 1-4. Distribution of Sample Saturation under Different Illuminations

At the same time, as shown in fig. 1-4, the tourmaline color saturation s decreases with the increase of illuminance. Saturation is the relative ratio of apparent chroma and apparent lightness in human vision, that is, saturation = apparent chroma/apparent lightness. Apparent brightness and apparent chroma both increase with the increase of illuminance, but the range of change of apparent brightness affected by illuminance is larger than apparent chroma, so the value of saturation decreases with the increase of illuminance. Based on the above analysis, it can be concluded that when the illuminance increases, the tourmaline color will be more bright, but the saturation will decrease, that is, the difference between the color brightness of tourmaline will become larger.

3.2. Color Appearance Prediction of Tourmaline Color under Different Environmental Changes

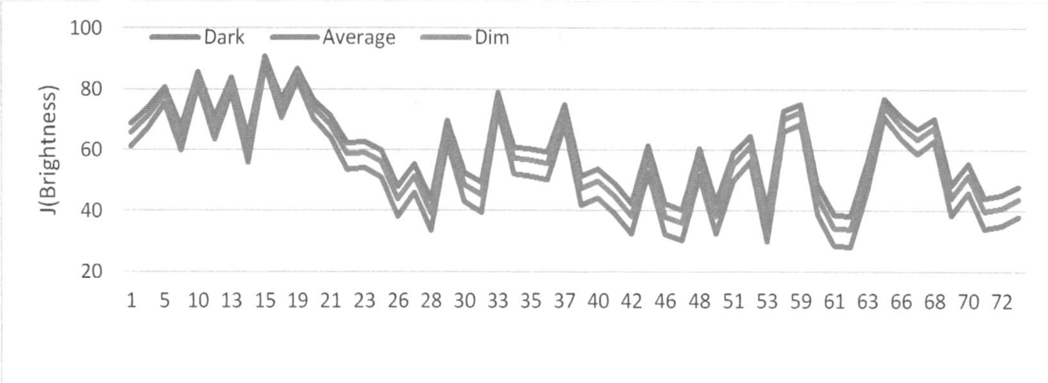

Figure 1-5. Distribution of Sample Brightness in Different Environments

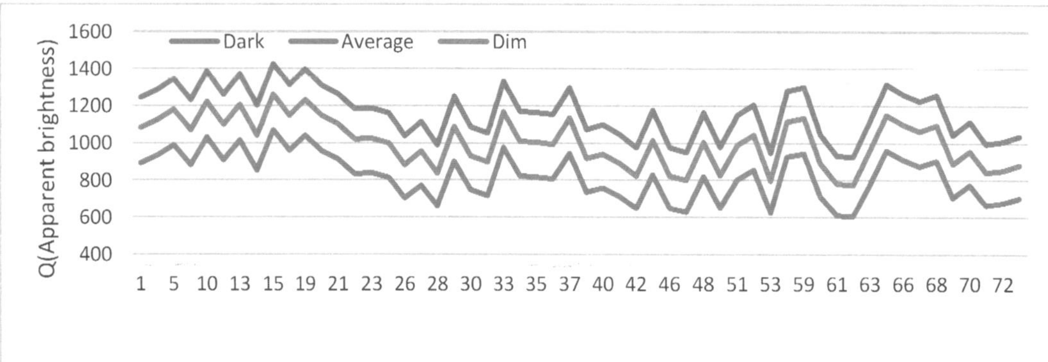

Figure 1-6. Distribution of Sample Apparent Value in Different Environments

On the premise of the same illumination and type of light source, changing the surrounding environment parameters in the model and simulating the Average→Dim→Dark environment respectively. As shown in figures 1-5 and 1-6, the abscissa is the specimen number and the ordinate is the corresponding value. It can be concluded that the lightness J and the apparent lightness Q of tourmaline increase in the process of environmental change. The color appearance model simulates that the relative brightness of tourmaline is increased compared with the brightest light source in the response environment, and the brightness stimulus intensity of tourmaline itself is also increased in human vision, which is due to the brightness compensation phenomenon in the process of environmental change.

Table 2-3. Table of Average Value and Increase of Green Tourmaline Color Appearance Attribute Luminosity J in Different Environments

	Average	Dim	Dark
Average of J	52.86	57.65	61.06
Increase (based on Average)	——	9.07%	15.52%
Increase (Increase progressively)	——	9.07%	5.91%

CIECAM16 color appearance model simulates the changes of results of lightness J and apparent lightness Q in different surrounding environments, and verifies Bartleson-Breneman effect in color appearance phenomenon. Bartleson-Breneman effect points out that in the surrounding environment, when the brightness of the surrounding environment changes from bright to dark, the perceived contrast will decrease, the change of bright area is not obvious, but the dark area will appear brighter. However, when the surrounding environment becomes brighter, the perceived contrast will increase because the change in bright areas is not obvious, while the dark areas will look darker. Because tourmaline lightness belongs to medium and low lightness, it is in dark area in environmental transformation. Therefore, in the process of environment transformation from Average→Dim→Dark, the brightness of the environment becomes dark, thus tourmaline. As a result, the green tourmaline that appear as dark areas are brighter.

Figure 1-7. Distribution of Sample Chromaticity in Different Environments

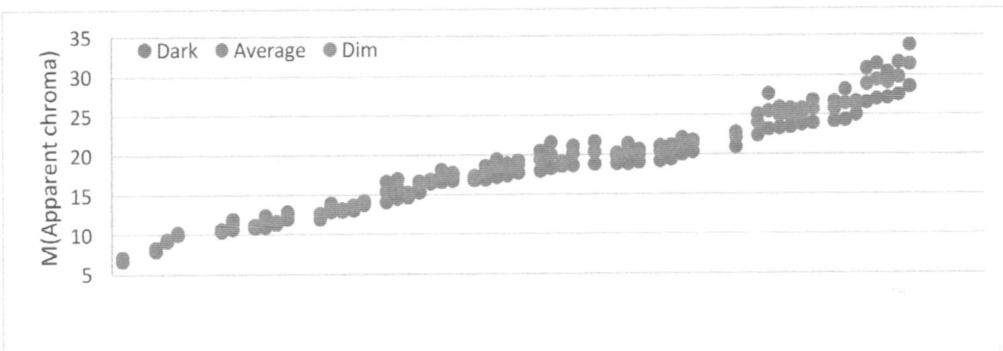

Figure 1-8. Distribution of Sample Apparent Colorimetry in Different Environments

As shown in figures 1-7 and 1-8, the chroma C and apparent chroma M of tourmaline have the same trend of change and both values decrease during the transformation of the surrounding environment from Average→Dim→Dark. This shows that as the surrounding environment darkens, the changes in the surrounding environment have a significant impact on the color of tourmaline.

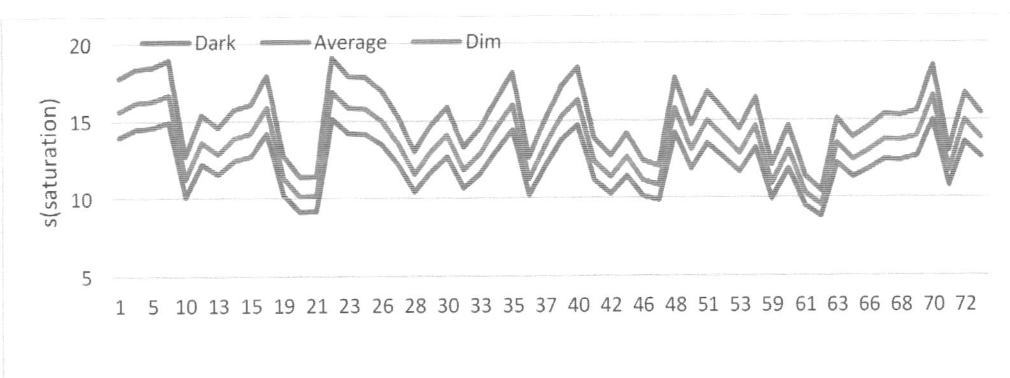

Figure 1-9. Distribution of Sample Saturation in Different Environments

As shown in figures 1-9, the color saturation s of tourmaline shows a clear and regular change trend under the transformation of different environments. When the surrounding environment changes from light to dark, the saturation s gradually decreases. The Average value of saturation s in the Average environment is 15.24, while the average value of saturation s in the Dim environment is 13.53, a decrease of 11.25%. Under Dark environment, the Average saturation S is 12.21, which is 19.88% lower than the average.

Table 2-4. Average and Increase of Color Appearance Attribute Saturation S of Green Tourmaline in Different Environments

	Average	Dim	Dark
Average of S	15.24	13.53	12.21
Increase (based on Average)	-	-11.25%	-19.88%
Increase (Increase progressively)	-	-11.25%	-9.73%

Figure 1-10. Distribution of Sample Tone Angle in Different Environments

Under other conditions being the same, as shown in figures 1-10, the Average value of hue angle h of tourmaline in average environment is 138.58, the average value of hue angle h in Dim environment is 140.24, and the average value of hue angle h in Dark environment is 141.85. Therefore, it can be seen that the hue angle of tourmaline does not change obviously with the change of light and shade in the surrounding environment. Generally, the hue angle increases in the process of transformation from Average→Dim→Dark in the surrounding environment.

3.3. Color Appearance Prediction of Tourmaline Color under Different Light Sources

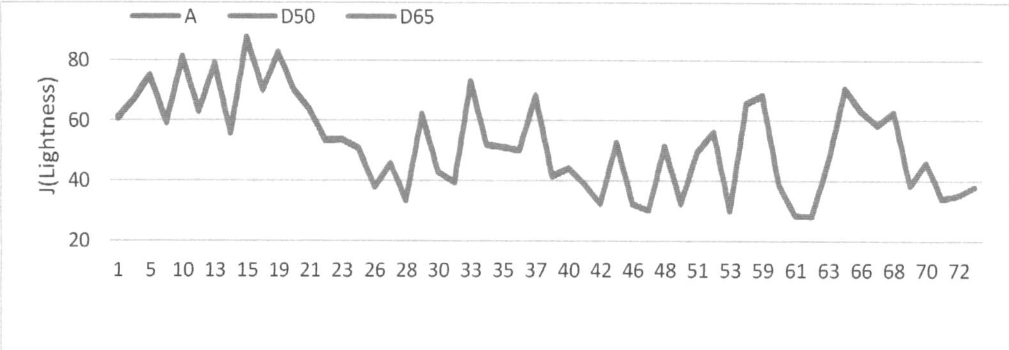

Figure 1-11. Distribution of Sample Brightness under Different Light Sources

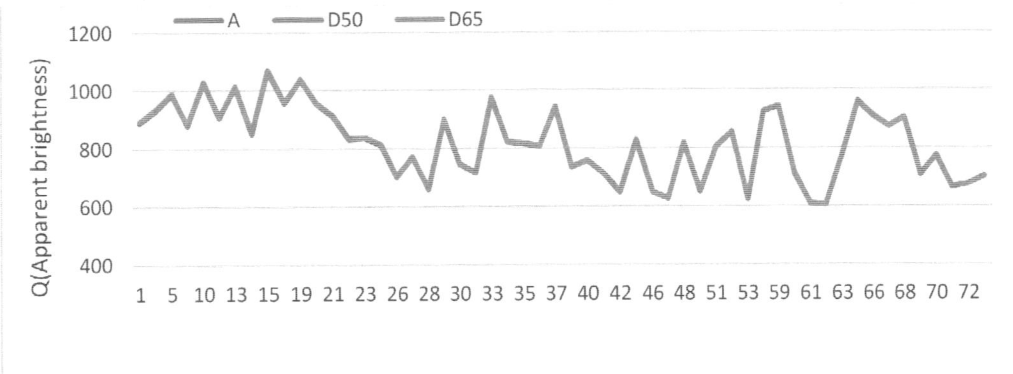

Figure 1-12. Distribution of Sample Visibility under Different Light Sources

When other observation conditions are the same, change different illumination sources, as shown in fig. 1-11 and 1-12, and the lightness of tourmaline specimen changes little. Under D65 light source, the lightness value is 52.86 and the apparent lightness is 820.24. Under D50 light source, the lightness value is 52.77 and the apparent lightness value is 819.63. Under A light source, the lightness value is 52.34 and the apparent lightness value is 816.99. It shows that the illumination source has little effect on brightness.

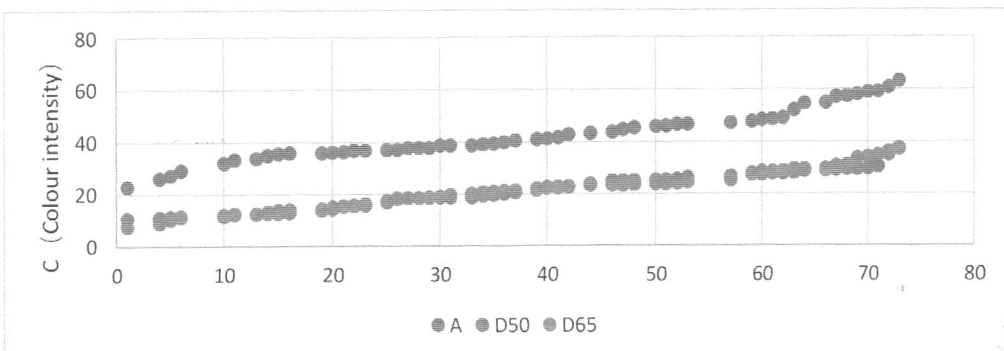

Figure 1-13. Distribution of sample chroma under different light sources

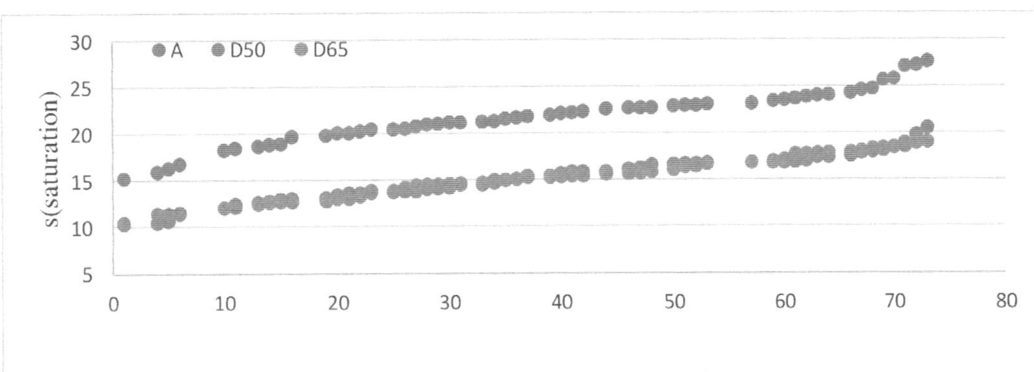

Figure 1-14. Distribution of Sample Saturation under Different Light Sources

Comparing the color intensity of tourmaline specimens under three different light sources, we can judge by the chroma C and saturation s. As shown in Figures 1-13 and 1-14, under A light source, the chroma C and saturation s of tourmaline color are the highest, indicating that tourmaline color is more vivid when viewed under A light source. The next is D50 light source and D65 light source, which have no obvious difference.

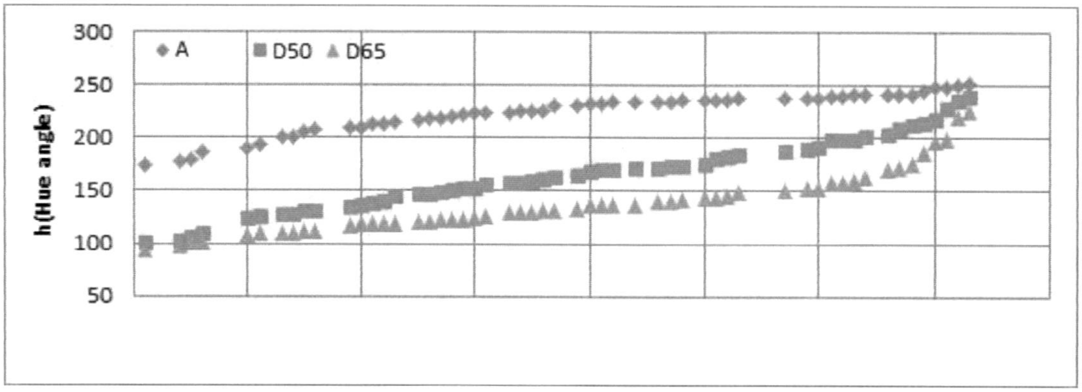

Figure 1-15. Distribution of Sample Tone Angle under Different Light Sources

Under the three kinds of light sources, the hue H and hue angle h of tourmaline specimens show obvious differences. The hue angle h is used to describe the hue of the color, and the hue is used to describe the visual attribute of the representation of the main color. As shown in fig. 1-15, under D65 light source, the average value of hue H is 171.82 and the average value of hue angle h is 138.58; Under D50 light source, the average value of hue H is 206.85 and the average value of hue angle H is 165.95; Under A light source, the average value of hue H is 282.54 and the average value of hue angle H is 224.41. In CIECAM16 color appearance model, the standard blue color H=300, h=237.53; Green H=200, h=164.25; Red H=400, h=380.14. With the decrease of color temperature, the hue angle of tourmaline samples adapting to the field should shift from green to yellow to green to blue under experimental conditions. The higher the color temperature, the more blue-tinted the light source. For example, the D65 light source for simulating blue sky sunlight, the A light source is a warm light source with yellow-tinted. When using the D65 light source to illuminate the tourmaline green sample, the green color should be blue-tinted, while when using the A light source to illuminate, the tourmaline green color should be yellow-tinted. The data calculated by CIECAM16 color appearance model is contrary to this result. The data show that when the illumination light source is changed from D65 light source with blue hue to A light source with yellow hue, the hue angle of tourmaline sample shifts to blue region instead, and the variable amplitude of hue angle under A light source decreases with the increase of hue angle. Because the CIECAM16 color appearance model predicts the Helson-Judd effect, that is, when the sample is illuminated by a colored light source, if the sample is brighter than the background, the hue of the sample tends to be the same as that of the light source, and if the sample is darker than the background, the hue of the sample tends to assume the complementary color of the light source. In the experimental conditions, the brightness of the background is greater than the sample itself, so when the blue light source D65 irradiates the sample, the final color appearance of tourmaline tends to be orange yellow, which is the complementary color of the light source color, making the hue angle of the sample yellow. When the light source A with yellow tone irradiates the sample, the final color appearance of the sample tends to present the complementary color of the light source A-blue, so the hue angle of the sample shifts to the blue region.

The brightness and apparent brightness of tourmaline under A light source do not change much with D65 light source and D50 light source, but they have better chroma and saturation, and the human vision is more vivid. This is the reason why many businesses choose A light source in their sales places. However, due to the Helson-Judd effect, the hue of the tourmaline is obviously shifted to the blue area, and the color of the green tourmaline cannot be directly reflected under the A light source. Therefore, D65 light source is more suitable for the sale and display of the green tourmaline under the comprehensive evaluation, while D50 light source is suitable for color comparison and inspection of tourmaline specimens.

4. Conclusion

1) The change of illuminance under D65 light source has little effect on the lightness J of green tourmaline, but the apparent lightness Q increases with the increase of illuminance, and the changing amplitude increases. The apparent chroma M increases with the increase of illuminance, and the saturation decreases. When the illuminance increases, the tourmaline will be brighter and brighter, and the difference in color and brightness between specimens will become larger.

2) When the environment changes from Average→Dim→Dark, the lightness J and apparent lightness Q of tourmaline increase, while the chroma C and apparent chroma M decrease. The saturation s of tourmaline gradually decreases. It has little influence on the hue angle h of tourmaline, and the value of h increases slightly.

3) Changing different light sources (D65, D50, A) has no obvious influence on tourmaline brightness and visual brightness. Under the A light source, the chroma C and saturation S of tourmaline color are the largest, and the tourmaline color is more brightly observed. However, due to the Helson-Judd effect. Therefore, the A light source cannot well display the color of tourmaline itself. On the whole, D65 light source is more suitable for the sale and display of green tourmaline. D50 light source is suitable for color comparison and inspection of tourmaline specimens.

5. References

[1] Zou Tianren, Yang Yueqing. Color and Composition of Chinese Tourmaline (Tourmaline) [J]. Deposit Geology, 1996, 15 (S1):65-68.
[2] Wang Jinjun, Tao Xiaofeng, Wang Wujun. Study on the Color Genesis of Xinjiang Green Tourmaline [J]. Journal of Rock Mineralogy, 2005 (04):319-323.
[3] Yang Yueqing, Wang Yong, Lu Qingtian, Zhao Jinhua. Study on Tourmaline in Nanping Granite Pegmatite in Fujian [J]. Journal of Rock Mineralogy, 2010, 29 (03):235-242.
[4] Harald G. Dill, Mirta M. Garrido, Frank Melcher, Maria C. Gomez, Liliana I. Luna. Depth-related variation of tourmaline in the breccia pipe of the San Jorge porphyry copper deposit, Mendoza, Argentina [J]. Ore Geology Reviews, 2012, 48.
[5] S. O. Olabanji, O. A. Ige, C. Mazzoli, D. Ceccato, J. A. Akintunde, M. De Poli, G. Moschini. Accelerator-based analytical technique in the evaluation of some Nigeria's natural minerals: Fluorite, tourmaline and topaz [J]. Nuclear Inst. and Methods in Physics Research, B, 2005, 240 (1-2).
[6] He dengliang, dong faqin. environmental functional properties of tourmaline and its application research trends [J]. China nonmetallic mineral industry guide, 2006 (01):10-14.
[7] Mariano Mercurio, Manuela Rossi, Francesco Izzo, Piergiulio Cappelletti, Chiara Germinario, Celestino Grifa, Maurizio Petrelli, Alessandro Vergara, Alessio Langella. The characterization of natural gemstones using non-invasive FT-IR spectroscopy: New data on tourmalines [J]. Talanta, 2018, 178.
[8] Nilo F. Cano, T. K. Gundu Rao, Jorge S. Ayala-Arenas, Carlos D. Gonzales-Lorenzo, Letícia M. Oliveira, Shigueo Watanabe. TL in green tourmaline: Study of the centers responsible for the TL emission by EPR analysis [J]. Journal of Luminescence, 2018.
[9] K. Krambrock, M. V. B. Pinheiro, S. M. Medeiros, K. J. Guedes, S. Schweizer, J.-M. Spaeth. Investigation of radiation-induced yellow color in tourmaline by magnetic resonance [J]. Nuclear Inst. and Methods in Physics Research, B, 2002, 191 (1).
[10] Zhang Lili, Yuan Xinqiang. Green Emerald Color Grading Technology Based on HSL Color Space Model [J]. Science, Technology and Engineering, 2017, 17 (26) 268-273.
[11] Ying GUO, Huan WANG, Xiang LI, Shirong DONG. Metamerism Appreciation of Jadeite-Jade Green under the Standard Light Sources D₆₅, A and CWF [J]. Acta Geologica Sinica - English Edition, 2016, 90 (6).
[12] Yin Ke, Tian Jian, Ma Yubo, Wu Yu, Wang Yang. Mineralogical Characteristics and Color Genesis of "Youqing Species" Jadeite [J]. Spectroscopy and Spectral Analysis, 2014, 34 (12):3323-3326.
[13] Wang Yongya, Gan Fuxi. Chromogenic mechanism and colorimetry of xiuyan jade, China [J]. Spectroscopy and Spectral Analysis, 2012, 32 (09): 2305-2310.

[14] Guo Ying, Sha Sha, Zhang Chengcheng. Effect of different achromatic background brightness on emerald green brightness [J]. Journal of Minerals, 2010, 30 (S1): 41-42.

[15] Luo Hongyu, Peng Mingsheng, Liao Shangyi, Gao Lisheng. Color-forming Mechanism of Emerald and Metamorphic Stone [J]. Modern Geology, 2005 (03): 355-360.

[16] Florian Schiller, Matteo Valsecchi, Karl R. Gegenfurtner. An evaluation of different measures of color saturation [J]. Vision Research, 2017.

[17] Wu Jiahui, Guo Ying, Deng Xia, Fang Yibin. CIE1976 L~*a~*b~* Evaluation of Illuminance on Manganese Aluminide Color in Uniform Color Space [J]. Silicate Bulletin, 2014, 33 (10): 2752-2760.

[18] Guo Ying, Zhang Jun, Mo Tao. Quality Evaluation of Emerald Green Brightness Based on CIE 1976 L~*a~*b~* Color Space [J]. Silicate Bulletin, 2010, 29 (03): 560-566.

[19] Mark D. Fairchild. Color appearance models and complex visual stimuli [J]. Journal of Dentistry, 2010, 38.

[20] Li Jincheng, Serina Liu. Research Progress of CIECAM02 Color Appearance Model [J]. imaging science and photochemistry, 2009, 27 (05):391-398.

[21] CIE Publication 159: 2004. A colour appearance model for colour man-agement systems: CIECAM02. Vienna: CIECentral Bureau; 2004.

[22] Snjezana Soltic, Andrew N. Chalmers, Radhika Jammalamadaka. Application of the CIECAM02 colour appearance model to predict the effect of gamma on the colours viewed on CRT monitors [J]. Signal Processing: Image Communication, 2004, 19 (10).

[23] Andrew Thwaites, Cai Wingfield, Eric Wieser, Andrew Soltan, William D. Marslen-Wilson, Ian Nimmo-Smith. Entrainment to the CIECAM02 and CIELAB colour appearance models in the human cortex [J]. Vision Research, 2018, 145.

[24] Yang Xiaoli, Yang Weiping, Yang Weiguo, Dai Zhifu, Sun Bo, Feng Jie, Shi Junsheng. Comparative Study on Color Space Uniformity of CIELAB and CIECAM02 [J]. Optical Technology, 2008 (04): 620-622+625.

[25] Dong Shuwen, Liao Ningfang, Song Li, Liang Jing, Lin Kai. CIECAM02 Model Predicts and Evaluates Chromatographic Color Appearance of Printing with Different Gloss [J]. journal of beijing institute of technology, 2014, 34 (08): 848-852.

[26] LuoMR, CuiGH, LiCJ. Uniform colour spaces based on CIECAM02 colour appearance model. ColorResAppl. 2006; 31: 320–330.

[27] Kuo C, Zeise E, Lai D. Robust CIECAM02 implementation and numerical experiment within an international color consortium workflow. J Imaging Sci Technol. 2008; 52:61.

[28] Kuo C, Zeise E, Lai D. Robust CIECAM02 implementation and numerical experiment within an ICC workflow. In: Proceeding of Fourteenth Color Imaging Conference (CIC14); 2006; Scottsdale AZ: IS&T. 215–219.

[29] Li CJ, Luo MR. Testing the robustness of CIECAM02. Color Res Appl. 2005; 30:99–106.

[30] Li CJ, Luo MR, Wang ZF. Different matrices for CIECAM02.Color Res Appl. 2014; 39:143–153.

[31] Li CJ, Luo MR, Sun PL. A modification of CIECAM02 based on the Hunt Pointer Estevez matrix. J Imaging Sci Technol. 2013; 57:1–8.

IOP Conf. Series: Materials Science and Engineering **746** (2020) 012006 doi:10.1088/1757-899X/746/1/012006

Analysis of Stress Distribution of Double-layer Shoulder Beam of Concrete-filled Steel Tubular

Yang Wang[1], Baofeng Qi[2] and Yao Peng[2]

[1]China Shipbuilding Industry Corporation International Engineering Co. Ltd, Beijing 100121, China
[2]Unit 91053 of Navy, Beijing 100071, China
Email: wangyang@csic602.com.cn

Abstract: At present, single-layer shoulder beam can not meet the operation of large-tonnage bridge cranes with different elevations. However, due to the complexity of the bearing forms of double-layer shoulder beam, there is no corresponding design formula in the code. In order to study the one-way pushing force performance of concrete filled steel tubular (CFST) double-layer shoulder beam with mid-column, combined with the prototype of the shoulder beams in a shipbuilding building, the scale of the beam was analyzed by ABAQUS software, and the P-Δ curve was calculated, the failure process and the stress mechanism were described, and the stress distribution law of lower shoulder beams in elastic stage was emphatically analyzed.

1. Introduction

Shoulder Beam is a transfer member that transmits the internal force form the upper column to the lower column and acts as a crane beam support, which is widely used in heavy factory. Because the shoulder Beam is an important part of the design of industrial plants, it has the characteristics of high bearing capacity and large lateral stiffness, which can effectively ensure the safe operation of the bridge crane [1]. As shown in Figure 1, the traditional double-story shoulder beam adopts a variable-section column to place the crane beam of the upper crane, its calculation model is also based on the calculation model of the single-layer shoulder beam. However, due to the complexity of the bearing forms of double-layer shoulder beam [2], it cannot be applied to existing design methods.

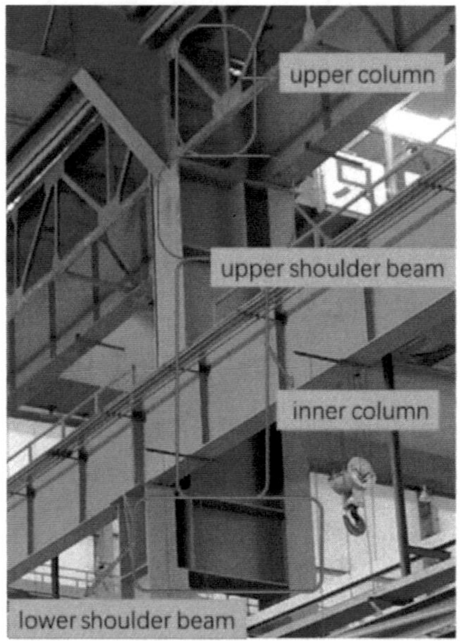

Figure 1. Double-layer shoulder beam of a shipbuilding building

Domestic scholar Jin Tiande, Rao Zhiying [3] made a finite element analysis of the single-layer shoulder beam of the concrete filled steel tubular in 1997. The experiment shows that the distribution of the normal stress of the web of the shoulder beam is not in accordance with the flat section assumption. In the form of failure, the shear failure of the shoulder web causes damage to the overall failure of the member. In this year, Shen Zuyan and Zheng Yi [4, 5] of Tongji University studied the ultimate bearing capacity and stiffness of concrete-filled steel tubular single-slab shoulder beam, and obtained the theoretical formula for calculating the ultimate bearing capacity of shoulder beam. In 2017, Lan Tao, Hu Weizhong, Men Jinjie [1, 6] and others established the finite element model of column shoulder beam of concrete filled steel tubular double-layer double-limb, and gave the calculation model and formula of double-layer shoulder beam.

2. Establishment of Finite Element Model

2.1. Model Design

The lower column of the double-layer shoulder beam studied in this paper adopts a circular steel tubular concrete column and it is inserted into the shoulder beam web range. In addition, adds a long stiffener at the shoulder beam concentration. As shown in picture 2. The upper column and the inner column are made of steel columns with a section of H 220×140×4×8 and a height of 300 mm on the upper column. Two stiffeners are added to the inner pillar height, and the inner limb column is separated from the right concrete tubular column. The upper shoulder beam is H 230×180×4×6, the lower shoulder beam is H 260×180×4×6, the height of the lower column is 600 mm, the web is Φ50×4 and the tilt angle is 450.

IOP Conf. Series: Materials Science and Engineering **746** (2020) 012006 doi:10.1088/1757-899X/746/1/012006

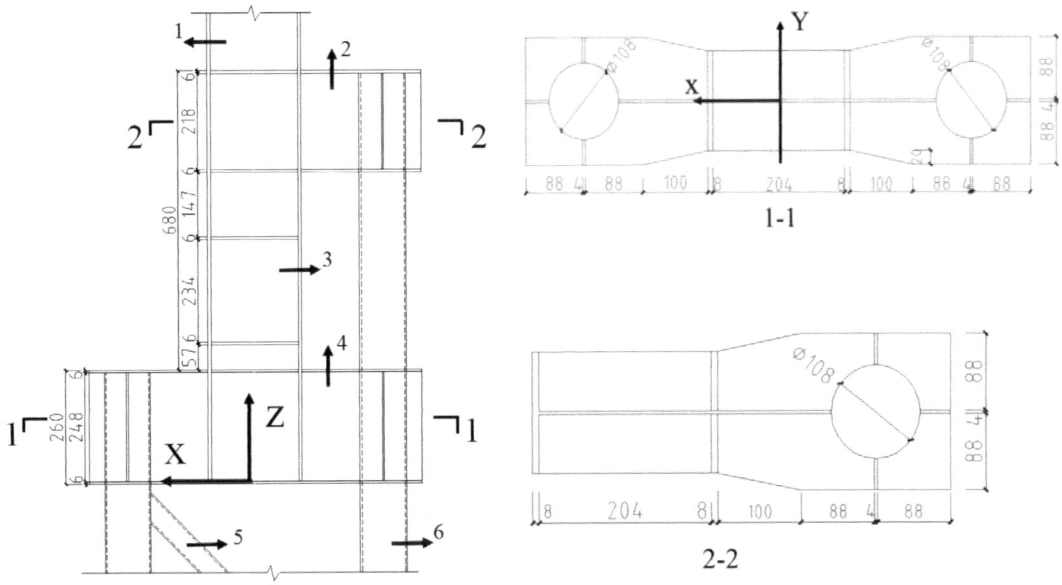

Figure 2. Double-layer shoulder beam (1-upper column; 2- upper shoulder beam; 3- inner column; 4- lower shoulder beam; 5- web; 6- lower column)

2.2. Unit type Selection and Meshing

C3D8R (8-node linear reduction integral solid element) is used for Steel pipe, diagonal bracing, concrete use, and this unit is not sensitive to the shape of the unit. Compared with the complete integral unit it can effectively reduce the calculation cost [7]. And C3D20R (20-node secondary reduction integral physical unit) is used for the upper column, the upper shoulder beam, the inner column, and the lower shoulder beam. At the opening of the shoulder beam the mesh is reasonably refined [6].as shown in Figure 3.

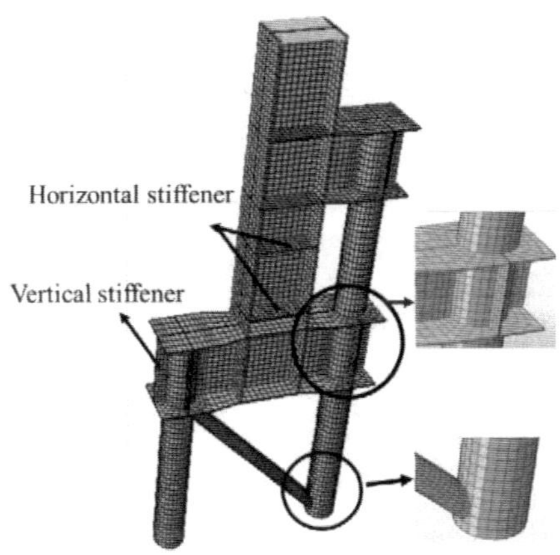

Figure 3. Meshing of finite model

2.3. The Constitutive Relationship of Materials

Selection of steel Q235B, Elastic modulus E=210 GPa, Poisson's ratio γ=0.3, yield stress Fy=230 Mpa.

Selection of Concrete C40. The Elastic modulus E=32.5 GPa, Poisson's ratio γ=0.2. The uniaxial stress-strain relationship of concrete is recommended by GB50010-2002 "Concrete Structure Design Code" [8].

2.4. Boundary Conditions and Interactions
The coupling point of the bottom of the concrete-filled steel tubular column is consolidated by simulation. The upper surface of the upper column and the horizontal load point of the lower layer are constrained by the out-of-plane displacement (Uy) and the in-plane rotation (URz).

2.5. Loading Method
The load is loaded in three steps, first applies the vertical load once and keeps invariant. The second step applies the lower horizontal load once and keeps invariant. The third step applies the horizontal load of the upper crane and uses displacement control until the component is destroyed.

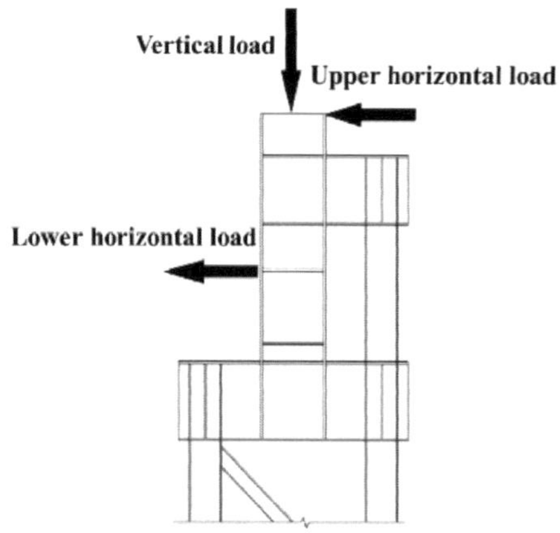

Figure 4. Location and direction of load action

3. Finite Element Analysis and Force Performance of Shoulder Beam Failure Process
After the vertical load is applied, most of it is transmitted to the lower shoulder beam. Due to the structural asymmetry, the left shoulder stress of the lower shoulder beam increases faster. As shown in Figure 5, the load is transmitted through the baroclinic belt of the left zone web. After the application of the lower horizontal load, the strain of the baroclinic belt of the left shoulder of the lower shoulder beam continues to develop. As the upper horizontal load is applied and gradually increases, the yield range expands rapidly until the full section of the left zone web yields. As shown in Figure 6. As the displacement of the top of the column continues to increase, the bearing capacity of the member reaches a maximum value. After that, the bearing capacity of the member decreases slowly. When the displacement of the top of the column reaches about 60 mm, the surface of the web is buckling and the bearing capacity drops rapidly. Then the lateral buckling pattern is shown in Figure 7 (the black line specifically indicates the out-of-plane buckling), the overall stiffness of the component is seriously degraded and cannot be carried. The overall failure mode of the member and the top load-displacement curve are shown in Figure 8 and Figure 9, respectively.

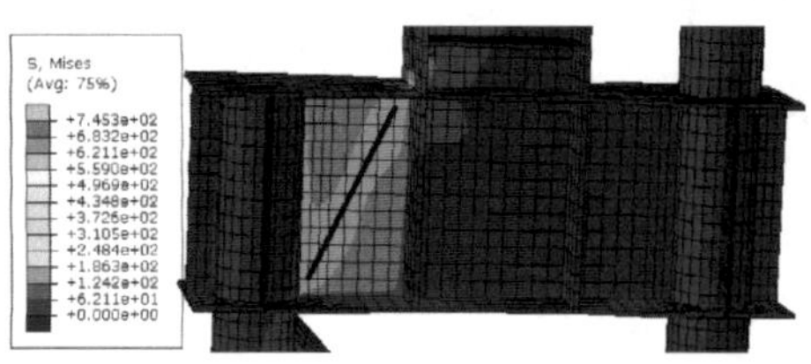

Figure 5. Baroclinic Belt of Lower Shoulder Beam

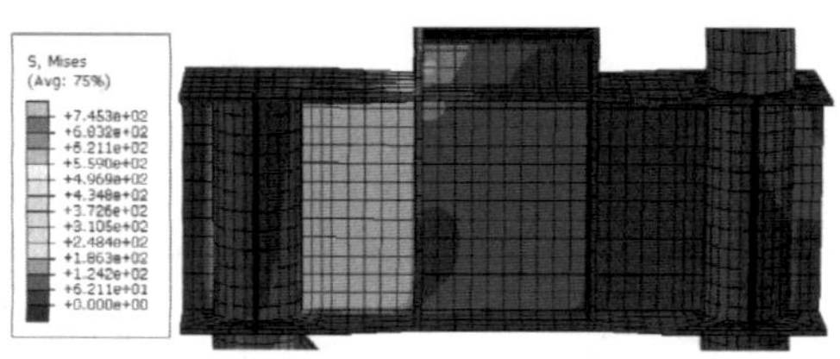

Figure 6. Full Section Yield of Lower Shoulder Beam

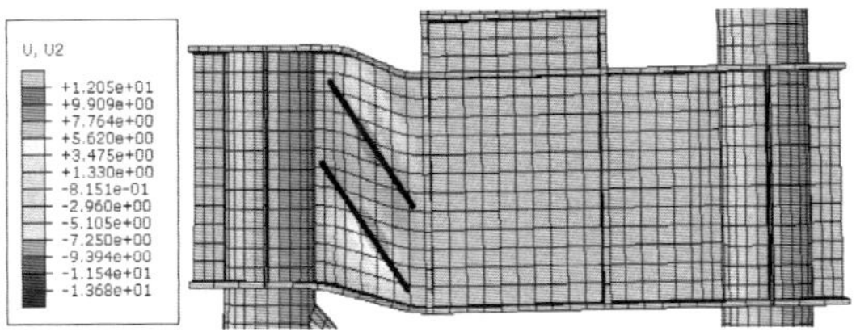

Figure 7. Outside buckling of web of lower shoulder beam

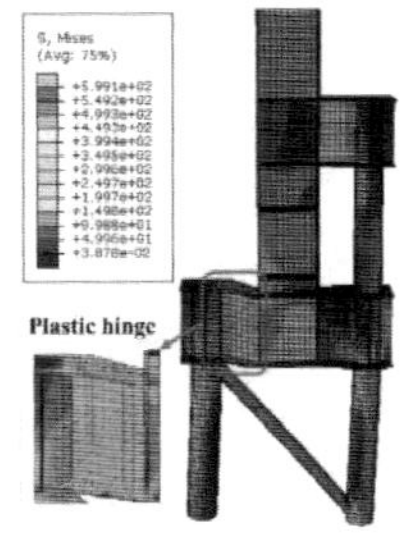

Figure 8. Overall failure mode of the component

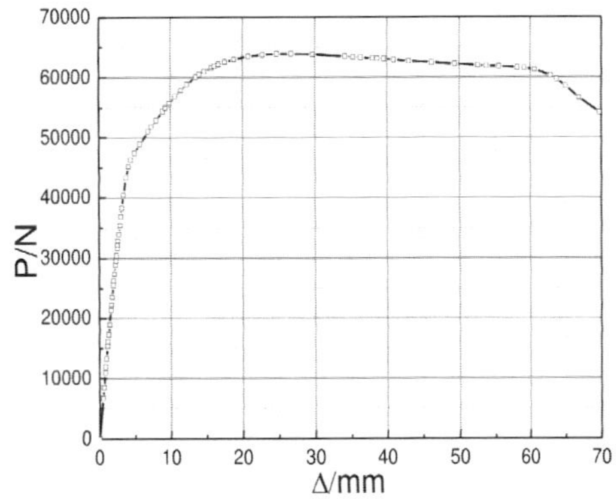

Figure 9. Load-displacement curve of column top

4. Finite Element Analysis of the Stress of the Lower Shoulder Beam

Stress analysis of the lower shoulder web at x=178 mm. As shown in Figure 10a, after the vertical load is applied, the compressive stress is small on both sides. When the horizontal displacement reaches 4 mm, the stress in the middle of the web suddenly drops below 5 MPa, and the distribution is irregular. As shown in Fig. 10b, after the vertical load is applied, the shear stress of the web is compressive stress, and most of the high-level stress is large. When the horizontal displacement reaches 4 mm, the shear yield stress is 135 MPa in the middle part of the web. In this time, the shear stress tends to be the same at all points of the web height. As shown in Fig. 10c, the general rule of the bending normal stress is that the upper portion of the upper portion is pulled. The pre-loading stress conforms to the flat section assumption, and the tensile stress value demarcation point is approximately at the center. After that, the upper compressive stress increases and the compression zone expands downward, which is seriously inconsistent with the flat section assumption.

IOP Conf. Series: Materials Science and Engineering **746** (2020) 012006 doi:10.1088/1757-899X/746/1/012006

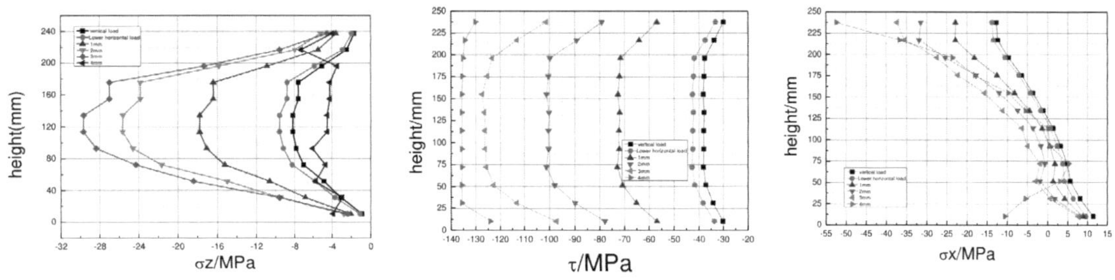

a. Extrusion stress b. Shear stress c. Bending stress

Figure 10. Stress Distribution of Lattice Web in Left Zone of Lower Shoulder Beam

5. Conclusion

(1) When the working conditions described in this paper are used, the lower shoulder beam is the main force-bearing part, and the lower shoulder beam web transmits the load through the diagonal pressure belt. The failure of the component is due to the shear failure of the lower shoulder web, the buckling occurs outside the surface, the member loses rigidity, and the bearing capacity decreases rapidly.

(2) The shoulder beam web is in a plane stress state, and the bending stress and the compressive stress value are low. In the late loading stage the bending stress does not conform to the flat section assumption; the shear stress distribution law is large in the middle part, and the stress on the two sides away from the oblique pressing belt is small.

6. References

[1] Hu Weizhong. Research on bearing capacity and design method of concrete filled steel tubular double shoulder beam [D]. Xi'an University of Architecture and Technology, 2017.

[2] Steel Structure Manual Editorial Board. Steel Structure Design Manual (Volume 1, Third Edition) [M]. Beijing: China Building Industry Press, 2004.

[3] Jin Tiande, Rao Zhiying. Stress Analysis of Concrete Filled Steel Tubular Column Shoulder Beams [J]. Zhejiang Architecture, 1997 (05):13-15+33.

[4] Zheng Yi. Analysis of strength and stiffness of concrete-filled steel tubular column shoulder beam [D]. Shanghai: Tongji University Master's Thesis, 1997.

[5] Shen Zuyan, Zheng Yi. Analysis of the stiffness of multi-limb shoulder beam [J]. Building Structure, 1999 (07): 50-52.

[6] Lan Tao, Hu Weizhong, Men Jinjie. Analysis of the mechanical behavior of concrete-filled steel tubular double-shoulder beam [J]. Steel Structures, 2017, 32 (06): 39-45.

[7] Shi Yiping, Zhou Yurong. Detailed examples of ABAQUS finite element analysis [M]. Beijing, Mechanical Industry Press, 2006.

[8] GB50010-2002, Specification for Concrete Structure Design [S]. Beijing, China Construction Industry Press. 2002.

IOP Conf. Series: Materials Science and Engineering **746** (2020) 012007 doi:10.1088/1757-899X/746/1/012007

A Concrete Pouring Temperature Forecast Method for Hydraulic Structures

Zhenyang Zhu[1,*], Yi Liu[1], Yaosheng Tan[2] and Mengxia Zhou[2]

[1]State Key Laboratory of Simulation and Regulation of Water Cycle in River Basin, China Institute of Water Resources and Hydropower Research, Beijing 100038, P.R. China
[2]China Three Gorges Projects Development Co., Ltd, Chengdu 610041, P.R. China
*Email: 1219921552@qq.com

Abstract. To address problems with the existing concrete pouring temperature forecast methods, an improved forecast method is presented in this paper. A concrete temperature variation calculation method during pouring is proposed. In this method, the effect of the external environmental temperature on the concrete pouring temperature for various measures is calculated using a finite element model. The relationships between heat conductivity, thermal diffusivity, surface heat transfer coefficient and the concrete temperature rise coefficient are obtained via fitting, and the reliability of these relationship equations are verified.

1. Introduction

Concrete temperature stress control measures include pouring temperature control, water cooling, surface heat insulating and dam body parting [1-5]. Pouring temperature control is one of the most effective hydraulic concrete temperature stress control measures. Concrete pouring temperature should meet design requirements. During the period from concrete production to pouring, its temperature changes continuously. Temperature variation during the process should be accurately calculated to obtain the temperature at the mixer outlet, to ensure it satisfies the temperature control requirement. In the current version of the Hydraulic Structure Concrete Standard of the People's Republic of China, the definition of pouring temperature is the temperature of the concrete located 10 cm beneath the paving layer surface before it is covered by additional layers of concrete [6]. Although there are studies on concrete pouring temperature, the solution process is complex and the definition of pouring temperature is significantly different from the definition in the standard.

Currently, the achievements and pending issues related to pouring temperature estimation include:

(1) The calculation of the mixer outlet temperature. The mixer outlet temperature is the temperature of concrete when mixing is completed. The mixer outlet temperature is related to concrete raw material temperature and the mixer outlet temperature control. The method for calculating the mixer outlet temperature from the temperature of the raw materials is mature.

(2) The calculation of concrete temperature variation during transport. After production, concrete must be transported to a storehouse surface for pouring. During this process, the concrete temperature is affected by the environmental temperature and changes constantly. The method for calculating the placement temperature from the mixer outlet temperature is mature.

(3) The temperature rise during pouring, including the temperature rise during concrete spreading, the temperature rise during paving and the self-hydration heat induced temperature rise during pouring. The temperature rise during concrete spreading has been studied extensively. The temperature rise

during paving and the hydration heat induced temperature rise during pouring should be investigated further.

Based on the results obtained from a finite element calculation, equations governing the relationships for parameters such as thermal diffusivity, paving layer thickness, external temperature, paving layer pouring duration, heat conductivity, surface heat transfer coefficient and temperature rise during paving layer pouring are derived.

2. The Existing Methods to Calculate the Concrete Temperature Rise During Pouring

The calculation of temperature during the period from concrete production to pouring includes the following items: mixer outlet temperature calculation, temperature variation calculation during transport, temperature variation calculation during spreading and vibration and temperature variation calculation during pouring. Currently, there are some achievements in temperature calculation during pouring:

The concrete pouring temperature T_p is calculated via the following equation [7]:

$$T_p = T_i + (T_h + R/\beta_a - T_i)(\varphi_1 + \varphi_2) \tag{1}$$

Where T_p is the pour temperature; T_i is the placement temperature; T_h is the air temperature; R is the solar radiation heat; β_a is the surface heat transfer coefficient (the value recommended by Zhu Bofang is 80 kJ/m^2·d·°C); φ_1 is the concrete spreading and vibration influence coefficient; and φ_2 is the paving layer interval influence coefficient.

The coefficient φ_1 is calculated via the following formula:

$$\varphi_1 = kt + \varphi' \tag{2}$$

Where t is the duration of concrete spreading and k is an empirical coefficient for spreading, and φ' is the coefficient for concrete vibration.

The existing methods only consider those projects using manual vibrating concrete, in which, spreading and vibration cannot be separated. Then, only k can be under consideration. According to previous research results, for those projects, when actual measurement data are unavailable, k=0.003(1/min), or k=0.18(1/h).

3. A New Method to Calculate the Temperature Variation During Concrete Pouring

The temperature variation during concrete pouring is the difference between the pouring temperature and the placement temperature. The following factors influence the temperature during concrete pouring: the environmental temperature, the concrete hydration heat and the old paving layer heat conduction. The pouring process is divided into two phases: the spread vibration phase and the paving layer interval phase. The calculation procedure is as follows:

(1) To calculate the effect of the environmental temperature on the concrete pouring temperature during pouring, assume the newly poured concrete and the old concrete have identical temperatures. A paving layer with arbitrary thickness is calculated via FEM. Additionally, length of the layer is set to a very large value, and the old paving layer heat conduction effect is ignored. The concrete temperature rise at a location 10 cm beneath the paving layer surface during the paving layer interval is calculated.

(2) Next, determine the effect of the old paving layer heat conduction on the concrete pouring temperature.

(3) The temperature variation during pouring is the combined result of the environmental temperature and the old paving layer heat conduction.

3.1. Calculating the Effect of Environmental Temperature on the Pouring Temperature during Pouring

During concrete pouring, the external-temperature-induced pouring temperature rise is calculated by iteration. The initial value is calculated using the following equation:

$$T_{p0} = T_i + \theta(\overline{\Delta\tau}) + (T_a - \theta(\overline{\Delta\tau}) - T_i)(\varphi_1 + \varphi_2) \tag{3}$$

Where T_a is the environmental temperature in °C; $\theta(\overline{\Delta\tau})$ is the hydration heat in °C; T_i is the placement temperature in °C; φ_1 is the concrete spreading and vibration influence coefficient; and φ_2 is the paving layer interval influence coefficient.

Methods for calculating other parameters in Eq. (3) are elaborated in the following sections.

The pouring temperature is calculated via iteration. The iteration recursion equation is as follows:

$$\begin{cases} T_{p1} = T_i + \theta(\overline{\Delta\tau}) + \left[T_a - \dfrac{(T_{p0} + T_1)}{2} \right](\varphi_1 + \varphi_2) \\ T_{p2} = T_i + \theta(\overline{\Delta\tau}) + \left[T_a - \dfrac{(T_{p1} + T_1)}{2} \right](\varphi_1 + \varphi_2) \\ \cdots\cdots\cdots \\ T_{pn} = T_i + \theta(\overline{\Delta\tau}) + \left[T_a - \dfrac{(T_{pn-1} + T_1)}{2} \right](\varphi_1 + \varphi_2) \end{cases} \tag{4}$$

After analysis, Eq. (4) has converges rapidly. When $n \geq 3$, the pouring temperature can be calculated accurately.

3.2. Calculating the Coefficient

The method for calculating φ_1 is similar to Eq.(2). For the mechanized vibration engineering, based on measured results, when actual measurement data are unavailable, $\varphi' = 0.012$. The calculation method of k in Eq. (2) is same as φ_2, both of which can be fitted by based on the much calculate result of finite element.

3.2.1. Determining the basic format

Based on the finite element analysis, k and φ_2 values for various surface heat transfer coefficients, heat conductivities and thermal diffusivities are calculated. Based on extensive data analysis, k and φ_2 is represented in the following format:

$$\begin{cases} k = \varphi_{21}\varphi_{22} \\ \varphi_2 = k\Delta\tau \end{cases} \tag{5}$$

Where φ_{21} is the surface heat exchange influence coefficient; and φ_{22} is the internal heat conduction influence coefficient.

The surface heat exchange influence coefficient is related to λ / β. Assume:

$$\varphi_{21} = a\left(\lambda / \beta\right)^b \tag{6}$$

Where a and b are coefficients to be determined.

The internal heat conduction influence coefficient is related to the thermal diffusivity. Considering that the concrete density is similar, we have:

$$\varphi_{22} = \left(dc / \lambda\right)^e \tag{7}$$

Where d and e are coefficients to be determined.

3.2.2. Determining the surface heat exchange influence coefficient φ_{21}

In this section, the coefficient in Eq. (6) is obtained via finite element calculation and analysis. The calculation conditions are as follows.

The pouring temperature is set to 0°C; and the external temperature is set to 10°C. The concrete pouring model top surface dissipates heat, and the other surfaces are adiabatic. The concrete pouring model height is 0.5 m; the heat conductivity is 164 kJ/m·d·°C; the specific heat is 0.9 kJ/kg·°C. When the surface heat transfer coefficient is 100-1200 kJ/m^2·d·°C, a unit time, a unit environmental temperature and a pouring-temperature-difference-induced concrete temperature change are calculated.

To facilitate calculation analysis, the coefficient *d* is set to 182.2, and we have $\varphi_{22}=1$. Based on the calculation result, when the specific heat is 0.9 kJ/kg·°C and the heat conductivity is 164 kJ/m·d·°C, the unit temperature difference induced per unit time concrete temperature rise under varying surface heat transfer coefficients are calculated, and the coefficient φ_{21} is obtained.

When the specific heat is 0.9 kJ/kg·°C, the heat conductivity is 164 kJ/m·d·°C, $\varphi_{22}=1$ and $\varphi_{21}=0.01127\left(\lambda/\beta\right)^{-0.844}$,The calculation results and the finite element results are compared via formula fitting, as shown in Fig. 1. The fitting value using Eq. (6) matches the finite element calculation result well.

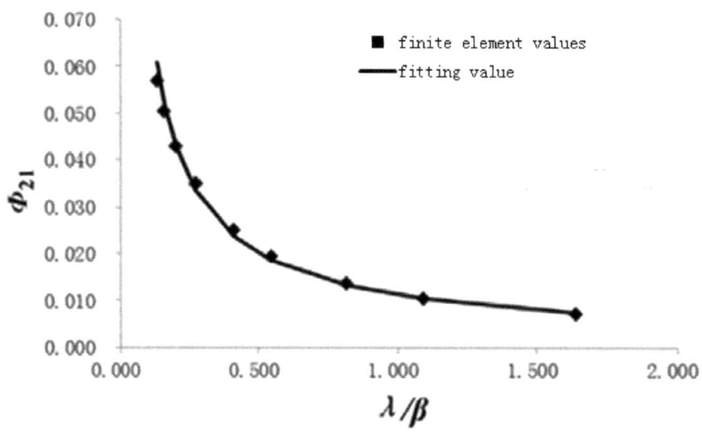

Figure 1. Relationship between φ_{21} and λ/β

3.2.3. Determining the internal heat conduction influence coefficient, φ_{22}

In this section, the coefficient in Eq. (7) is obtained via finite element calculation analysis. The calculation conditions are as follows:

The pouring temperature is set to 0°C; the external temperature is set to 10°C. The concrete pouring model top surface dissipates heat, while the other surfaces are adiabatic. The heat conductivity is 164 kJ/m·d·°C, the surface heat transfer coefficient is 600 kJ/m^2·d·°C, and the specific heat is 0.6-1.2 kJ/kg·°C. The unit-temperature-difference-induced concrete temperature rise is calculated, and φ_{22} is obtained.

When the heat conductivity is 164 kJ/m·d·°C, the surface heat transfer coefficient is 600 kJ/m^2·d·°C. The calculation result shows that φ_{22} is unrelated to the pouring interval and is only related to the specific heat.

When the heat conductivity is 164 kJ/m·d·°C, the surface heat transfer coefficient is 600 kJ/m^2·d·°C and $\varphi_{22}=\left(182.2\ c/\lambda\right)^{-0.861}$. The calculation results are compared with the finite element results using a fit with Eq. (7), as shown in Fig. 2.

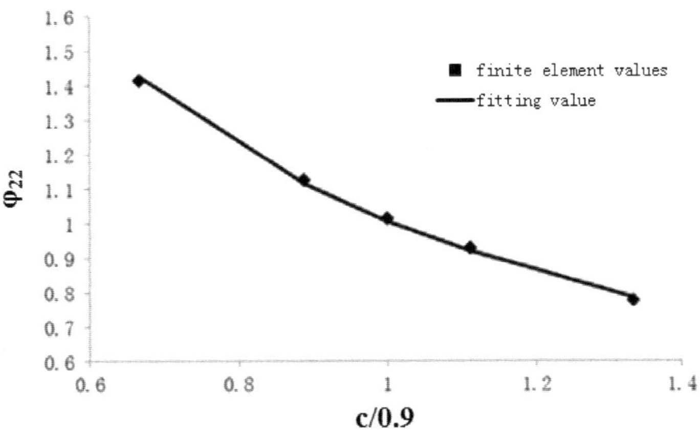

Figure 2. Relationship between φ_{22} and specific heat

3.2.4. Validation of k and φ_2 application scope

Based on Eqs. (5-7), φ_2 is determined:

$$\begin{cases} k = 0.01127\left(\lambda / \beta\right)^{-0.844}\left(182.2c / \lambda\right)^{-0.861} \\ \varphi_2 = k\Delta\tau \end{cases} \tag{8}$$

In this paper, Eq. (8) is validated using different concrete material parameters. Tables 1 and 2 list $\varphi_{21}\varphi_{22}$ and φ_{22} for various values of specific heat when the surface heat transfer coefficient is 900 kJ/m^2·d·°C and the heat conductivity is 164 kJ/m·d·°C. When the surface heat transfer coefficient is 900 kJ/m^2·d·°C, it represents the scenario for which heat insulating material is not applied.

Tables 3 and 4 list $\varphi_{21}\varphi_{22}$ and φ_{22} for various values of specific heat when the surface heat transfer coefficient is 300 kJ/m^2·d·°C and the heat conductivity is 164 kJ/m·d·°C. Based on the calculation results, $\varphi_{21}\varphi_{22}$ and φ_{22} are unrelated to the paving layer pouring interval and can be accurately calculated by fitting Eq. (8). Extensive on-site investigation shows that when the surface heat transfer coefficient is 300 kJ/m^2·d·°C, it represents the scenario for which heat insulating material is applied.

Apart from these two typical scenarios (for surface heat dissipation coefficients of 300 kJ/m^2·d·°C and 900 kJ/m^2·d·°C) this study has also investigated various φ_{22} values when λ / β is in the range of 0.164 to 1.64. The finite element calculation results are consistent with the fitting result of Eq. (8), indicating that this equation applies to all scenarios.

Table 1. $\varphi_{21}\varphi_{22}$ for various values of specific heat (surface heat transfer coefficient: 900 kJ/m^2·d·°C)

specific heat	3 hours	4 hours	5 hours	6 hours	7 hours	8 hours
0.8	0.055	0.055	0.053	0.051	0.049	0.048
0.9	0.049	0.049	0.048	0.047	0.045	0.044
1.0	0.044	0.045	0.044	0.043	0.042	0.041
1.2	0.036	0.037	0.037	0.037	0.036	0.035

Table 2. φ_{22} for various values of specific heat (surface heat transfer coefficient: 900 kJ/m^2·d·°C)

specific heat	3 hours	4 hours	5 hours	6 hours	7 hours	8 hours
0.8	1.13	1.11	1.11	1.09	1.10	1.08
0.9	1.01	1.01	1.01	1.00	1.01	1.00
1.0	0.90	0.91	0.92	0.92	0.93	0.92
1.2	0.74	0.76	0.78	0.79	0.80	0.80

Table 3. $\varphi_{21}\varphi_{22}$ for various values of specific heat (surface heat transfer coefficient: 300 kJ/m^2·d·°C)

specific heat	3 hours	4 hours	5 hours	6 hours	7 hours	8 hours
800	0.022	0.022	0.022	0.021	0.021	0.020
900	0.020	0.020	0.020	0.019	0.019	0.019
1000	0.017	0.018	0.018	0.018	0.017	0.017
1200	0.014	0.015	0.015	0.015	0.015	0.015

Table 4. φ_{22} for various values of specific heat (surface heat transfer coefficient: 300 kJ/m^2·d·°C)

specific heat	3 hours	4 hours	5 hours	6 hours	7 hours	8 hours
800	1.10	1.11	1.10	1.13	1.10	1.07
900	0.98	0.99	0.99	1.02	1.00	0.98
1000	0.87	0.89	0.90	0.93	0.92	0.90
1200	0.70	0.73	0.75	0.79	0.78	0.77

4. Acknowledgments

This research was supported by the National Key Research and Development Project of China (Grant No. 2018YFC0406703), the National Natural Science Foundation of China (Grant No. 51779277, 51579252, 51439005), the Special Scientific Research Project of the State Key Laboratory of Simulation and Regulation of Water Cycles in River Basins (2016ZY10), a Special Scientific Research Project of the China Institute of Water Resources and Hydropower Research (Grant No. SS0145B392016, SS0145B612017), and the Special Scientific Research Project of the China Institute of Water Resources and Hydropower Research (KY1799).

5. References

[1] Cervera M., Faria J., Oliver J., et al, Comput. Struct., 2002 **80** 1511-1521.
[2] Zhu B.F., Water Power 2003 **4** 29-32.
[3] Zhu Z.Y., Qiang S., Chen W.M., Comput. Concrete 2013 **11** 441-462.
[4] Nasir M, Al-Amoudi O S B, Maslehuddin M, Construction and building materials 2017 **152** 943-953.
[5] Zhu ZY, Zhang GX, Yi L, et al, Struct Eng Mech 2019 **69** 33-42.
[6] Sun Z.Y., Zhou H.G. Chen X.M. and et al 2015 *Specifications for hydraulic concrete construction (DL/T 5144-2015)* (China electric power press).
[7] Zhu B.F. 1999 *Thermal stresses and temperature control of mass concrete* (China electric power press).

IOP Conf. Series: Materials Science and Engineering **746** (2020) 012008 doi:10.1088/1757-899X/746/1/012008

Application of Matlab to Data Processing of Buckling Experiments

Jiahui Zhou, Peng Yang, Jie Sun, Chenyu Shi and Jianghong Xue*

School of Mechanics and Construction Engineering, Jinan University, Key Lab of Disaster Forecast and Control in Engineering Ministry of Education, Guangzhou 510632, Guangdong, China

Email: txuej@jnu.edu.cn

Abstract. In this paper, matlab programs were written to efficiently process the experiment data for buckling and post-buckling of delaminated composite beams. In the experiments, data collected from 810 Material Test System (MTS) and from dial indicator were sampled at different frequencies, thus were not matched with each other in time. To solve this problem matlab programs were developed. It shows that using the developed program, the experiment data from the two testing systems are corresponding and truly reflect the changes of mechanical parameters in the buckling and post-buckling process of specimens in the experiments.

1. Introduction

In mechanical experiments, the sampling frequencies of several machines working at the same time may be different due to mechanical failure or operational errors. As a result, the different quantities collected by different machines are not corresponding. Under the circumstances, data processing is undoubtedly required. However, it is very troublesome and time-wasting for us to pick the useful data from such vast amounts of that. Computer programs can be used to solve the problem above very quickly and conveniently.

As we know, Matlab is a matrix-oriented high-level programming environment which gives access to fast algorithms for a large number of numerical tasks on many common computer platforms [1]. Matlab has been used to realize multi-channel synchronous data acquisition [2], to process many kinds of data such as compaction test data, NMR data, ocean chemistry data [1, 3, 4] and so on. But there is few application to non-corresponding buckling experiments data.

This paper introduces a matlab program to realize data processing mentioned above, and shows how to apply it to the data processing of some buckling experiments to study the nonlinear failure of damaged composite laminate. Applying the Matlab program, we can save time and simplify data processing.

2. Problem Solving and Programing Ideas

When we experiment on delaminated composite beams, there are three quantities need measuring: lateral deflection, longitudinal displacement and axial compressive force. Among them, longitudinal displacement and axial compressive force are measured by Material Test System, while axial compressive force is measured by the dial indicator. Normally, these three quantities should correspond with each other. Because of a problem with the dial indicator, its sampling frequency is not consistent with the set frequency and is mutative, which leads to different acquisition frequencies of these three quantities.

IOP Conf. Series: Materials Science and Engineering **746** (2020) 012008 doi:10.1088/1757-899X/746/1/012008

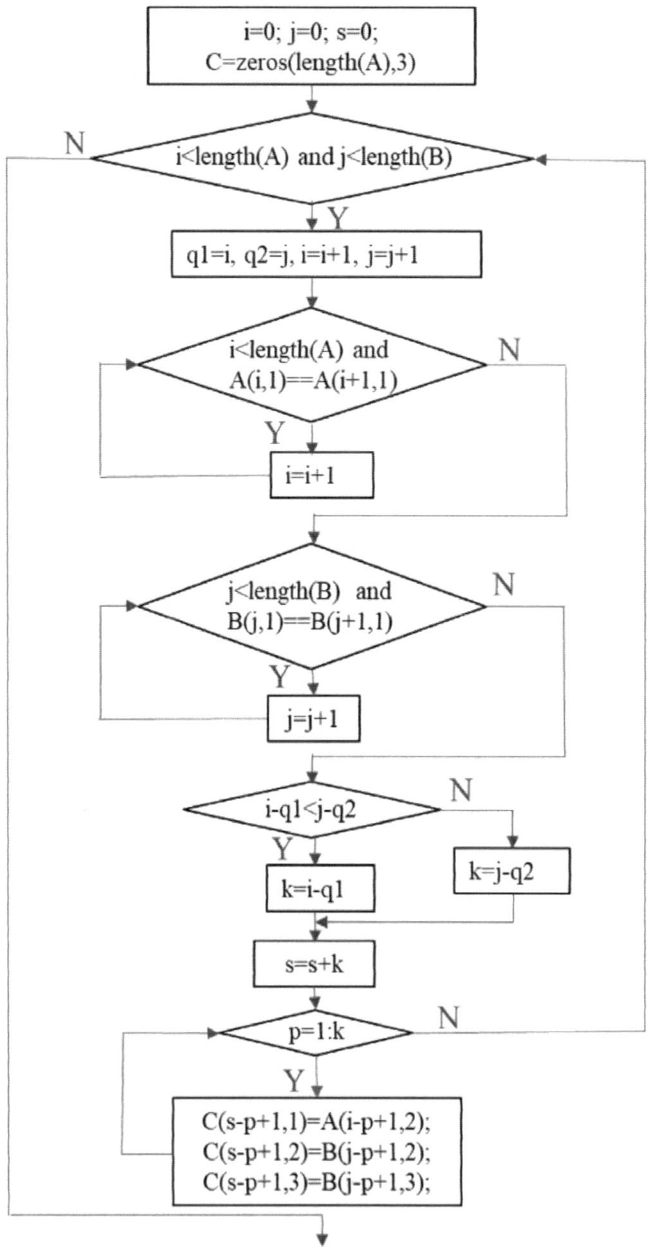

Figure 1. The program flow chart of the main part

As both the 810 Material Test System and the dial indicator record the time when they collect each sample, to make the three quantities matched, we let the number of samples of all the three quantities in a given second depends on the minimum number of samples in this second. For example, in a certain second, the MTS samples 5 times, while the dial indicator samples 3 times, so the number of samples of each quantity in this second is three (the first two extra samples from MTS will be discarded). The program flow chart of the main part is shown in figure 1. To take advantage of fast matrix operations in matlab most processing commands have been vectorized. The program code of the main part is shown in the appendix.

3. Results and Discussion

This section is going to compare the data before and after processing. Plots of force against deflection before and after data processing for two typical specimens are shown in the figure 2, which indicates the data processing is very effective. It can be found that the deflection of the specimen at yield is much smaller after data processing than before, and certainly the data processed can reflect the real situation more accurately.

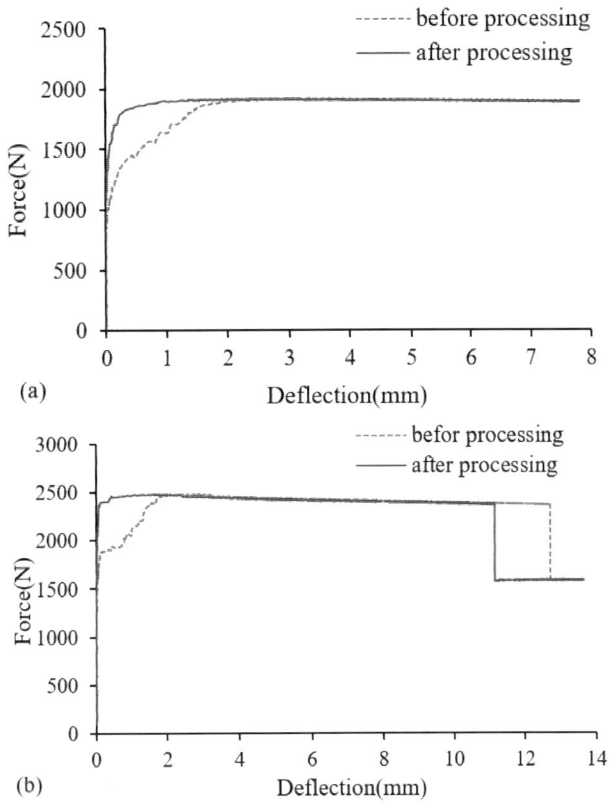

Figure 2. Plots of force against deflection before and after data processing for two typical specimens

If the original data is used to analyse the buckling and post-buckling of the delaminated composite beams, we are going to get wrong results. The program proposed in this paper can help to analysis experiment data correctly and distribute to follow-up work.

Although these quantities will not correspond very strictly after being processed, the errors are very small and are not large enough to affect the observation of the failure mode of the specimens because the samples were screened within every second.

Since the data of force and displacement are measured by the same testing system and there is little difference between the data before and after processing, no comparison is made here.

4. Conclusion

By using the matlab program provided in this paper, we can process experiment data much more conveniently and efficiently to deal with the data obtained from the buckling experiments so as to avoid manual and repeated processing of similar data. After applying the program, different quantities are matched with each other in time, so that the properties of the materials or structures can be analysed correctly and accurately. With this program, we lay a solid foundation for the subsequent research and offer a reference for similar experiment data processing.

5. Acknowledgements
The work presented herein was conducted with the financial support of Guangdong Natural Science Foundation (2017A030313013).

6. Appendices

```
i=0;j=0;
s=0;
%Create a new matrix. Pre-allocate memory for efficiency.
C=zeros (length(A),3);
while (i<length(A)) && (j<length(B))
%A refers to the matrix consists of the data collected by the dial indicator.
%B refers to the matrix consists of data collected by 810 Material Test System (MTS).
    q1=i;q2=j;
    i=i+1;j=j+1;
    while (i<length(A))&&(A(i,1)==A(i+1,1))
        i=i+1;
    end
    while (j<length(B))&&(B(j,1)==B(j+1,1))
        j=j+1;
    end
    if i-q1<j-q2
        k=i-q1;
    else
        k=j-q2;
    end
    s=s+k;
    for p=1:k
        C (s-p+1,1) = A (i-p+1,2);
        C (s-p+1,2) = B (j-p+1,2);
        C (s-p+1,3) = B (j-p+1,3);
    end
end
```

7. References

[1] Gunther U L, Ludwig C and Ruterjans H 2000 *J. Magn. Reson.* **145** 201-208.
[2] Yang Y, Tan Y, Huang J and Yuan N 2009 Realization of multi-channel synchronous data acquisition using matlab *Microcomputer Information* **25** 4-5, 84.
[3] Yu Z and Cheng S 2012 *Applied Mechanics and Materials* vol 170-173, *et al* (Yantai: China) p 611.
[4] Chen X, Lu C, Liu Z and Yang Y 2005 *Computers and Applied Chemistry* **22** 59-61.

IOP Conf. Series: Materials Science and Engineering **746** (2020) 012009 doi:10.1088/1757-899X/746/1/012009

Numerical Simulation of Thermal Performance in Hollow-structure Aluminum of Bullet Train

Weibing Shi[1], Li Li[1], Mengmeng Li[2,3] and Xiao Zeng[2]

[1]R&D Center OF CRRC QINGDAO SIFANG CO, LTD, Qingdao 266111, Shandong, China

[2]School of Energy science and engineering, Central South University, Changsha 410083, Hunan, China

Email: 2960944083@qq.com

Abstract. In order to know the characteristic of heat transfer performance in hollow-structure aluminium, the flow and heat transfer in it of bullet train "Fuxing" was numerical simulated. Changes for the equivalent thermal conductivity under the influence of four factors—the temperature difference, wall thickness, rid thickness and number of cavities, were studied and analysed separately adopting control variate numerical method. Numerical simulation results showed that the influence of temperature difference and wall thickness on the equivalent thermal conductivity of profiles were limited; the number of cavities and rib thickness were the main factors affecting the equivalent thermal conductivity of profiles, with the number of cavities and rib thickness increasing, the equivalent thermal conductivity of hollow aluminum increased, and viced versa. In the practical application, proper parameters of hollow-structure were selected to reduce heat transfer and energy consumption of train.

1. Research Background and Significance

It is important for lightweight car body to develop modern high-speed train. With the stainless steel material being replaced as extruded hollow-structure aluminum, it lowers the fees of train's operation and maintenance while reducing the weight of train [1]. And owning the good plasticity, easy extrusion, excellent sound insulation, heat preservation as well as recyclability, it gradually becomes the most important part of train's body. And aluminum alloy is the main material for train construction. Therefore, it is important to obtain equivalent thermal conductivity in complex coupled heat transfer process for hollow aluminum that composite structure. Nowadays, there have been large results of study on the extruded hollow-structure aluminum and technique engineering, but there is little study for heat transfer of it. Hui Yuan [2] and Jinlei Jiang [2] did experiment to measure thermal conductivity, Huasheng Xiong [3] measured thermal conductivity of aluminum extrusion by using steady flat plate method, Xiaojing Ma [4] studied the relationship between thermal conductivity and position of the hot wall related to the direction of gravity. Based on above study, the relationship between equivalent heat transfer coefficient on hollow-structure aluminum and temperature difference, rib thickness and cavity amount discussed in this essay.

2. Hollow-structure Aluminum Heat Transfer Model Establishment

2.1. Geometric Model

In accordance with the requirement, the hollow-structure aluminum is studied by using control variate method in this essay, and makes the simulation research for four influence factors, such as temperature difference (T), wall thickness (H), rib thickness (L) and number of cavities (N). The structure of hollow-structure aluminum is shown as figure 1.

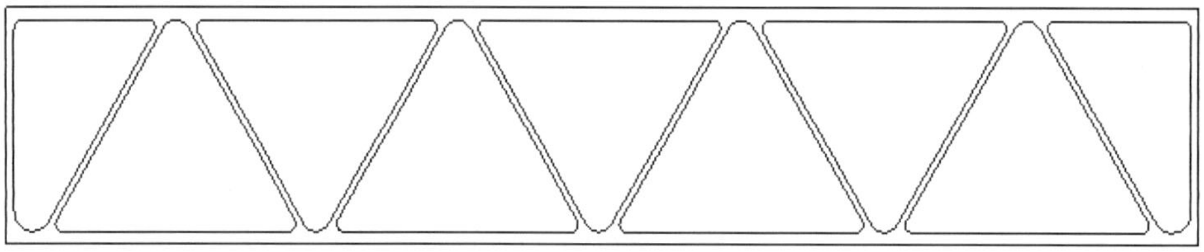

Figure 1.Hollow aluminum structure schematic

In this experiment, the cavity of hollow-structure aluminum is triangular and the solid walls of it are all made from aluminum alloy material. The control variables are shown in Table 1. The natural convection of air in the cavity is aroused by the buoyancy force which is made by density difference, the air density adopts the model of Boussinesq, as well as the average temperature of high and low wall temperature sets as heat expansion coefficient. The heat transfer of hollow-structure aluminum has been made through following three ways: (1) heat conduction of hollow-structure aluminum; (2) radiation heat transfer between cavity surfaces; (3) convective heat transfer of internal gases. In the process of heating, aluminum alloy is passed large amount of heat to make the average temperature of hollow wall equal, and radiation heat transfer in the wall of hollow metal is very weak [3, 4], so it ignores the radiant heat transfer of air inside the cavity.

Table 1. Variable list

temperature difference (K)	1, 2, 5, 10, 20
wall thickness (mm)	1, 2, 4, 6, 8
rib thickness (mm)	1, 2, 4, 6, 8
cavity number	7, 9, 11, 13, 15

2.2. Physical Model

Only the one-dimensional heat transfer of thickness direction for hollow-structure aluminum is considered of this model, the heat transfer of sided direction is ignored, and one-dimensional steady-state heat equation along the direction of hollow-structure aluminum.

$$\rho c\frac{\partial T}{\partial t}=\frac{\partial}{\partial z}(\lambda\frac{\partial T}{\partial z}) \tag{1}$$

Where, ρ is the effective density, c is the effective specific heat capacity, to be obtained by weighted average in accordance with the area ratio between air and aluminum alloy separately. The thermal property parameters of aluminum alloy and air in the cavity at the operating temperature are shown in Table 2.

Table 2. Thermophysical parameters of material

material	density ρ [kg/m3]	specific heat Cp [J/(kg·K)]	thermal conductivity λ [W/ (m·K)]	Viscosity μ [10-5kg/(m·s)]	Coefficient of Thermal Expansion [10-5K-1]
aluminum alloy	2730	873	202	-	2.117
air (281.65K)	1.2539	1005	0.02496	1.752	355

2.3. Calculation Condition Setting

In addition to the above computing, the initial condition and the boundary condition needs to be set. The first boundary condition is adopted for the hollow aluminum alloy heat transfer simulation in this essay. Taking the example of winter experiment with 2K of temperature difference, the external temperature outside the whole train is 280.65K as low temperature of outside wall, the upper and lower wall of hollow-structure aluminum is T_1=282.65K, T_2=280.65K, temperature load ΔT=2K . Only considering the heat transfer about internal and external temperature difference, its two-sided wall can be set as insulation condition, and cavity and interface of hollow-structure aluminum can be seen as coupled heat transfer.

The control volume method and the uncoupled steady-state implicit format are used to solve the governing equations, and the laminar flow model is used to calculate the simulated heat transfer and flow. The coupled pressure and speed adopt Simples algorithm. The convection term adopts a second-order upwind scheme. Except for the absolute value of energy calculation residue, the definition of convergence condition is less than 1×10^{-8}, the others are not less than 1×10^{-4}.

2.4. Calculation of Equivalent Thermal Conductivity

The heat performance of hollow aluminum is mainly characterized by equivalent thermal conductivity. The equivalent heat transfer coefficient of hollow aluminum can be shown as:

$$\lambda_{eq} = q\delta / (T_1 - T_2) \tag{2}$$

Where, λ is equivalent thermal conductivity, q is the total surface heat flow density under steady state, δ is the thickness of hollow aluminum, T_1 is the thickness of hot side, T_2 is the temperature of cold side. Through the final calculation result of above model, and combined above equation, the equivalent thermal conductivity of hollow-structure aluminum can be found.

3. Analysis of Calculation Results

3.1. Hollow-structure Aluminum Heat Transfer Characteristics

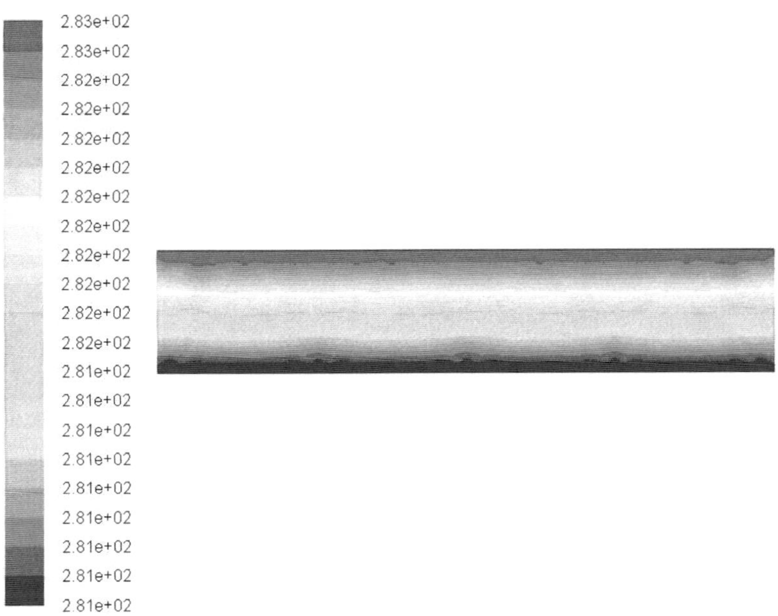

Figure 2. Temperature cloud diagram of hollow aluminum

The above figure shows temperature cloud distribution of the hollow-structure aluminum when temperature loading of hollow-structure aluminum is ΔT=2K, the high-temperature wall has same distribution trends corresponding to temperature in the cavity, so does the low temperature. It can be predicted that the natural convection is not so strong from the temperature fluctuation range and the natural convection has the limited impacts on equivalent heat transfer coefficient.

3.2. The Influence of Structure Parameters on the Heat Transfer Performance

In order to properly design the insulation of hollow-structure aluminum, the impact of each structural parameter on heat transfer is studied by adopting control variate. Figure3 to figure4 is context diagram of structure parameters of hollow aluminum and equivalent thermal conductivity.

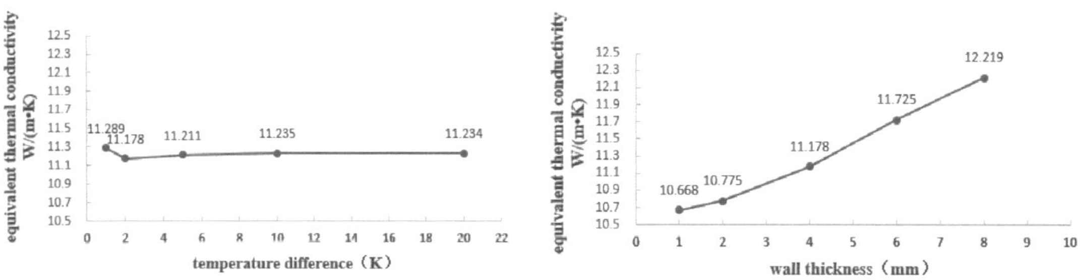

(a) (b)

Figure 3. The influence of temperature difference (a) and wall thickness (b) of hollow aluminum on equivalent thermal conductivity

From figure 3 (a), it is known that although there are some fluctuations for the equivalent thermal conductivity coefficient of hollow-structure aluminum under the different temperature, the numeric change is very little. The simulation result shows that there is little impact for temperature change of the equivalent thermal conductivity coefficient on hollow-structure aluminum, which shows it is very weak for natural convection in the cavity.

From the figure 3 (b), the impact of aluminum wall thickness on equivalent thermal conductivity coefficient shows approximately linear, the equivalent thermal conductivity coefficient of hollow-structure aluminum increases with the thickness of wall. Under the other conditions of non-changed parameters, with thickness for aluminum wall increasing, the proportion of aluminum alloy is heightened, the heat transfer of aluminum alloy is enlarged and the equivalent thermal conductivity coefficient is increased.

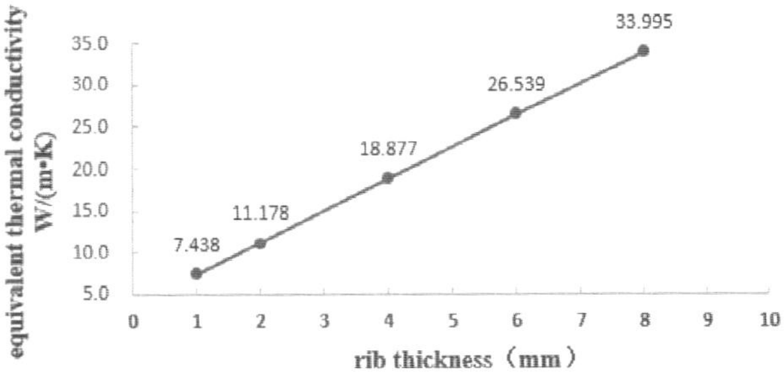

(a)

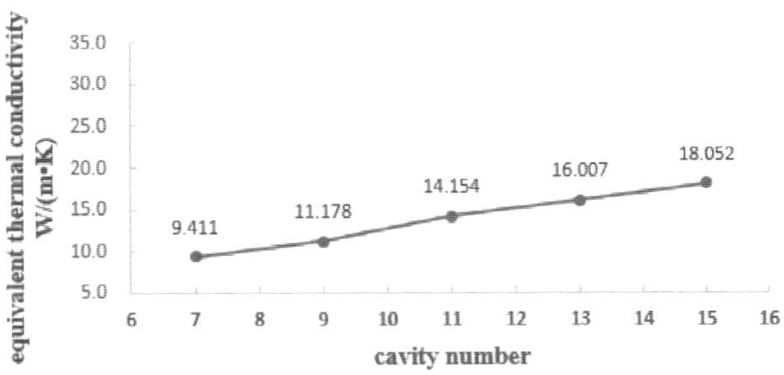

(b)

Figure 4. The influence of rib thickness (a) and cavity number (b) of hollow aluminum on equivalent thermal conductivity

It can be known on figure 4 (a), the impact of rib thickness on equivalent thermal conductivity coefficient of hollow-structure aluminum shows approximately linear, the equivalent thermal conductivity coefficient increases with the thickness of rib. Under the other conditions of non-changed parameters, with thickness for rib thickness increasing, it not only increases the proportion of aluminum alloy, but also it makes more heating transfer with the aluminum alloy in accordance with the principle of thermal-conduction resistance. On the whole, it makes the speed rate of equivalent thermal conductivity coefficient larger than the wall thickness of figure 3 (b) on its impacts.

It can be known from figure 4 (b), equivalent thermal conductivity coefficient increases with the number of cavities. There is little study on this aspect, which predicts that the interaction and composition between cavity leads to the increase of equivalent thermal conductivity coefficient while the number of cavities increases the proportion of aluminum alloy in the middle of total ingredient. The superimposed impact on these two reasons make hollow amount more obvious on heat transfer.

4. Conclusion and Prospect

1. With the change of load temperature on both sides of the hollow aluminum, the equivalent thermal conductivity keeps almost constant, it shows the natural convection in the cavity is very weak.

2. The number of cavities and rib thickness are the main factors that affect the heat transfer performance of hollow-structure aluminum. Wall thickness and temperature difference have limited influence on heat transfer performance.

3. Reasonable structure parameters should in accordance with specific conditions of the train in the actual engineering application, such as strength and load bearing, to reduce the energy consumption of train.

5. References

[1] Boming Wang 2014 *High-speed bullet train overall and bogies (2nd edition)* (Chengdu: Southwest Jiaotong University Press) p 330-331.

[2] Huasheng Xiong 2012 *Study on an emu's thermal insulation and the research of heat transfer coefficient K value* (Qingdao: Qingdao Technological University).
[3] Xiaojing Ma 2014 *Experimental and numerical studies of heat transfer characteristics of the layer combined with aluminum extrusion and multi insulation material layers* (Lanzhou: Lanzhou Jiaotong university).
[4] Hui Yuan, Jinlei Jiang 2013 The emulating calculation and experimental research of aluminum profiles' heat transfer coefficient of high-speed train *Railway vehicle*. 51 (6) 7-9.

IOP Conf. Series: Materials Science and Engineering **746** (2020) 012010 doi:10.1088/1757-899X/746/1/012010

Investigation on the Iron Oxide Behaviors in the SiO_2-based Slag

Qingchun Yu [1], Jiahao Zhang[1], Jing Chen[2] and Chunpei Yan[1]

[1]State Key Laboratory of Complex Nonferrous Metal Resources Clear Utilization, Faculty of Metallurgical and Energy Engineering, Kunming University of Science and Technology, Kunming 650093, P.R. China

[2]Faculty of Public Safety and Emergency Management, Kunming University of Science and Technology, Kunming 650093, P.R. China

Email: yqcy@163.com

Abstract. Various waste blast furnace slag not only occupies a large amount of land, but also seriously pollutes environment. Preparation of glass-ceramics using the waste blast furnace slag receives more and more attention. In this work, effect of iron oxide behaviour on the crystallization and physical and chemical properties in the SiO_2-based slag was studied. Results show that the presence of small amount of iron oxide in slag can promote the crystallization of CaO-MgO-Al_2O_3-SiO_2 quaternary basic glass. With the increase of Fe_2O_3, the crystalline phase content increases, and the glass phase content decreases. Meanwhile, the density of glass-ceramics increases, but the water absorption decreases. Alkali corrosion resistance of glass-ceramics increases with the increase of iron oxide content in range of 0.5 wt. % - 3 wt. %.

1. Introduction

With the rapid development of iron and steel industry, blast furnace slag discharge is increasing significantly. Over the years, slag has accumulated nearly 15 $\times 10^7$t [1]. Direct storage not only occupies a large amount of land, but also seriously pollutes environment. The comprehensive utilization of the waste blast furnace slag is an urgent problem. Glass-ceramics is a kind of polycrystalline solid material with homogeneous coexistence of glass phase and microcrystalline phase, which precipitates in a certain basic glass through a specific heat treatment system with addition of nucleating agent. Glass-ceramics [2] concentrates the characteristics of glass and ceramics, which have high mechanical strength, wear resistance, corrosion resistance, excellent electrical insulation, stable dielectric constant, and adjustable expansion coefficient. So, glass-ceramics can be widely used as structural and functional materials in high-tech fields such as optics, electronics, astronautics and biology. At the same time, glass-ceramics can also be widely used as decorative or protective materials in industrial and civil buildings.

The main components of glass-ceramics are SiO_2, CaO and Al_2O_3, which are similar to those of blast furnace slag. Therefore, researchers [3-6] pay more and more attention to the preparation of glass-ceramics using the waste slag. Agarwell developed wear-resistant microcrystalline ceramics [2] using blast furnace slag. Vecoglu obtained glass-ceramics directly from Turkish blast furnace slag with addition of 3%-5% titanium dioxide. It was found that the grain size changed significantly with the increase of the amount of nucleating agent. The wear resistance, hardness, and bending strength of the product increased with the increase of nucleating agent and crystallization temperature. Khater discovered that the volume crystallization of hedenbergite phase increased, but the volume crystallization degree decreased with the increase of fayalite phase.

Blast furnace slag contains a certain amount of iron, scholars have made a detailed study on the effect of iron on crystallization behaviour of glass-ceramics, magnetic properties and biological activity of products [7, 8]. Zhang [9] prepared glass-ceramics from tailings of a tungsten mine in Hunan Province. The effect of Fe_2O_3 on crystallization temperature was studied. It was found that when Fe_2O_3 content reached 6 wt. %, the crystallization temperature of glass-ceramics decreased significantly, and the crystallization ability of glass-ceramics enhanced effectively. In addition, Wu [10] drew a similar conclusion in the the preparation of glass-ceramics with stone powder. Li [11] prepared glass-ceramics with Baiyun Obo tailings and fly ash. The results show that the activation energy of crystallization first increases, then decreases with the increase of iron oxide content. The maximum is 426 kJ•mol-1. Rezvani [12] investigated the effect of Fe_2O_3 on the crystallization behaviour of glass-ceramics using Fe_2O_3, Cr_2O_3 and titanium dioxide as mixed nucleating agents. Alizadeh [13] studied the crystallization behaviour and sintering properties of $CaO\text{-}SiO_2\text{-}MgO\text{-}P_2O_5$ glass-ceramics by sintering method. It was found that the glass-ceramics had high density and mechanical strength with addition of 5.0wt% Fe_2O_3. The main crystalline phases were white phosphorite and diopside when sintering temperature was low. At high temperatures, the main crystalline phase is calcium silica. Ebisawa [14] found that magnetism and biological activity of the $CaO\text{-}SiO_2\text{-}FeO\text{-}Fe_2O_3$ glass-ceramics were obvious when the iron oxide content was about 45 wt. %-50 wt. %.

As discussed in the existing literature above for the preparation of glass-ceramics from industrial waste blast furnace slag, it can be seen that most of them are designed as $SiO_2\text{-}CaO\text{-}Al_2O_3$ ternary system, and their main crystalline phases are usually diopside, wollastonite or corresponding solid solutions, such as calcium-iron limestone and magnesium. The structure and properties of glass-ceramics depend on the mineral composition of crystal phase and the chemical composition of glass phase. However, the effect of iron on the basic glass structure, crystallization behaviour and physical and chemical properties of $SiO_2\text{-}CaO\text{-}MgO\text{-}Al_2O_3$ quaternary system has rarely been reported. In this work, effect of iron oxide on the crystallization and physical and chemical performance was investigated.

2. Experimental

2.1. Sample Preparation

In order to study the effect of Fe_2O_3 on the crystallization of glass-ceramics, five basic glass samples, with chemical compositions presented in Tab. 1, were prepared by mixing SiO_2, Al_2O_3, MgO, CaO and Fe_2O_3 to simulate the waste blast furnace slag. The melting of these glass samples was carried out in corundum crucible using a high temperature furnace with $MoSi_2$ heating element. The melts were kept at 1450 °C for 2 h, and were immediately poured on a pre-heated metallic mould with annealing at 500 °C for 2 h to obtain the basic glass samples.

Table1. Chemical composition of prepared glasses (wt.%)

Samples	CaO	MgO	Al_2O_3	SiO_2	Fe_2O_3
1#	21.000	15.000	10.000	54.000	-
2#	20.895	14.925	9.95	53.730	0.500
3#	20.790	14.850	9.900	53.460	1.000
4#	20.580	14.700	9.800	52.920	2.000
5#	20.370	14.550	9.700	52.380	3.000

2.2. Analysis Methods

Differential scanning calorimetry (DSC) analysis was made using a Netzsch STA 449F3 thermal balance from 200 to 1200°C. The particle size of the basic glass samples ranged from 74 to 104 mesh. Crystallization temperature was determined under condition of heating rates of 5, 10, 15, and 20 °C/min, respectively. Density of glass-ceramics product is calculated by (1), in which ρ is the density of glass-ceramics product, g/cm^3, and ρ_w is the density of water, g/cm^3. m and m_1 are the mass of glass-ceramics product in air and in water, respectively. Water absorption of glass-ceramics was tested by GB/3299-1996 [15]. Acid and alkali resistance of glass-ceramics product was tested in laboratory by the following procedures: Firstly, glass-ceramics were washed with 1% NaOH or 1% HCl solution to remove the surface impurities, then dried after rinsing with distilled water 3-5 times. Mass of glass-ceramics product m_1 was obtained. Secondly, the glass-ceramics product was immersed in 1% NaOH or 1% HCl solution at 60 °C for 3 hours. The pretreated glass-ceramics product was rinsed distilled water 3-5 times washing with, and dried in a vacuum drying chamber. The final mass m_2 was obtained. The acid-alkali corrosion resistance of glass-ceramics can be expressed as follows:

$$\rho = \frac{m}{m-m_1} \times \rho_w \quad (1)$$

$$C=\frac{m_1-m_2}{m_1}\times 100\% \quad (2)$$

3. Results and Discussion

3.1. *Effect of Iron Oxide on Crystallization*

Table 2 shows the DSC profiles of glass samples about crystallization peak temperature (T_p) with different Fe_2O_3 content heating rates of 5, 10, 15, and 20 °C/min, respectively. It is found that the glass crystallization peak temperatures T_p decrease with increasing the Fe_2O_3 content at the same heating rate.

Table 2. Crystallization peak temperature of samples at different heating rate (°C)

samples	Crystallization peak temperature			
	5°C/min	10°C/min	15°C/min	20°C/min
1#	994.4	1015.3	1026.5	1038.9
2#	972.4	991.2	1004.5	1018.6
3#	990.4	1008.1	1020.0	1030.6
4#	975.6	997.8	1010.5	1017.0
5#	969.9	986.6	992.2	999.5

Crystallization peak temperature of sample 2# is lower than that of 1#, 3# and 4#. This abnormal phenomenon may be caused by the decomposition of Fe_2O_3 in the basic glass system. The decomposition reaction of Fe_2O_3 occurs during the melting and clarifying process of basic glass as follows:

$$Fe_2O_3=FeO+1/2O_2 \quad (3)$$

$$3Fe_2O_3=2Fe_3O_4+1/2O_2 \quad (4)$$

The decomposition equilibrium is calculated by HSC software as shown in Figure 1. It can be found that mole ratio of Fe_2O_3 decreases with the increase of temperature, which coincides with the increase of mole ratio of Fe_3O_4 and FeO.

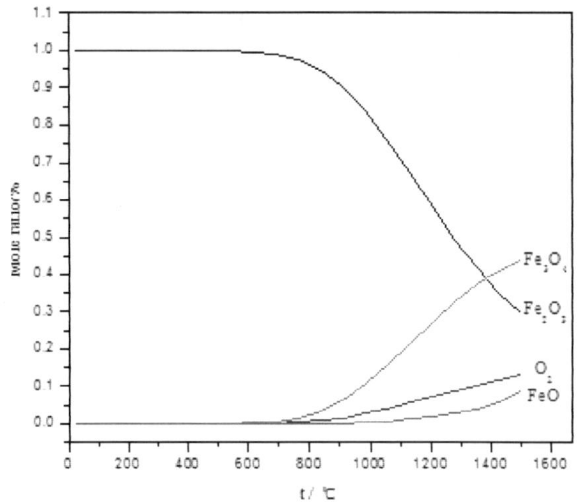

Figure 1. The decomposition equilibrium of iron oxide

In glass systems, iron generally exists in the valence states of Fe^{2+} and Fe^{3+}, but they play opposite roles in the glass grid [16]. Fe^{2+} can decrease the viscosity of molten basic glass and crystallization temperature. On the contrary, Fe^{3+} can increase the crystallization temperature by forming a grid structure when there are more monovalent (Na^+, K^+) or bivalent (Mg^{2+}, Ca^{2+}) ions in the molten basic glass. According to figure 1, 70 mol % Fe_2O_3 will decompose into FeO or Fe_3O_4 when the temperature rises to 1450 C. However, reactions (3) and (4) are affected by the viscosity of glass system and the oxidizing atmosphere in the melting environment. It can be inferred that when a small amount of Fe_2O_3 is introduced into the CaO-MgO-Al_2O_3-SiO_2 quaternary system, the decomposition reaction of Fe_2O_3 will occur. Fe^{3+} will be changed into Fe^{2+}. Ratio of Fe^{3+}/Fe^{2+} in the system decreases, thus reducing its crystallization temperature. When the Fe_2O_3 content is high, Fe^{3+} greatly increases the system viscosity, and produces more oxygen in melting environment. The decomposition reaction is inhibited by the atmosphere, leading the increase of crystallization temperature.

3.2. Effect of Iron Oxide Ondensity and Water Absorption

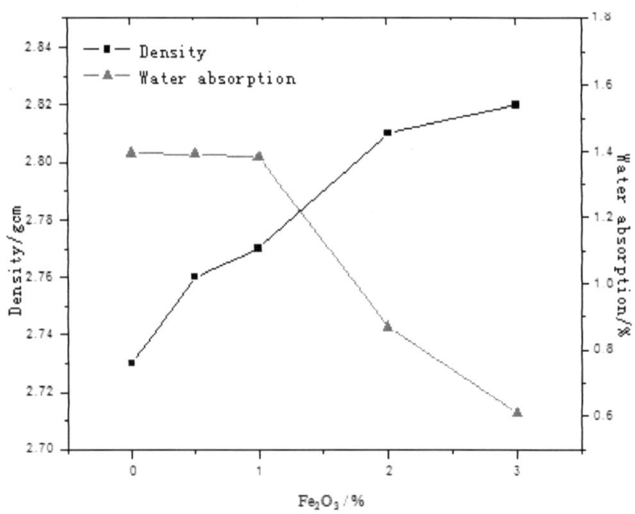

Figure 2. Relationship between Fe_2O_3 content and glass-ceramics density and water absorption

Figure 2 shows the patterns of density and water absorption versus iron oxide content on condition of the basic glass samples nucleated at 750 °C for 2 h and crystallized at 1050 °C for 2 h. The results are shown in table 3. From the data in table 3 and the patterns in figure 2, it can be seen that the density of glass-ceramics increases with the increase of iron oxide content. This is mainly due to the close relationship between the density of glass-ceramics and the composition and heat treatment process. The density of glass-ceramics increases with the addition of iron oxide. By precisely controlling the density of glass-ceramics, it can be sensitively detected the fluctuation of basic composition of glass-ceramics and the change of crystallization effect caused by heat treatment. To some extent, water absorption reflects the degree of crystallization and compactness of glass-ceramics. From Fig. 2, it can be seen that the change of water absorption is not significant when iron oxide is less than 1 wt. %. As the Fe_2O_3 content is more than 1 wt. %, the water absorption decreases, but all of the experimental date shown in figure 2 meet the requirements of national standards. This irregular change is caused by bubbles on the surface of glass-ceramics.

3.3. Effect of Iron Oxide on Acid and Alkali Resistance

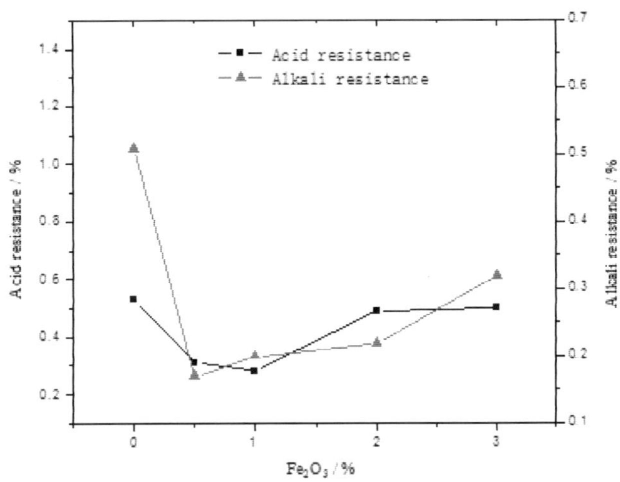

Figure 3. Relationship between Fe_2O_3 content and the acid and alkali resistance of glass-ceramics

Table 3. Results of glass-ceramics properties

Samples	Density g/cm^3	Water absorption %	Acid resistance %	Alkali resistance %
1#	2.73	1.395	0.53	0.51
2#	2.76	1.392	0.31	0.17
3#	2.77	1.384	0.28	0.2
4#	2.81	0.868	0.49	0.22
5#	2.82	0.611	1.36	0.32

Figure 3 shows the relationship between acid and alkali resistance of the glass-ceramics versus iron oxide content under conditions that the basic glass samples nucleated at 750 °C for 2 h and crystallized at 1050 °C for 2 h. The results are shown in table 3. It can be seen from the figure 3 that after soaking in 1% sodium hydroxide solution for 3 hours, the mass loss rate of glass-ceramics with iron oxide is lower than that of glass-ceramics without iron oxide. Alkali corrosion resistance of glass-ceramics varies gradually with the increase of Fe_2O_3 content. It can be inferred that the glass phase in glass-ceramics corrodes in alkaline solution. With the increase of Fe_2O_3, the crystalline phase content increases, and the glass phase content decreases. The alkali corrosion resistance of glass-ceramics increases. After soaking in 1% hydrochloric acid solution at constant temperature for 3 hours, the mass loss of glass-ceramics sample decreases slightly with the addition of iron oxide, which indicates that adding a small amount of iron oxide will enhance its acid corrosion resistance, but it will increase slightly with the increase of iron oxide content, indicating that the addition of excessive iron oxide will reduce the acid corrosion resistance.

4. Conclusions

(1) With the increase of iron oxide content, the crystallization peak temperature (T_p) of glass-ceramics decreases obviously. When Fe_2O_3 content is 0.5 wt. %, the crystallization peak temperature is lower than that without Fe_2O_3, and also lower than that with Fe_2O_3 content 1 wt. % and 2 wt. %, respectively.

(2) With the increase of iron oxide content, the density of glass-ceramics increases, but the water absorption decreases. Alkali corrosion resistance of glass-ceramics increases with the increase of iron oxide content in range of 0.5 wt. % - 3 wt. %. When iron oxide content is higher than 1 wt. %, the acid corrosion resistance decreases slightly.

(3) The presence of small amount of iron oxide less than 1 wt. % in slag can promote the crystallization of basic glass of CaO-MgO-Al_2O_3-SiO_2 quaternary system and improve the physical and chemical properties of glass-ceramics.

5. Acknowledgement

Project (51864025) supported by the National Natural Science Foundations of China.

6. References

[1] Yanfei Zou, Huiqing Jiang. Present situation of blast furnace slag treatment method and comprehensive utilization technology [J]. 2011 *Gold Mountain* **6** 42 (In Chinese).

[2] Jinshu Cheng, Hong Li, Liying Tang, Feng He. *Glass-ceramics* [M]. 2006 Beijing: Chemical Industry Press p 352 (In Chinese).

[3] Karamanov A, Pisciella P, Pelino M. The effect of Cr_2O_3 as a nucleating agent in iron-rich glass-ceramics [J]. 1999 *Journal of the European Ceramic Society* **19** 2641.

[4] Alexander Karamanov, Paola Pisciella, Carlo Cantalini, Mario Pelino. Influence of Fe^{3+}/Fe^{2+} Ratio on the Crystallization of Iron-Rich Glasses Made with Industrial Wastes [J]. 2000 *Journal of the American Ceramic Society* **83** 3153.

[5] Karamanov A, Pisciella P, Pelino M. The crystallisation kinetics of iron rich glass in different atmospheres [J]. 2000 *Journal of the European Ceramic Society* **20** 2233.

[6] Karamanov A, Pisciella P, Pelino M. The effect of Cr_2O_3 as a nucleating agent in iron-rich glass-ceramics [J]. 1999 *Journal of the European Ceramic Society* **19** 2641.

[7] Kai Zhang. Study on crystallization behavior of glass-ceramicsderived from simulated molten steel slag [D]. 2012 *Wuhan,Huazhong University of Science and Technology* p 32 (In Chinese).

[8] Hirose K, Honma T, Doi Y, et al. Mssbaucr analysis of Fe ion state in lithiumiron phosphate glasses and their glass-ceramics with olivinc-type $LiFePO_4$ crystals [J]. 2008 *Solid State Communications* **146** 273.

[9] Shugen Zhang, Qi Wei, Dawei Wang. The study on the crystallization of glass-ceramics with Fe_2O_3 as nucleation agent [J]. 1999 *Bulletin of the Chinese Ceramics Society* **4** 77 (In Chinese).

[10] Dandan Wu. Crystallization of Fe_2O_3 on glass from crystallization and microstructures in R_2O-CaO-MgO-Al_2O_3-Fe_2O_3-SiO_2 system and application of stone powder in glass-ceramics. [D]. 2007 *Overseas Chinese University* p 45 (In Chinese).

[11] Li Baowei, Du Yongsheng, Zhang Xuefeng, Jia Xiaolin, Zhao Ming, Chen Hua. Effects of Iron Oxide on the Crystallization Kinetics of Baiyunebo Tailing Glass-Ceramics [J]. 2013 *Transactions of the Indian Ceramic Society* **2** 119.

[12] Rezvani M, Eftekhari-Yekta B, Solati-Hashjin M, et al. Effect of Cr_2O_3, Fe_2O_3 and TiO_2 nucleants on the crystallization behaviour of SiO_2-Al_2O_3-CaO-MgO (R_2O) glass-ceramics [J]. 2005 *Ceramics international* **31**75.

[13] P. Alizadeh, B. Eftekhary Yekta, A. Gervei. Effect of Fe_2O_3 addition on the sinterability and machinability of glass-ceramics in the system MgO-CaO-SiO_2-P_2O_5 [J]. 2004 *Journal of the European Ceramic Society* **24** 3529.
[14] Ebisawa Y, Miyaji F, Kokubo T, et al. Bioactivity of ferromagnetic glass-ceramics in the system FeO-Fe_2O_3-CaO-SiO_2 [J]. 1997 *Biomaterials* **18** 1277.
[15] Standard test method for water absorption of domestic ceramic ware. 1996 *State Bureau of Technical Supervision* p 3.
[16] Z. J. Yang, Y. Li, A. Q. Cang, M. L. Diao, W. B. Guo, The influence of Fe^{2+} and Fe^{3+} on crystallization of CaO-Al_2O_3-SiO_2-MgO system glass-ceramic [J]. 2012 *Materials Science and Technology* **20** 45.

IOP Conf. Series: Materials Science and Engineering **746** (2020) 012011 doi:10.1088/1757-899X/746/1/012011

Exploring the Optimal Growth Parameters of Al_2O_3 Grown by Atomic Layer Deposition

Shuping Li[1] and Zeqian Ren[2]

[1]Suzhou Industrial Park Institute of Services Outsourcing, Jiangsu Suzhou 215123, China, No 99, Ruoshui Road, Suzhou industrial park
[2]Suzhou Institute of Nano-Tech and Nano-Bionics, Chinese Academy of Sciences, Suzhou 215123, China, No 398, Ruoshui Road, Suzhou industrial park
Email: lisp@siso.edu.cn

Abstract: High quality Al_2O_3 thin film is grown by atomic layer deposition (ALD). The influence of growth parameters such as growth temperature and purge time on the surface morphology, dielectric constant and so on have been investigated. It is found that Al_2O_3 thin film at a flow rate of 150 sccm, H_2O flow rate of 150 sccm, pulse time of 0.1 s, purge time of 6 s, cycle period of 450 times and growth temperature of 150°C, exhibits the optimal material quality with a growth rate of 0.79 Å /cycle, a root-square surface roughness (RMS) of 0.152 nm and a capacitance of 13.8 nf.

1. Introduction

Al_2O_3 thin film is considered to be the most promising candidate to traditional silicon dioxide gate dielectric materials due to the large band gap, high hardness, excellent electrical properties, chemical stability and thermal stability. In recent years, the research on the deposition technology of Al_2O_3 film has attracted numerous attentions. The methods of film deposition include physical vapor deposition (PVD), chemical vapor deposition (CVD), and atomic layer deposition (ALD). Among them, the atomic layer deposition technology can precisely control the growth thickness (the thickness unit is accurate to the atomic level) and has a good step coverage ability due to its special reaction mechanism. It is widely used in the preparation of high-k dielectric films.

This paper introduces a method for preparing high mobility Al_2O_3 films by atomic layer deposition. The surface morphology, dielectric properties, etc. of alumina films prepared under different deposition temperatures and purge time are analyzed. The variation in characteristics is to obtain suitable parameters and conditions for depositing high-quality Al_2O_3 thin film.

2. Experiment

2.1. Experimental Principle

Atom layer deposition (ALD) is a vapor-phase technique for preparing dense and ultrathin films with precise growth control. In a typical ALD procedures for depositing a binary compound film, sequential alternating pluses of gas precursors are introduced into a reaction chamber with an inert purge step in between until the appropriate thickness is reached. Precursors are then absorbed onto the substrate and react with each other to form the desired film. The reaction between precursors and the substrates is crucial for depositing films. Trimethylaluminum and high-purity H_2O were used as precursors and the carrier gas was nitrogen (N_2).

2.2. Experimental Chemicals

Chemicals used in this paper included acetone (CH_3COCH_3), isopropyl alcohol ($(CH_3)_2CHOH$) and TMA and so on.

2.3. Experimental Equipments

Al_2O_3 thin film was grown on silicon substrates in an R-200 Advanced ALD system. Prior to the growth of Al_2O_3 thin film, substrates were pre-treated using ultrasonic cleaner to achieve better surface morphology. For material characterization, its thickness was measured by ellipsometer (M-2000DI J. A. Woollam). Atom force microscopy (AFM, Brucker Dimension ICON) was applied for surface inspection. Its electrical properties were evaluated by parameter analyzer (Keithley 4200). The specific models and manufacturers are shown in table 1.

Table 1. The list of experimental equipments

Experimental equipments	Model	Manufacturer	Application
Ultrasonic cleaner	KQ-600E	Kunshan Ultrasonic Instrument Co., Ltd.	Sample clean
Atomic layer deposition system	R-200 Advanced	Picosun	Film deposition
Atomic force microscope	Burker	Dimension ICON	Analysis of Surface morphology
Spectral ellipsometer	M-2000DI	J.A. Woollam	Thickness test
Semiconductor characteristic analysis system	Keithley4200, Suss PM8	Keithley, Suss	Electrical performance test

3. The Influence of Growth Temperature on the Properties of Al_2O_3 Thin Film

Three Al_2O_3 thin film samples were grown at 100°C, 120°C and 150°C to study the influence of growth temperature. The flow rate of TMA and H_2O were both set to be 150 sccm. The timing sequence (TMA exposure, purge, water exposure, purge) during these TMA/H_2O cycles was (0.1 s, 1 s, 0.1 s, 1 s).

3.1. The Influence of Deposition Temperatures on Surface Morphology

Fig. 1 shows the surface morphology of these Al_2O_3 thin film samples measured by AFM in contact mode and the root mean square (RMS) roughness for each sample grown in different temperatures were displayed in table 2.

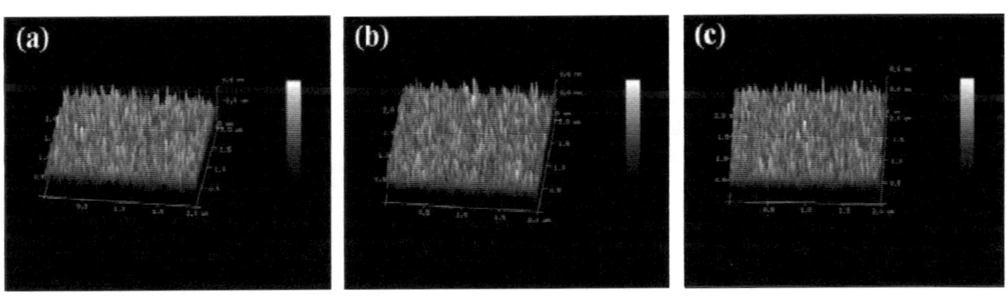

(a) 100°C (b) 120°C (c) 150 °C

Figure 1. The surface morphology of different deposition temperature

Table 2. The film RMS at different deposition temperature

Deposition temperature (°C)	RMS (nm)
100	0.189
120	0.176
150	0.152

The surface was shown to be smoother with evaluated growth temperature and the RMS roughness has reached to the smallest of 0.152 nm for the sample grown at 150°C. Weakened chemical reaction and decreased growth rate mainly account for the rougher surface at lower temperature. TMA may not fully decompose so that other substances might be incorporated into the film, leading to more uneven surface. Enhanced physical absorption at lower temperature is another reason for the rougher surface. In that case, multi-layers tend to form on the growing surface and then interfere the next cycle, resulting in the undulated surface condition.

3.2. The Influence of Deposition Temperatures on Growth Rate

In order to study the influence of different deposition temperatures on the growth rate of the film, the thickness of the film at different deposition temperatures were tested by ellipsometer. The results are shown in table 3.

Table 3. The growth rate of different deposition temperature

Deposition temperature (°C)	Thickness (nm)	Growth rate (Å /cycle)
100	30.7	0.68
120	32.8	0.73
150	35.6	0.79

As can be seen from the table 3, the deposition temperature has little influence on the growth rate. The growth rate of 100°C, 120°C, 150°C is 0.68 Å /cycle, 0.73 Å /cycle, 0.79 Å /cycle, respectively. The largest growth rate is achieved at 150°C.

3.3. The Influence of Deposition Temperatures on Dielectric Constant

The relationship between the dielectric constant and the capacitance of the film can be expressed by Equation 1-1, the dielectric constant of the film is proportional to the capacitance of the film.

$$C = \varepsilon_0 \varepsilon_r A / t_{ox} \tag{1-1}$$

Where C is gate capacitance, ε_r is relative permittivity, ε_0 is Vacuum dielectric constant, A is the area of Al_2O_3 thin film and t_{ox} is thickness of Al_2O_3 thin film.

In order to study the influence of different deposition temperatures on the dielectric constant, a semiconductor analysis system was used to measure the film capacitance obtained at different deposition temperatures. The test results are shown in table 4.

Table 4. The capacitance of different deposition temperature

Deposition temperature (150 °C)	Thickness (nm)	Capacitance (nf)
100	30.7	0.20
120	32.8	0.25
150	35.6	13.8

When the deposition temperature is changed from 100°C to 150°C, the thickness of the film is nonlinear with the film capacitance. The Al_2O_3 thin film prepared at the deposition temperature of 100°C and 120°C has a low capacitance and a low dielectric constant. When the deposition temperature reaches 150°C, the capacitance of the film changes significantly, increasing to 13.8 nf. The Al_2O_3 thin film prepared at the deposition temperature of 150°C has the highest capacitance and dielectric constant, and the dielectric properties of the film are the best.

4. The Influence of Purge Time on the Properties of Al_2O_3 Thin Film

The influence of purging time on the film quality of Al_2O_3 was studied. The experimental conditions were designed as follows: carrier gas flow rate TMA is 150 sccm, H_2O flow rate is 150 sccm, pulse time is set to 0.1 s, deposition temperature is 150°C, and cycle period is 450 times. The film preparation was carried out by setting the purge time to 3 s, 6 s, and 9 s, respectively.

4.1. The Influence of Purge Time on Surface Morphology

In order to study the influence of the purge time on the surface morphology of the Al_2O_3 thin film, the morphology of the film is obtained by AFM as shown in Fig. 2.

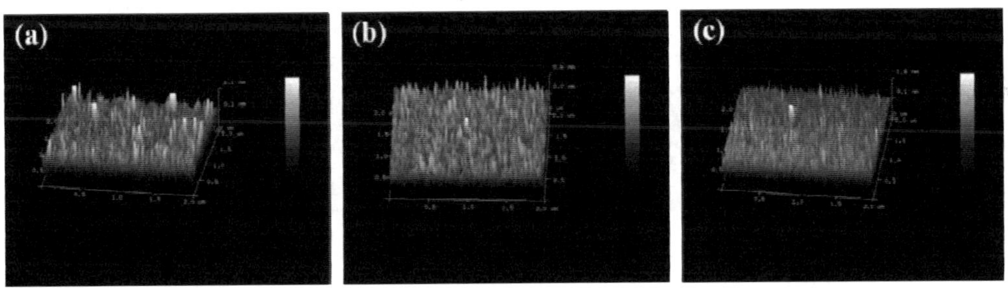

(a) 3 s (b) 6 s (c) 9 s

Figure 2. The surface morphology of different purge time

The RMS of the film measured under different purge time is shown in table 5.

Table 5. The RMS of different purge time

Purge time (s)	RMS (nm)
3	0.326
6	0.152
9	0.398

The surface morphology of the film is significantly changed with different purge time. The RMS is 0.326 nm at a purge time of 3 s, a roughness of 0.152 nm at a purge time of 6 s, and a roughness of 0.398 nm at a purge time of 9 s. At the purge time of 6 s, the surface roughness is the lowest, which is 0.152 nm.

4.2. The Influence of Different Purge Time on Growth Rate

In order to study the influence of different purge time on film thickness, the growth rate of the film under different purge time was tested by ellipsometer as shown in Table 6.

Table 6. The growth rate of different purge time

Purge time (s)	Thickness (nm)	Growth rate (Å /cycle)
3	35.1	0.78
6	35.6	0.79
9	35.5	0.79

It is known from the data that the influence of the purge time on the growth rate of Al_2O_3 thin film is not very significant, and the difference between the thicknesses caused by the different purge times is ignorable. However, it can be seen that after the purge time is more than 6 s, the thickness uniformity of the Al_2O_3 thin film is relatively better, and the rate of deposition of the film is relatively stable. Therefore, we choose the purge time of 6 s as the better growth condition when the pulse time of TMA is 0.1 s. The reason for the difference is mainly due to the fact that when the purge time is relatively short, the TMA entering the reaction chamber is not completely purged, and residual TMA will react with other substances in the chamber.

4.3. The Influence of Different Purge Time on Dielectric Constant

In order to study the influence of different purge time on the dielectric constant of the film, the capacitance of the prepared film was tested by a semiconductor analysis system. The test results are shown in table 7.

Table 7. The capacitance of different purge time

Purge time (s)	Thickness (nm)	Capacitance (nf)
3	35.1	8.2
6	35.6	13.8
9	35.5	3.0

With the increasing of purge time, the capacitance of the film does not change regularly. From table 7, at the purge time of 3 s, 6 s, 9 s, the capacitance is 8.2 nf, 13.8 nf, 3.0 nf, respectively. When the purge time is 6 s, the dielectric properties of the film are optimal compared to the capacitance and dielectric constant obtained at 3s and 9s.

5. Summary

In summary, we have studied the influence of deposition temperature and purge time on the surface morphology, film thickness, deposition rate and dielectric constant of the Al_2O_3 thin film which is deposited by ALD. By the optimization of growth parameters, we obtained a better Al_2O_3 thin film with growth rate of 0.79 Å /cycle, RMS of 0.512 nm, capacitance of 13.8 nf, which the flow rate of TMA is 150 sccm, the H_2O flow rate is 150 sccm, the pulse time is 0.1 s, purge time is 6 s and cycle period is 450 times.

6. Reference

[1] Li Chunya, Zhang Hao, Song Jiantao, Ding Xingwei, Wei Bin. Effects of trimethylaluminium pulse duration and purging time on the performance of Al_2O_3 film by atomic layer deposition [J]. Journal of Functional Materials, 2017, 48 (07): 7178-7181.

[2] Shaeri M R, Jen T C, Yuan C Y. Improving atomic layer deposition process through reactor scale simulation [J].International Journal of Heat and Mass Transfer, 2014, 78:1243-1253.

[3] Puurunen R L.A short history of atomic layer deposition: Tuomo Suntola's atomic layer epitaxy [J].Chemical Vapor Deposition, 2014, 20 (10-12): 332-344.

[4] Zhang Li-Rong, Ma Xue-Xue, Wang Chun-Fu, High speed gate driver circuit basd on metal oxide thin film transistors. Acta Physica Sinica, 2016, 65(02): 356-363.

[5] Li Yanying. Preparation and Dielectric Properties of Aluminlum Oxide Thin Films Grown by Atomic Layer Deposition [D].Harbin Institute of Technology, 2017.

[6] Robertson J, Wallace R M. High-k Materials and Metal Gates for Cmos Applications [J]. Materials Science & Engineering R-Reports, 2015, 88: 1-41.

[7] K.L. Jarvis, P.J. Evans, A. Nelson, G. Triani. Comparisons of alumina barrier films deposited by thermal and plasma atomic layer deposition [J]. Materials Today Chemistry, 2019, 11.

[8] Wei Hu, Cheng Li, D.G. Shchukin. Ceramic honeycomb-like alumina film as corrosion inhibitor carrier and mechanism analysis [J]. Colloids and Surfaces A: Physicochemical and Engineering Aspects, 2018, 555.

IOP Conf. Series: Materials Science and Engineering **746** (2020) 012012 doi:10.1088/1757-899X/746/1/012012

Analysis of Climatic Characteristics and Multiple Corrosion Factors of Salty Spray Environment in Shenzhen

Jun Liu [1], Lu Hu[1] and Yingqiu Song[2,*]

[1]Guangdong Provincial Key Laboratory of Durability for Marine Civil Engineering, Shenzhen University, Shenzhen, Guangdong, China 518060
[2]Normal university, Shenzhen University, Shenzhen, Guangdong, China 518060
Email: wookee2012@163.com, [*]Corresponding author: Yingqiu Song

Abstract: In the coastal salt fog area, there are many sources of atmospheric corrosion in the ocean. For the concrete structure, the effects of carbonization, sulfate, chloride and other complex effects should be considered. The experiment was carried out by setting up a station in the coastal salt fog area of Shenzhen, and the carbon dioxide and chloride ion content in the atmosphere were observed on the spot. The study found that the coastal environment of Shenzhen has the characteristics of high temperature and high humidity, and the rainfall is large. The average temperature in the past five years has reached 23.5 degrees Celsius and the average humidity has reached 75.4%. The atmospheric carbon dioxide content in the coastal salt spray area is higher in spring and winter, and lower in summer and autumn. The atmospheric chloride ion concentration in the coastal salt spray area is characterized by high winter and low summer temperatures throughout the year, and the annual chloride ion concentration fluctuates greatly.

1. Introduction

From the global perspective, there are a large number of coastal areas. In the coastal environment, there are many corrosive factors in the atmosphere, which will adversely affect the durability of the reinforced concrete structure, which may lead to structural safety problems. Premature failure of concrete structures is caused by a certain chemical reaction between the harmful medium in the environment and the calcium hydroxide and cement hydration products in the concrete, resulting in changes in the internal components of the concrete [1]. Concrete may gradually lose gelation.

There are many corrosive factors in the coastal environment. Concrete is a non-uniform, porous solid, liquid and gas three-phase composite material. Corrosion ions in the environment will enter the structure through the pores of the concrete [2, 3]. Shenzhen is located in the south of China and belongs to the typical marine atmosphere. There are a large number of reinforced concrete buildings located in the coastal salt fog area. It is necessary to consider the effects of carbonation and chloride ions. When analyzing these factors, it is necessary to understand the main areas of the coastal salt spray area, the content of corrosive ions. In this paper, a field observation station is set up in the coastal salt fog area to measure the content of chloride ion and carbon dioxide in the salt spray zone. It has certain practical significance for analyzing the durability of the concrete structure in the coastal salt spray area.

2. Experiment

An environmental observation station was set up in Shenzhen to measure the content of chloride ions and carbon dioxide in the atmosphere. The distance between the measuring points and the coastline is 200 meters.

3. Results and Analysis

3.1. Shenzhen Coastal Environmental Climate Characteristics

Shenzhen is located south of the Tropic of Cancer, east longitude 113°46' to 114°37', north latitude 22°27' to 22°52'. It is located in the south of Guangdong Province, east to Daya Bay and Dapeng Bay, west to the Pearl River Estuary and the Boyang Sea. It is a subtropical maritime climate zone with mild climate, abundant rainfall and long sunshine hours. This paper collected the main meteorological data from 2014 to 2018 from the website of Shenzhen Meteorological Bureau.

Table 1. 2014-2018 Shenzhen meteorological data

Year	Average temperature /°C	Maximum temperature /°C	Minimum temperature /°C	Total rainfall /mm	Rainy days /d	Average relative humidity /%
2014	23.2	35.0	4.4	1725.5	129	73
2015	23.9	36.2	8.6	1500.8	123	72
2016	23.2	36.7	1.7	2490.6	153	79
2017	23.6	36.9	7.4	1967.1	127	77
2018	23.4	35.6	4.2	1957.2	121	76

3.1.1. Temperature and Humidity

From 2014 to 2018, the average temperature is 23.5°C, the maximum temperature is 36.9°C, the minimum temperature is 1.7°C, and the annual average humidity is 75.4%. It can be seen from Fig. 1 and Fig. 2 that Shenzhen has the characteristics of high temperature and high humidity, and the temperature and humidity changes throughout the year are relatively stable, almost at a relatively high level.

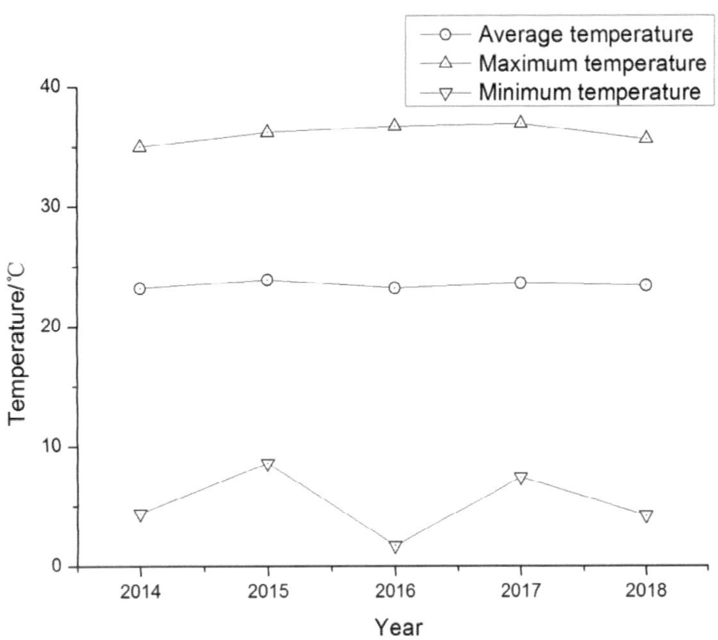

Figure 1. Temperature change in Shenzhen coastal environment

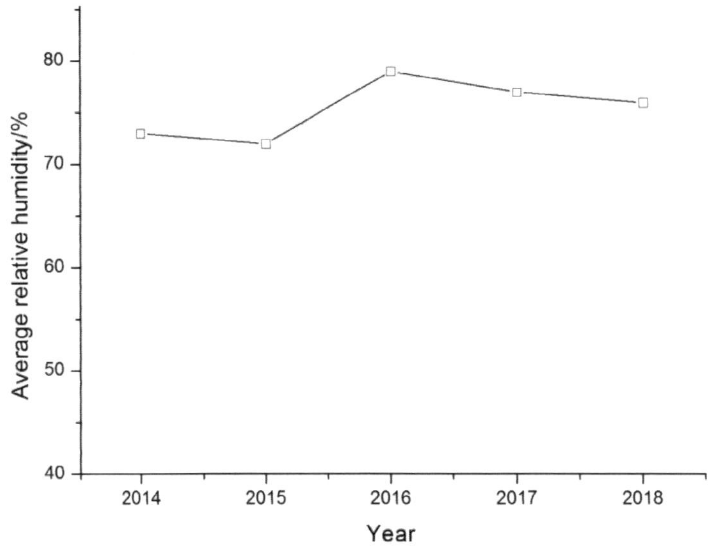

Figure 2. Humidity change in Shenzhen coastal environment

3.1.2. Precipitation

The precipitation in Shenzhen is quite abundant. The average precipitation from 2014 to 2018 is 1928.2 mm. The annual rainfall varies greatly, with the most years 2490.6 mm, mainly due to the strong weather and climate in Shenzhen due to the strong El Niño in 2016. It is characterized by "winter cold, hot summer, heavy rainfall, and typhoon and rain". The accumulated rainfall is nearly 30% higher than the average climate of the same period, 66.0% more than in 2015, 50% more than the past five years, and 68.4% more than last year. A total of 12 heavy rains, 3 heavy rains, and strong convection were recorded. The year with the least rainfall is only 1500.8 mm (2015). In general, 85% of the annual rainfall occurs from April to September, 48% of which is distributed from July to September, with an average rainfall of more than 900 mm; the average rainfall from April to June is more than 700 mm, mainly from cold air and the tropical warm and humid air flow is formed together. The monthly rainfall changes during the year are unimodal, up to August, at least December.

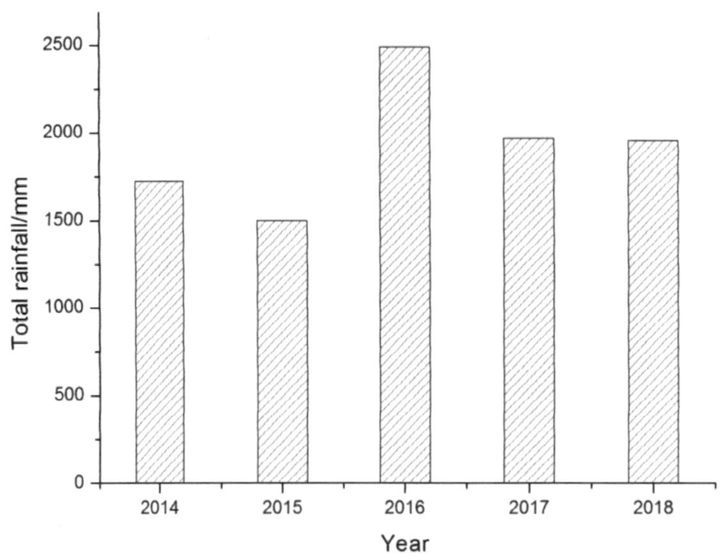

Figure 3. Precipitation change in Shenzhen coastal environment

3.2. Carbon Dioxide Content in Coastal Salt Spray Area

The cement hydration product is alkaline, and the presence of alkaline components ensures the stability of other hydration components. Carbon dioxide and other acid gas pollutants from the atmosphere react with hydroxides formed in concrete, and these hydroxides are reacted. Conversion to non-alkaline carbonates, causing the pH in the concrete to drop [4, 5], this may damage the passivation state of the reinforcement. Calcium carbonate is deposited in the pores of the concrete, blocking part of the pores. As the reaction continues, the pH in the pores drops to 8.3, and the silicate, aluminate and ferrite become unstable and begin to decompose. It is no longer passivated, and as oxygen and moisture enter the pores, the steel begins to corrode, and carbon dioxide has a greater impact on the durability of the concrete structure.

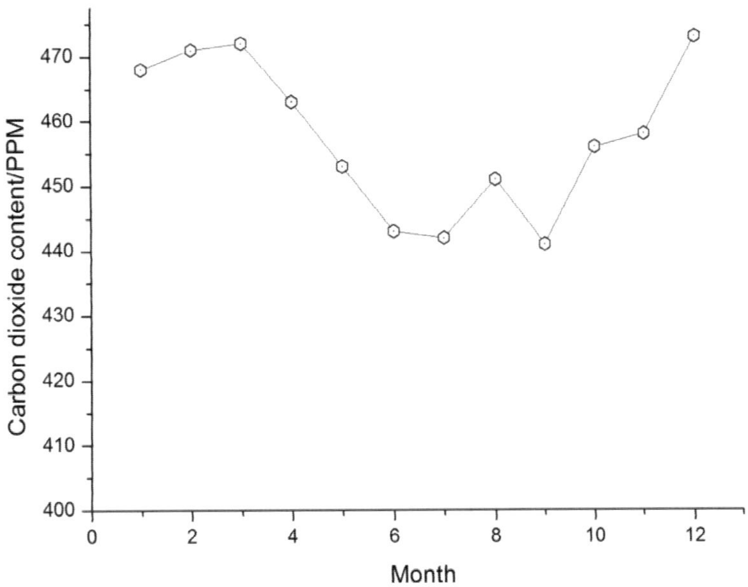

Figure 4. Carbon dioxide content in the survey area changes with the month

In the observation area of Shenzhen Binhai, it can be seen from Figure 4 that the carbon dioxide content is higher in spring and winter, and the carbon dioxide content in summer and autumn is lower. The carbon dioxide content in each month does not change much, and it varies from 440 to 470 ppm.

3.3. Chloride Ion Content in The Atmosphere of Coastal Salt Spray Area

In the atmosphere near the ocean, seawater droplets containing sodium chloride are mixed into the atmosphere when the water droplets are scattered. The sea breeze and the mist contain a large amount of particles of chloride or NaCl, and the chloride ions accelerate the corrosion of the steel [6]. In general, the high alkalinity of cement hydration produces a dense passivation film on the surface of the steel in the concrete. Previous studies have suggested that the passivation film is composed of iron oxide, but recent studies have shown that the passivation the film contains Si-O bonds, which have strong protection for steel bars. Chloride ions are extremely strong depassivating agents. When chloride ions are added to the concrete to reach the surface of the steel bar and adsorbed to the local passivation film, the place can be the pH value is rapidly lowered and the pH of the steel surface is lowered to below 4, thereby destroying the passivation film on the surface of the steel.

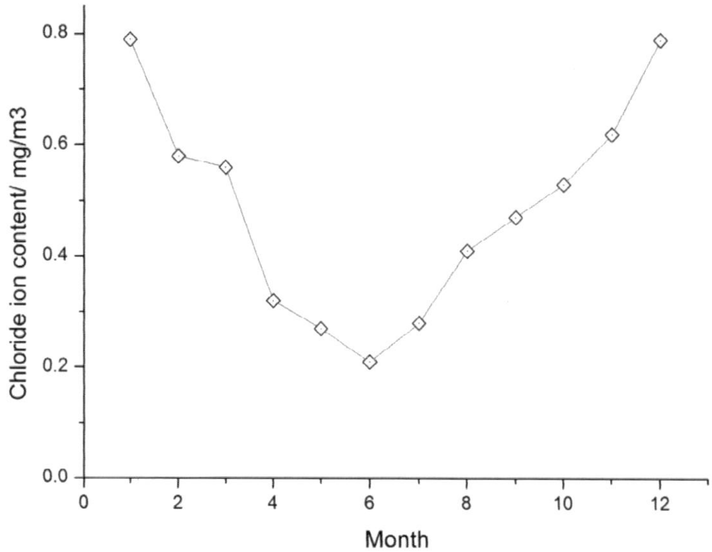

Figure 5. Chloride ion content in the measurement area changes with the month

Figure 5 shows that the chloride ion concentration is characterized by high winter and low summer temperatures throughout the year. In general, high clustering in 10, 11 and 12 months, low clustering in 4, 5, 6, 7, 8, and September, the annual chloride ion concentration fluctuated greatly.

4. Conclusion

1. In the coastal salt fog area, there are many sources of atmospheric corrosion in the ocean. For the concrete structure, the effects of carbonization, sulfate and chloride ions should be considered.

2. Shenzhen coastal environment has the characteristics of high temperature and high humidity. The average temperature in the past five years has reached 23.5 degrees Celsius and the average humidity has reached 75.4%.

3. The atmospheric carbon dioxide content in the coastal salt spray area is higher in spring and winter, lower in summer and autumn, and little change in carbon dioxide content in each month.

4. The concentration of atmospheric chloride ions in the coastal salt spray area is high in winter and low in summer. The chloride ion content in October, November and December is high, and the chloride ion concentration fluctuates greatly throughout the year.

5. Acknowledgments

The financial supports from Natural Science Foundation of China (No. 51578340), Shenzhen city science and technology project (No. JCYJ20170302143133880, JCYJ20180305124008155) and Shenzhen International Cooperation Research Project (No. GJHZ20180928155602083).

6. References

[1] Jun Liu, Guangfeng Ou, Qiwen Qiu, Feng Xing, Kaifeng Tang, Jiaxin Zeng. Atmospheric chloride deposition in field concrete at coastal region. *Construction and Building Materials*, 2018, 190: 1015-1022.

[2] Borges PHR, Costa JO, Milestone NB, Lynsdale CJ, Streatfield RE. Carbonation of CH and C–S–H in composite cement pastes containing high amount of BFS. *Cement Concrete Research.* 2010, 40: 284-92.

[3] Jun Liu, Feng Xing, Biqin Dong, Hongyan Ma, Dong Pan. Study on water sorptivity of the surface layer of concrete. *Materials and Structures*, 2014, 47: 1941–1951.

[4] Pu Q, Jiang LJ, Chu H, Xu Y, Zhang Y. Evolution of pH and chemical composition of pore solution in carbonated concrete. *Construction and Building Materials*, 2012, 28: 519-24.

[5] Thiery M, Villian G, Dangla P, Platret G. Investigation of the carbonation front shape on cementitious materials: effect of the chemical kinetics. *Cement Concrete Research*, 2007, 37: 1047-58.

[6] J. Williamson, O.B. Isgor. The effect of simulated concrete pore solution composition and chlorides on the electronic properties of passive films on carbon steel rebar. *Corrosion science* [J]. 2016, 106: 82-95.

IOP Conf. Series: Materials Science and Engineering **746** (2020) 012013 doi:10.1088/1757-899X/746/1/012013

Effect of Particle Flow Rates on Erosion Ratio in Particle Impingement Experiments

Nan Lin[1], Renyang He[1], Hao Zhang[2], Junqiang Wang[1], Honglian Ma[1] and Hui Huang[1]

[1]China Special Equipment Inspection and Research Institute, Beijing 100029, China
[2]Key Laboratory of Vehicle Advanced Manufacturing, Measuring and Control Technology (Ministry of Education), Beijing Jiaotong University, Beijing 100044, China
Email: sy_linnan@hotmail.com

Abstract. Erosion caused by solid particles could not be neglected in oil and gas transportation industry, especially in gathering gas pipelines. Particles with small size may also erode the inner wall of pipelines after cumulative impacting and even lead to the leakage. Normally, a lot of solid particles are carried in the gathering gas before filtrating. Additionally, corrosion productions may fall off the wall as the gas flowing. Researches on erosion of pipeline caused by solid particles are important for predicting the remaining wall thickness, which can be a guide for pipeline inspection and safety protection. However, effect of particle flow rates on erosion ratio is normally omitted in studies on erosion ratio calculations. It was observed that particle flow rates have a great influence on the mass loss of specimens in the impingement experiments of this work. According to groups of Particle Tracking Velocimetry (PTV) results obtained under various conditions, the particle velocity decreased when the particle flow rate was larger under the same experimental condition and the mass loss of the specimen became lower. Relations between particle velocity and particle flow rate under different experimental conditions were calculated according to the PTV results. Further, relations between erosion ratio and particle flow rate were purposed in this work. It is better to consider effect of particle flow rate while calculating solid particle erosion of pipelines in industry.

1. Introduction

Failures of pipelines and facilities caused by particle erosion are normal in the oil and gas industry. Serious erosion may even lead to leakage accidents or hidden danger for the society and companies [1, 2]. Even through the gathering gas is filtered to omit particles, some particles with small sizes are still there carried by the natural gas. In addition, corrosion productions may also fall off the inner wall as the gas flowing. Erosion occurs under impacting of these small particles. It is well known that particle velocity, size and impact angle have major contributions to erosion [1, 3, 4]. Levy and Sapate investigated the effects of particle shape, hardness and angle on erosion for a range of size of particles [5, 6] to propose simple relations between them. Kumar used a high speed impingement experimental rig to obtain erosion behaviors on two different kinds of coating under 90° [7]. Ballout and Lindsley used rotating devices to investigate the velocity exponent in erosion calculations [8, 9]. Currently, effect of particle size distributions in solid-gas and solid-liquid two-phase flow were discussed based on comparing experimental results by Zamani and Nguyen separately in their works [10, 11]. In researches about simulations, Wong investigated the erosion on pipes with small diameter by simulating [12]. Arabnejad and Mansouri proposed a better calculation method on erosion ratio calculation [13, 14] in their papers. However, little researches mentioned effect of particle flow rate on erosion and particle flow rate is rarely considered in particle impingement experiments. Actually, particle flow rate has a great influence on erosion ratio. In this work, effect of particle flow rates on erosion for gas-solid two phase flow were discussed. In order to make the experimental results more comparable, particle flow rate should be well controlled while studying the effect of particle parameters (such as size and shape) on erosion ratio [15].

2. Effect of Particle Flow Rates on Impingement Experimental Results

An experimental rig was designed to investigate relations between the erosion ratio and particle parameters. In order to obtain better comparable experimental results, experiments mentioned in this paper are similar to those finished before in Erosion Corrosion Research Center (E/CRC) [15]. The connection of test rig is shown as Fig. 1. Particles and compressed gas came together from two different tubes and mixed in the nozzle. The specimen was fixed on the holder and the impact angle was controlled by the holder. The distance between the nozzle exit center and the specimen surface was 1.27 cm. The 316 stainless steel (SS316) specimens were used in these experiments and have dimensions of 75×50×1.5mm. Air and silica sands were used in the impingement experiments.

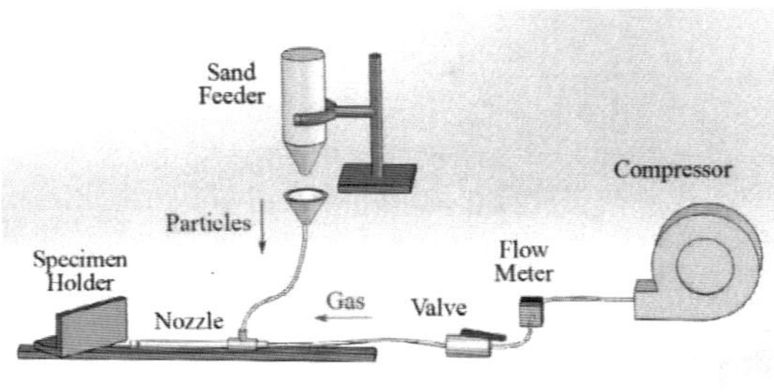

Figure 1. Connection of test rig

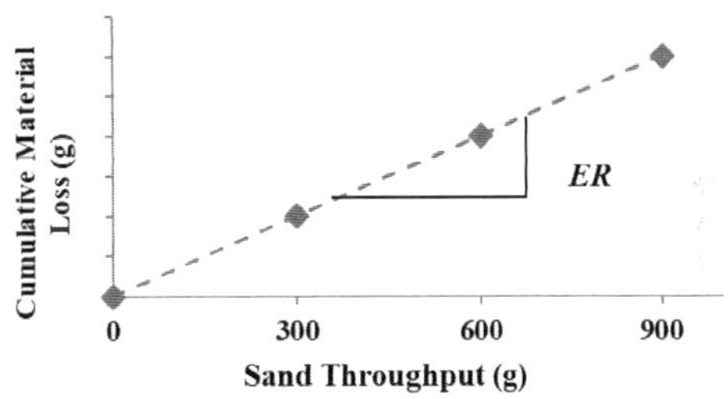

Figure 2. Method of erosion ratio calculation

The gas pressure near the specimen surface was measured before each group of impingement experiment to calculate the gas velocity. The erodent specimen should be swiped to omit possible broken particles which are attached on the surface. The material loss were calculated by measuring the weight of the specimen before and after particle impacting. Erosion ratio (ER) was defined as Fig. 2 to describe the impingement experimental results. The vertical axis represents the cumulative material loss (g) and the horizontal axis is the sand throughput (g) in experiments. The experimental results can be calculated according to equation 1.

$$ER=\frac{m_{\text{loss}}}{m_{\text{particle}}} \tag{1}$$

Where, ER is the erosion ratio, m_{loss} is the mass loss during the impingement experiment, m_{particle} is the mass of erodent particles used in experiments.

Groups of experiments under different particle flow rates and incidence angles were design in this work. Parameters are shown in Tab. 1, the particle size is 150 μm and the throughput of erodent particles for each group is 300 g.

Table 1. Parameters of experiments

Parameter	Value	
Particle size	150 μm	
Throughput	300 g	
Gas velocity	94 m/s	
Impact angle	15°	90°
Particle flow rate	0.53 g/s	2.83 g/s

When the particle flow rate is high, more particles come out from the nozzle, so they are more likely to impact other particles while moving to reduce the particle erosion. On the other side, more particles will impact the specimen surface at the same time. Thus, it is hard to predict erosion ratio under various particle flow rates without analyzing the impingement experimental results.

Experimental results under different particle flow rates (0.53 g/s and 2.83 g/s) and different impact angels (15° and 90°) were compared in Fig. 3. According to Fig. 3, the cumulative material loss of specimens are larger when the particle flow rate is lower under those two impact angles. Thus, erosion ratio is higher when the particle flow rate is lower when using the particle throughputs are the same in impingement experiments. It is necessary to obtain more experiments results under various conditions for further investigation on relation between erosion and particle flow rate.

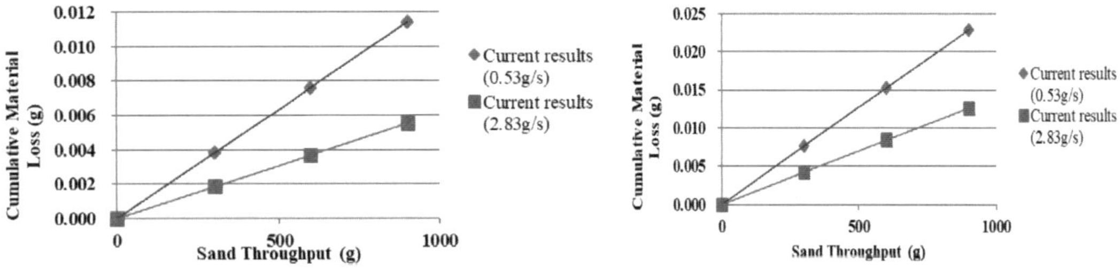

(a) Impact angle: 90° (b) Impact angle: 15°

Figure 3. Comparison of erosion test results

3. Effect of Particle Flow Rates on Particle Velocity

PTV was used to measure and calculate particle velocity in this work. Particle velocities were caught by PTV with two different sand flow rates to study effect of particle flow rate on particle velocity. PTV system includes a double-pulsed laser, a CCD camera, a synchronizer and a processor [16]. The schematics of PTV is shown in Fig. 4. The gas velocity (1.27 cm away from the nozzle exit) and particle flow rate were measured before every group of experiments.

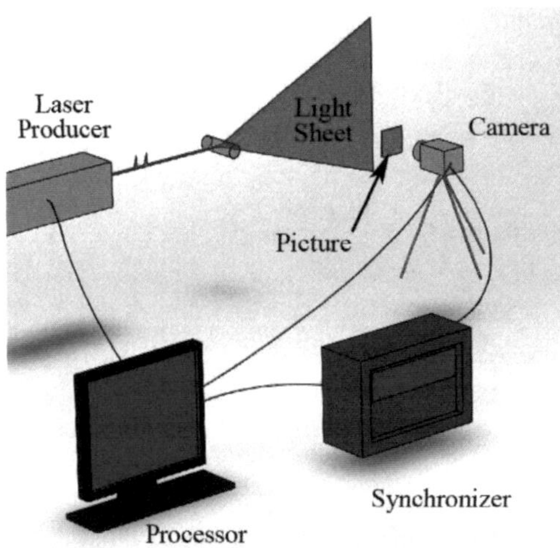

Figure 4. Schematics of particle image velocimetry

Particles caught by PTV are shown in Fig. 5(a). Different velocities of particles near the plane 1.27 cm away from the nozzle exit were calculated by the two contiguous images obtained from the CCD camera. The count of particle with various velocities are shown in Fig. X (b). The vertical and horizontal axis represent the count of particles and particle velocities caught through PTV separately.

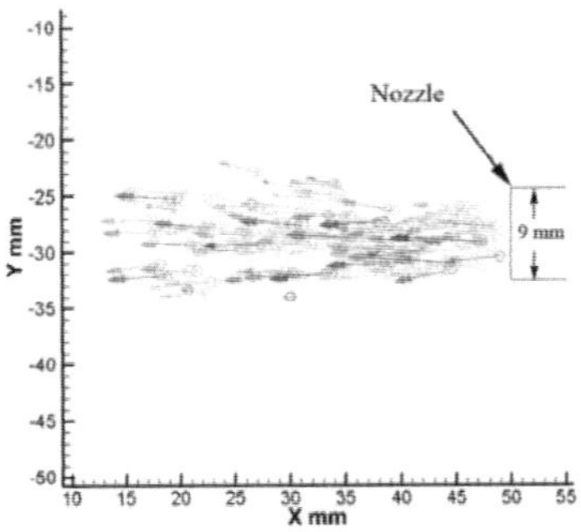

(a) Particles capture

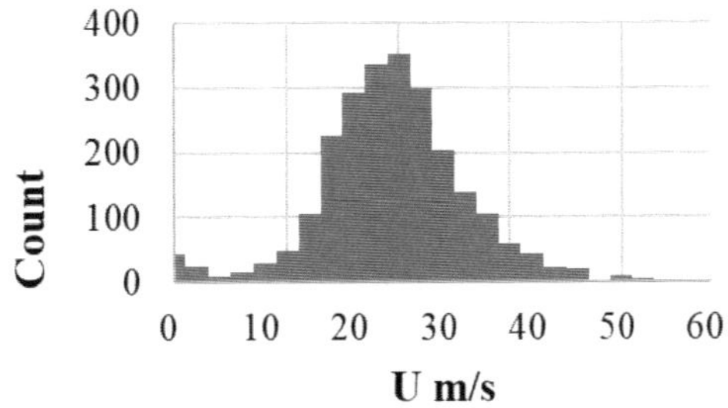

(b) Particle velocity

Figure 5. Particles capture and particle velocity obtained from particle tracking velocimetry

Relations between particle velocity and gas velocity were obtained by PTV. In these tests, the particle size is 150 μm and the particle flow rates are 0.53 g/s and 2.83 g/s separately. Particle velocities of these two kinds of particles in front of the nozzle are shown in Fig. 6 under different gas velocities. The vertical axis is the particle velocity (m/s) and the horizontal axis shows the gas velocity (m/s). It is observed that solid particles are more likely to be accelerated by the flowing gas when the particle flow rate is low. So particles have a higher velocity than those under high particle flow rate even when the gas velocities are the same, and they may cause more erosion on specimen surfaces.

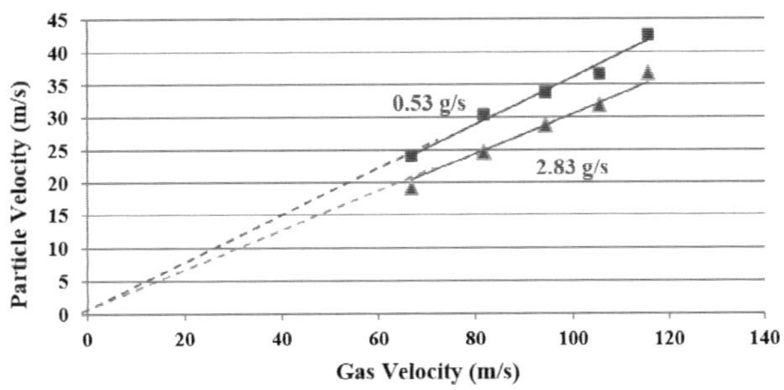

Figure 6. Relation between particle velocity and gas velocity

In order to obtain better comparable experiment results, particle flow rate should be well controlled during impingement experiments for investigating effect of particle parameters on erosion. A vibratory feeder was designed in E/CRC [15] to control the particle flow rate in impingement experiments, which can control most small sand drop rates required in experiments. As shown in Fig. 7, particles are put in the particle inlet and go out from the outlet through the transparent tube. The particle drop rates are controlled by setting the frequency of the vibrator at the end of the device. The particle flow rates were test before every groups of experiments to keep them constant.

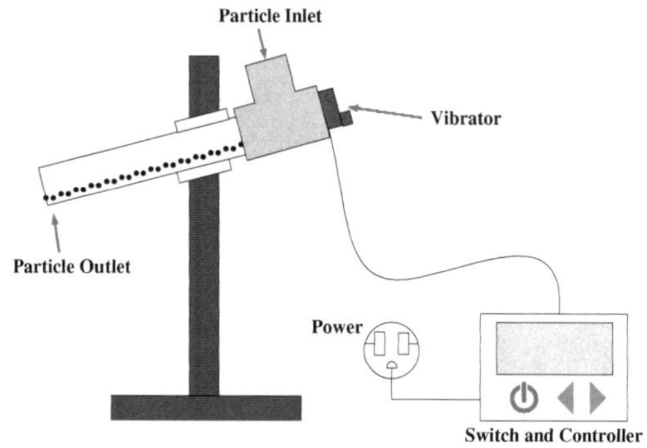

Figure 7. Vibratory feeder

Four different particle flow rates (such as 0.53 g/s, 1.02 g/s, 1.52 g/s and 2.83 g/s) were set in the impingement experiments to investigate effect of particle flow rates on erosion. As shown in Tab. 2, relations between the particle velocity and gas velocity for particles with various particle sizes has been calculated by E/CRC [17].

Table 2. Relation between the particle velocity and gas velocity

Particle diameter	Relation between V_p and V_g
75 μm	V_p=0.415V_g
150 μm	V_p=0.3063V_g
300 μm	V_p=0.3171V_g
600 μm	V_p=0.01709V_g

According to Tab. 2, the relation between the particle velocity and gas velocity is simply accord with equation 2.

$$V_{\mathrm{p}} = C_{\mathrm{p}/g} V_{\mathrm{g}} \tag{2}$$

Where, $C_{\mathrm{p}/g}$ is the relative coefficient between particle velocity and gas velocity. Based on Tab. 2, sands with the size of 75 μm have a better motion performance. Normally, smaller particles have better motion performance with flow in gas-solid two-phase flow while flowing, so the relative coefficient is bigger for smaller particles. However, because of the densities of particles are not constant, changing of particle velocities are similar when the sand sizes are 150 μm and 300 μm.

It was found that the relative coefficient is not constant for different particle flow rates, even when the particle size are the same. Relative coefficients for sands with various sizes were calculated according to the PTV test results while the range of the gas velocities is from 66 m/s to 115 m/s. Relations between particle velocity and gas velocity of different particles under various particle flow rates (FR) are shown in Tab. 3, Tab. 4, Tab. 5 and Tab. 6. It is obvious that the relative coefficient is bigger for smaller particles and the relative coefficients are not constant for each particle size. Because the particle interaction, volume fraction and acceleration changes when the particle flow rates are different. Thus, the particle flow rates should be well controlled in particle impingement experiments to obtain more comparable experimental results.

Table 3. Relation between particle velocity and gas velocity (FR=0.53 g/s)

Particle diameter	Relation between V_p and V_g
75 μm	V_p=0.5267V_g
150 μm	V_p=0.3612V_g
300 μm	V_p=0.2959V_g
600 μm	V_p=0.2633V_g

Table 4. Relation between particle velocity and gas velocity (FR=1.02 g/s)

Particle diameter	Relation between V_p and V_g
75 μm	V_p=0.4581V_g
150 μm	V_p=0.3421V_g
300 μm	V_p=0.2816V_g
600 μm	V_p=0.2448V_g

Table 5. Relation between particle velocity and gas velocity (FR=1.52 g/s)

Particle diameter	Relation between V_p and V_g
75 μm	V_p=0.4403V_g
150 μm	V_p=0.3151V_g
300 μm	V_p=0.2379V_g
600 μm	V_p=0.2198V_g

Table 6. Relation between particle velocity and gas velocity (FR=2.83 g/s)

Particle diameter	Relation between V_p and V_g
75 μm	V_p=0.4372V_g
150 μm	V_p=0.3054V_g
300 μm	V_p=0.2131V_g
600 μm	V_p=0.2047V_g

4. Effect of Particle Flow Rates on Erosion Ratio

Because the particle diameter is bigger, the flow rate of 600 μm sands is unsteady when it is low. So particle erosion experiments are compared only for particles with the size of 75 μm, 150 μm and 300 μm. In order to investigate the effect of particle flow rates on erosion, the particle velocities are controlled at about 24 m/s for all different particles. Relations between the particle velocity and gas velocity of these three kinds of particles are obtained by PTV tests, so the gas velocities can be calculated for different experimental conditions. The throughput of sands are 300 g for each particle impingement experiments under different conditions and the average values of mass loss on specimens were measured and calculated. The experimental results are shown in Fig. 8, where the vertical axis is the erosion ratio (g/g) calculated by equation 1 and the horizontal axis represents particle flow rate (g/s).

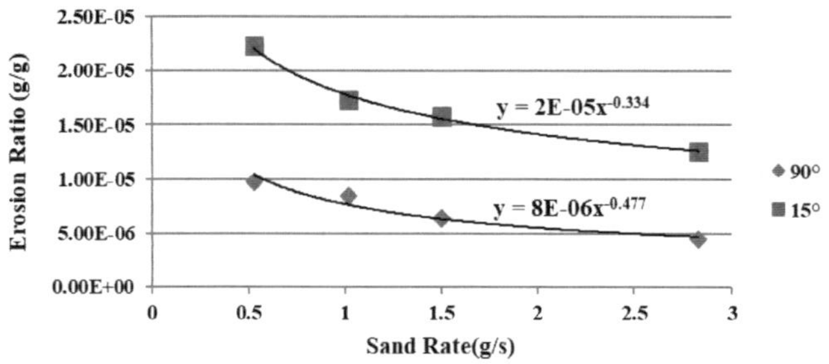

(a)Specimen: SS316, Velocity: 24 m/s, Sand Size: 75 μm

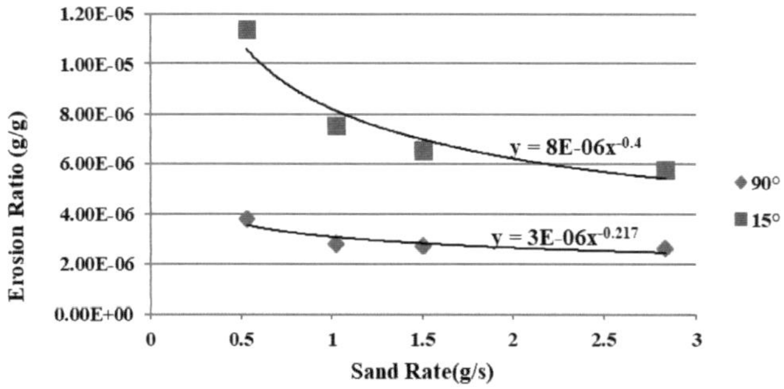

(b) Specimen: SS316, Velocity: 24 m/s, Sand Size: 150 μm

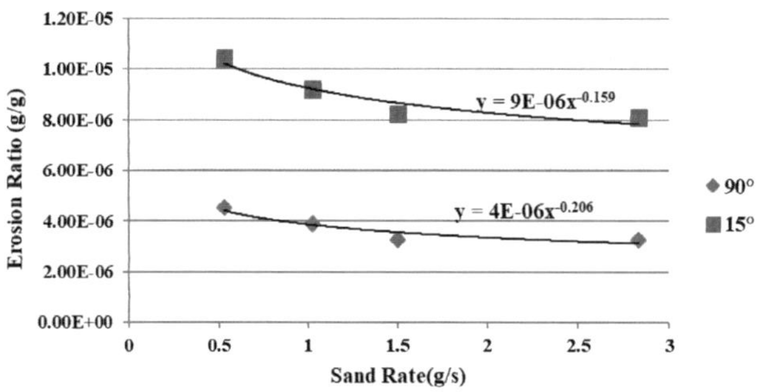

(c) Specimen: SS316, Velocity: 24 m/s, Sand Size: 300 μm

Figure 8. Erosion ratio of particles with different sizes under different sand rates

According to Fig. 7, interactions between particles are more obvious when the particle flow rates are higher, so the particle erosion ratio decreases as the particle flow rate increases and the relation is complicated. The influence of particle flow rate on erosion is stronger when the impact angle is 90°, because the collisions between particles and the specimen surface are mainly in the normal direction when the impact angle tending to 90° and the rebound particles are easy to collide with each other [18-20]. And when the particle flow rate increases, the collision probability between particles increases obviously, which affects the erosion ratio on specimens. Additionally, the erosion ratio distresses faster as the particle flow rate increases for smaller particle, since the relative coefficients for smaller particles are bigger.

5. Effect of Particle Flow Rates in Erosion Ratio Calculations

Particle interactions are usually ignored in particle erosion calculations, however, particle flow rate have a great influence on particle collisions and even erosion ratio in such impingement experiments mentioned in this paper. So it is better to take particle flow rates into consideration when calculating particle erosion ratio in solid-gas two-phase flow. According to Fig. 7, relation between erosion ratio and particle flow rate conform to the equation 3.

$$ER \propto \mathrm{a}Q_{\mathrm{p}}^{\ \mathrm{b}} \tag{3}$$

In order to improve the particle erosion calculating method, the experimental results are combined with the erosion calculating equation shown as equation 4 [21, 22].

$$ER = \frac{m_{\mathrm{p}} c(d_{\mathrm{p}}) f(\alpha) v^{b(u)}}{A_{\mathrm{face}}} \tag{4}$$

Where ER is the erosion ratio of the material, m_p is the particle mass, $c(d_{\mathrm{p}})$ is a function of the particle diameter d_{p}, $f(\alpha)$ is a function of the impact angle α, $v^{b(u)}$ is a factor about velocity. Varying the formation of functions of particle mass and diameter as follows.

$$ER = \frac{f(\alpha) v^{b(u)} f(Q_{\mathrm{p}}) D^{f_1^*(\alpha)}}{A_{\mathrm{face}} f_2^*(\alpha)} \tag{5}$$

Where Q_{p} is the particle flow rate and $f(Q_{\mathrm{p}}) = \mathrm{a}Q_{\mathrm{p}}^{\ \mathrm{b}}$. The factor a and b for different particles can be confirmed by experimental results. More experimental investigations about effect of particle sizes, shapes and flow rate on erosion ratio under well controlled particle flow rates are necessary to better modify the erosion calculation method. The area of the nozzle exit, accelerate distance and accelerated factor should also be taken into consideration in further researches to propose a better equation.

6. Summary and Conclusions

Effect of particle flow rates on erosion were discussed based on the impingement experimental results for particles with different sizes of 75 μm, 150 μm, 300 μm and 600 μm. The conclusions are summarized as follows.

1. According to the test results, particles with lower flow rates caused more erosion on the surface of the target material when the particle diameter, gas velocity and particle throughput are the same under the experimental condition mentioned in this paper. Based on PTV test, particles are more likely to be accelerated by the gas when the particle flow rate is lower.
2. Because of the effect of particle flow rates on particle velocity, it should be well controlled during impingement experiments. Relation between the changing particle velocity and particle flow rates for four different sizes were obtained based on PTV test results.
3. The erosion ratio is affected by the particle flow rate. If the interaction between particles is neglected in simulations, the relative coefficient of particle flow rate should be taken into consideration while calculating erosion ratio.
4. The area of the nozzle exit, accelerate distance and accelerated factor should also be taken into consideration in further researches to propose a better equation.

7. Acknowledgment

This work was supported by National Key Research and Development Project (2018YFF0215000) and Youth Foundation of China Special Equipment Inspection and Research Institute (2018 Youth 20).

8. References

[1] Arabnejad H, Mansouri A, Shirazi S A et al., Development of mechanistic erosion equation for solid particles, *Wear*, **332-333** (2015) 1044–1050.

[2] Paris M, Najmi K, Najafifard F et al., 2014. A comprehensive review of solid particle erosion modeling for oil and gas wells and pipelines applications. *Journal of Natural Gas Science and Engineering*, **21**, 850-873.

[3] Bellman R, Levy A, Erosion mechanism in ductile metals, *Wear*, **70** (1981) 1-28.
[4] Srinivasan S, Scattergood R. O, Effect of erodent hardness on erosion of brittle materials. *Wear*, **128** (2) (1988) 139-152.
[5] Sapate S. G, RamaRao A V, Effect of Erodent Particle Hardness on Velocity Exponent in Erosion of Steels and Cast Irons, *Materials and manufacturing processes*, **18** (2003) 783-802.
[6] Levy A, Chik P, The effects of erodent composition and shape on the erosion of steel, *Wear* **89** (1983) 151-162.
[7] Kumar R K, Kamaraj M, Seetharamu S, Effect of Spray Particle Velocity on Cavitation Erosion Resistance Characteristics of HVOF and HVAF Processed 86WC-10Co4Cr Hydro Turbine Coatings, *Journal of Thermal Spray Technology*, (2016) 1059-9630.
[8] Ballout Y A, Mathis J A, Talia J E, Effect of particle tangential velocity on erosion ripple formation, *Wear*, **184** (1995) 17-21.
[9] Lindsley B A, Marder A R, The effect of velocity on the solid particle erosion rate of alloys, *Wear*, **225-229** (1999) 510-516.
[10] Zamani S, Mahmoodabadi M, Effect of particle-size distribution on wind erosion rate and soil erodibility, *Archives of Agronomy and Soil Science*, **59** (2013) 1743-1753.
[11] Nguyen V B, Nguyen Q B, Zhang Y W, Effect of particle size on erosion characteristics, *Wear*, **348-349** (2016) 126-137.
[12] Wong C Y, Boulanger J, Short G, Modelling the effect of particle size distribution in multiphase flows with computational fluid dynamics and physical erosion experiments, *Advanced Materials Research*, **891-892** (2014) 1615-1620.
[13] Arabnejad H, Mansouri A, Shirazi S A et al., Evaluation of Solid Particle Erosion Equations and Models for Oil and Gas Industry Applications, in: *Proceedings of the SPE Annual Technical Conference and Exhibition*, Houston, 2015.
[14] Mansouri A, Arabnejad H, Shirazi S A et al., A combined CFD/experimental methodology for erosion prediction, *Wear*, **332-333** (2015) 1090-1097.
[15] Lin N, Arabnejad H, Shirazi S A et al. Experimental Study of Particle Size, Shape and Particle Flow Rate on Erosion of Stainless Steel, *Powder Technology*, **226** (2018) 70-79.
[16] Arabnejad H, Development of erosion equations for solid particle and liquid droplet impact, PhD Thesis, *The University of Tulsa*, 2015.
[17] Li Q, Erosion prediction in contractions and expansions based on computational fluid dynamics (CFD), Master Thesis, *The University of Tulsa*, 2015.
[18] Oka Y I, Okamura K, Yoshida T, Practical estimation of erosion damage caused by solid particle impact: Part 1: effects of impact parameters on a predictive equation, *Wear*, **259** (1) (2005) 95-101.
[19] Tilly G, A two stage mechanism of ductile erosion. *Wear*, **23** (1) (1973) 87-96.
[20] Sheldon G, Similarities and differences in the erosion behavior of materials, *Journal of Fluid Engineering*, **92** (3) (2015) 619-626.
[21] Lee B E, Tu J Y, On numerical modeling of particle-wall impaction in relation to erosion prediction in relation to erosion prediction: Eulerian versus langrangian method. *Wear*, 2002, **252**: 178-188.
[22] Edwards J K, Mclaury B S, Shirazi S A, Evaluation of alternative pipe bend fittings in erosive service. In ASME 2000 Fluids Eng. *Division Sumer Meet.*, Boston, pp. 959-966.

IOP Conf. Series: Materials Science and Engineering **746** (2020) 012014 doi:10.1088/1757-899X/746/1/012014

Effect of Scratches on Passive Anti-corrosion in Cr-Mn Ferroalloy

Junlong Li [1, 2], Jianrui Feng[1, 3], Dingbo Li [1], Xiuwen Wu [1, *] and Yunhang Ling [4]

[1]School of Science, China University of Geosciences, Beijing 100083, PR China
[2]Beijing Synchrotron Radiation Laboratory, Institute of High Energy Physics, Chinese Academy of Science, Beijing 100049, PR China
[3]State Key Laboratory of Fire Science, University of Science & Technology of China, Hefei 230026, PR China
[4]Lab of Advanced Materials, School of Materials Sciences and Engineering, Tsinghua University, Beijing 100084, PR China
[*]wuxw@cugb.edu.cn

Abstract. To study the scratch effect on the metal corrosion, some special scratches on the Cr-Mn ferroalloy were formed by rubbing a sandpaper on the sample surface. Two environmental cases are considered, that is, a humidity air, and a humidity air bottomed with alkaline–organic solution successively. The micro-surfaces of the samples are observed by Energy dispersive spectrometer, and the elementary contents are analyzed on an area and a line. Special scratches on the surface of Cr-Mn ferroalloy do not accelerate the metal corrosion in a humid organic alkaline environment. On the contrary, it would play a positive role in the sample passive anti-corrosion process. Chromium oxide and manganese oxide particles are produced by the oxidation of metals in a humidity air would gather in the grooves caused by the scratches. The relative study is less reported till now.

1. Introduction

Owing to the huge economic losses caused by the destruction of metal structures, the scientific community has devoted increasing attention to the corrosion phenomenon in recent decades. The impact of corrosion on the domestic economy of countries is as high as 3% of their GDP [1]. Ferroalloy containing Cr are widely used in industrial production and daily life, and they may be corroded more quickly when there are scratches and other damage to their surfaces [2-4]. When a metal is in a non-neutral environment, the corrosion process of the metal mainly involves H^+ or OH^- [5], and the metal is oxidized very quickly [6, 7]. However, when there are oxide films or grooves on the metal surface that can hold Cr or Mn, and the metal system contains organic matter that can be oxidized, the metal will be corroded more slowly by OH^- [8]. This is common in wet, alkaline soil, in which there is always organic matter that can be easily oxidized [9, 10], and when metal comes into contact with it, the above passive anti-corrosion system is established. Though there some studies have been reported about the corrosion of metals with scratches on their surfaces [6, 7], there less studies were done about the effect of scratches or grooves on anti-corrosion of metals in an alkaline–organic humidity environment.

The focus of this study is to explore the scratch effect on the Cr-Mn ferroalloy anti-corrosion. Two environmental cases were considered: a humidity air, and a humidity air and alkaline–organic solution successively.

IOP Conf. Series: Materials Science and Engineering **746** (2020) 012014 doi:10.1088/1757-899X/746/1/012014

2. Mechanism

The corrosion mechanism was studied after two different stage treatments. The samples were treated in a humidity air and alkaline–organic solution. After some preliminary experiments, we found that when the above treatments were carried out alternately, the metal corrosion rate on the sample surfaces could be slowed down.

If the samples were only treated in a humidity air, the oxide films would form on the Cr-Mn ferroalloy surfaces, and the reactions are as follows.

$$4Cr + 3O_2 \rightarrow 2Cr_2O_3 \quad (1)$$

$$2Mn + O_2 \rightarrow 2MnO \quad (2)$$

As there are loose ferric hydroxide, ferrous hydroxide, and ferric oxide in the oxide film, the oxide film on the sample surfaces would quickly fall off, and then the metal surfaces would be exposed. The oxide film in a groove would be retained by the groove protection. EDS analysis around some grooves of the sample surfaces after a humidity air treated found that the relative content of element O in the grooves is higher than that in other areas, and element O should be belong to Cr2O3 or MnO, meaning that the grooves have the ability to retain the Cr, Mn oxide films or their oxide aggregates.

If the samples were treated in a humidity air and alkaline–organic solution successively, the following phenomena would occur. When the Cr-Mn oxides retained in the grooves met the organic functional groups in the alkaline–organic solution, the Cr-Mn oxides would catalyze the organic functional groups of the alkaline–organic solution. The organic functional groups were oxidized to release free H^+, and the free H^+ would react with the OH^- to form H_2O. The pH value of the liquid films near the grooves tended to neutral for consuming the OH^- of the alkaline liquid –organic solution, which resulted in the corrosion rate around the grooves being slowed down compared with other surface areas without grooves. The reactions mentioned above were presented in Equation 3 and 4. The anti-corrosion process in the grooves was shown in Figure 1.

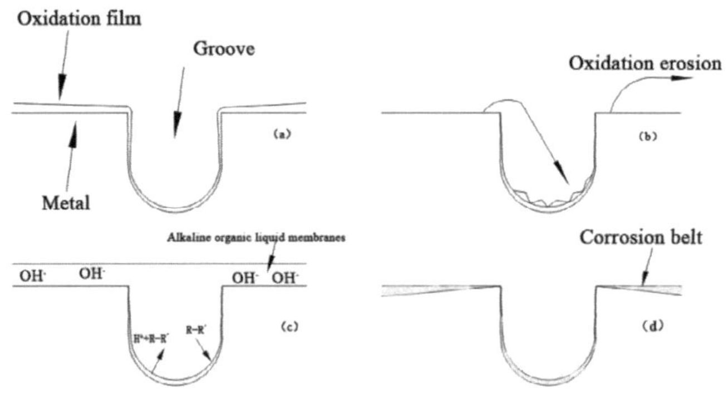

Figure 1. Schematic diagram of passive protection mechanism (a) Formation process of oxide film; (b) Oxide film was denuded and detached on the surface of samples except groove; (c) Corrosion process of alkaline organic liquid film; (d) Gradient thickness oxide films were formed on sample surface except groove

$$Cr_2O_3 + R\text{-}R' \rightarrow R\text{-}R'O^- + H^+ + Cr^{2+} \quad (3)$$

$$MnO + R\text{-}R' \rightarrow R\text{-}R'O^- + H^+ + Mn^{2+} \quad (4)$$

3. Materials and Methods

Model 202 stainless-steel sheet with a size 1 cm × 1 cm × 1 mm was used in the experiment. A humidity air treatment stage was described as follows. Firstly, the samples were soaked in dilute nitric acid (0.05 mol/L) for 1 h to remove the original oxide film on the sample surface. Secondly, after the sample surfaces got dry, scratches on the sample surfaces were made by shifting a sandpaper (surface particle size of 1 char) on the sample surfaces for 20 cm under 100 N pressure. Thirdly, the samples were laid aside for 48 h at a 95% humidity and 300 K atmosphere. Finally, the samples were dried at 330 K for 24 h. An alkaline–organic solution treatment stage was described as follows. Firstly, preparing the alkaline organic solution: Tartaric acid (0.03 g), sodium ascorbic acid (0.038 g), glucose (0.015 g), sucrose (0.03 g), citric acid (0.039 g), and ethanol (1 ml) were added to a distilled water (200 mL), using sodium carbonate to adjust the aqueous solution pH value to 9. Secondly, a skimmed cotton pad was immersed in the alkaline organic solution until it reached a saturation adsorption. Thirdly, the samples were laid side by side on the skimmed cotton pad for 168 h in a dustless environment (Humidity 100%; Temperature 300 K). Fourthly, the samples were immersed in a distilled water for 1 h. Finally, the samples were dried in a vacuum drying oven at 300 K for 0.5 h.

Energy dispersive spectrometer (EDS) analysis was carried out on the surface of the samples treated by a humidity air, or treated in a humidity air and alkaline–organic solution successively. The EDS analysis carried out in a line (7 μm – 10 μm), which crossed the grooves on the sample surfaces, and the step length used is 30 nm. The EDS analysis carried out in an area about 225 μm2 –250 μm2. The elemental contents of the flat surface and the groove surface were analyzed separately, and they were calculated in an average based on two micro area elemental analysis of a sample.

4. Results and Discussion

Figure 2 shows the EDS analysis result of the samples that was treated in a humidity air. The average mass percentage of element O in the groove was 0.95%, element Cr 16.77%, and element Mn 1.33%. The average mass percentage of element O was 0.84%, element Cr17.25%, and element Mn 1.37% on the flat surface. It was found that the relative content of O increased while the relative contents of Cr and Mn decreased. It was further found that the relatively flat surface in the groove has more Cr-Mn oxide. Considering the relative contents of elements due to O in the magnitude of the absolute value range from 0.01% to 0.05%, carbon and silica refractory erosion of inert-element relative content conforms to the trend, but the relative contents of elemental Cr and Mn are greater than 0.05%; therefore, the grooves in the process of retaining Cr-Mn oxide are slightly lost.

Next, EDS line analysis was carried out for the samples that were treated in a humidity air and alkaline–organic solution successively, and the results were shown in Figure 3. The mass percentage of oxygen at the groove is 1.60%, and 2.89 – 3.85% on the flat surface. The mass percentage of oxygen in the groove is lower than that in the other surface, and showed an increase tendency in the flat surface as the analysis point more and more distancing the grooves. But the increase tendency ended as the distance of the analysis point and the groove is too far. Based on the above oxygen distribution characteristic, the existing of the grooves on the sample surface made that the samples have a better anti-corrosion ability compared with the samples without grooves on their surface in an alkaline-organic humidity environment. For no relative studies were reported till now based on our knowledge, we did not give comparative discussion with others.

Comparing the oxygen mass percentages of the samples treated by a humidity air, or treated in a humidity air and alkaline–organic solution successively, they are higher when the samples only treated by a humidity air, indicating that the grooves on the sample surfaces can improve the sample anti-corrosion ability in the alkaline–organic humidity environment.

IOP Conf. Series: Materials Science and Engineering **746** (2020) 012014 doi:10.1088/1757-899X/746/1/012014

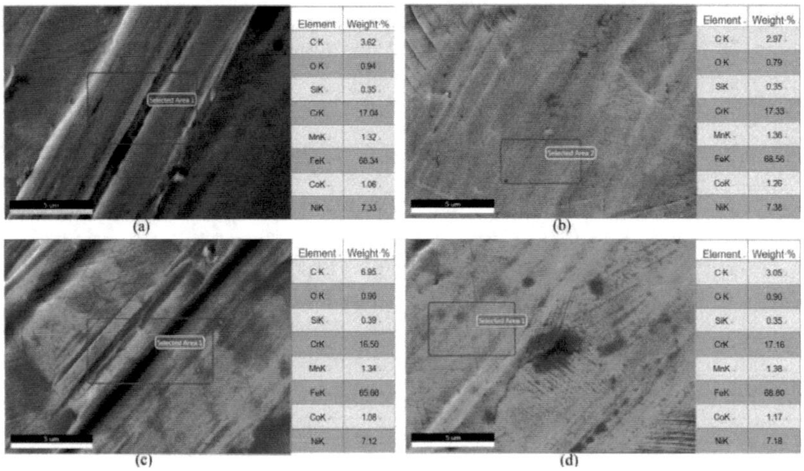

Figure 2. EDS photos and elemental spatial distribution analysis of micro areas on sample surfaces (Samples here were treated in a humidity air)

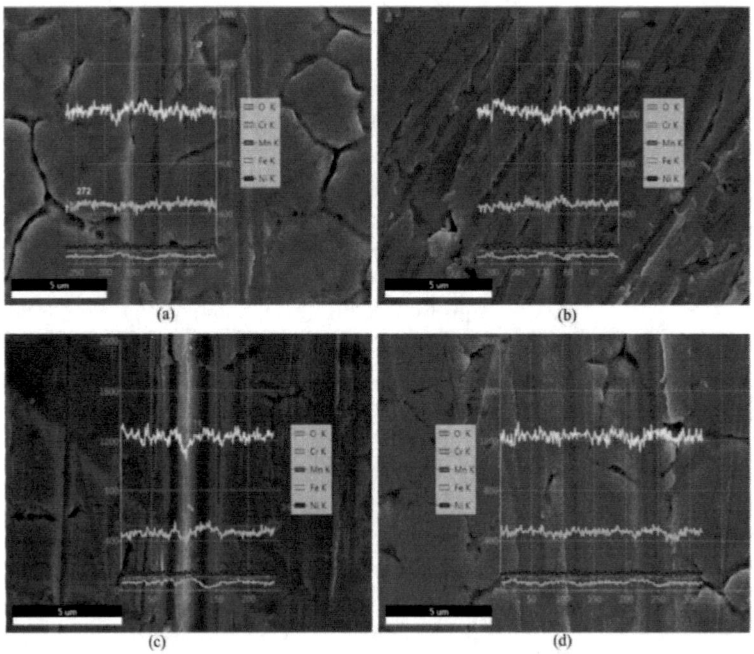

Figure 3. EDS photos and elemental spatial distribution analysis of some lines spanning grooves on sample surfaces (Samples here were treated in a humidity air and alkaline–organic solution successively)

5. Conclusions

The existing of the grooves on the sample surface would accelerate the sample surface corrosion in a moist environment. The existing of the grooves on the sample surface improved the sample anti-corrosion ability in an alkaline organic moist environment. The existing of the organic functional groups would tune the pH value of the alkaline liquid films near the grooves to neutral, which resulted in the anti-corrosion ability of the samples was improved.

6. Acknowledgements

Jun-Long Li and Jian-Rui Feng contributed equally to this work. This work was supported by the National Nature Science Foundation of China (No. U1430118, 51771098), and National Undergraduate Innovation Project of China University of Geosciences, Beijing (NO. S201911415064).

7. Reference

[1] Britton SC 2013 *Br. Corros.* **16** 62–62.
[2] Engelhardt G and Macdonald DD 2004 *Corros. Sci.* **46** 2755–2780.
[3] Hara T, Aaahi H and Kaneta H 2000 *Corros.* **56** 183–188.
[4] Tarragó JM, Fargas G, Isern L, Dorvlo S, Tarres E, Müller CM, Jiménez-Piqué E and Llanes L 2016 *Mater. Des.* **111** 36–43.
[5] Smiyan OD, Grigorenko GM and Vainman AB 2002 *Int. J. Hydrogen Energy* **27** 801–812.
[6] Higgs GB, Hanzlik JA, Macdonald DW, Gilbert JL, Rimnac CM and Kurtz SM 2013 *J. Arthroplasty* **28** 2–6.
[7] Landolfo R, Cascini L and Portioli F 2010 *Sustainability* **2** 2163–2175.
[8] Umoren SA and Solomon MM 2017 *J. Environ. Chem. Eng.* **5** 246–273.
[9] Avdeev YG and Kuznetsov YI 2012 *Russ. Chem. Rev.* **81** 1133.
[10] Salz D, and Vissing KD 2012 *US Patent* 0003483.

IOP Conf. Series: Materials Science and Engineering **746** (2020) 012015 doi:10.1088/1757-899X/746/1/012015

General Stress Fields of Orthotropic Plate with Mode II Crack

Long Zhang, Purong Jia*, Yongyong Suo and Cheng Jia

School of Mechanics and Civil-Engineering & Architecture, Northwestern Polytechnical University, Xi'an 710129, PR China
*Corresponding author E-mail: prjia@nwpu.edu.cn

Abstract. The plane crack problem of the orthotropic plate is discussed for the shear loading. The solution of this problem is mainly achieved by the complex variable method. Essential equations are introduced fully on the basis of linear elastic mechanics. The stress functions have been given to meet the needs of shear mode II crack states for orthotropic materials. The formulae for stress fields are derived by the complex functions in two cases. With the simplification of relative functions, the expressions of the singular stress fields in the crack-tip are determined by the real functions.

1. Introduction

Conventional fracture mechanics discusses on the homogeneous materials in great detail and has been highly successful. Nowadays the fracture mechanics of anisotropic materials may be relative to many engineering structures [1, 2]. The orthotropic plates must be for the base of composite materials in common use. The valid method to solve crack tip field problems in anisotropic materials may be in using complex analytic function theory, and the results have been reported [3~5]. But the general solutions for composite materials have not given completely or perfectly. Therefore, it is necessary to make up a new solution for the singular stresses of cracked composite materials [6~8]. Particularly, the study of the shear mode II crack problem as shown in Figure 1 must be very necessary for the plane fracture mechanics of orthotropic materials.

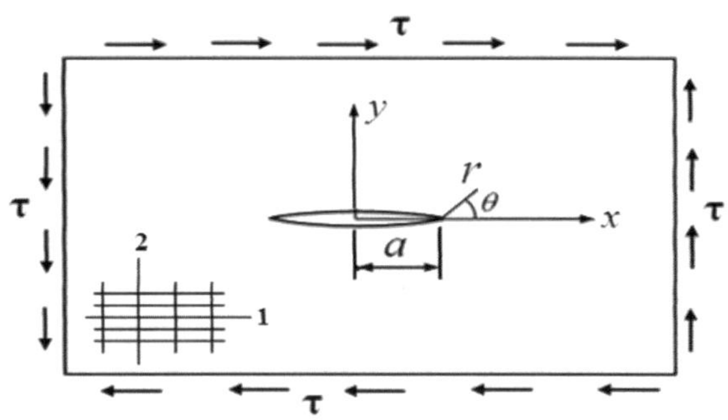

Figure 1. Model of the plate with mode II crack

2. Elastic Equations and Stress Function

To solve the plane stress problems of anisotropic materials, the body forces can not be taken into consideration, and the equilibrium equations are in the following form:

$$\frac{\partial \sigma_x}{\partial x} + \frac{\partial \tau_{xy}}{\partial y} = 0 \quad , \quad \frac{\partial \tau_{xy}}{\partial x} + \frac{\partial \sigma_y}{\partial y} = 0 \tag{1}$$

In order to satisfy the equilibrium equations, the plane stress components are defined by the stress function $F = F(x, y)$ as follows:

$$\sigma_x = \frac{\partial^2 F}{\partial y^2} , \quad \sigma_y = \frac{\partial^2 F}{\partial x^2} , \quad \tau_{xy} = -\frac{\partial^2 F}{\partial x \partial y} \tag{2}$$

The orthotropic coefficients are often expressed as E_1, E_2, ν_{12}, G_{12}. For the convenience of the below explanation, we use the characters, B and C, as the material parameters, they are:

$$B = \frac{E_1}{2G_{12}} - \nu_{12} , \quad C = \frac{E_1}{E_2} \tag{3}$$

The governing equation of the strain compatibility condition can be expressed by

$$\frac{\partial^4 F}{\partial y^4} + 2B \frac{\partial^4 F}{\partial x^2 \partial y^2} + C \frac{\partial^4 F}{\partial x^4} = 0 \tag{4}$$

To determine the solution of the partial differential equation, it is very advantageous to use complex variables. Thus, we introduce the complex variable (w) and its conjugate ($\overline{w}$), they are defined by: $w = x + ihy$, $\overline{w} = x - ihy$, where h is a real arbitrary constant, and also we suppose that $h > 0$. By above definition, the real stress function $F(x, y)$ can be expressed by the complex variables. The partial derivations of F with x or y can be transformed by following expressions:

$$\frac{\partial F}{\partial x} = \frac{\partial F}{\partial w}\frac{\partial w}{\partial x} + \frac{\partial F}{\partial \overline{w}}\frac{\partial \overline{w}}{\partial x} = \frac{\partial F}{\partial w} + \frac{\partial F}{\partial \overline{w}} , \quad \frac{\partial F}{\partial y} = \frac{\partial F}{\partial w}\frac{\partial w}{\partial y} + \frac{\partial F}{\partial \overline{w}}\frac{\partial \overline{w}}{\partial y} = ih(\frac{\partial F}{\partial w} - \frac{\partial F}{\partial \overline{w}}) \tag{5}$$

The important relations of partial derivations are as

$$\frac{\partial^2 F}{\partial x^2} + \frac{1}{h^2}\frac{\partial^2 F}{\partial y^2} = 4\frac{\partial^2 F}{\partial w \partial \overline{w}} \quad , \quad (\frac{\partial^2}{\partial x^2} + \frac{1}{h^2}\frac{\partial^2}{\partial y^2})^2 F = 16\frac{\partial^4 F}{\partial w^2 \partial \overline{w}^2}$$

Using above relations, the partial derivative equation (4) can be written by

$$D_1(\frac{\partial^4 F}{\partial w^4} + \frac{\partial^4 F}{\partial \overline{w}^4}) + D_2(\frac{\partial^2}{\partial w^2} + \frac{\partial^2}{\partial \overline{w}^2})\frac{\partial^2 F}{\partial w \partial \overline{w}} + D_3\frac{\partial^4 F}{\partial w^2 \partial \overline{w}^2} = 0 \tag{6}$$

Where, $D_1 = h^4 - 2Bh^2 + C$, $D_2 = 4C - 4h^4$, $D_3 = 6h^4 + 4Bh^2 + 6C$.
In the following, we shall solve this equation (6).

Case 1: $\frac{\partial^4 F}{\partial w^2 \partial \overline{w}^2} = 0$, $D_1 = D_2 = 0$

In this case, two parameters must be given as follows

$$C = h^4 , \quad B = h^2$$

We know that, $h > 0$, the solution can be determined by

$$h = \sqrt{\frac{E_1}{2G_{12}} - \nu_{12}} = \sqrt[4]{\frac{E_1}{E_2}} \tag{7}$$

By the equation $\dfrac{\partial^4 F}{\partial w^2 \partial \overline{w}^2} = 0$, we can determine the stress function F, that is:

$$F = A_1\Psi_1 + A_2\overline{\Psi}_1 + A_3\overline{w}\Psi_2 + A_4 w\overline{\Psi}_2 \tag{8}$$

Where, $\Psi_1 = \Psi_1(w)$ and $\Psi_2 = \Psi_2(w)$ are the analytic functions.
For Mode II crack problem, the constants and complex functions can be selected as:

$$A_1 = A_4 = -iA_0\ ,\quad A_2 = A_3 = iA_0\ ,\qquad \Psi_1 = w\Phi\ ,\quad \Psi_2 = \Phi$$

Then, the stress function can be determined by

$$F = A_0 i\cdot(\overline{w} - w)(\Phi + \overline{\Phi}) = 2A_0 hy(\Phi + \overline{\Phi}) \tag{9}$$

The partial derivations can be given as follows

$$\frac{\partial F}{\partial x} = 2A_0 hy(\Phi' + \overline{\Phi}')\quad ,\qquad \frac{\partial F}{\partial y} = 2A_0 h[\Phi + \overline{\Phi} + ihy(\Phi' - \overline{\Phi}')]$$

$$\frac{\partial^2 F}{\partial x^2} = 2A_0 hy(\Phi'' + \overline{\Phi}'')\quad ,\qquad \frac{\partial^2 F}{\partial y^2} = 2A_0 h^2[2i(\Phi' - \overline{\Phi}') - hy(\Phi'' + \overline{\Phi}'')]$$

$$\frac{\partial^2 F}{\partial x \partial y} = 2A_0 h[\Phi' + \overline{\Phi}' + ihy(\Phi'' - \overline{\Phi}'')]$$

According to the stress expressions (2), the plane stress components can be expressed as:

$$\begin{cases} \sigma_x = -4A_0 h^2(2\,\mathrm{Im}\,\Phi' + hy\,\mathrm{Re}\,\Phi'') \\ \sigma_y = 4A_0 hy\,\mathrm{Re}\,\Phi'' \\ \tau_{xy} = -4A_0 h(\mathrm{Re}\,\Phi' - hy\,\mathrm{Im}\,\Phi'') \end{cases} \tag{10}$$

Case 2: $\dfrac{\partial^2 F}{\partial w \partial \overline{w}} = 0\ ,\quad D_1 = 0$

Under the case, the characteristic equation is given as

$$h^4 - 2Bh^2 + C = 0$$

Obviously, the solution can be determined by

$$h^2 = B \pm \sqrt{B^2 - C} \qquad (B^2 > C)$$

Where, we suppose that: $B^2 > C$, and also let $h_1 > h_2 > 0$. Then, two solutions are:

$$\begin{cases} h_1 = \sqrt{\dfrac{E_1}{2G_{12}} - \nu_{12} + \sqrt{(\dfrac{E_1}{2G_{12}} - \nu_{12})^2 - \dfrac{E_1}{E_2}}} \\ h_2 = \sqrt{\dfrac{E_1}{2G_{12}} - \nu_{12} - \sqrt{(\dfrac{E_1}{2G_{12}} - \nu_{12})^2 - \dfrac{E_1}{E_2}}} \end{cases} \tag{11}$$

By the equation $\dfrac{\partial^2 F}{\partial w \partial \overline{w}} = 0$, the stress function F may be in the following form

$$F = A_1\Psi_1 + A_2\overline{\Psi}_1 + A_3\Psi_2 + A_4\overline{\Psi}_2 \tag{12}$$

Where, $\Psi_1 = \Psi_1(w_1)$ and $\Psi_2 = \Psi_2(w_2)$ are the analytic functions. The complex variables are

$$w_1 = x + ih_1 y \quad , \quad w_2 = x + ih_2 y$$

The constants can be given by: $A_1 = A_4 = iA_0$, $A_2 = A_3 = -iA_0$. Therefore, we have:

$$F = A_0 i \cdot (\Psi_1 - \overline{\Psi}_1 - \Psi_2 + \overline{\Psi}_2) \tag{13}$$

The partial derivations can be given as follows

$$\frac{\partial F}{\partial x} = A_0 i \cdot (\Psi_1' - \overline{\Psi}_1' - \Psi_2' + \overline{\Psi}_2') \quad , \quad \frac{\partial F}{\partial y} = A_0 i \cdot [ih_1(\Psi_1' + \overline{\Psi}_1') - ih_2(\Psi_2' + \overline{\Psi}_2')]$$

According to the stress expressions (2), the plane stress components can be expressed as:

$$\begin{cases} \sigma_x = -A_0 i \cdot [h_1^2(\Psi_1'' - \overline{\Psi}_1'') - h_2^2(\Psi_2'' - \overline{\Psi}_2'')] = 2A_0 \operatorname{Im}(h_1^2\Psi_1'' - h_2^2\Psi_2'') \\ \sigma_y = A_0 i \cdot (\Psi_1'' - \overline{\Psi}_1'' - \Psi_2'' + \overline{\Psi}_2'') = -2A_0 \operatorname{Im}(\Psi_1'' - \Psi_2'') \\ \tau_{xy} = A_0[h_1(\Psi_1'' + \overline{\Psi}_1'') - h_2(\Psi_2'' + \overline{\Psi}_2'')] = 2A_0 \operatorname{Re}(h_1\Psi_1'' - h_2\Psi_2'') \end{cases} \tag{14}$$

Evidently, the stress expressions are given with the application of complex variable method.

3. Solution of Mode II Crack

To solve the Mode II crack boundary problem as shown in Figure 1, we must consider the stress boundary conditions as follows:

$$\text{At } |x| < a \;,\; y = 0 : \; \sigma_y = \tau_{xy} = 0 \;. \qquad \text{At } x^2 + y^2 \to \infty : \; \sigma_x = \sigma_y = 0 \;,\; \tau_{xy} = \tau \;.$$

In the following, we discuss the solutions in above section for case I and case II.

For the first case, the stress components are expressed as equation (10). We select the constant and complex function as:

$$4A_0 h = -\tau \quad , \quad \Phi' = \sqrt{\frac{w^2}{w^2 - a^2}} \quad , \quad \text{then: } \Phi'' = -\frac{a^2}{w}\sqrt{\frac{w^2}{(w^2 - a^2)^3}}$$

Thus, the stresses can be determined by

$$\begin{cases} \dfrac{\sigma_x}{\tau} = 2h \operatorname{Im}\sqrt{\dfrac{w^2}{w^2 - a^2}} - h^2 y \operatorname{Re}[\dfrac{a^2}{w}\sqrt{\dfrac{w^2}{(w^2 - a^2)^3}}] \\ \dfrac{\sigma_y}{\tau} = y \operatorname{Re}[\dfrac{a^2}{w}\sqrt{\dfrac{w^2}{(w^2 - a^2)^3}}] \\ \dfrac{\tau_{xy}}{\tau} = \operatorname{Re}\sqrt{\dfrac{w^2}{w^2 - a^2}} + hy \operatorname{Im}[\dfrac{a^2}{w}\sqrt{\dfrac{w^2}{(w^2 - a^2)^3}}] \end{cases} \tag{15}$$

Obviously, the stresses can meet the needs of the boundary conditions. The solution is suitable to whole plate as shown in Figure 1. Nevertheless, the local region near the crack tip is more important. So we select one small region at the crack tip on the right. By using the polar coordinate system, the functions in above equation can be simplified as follows ($r << a$):

$$x = a + r\cos\theta \quad , \quad y = r\sin\theta \quad , \quad w = a + r\cos\theta + ihr\sin\theta$$

$$w \approx a \ , \ \ w + a \approx 2a \ , \ \ \frac{w^2}{w^2 - a^2} = \frac{a}{2r(\cos\theta + ih\sin\theta)}$$

Then the stresses at near crack-tip can be simplified as:

$$\begin{cases} \sigma_x = \dfrac{K_{II}}{\sqrt{2r}}[2h\,\mathrm{Im}\sqrt{\dfrac{1}{\cos\theta + ih\sin\theta}} - h^2\dfrac{\sin\theta}{2}\mathrm{Re}\sqrt{\dfrac{1}{(\cos\theta + ih\sin\theta)^3}}] \\ \sigma_y = \dfrac{K_{II}}{\sqrt{2r}}\dfrac{\sin\theta}{2}\mathrm{Re}\sqrt{\dfrac{1}{(\cos\theta + ih\sin\theta)^3}} \\ \tau_{xy} = \dfrac{K_{II}}{\sqrt{2r}}[\mathrm{Re}\sqrt{\dfrac{1}{\cos\theta + ih\sin\theta}} + h\dfrac{\sin\theta}{2}\mathrm{Im}\sqrt{\dfrac{1}{(\cos\theta + ih\sin\theta)^3}}] \end{cases} \tag{16}$$

Where, $K_{II} = \tau\sqrt{a}$ is called the stress intensity factors for mode II crack. The stress expressions show that the stress components tend to infinity at the crack tip (when, $r \to 0$), this is so-called the stress singularity. As yet, the functions in above equation remain the complex form.

For the second case, the stress components are expressed by equation (14). To meet the stress boundary conditions, the complex functions can be selected as:

$$\Psi_1'' = \sqrt{\frac{w_1^2}{w_1^2 - a^2}} \ , \ \ \Psi_2'' = \sqrt{\frac{w_2^2}{w_2^2 - a^2}} \ . \ \ \text{And} \ \ 2A_0 = \frac{\tau}{h_1 - h_2}$$

Then, the stresses in equation (14) can be transformed as follows

$$\begin{cases} \sigma_x = \dfrac{\tau}{h_1 - h_2}\mathrm{Im}(h_1^2\sqrt{\dfrac{w_1^2}{w_1^2 - a^2}} - h_2^2\sqrt{\dfrac{w_2^2}{w_2^2 - a^2}}) \\ \sigma_y = -\dfrac{\tau}{h_1 - h_2}\mathrm{Im}(\sqrt{\dfrac{w_1^2}{w_1^2 - a^2}} - \sqrt{\dfrac{w_2^2}{w_2^2 - a^2}}) \\ \tau_{xy} = \dfrac{\tau}{h_1 - h_2}\mathrm{Re}(h_1\sqrt{\dfrac{w_1^2}{w_1^2 - a^2}} - h_2\sqrt{\dfrac{w_2^2}{w_2^2 - a^2}}) \end{cases} \tag{17}$$

In the vicinity region of the crack tip ($r << a$), the complex functions can be simplified by using the polar coordinate system ($x = a + r\cos\theta$, $y = r\sin\theta$): $w_1 \approx a$, $w_1 + a \approx 2a$

$$w_2 \approx a \ , \ \ w_2 + a \approx 2a \ , \ \ w_1 - a = r(\cos\theta + ih_1\sin\theta) \ , \ \ w_2 - a = r(\cos\theta + ih_2\sin\theta)$$

Then the stresses at near crack-tip can be expressed as ($K_{II} = \tau\sqrt{a}$):

$$\begin{cases} \sigma_x = \dfrac{1}{h_1 - h_2}\dfrac{K_{II}}{\sqrt{2r}}\mathrm{Im}(\dfrac{h_1^2}{\sqrt{\cos\theta + ih_1\sin\theta}} - \dfrac{h_2^2}{\sqrt{\cos\theta + ih_2\sin\theta}}) \\ \sigma_y = \dfrac{-1}{h_1 - h_2}\dfrac{K_{II}}{\sqrt{2r}}\mathrm{Im}(\dfrac{1}{\sqrt{\cos\theta + ih_1\sin\theta}} - \dfrac{1}{\sqrt{\cos\theta + ih_2\sin\theta}}) \\ \tau_{xy} = \dfrac{1}{h_1 - h_2}\dfrac{K_{II}}{\sqrt{2r}}\mathrm{Re}(\dfrac{h_1}{\sqrt{\cos\theta + ih_1\sin\theta}} - \dfrac{h_2}{\sqrt{\cos\theta + ih_2\sin\theta}}) \end{cases} \tag{18}$$

As yet, the functions in above equation remain the complex form.

4. Function Simplification and Stresses

In order to simplify above stress expressions, it is necessary to make some function transformation. Now, we let:

$$\cos\theta = g\cos\beta \ , \qquad h\sin\theta = g\sin\beta \tag{19}$$

We can obtain that:

$$h\tan\theta = \tan\beta \ , \qquad \cos^2\theta + h^2\sin^2\theta = g^2(\cos^2\beta + \sin^2\beta) = g^2$$

$$\beta = \arctan(h\tan\theta) \ , \qquad g = \sqrt{\cos^2\theta + h^2\sin^2\theta} \tag{20}$$

Then, the complex function can be simplified by:

$$\frac{1}{\sqrt{\cos\theta + ih\sin\theta}} = \frac{1}{\sqrt{g\cos\beta + ig\sin\beta}} = \frac{1}{\sqrt{g}}e^{-i\beta/2} = \frac{1}{\sqrt{g}}(\cos\frac{\beta}{2} - i\sin\frac{\beta}{2})$$

$$\sqrt{\frac{1}{(\cos\theta + ih\sin\theta)^3}} = \frac{1}{\sqrt{g^3}}(\cos\frac{3\beta}{2} - i\sin\frac{3\beta}{2})$$

For Case 1, the stresses in equation (16) can be transformed as

$$\begin{cases} \sigma_x = \dfrac{K_{II}}{\sqrt{2r}}[2h\,\mathrm{Im}\dfrac{1}{\sqrt{g}}(\cos\dfrac{\beta}{2} - i\sin\dfrac{\beta}{2}) - h\dfrac{g\sin\beta}{2}\mathrm{Re}\dfrac{1}{\sqrt{g^3}}(\cos\dfrac{3\beta}{2} - i\sin\dfrac{3\beta}{2})] \\ \sigma_y = \dfrac{K_{II}}{\sqrt{2r}}\dfrac{g\sin\beta}{2h}\mathrm{Re}\dfrac{1}{\sqrt{g^3}}(\cos\dfrac{3\beta}{2} - i\sin\dfrac{3\beta}{2}) \\ \tau_{xy} = \dfrac{K_{II}}{\sqrt{2r}}[\mathrm{Re}\dfrac{1}{\sqrt{g}}(\cos\dfrac{\beta}{2} - i\sin\dfrac{\beta}{2}) + \dfrac{g\sin\beta}{2}\mathrm{Im}\dfrac{1}{\sqrt{g^3}}(\cos\dfrac{3\beta}{2} - i\sin\dfrac{3\beta}{2})] \end{cases}$$

Evidently, the functions can be simplified easily. And then, the singular stress fields in the crack-tip for the first case must well be determined by

$$\begin{cases} \sigma_x = -\dfrac{K_{II}}{\sqrt{2r}}\dfrac{h}{\sqrt{g}}\sin\dfrac{\beta}{2}(2 + \cos\dfrac{\beta}{2}\cos\dfrac{3\beta}{2}) \\ \sigma_y = \dfrac{K_{II}}{\sqrt{2r}}\dfrac{1}{h\sqrt{g}}\sin\dfrac{\beta}{2}\cos\dfrac{\beta}{2}\cos\dfrac{3\beta}{2} \\ \tau_{xy} = \dfrac{K_{II}}{\sqrt{2r}}\dfrac{1}{\sqrt{g}}\cos\dfrac{\beta}{2}(1 - \sin\dfrac{\beta}{2}\sin\dfrac{3\beta}{2}) \end{cases} \tag{21}$$

When $h = 1$, then $g = 1$, $\beta = \theta$.

For Case 2, it is similar to have the transformation as the first case. We let:

$$\cos\theta = g_1\cos\beta_1 = g_2\cos\beta_2 \,,\ h_1\sin\theta = g_1\sin\beta_1 \,,\ h_2\sin\theta = g_2\sin\beta_2 \tag{22}$$

We can obtain that:

$$h_1\tan\theta = \tan\beta_1 \,,\ h_2\tan\theta = \tan\beta_2 \,,\ \cos^2\theta + h_1^2\sin^2\theta = g_1^2 \,,\ \cos^2\theta + h_2^2\sin^2\theta = g_2^2$$

$$\begin{cases} \beta_1 = \arctan(h_1\tan\theta) \ , \quad g_1 = \sqrt{\cos^2\theta + h_1^2\sin^2\theta} \\ \beta_2 = \arctan(h_2\tan\theta) \ , \quad g_2 = \sqrt{\cos^2\theta + h_2^2\sin^2\theta} \end{cases} \tag{23}$$

Then, the complex functions can be simplified by:

$$\frac{1}{\sqrt{\cos\theta + ih_1 \sin\theta}} = \frac{1}{\sqrt{g_1 \cos\beta_1 + ig_1 \sin\beta_1}} = \frac{1}{\sqrt{g_1}}(\cos\frac{\beta_1}{2} - i\sin\frac{\beta_1}{2})$$

$$\frac{1}{\sqrt{\cos\theta + ih_2 \sin\theta}} = \frac{1}{\sqrt{g_2}}(\cos\frac{\beta_2}{2} - i\sin\frac{\beta_2}{2})$$

The stresses in equation (18) can be transformed as

$$\begin{cases} \sigma_x = \frac{1}{h_1 - h_2}\frac{K_{II}}{\sqrt{2r}}\mathrm{Im}[\frac{h_1^2}{\sqrt{g_1}}(\cos\frac{\beta_1}{2} - i\sin\frac{\beta_1}{2}) - \frac{h_2^2}{\sqrt{g_2}}(\cos\frac{\beta_2}{2} - i\sin\frac{\beta_2}{2})] \\ \sigma_y = \frac{-1}{h_1 - h_2}\frac{K_{II}}{\sqrt{2r}}\mathrm{Im}[\frac{1}{\sqrt{g_1}}(\cos\frac{\beta_1}{2} - i\sin\frac{\beta_1}{2}) - \frac{1}{\sqrt{g_2}}(\cos\frac{\beta_2}{2} - i\sin\frac{\beta_2}{2})] \\ \tau_{xy} = \frac{1}{h_1 - h_2}\frac{K_{II}}{\sqrt{2r}}\mathrm{Re}(\frac{h_1}{\sqrt{g_1}}(\cos\frac{\beta_1}{2} - i\sin\frac{\beta_1}{2}) - \frac{h_2}{\sqrt{g_2}}(\cos\frac{\beta_2}{2} - i\sin\frac{\beta_2}{2})) \end{cases}$$

Evidently, the complex functions can be simplified easily. And at last, the singular stress fields in the crack-tip for the second case must well be determined as follows:

$$\begin{cases} \sigma_x = -\frac{1}{h_1 - h_2}\frac{K_{II}}{\sqrt{2r}}(\frac{h_1^2}{\sqrt{g_1}}\sin\frac{\beta_1}{2} - \frac{h_2^2}{\sqrt{g_2}}\sin\frac{\beta_2}{2}) \\ \sigma_y = \frac{1}{h_1 - h_2}\frac{K_{II}}{\sqrt{2r}}(\frac{1}{\sqrt{g_1}}\sin\frac{\beta_1}{2} - \frac{1}{\sqrt{g_2}}\sin\frac{\beta_2}{2}) \\ \tau_{xy} = \frac{1}{h_1 - h_2}\frac{K_{II}}{\sqrt{2r}}\mathrm{Re}(\frac{h_1}{\sqrt{g_1}}\cos\frac{\beta_1}{2} - \frac{h_2}{\sqrt{g_2}}\cos\frac{\beta_2}{2}) \end{cases} \tag{24}$$

Above equations are the singular stress fields to be expressed as basic real functions.

5. Acknowledgments

The authors would like to acknowledge the financial support of the Natural Science Foundation of China (NSFC Grant No: 51475372).

6. References

[1] Sih G C 1991. Mechanics of fracture initiation and propagation. Kluwer Academic Publishers.
[2] Friedrich K 1989. Application of fracture mechanics to composite materials. Elsevier Sci. Pub.
[3] Jia P, Suo Y, Jia C, Wang Q 2019. Stress analysis of orthotropic wedge loaded on the apex. IOP Conference Series: Materials Science and Engineering, 585: 448-453.
[4] Wang G, Jia P, Suo Y, et al 2019. Elasticity solution of composite material wedge loaded with a concentrated moment. Journal of Materials Science and Chemical Engineering, 7: 77-85.
[5] Fakoor M 2017. Augmented strain energy release rate (ASER): A novel approach for investigation of mixed-mode I/II fracture of composite materials. Eng Fract Mech; 179: 177-189.
[6] Fallah N, Nikraftar N 2018. Meshless finite volume method for the analysis of fracture problems in orthotropic media. Eng Fract Mech; 204: 46-62.
[7] Zhang H, Qiao P 2019. A state-based peridynamic model for quantitative elastic and fracture analysis of orthotropic materials. Eng Fract Mech; 206: 147-171.
[8] Jia P, Wang G, Zhang L, Suo Y 2019. Stress and strain of cracked orthotropic sheet. IOP Conference Series: Materials Science and Engineering, 484: 170-175.

IOP Conf. Series: Materials Science and Engineering **746** (2020) 012016 doi:10.1088/1757-899X/746/1/012016

Application Status of Chlorophyll-a Determination Method in Seawater

Li Zhang, Qian Shi and Zhaoyu Wang

Institute of oceanographic Instrumentation, Qilu University of Technology (Shandong Academy of Sciences), Qingdao 266061, China
zhang1222li@163.com

Abstract: As an important indicator of water eutrophication and water ecological health, the monitoring methods of Chlorophyll-a are being increasingly valued by environmental supervision departments and relevant scientific research institutions. This paper reviews several commonly used monitoring methods, and systematically analyzes the advantages and disadvantages of each method. It is found that each monitoring method has certain restrictions. Therefore, when the environmental protection department conducts routine monitoring or scientific research institutions conduct long-term tracking monitoring, they should choose the appropriate method according to the different monitoring objects, or even combining different monitoring methods to determine chlorophyll-a, so as to provide reliable data support for water environmental management and scientific research work.

1. Introduction

Chlorophyll a is an indispensable investigation item in water ecological survey, and is the most direct and effective indicator reflecting the biomass of phytoplankton and even the degree of eutrophication of water body [1].Chlorophyll a is an essential measurement parameter in field observation [2].At present, the monitoring method of phytoplankton chlorophyll-a mainly include spectrophotometry, chromatography, fluorescence method, due to the different research and accuracy for determination of chlorophyll a, time, adopts the research methods are also different, furthermore, due to the current research method is numerous, often bring confusion for researchers in terms of methods, therefore, this article will focus on the phytoplankton chlorophyll-a monitoring methods were summarized that reference and reference for the related research work in the future.

2. Spectrophotometry

The maximum absorption peak of chlorophyll a is located at 663nm. According to Lambert Beer law, within a certain concentration range, the absorbance value of chlorophyll a is directly proportional to its concentration. After the water sample is filtered, concentrated, extracted and constant volume, its absorbance value is determined at the maximum absorption wavelength, interference value is removed, and the concentration of chlorophyll a can be calculated according to the standard curve of chlorophyll a. Currently, the determination of chlorophyll generally adopts spectrophotometry for accurate quantification. Spectrophotometry is an analytical method based on the selective absorption of light by substances and molecules, which belongs to spectrophotometry [1]. The most commonly used method for determination of phytoplankton chlorophyll in domestic water bodies is the spectrophotometric method recommended by water and wastewater monitoring and analysis methods (the fourth edition) [2]. Spectrophotometry is the most widely used method because of its low price, simple operation and stable results.

However, this method has the disadvantages of long time consumption, unstable operation and so on, which leads to a large error in the result. Moreover, contact with acetone, an organic reagent, is easy to cause harm to human body.Xiao-yan zheng [3], etc., on the basis of in the method, put forward the chlorophyll a and the determination of magnesium content of chlorophyll a, and the method of acetone extraction, grinding method of cell disruption was improved, using extraction instead of grinding, not only avoid the acetone in the process of grinding damage to people, reducing the chlorophyll-a in light exposure, and to reduce the repeated transfer in the process of the loss of chlorophyll a, greatly improve the efficiency of the experiment; Li Huiling et al. simplified and improved the determination method of chlorophyll a in water and waste water monitoring and analysis method (the fourth edition) [2], and extracted chlorophyll a by grinding free method on the basis of the original method. The experimental results showed that the extraction rate and stability of chlorophyll a were improved, and the accuracy of the experimental results was ensured [4].

The current spectrophotometric method is mainly divided into acetone method and ethanol method according to different extraction agents, among which, acetone method is less used at present due to the larger toxicity of acetone, and the use of ethanol method is gradually increasing. According to the different methods, the method can be divided into monochrome method and tricolor method. As a conventional method for the determination of phytoplankton chlorophyll a, spectrophotometry is mostly applied to the routine monitoring of scientific research and environmental protection surface water, and suitable for the determination of indoor water environment samples in large quantities. In order to improve the accuracy of the method, the spectrophotometry can be improved from the following aspects: changing the way of cell fragmentation, such as ultrasound, repeated thawing; Select better extraction reagent, such as mixed reagent; Optimize extraction methods, such as delayed or heated extraction; Select better filter membrane, such as glass fiber filter membrane, freeze and store in dark, etc. [5].

3. Chromatography

Chromatography is the principle of the separation of material in the mobile phase and stationary phase adsorption force, when the samples through the chromatographic column, separated material takes place continuously in the current phase and fixed phase adsorption, desorption, of the different time to components through chromatographic column, different material will successively outflow of chromatographic column, enter into the detector is analyzed, so as to achieve the purpose of separation [6].Many scholars, including Mantoura [7], Suzuki and Wright [8], have successively established an HPLC method for the analysis of photosynthetic pigments in Marine phytoplankton. High performance liquid chromatography (HPLC) has been widely used in the determination of phytoplankton pigment characteristics, the method is reproducible, the instrument is relatively cheap, and can meet most of the field sample detection. However, there are still three deficiencies in HPLC method, which affect the accuracy of pigment qualitative and quantitative analysis: pigments with the same or similar spectral characteristics are difficult to be distinguished by HPLC [9], and false negative or false positive are likely to occur. In quantitative analysis, chromatographic peaks with low concentration are easily disturbed by their isomers, pigment degradants and adjacent high concentration chromatographic peaks [10]. It is impossible to distinguish different pigments in the co-efflux components, and the co-efflux unknown pigment components are often mistaken for known pigments [11].The detector was not sensitive enough, and some important pigments with low content, such as Allo indicating micro phytoplankton and zeaxanthin indicating micro phytoplankton (Zea), tended to be near the detection limit of HPLC method with low quantitative accuracy.

4. Remote Sensing Imaging

Based on geographic information system of remote sensing technology as a new type of regional water environmental investigation and monitoring method, has a strong advantage in space and time continuity, commonly used method is through the study of remote sensing image spectral characteristics and the relationship between the chlorophyll a concentration, building mathematical model inversion chlorophyll a concentration [12].

Chen Chuqun et al. [13] selected several combinations of bands from TM1 to TM4 as the dependent variable, analyzed the correlation degree between combinations of bands and chlorophyll mass concentration, modeled the combinations of bands with the highest correlation degree, and obtained five inversion models for estimating chlorophyll a mass concentration in surface seawater. Andreo et al. [14] used MODIS chlorophyll product data and GIS to discuss and analyze chlorophyll a mass concentration in the gulf of Patagonia, Argentina, and obtained the maximum and minimum variation range of phytoplankton distribution, which was used to monitor the changes of water environment. Watanabe et al. [15] used Landsat8 OLI images to combine band values to form a single factor, establish a regression equation, and invert chlorophyll a mass concentration. Dalu et al. [16] used tm1-tm3 band data and the verified 3-band empirical formula to calculate the mass concentration of chlorophyll a and suspended solids, and discussed the correlation between the mass concentration of chlorophyll a in urban reservoirs and the distribution of the mass concentration of suspended solids and the change of reservoir water level.

5. The Fluorescence Method

Chlorophyll molecules will absorb the energy of photons when exposed to excited light, and transition from the most stable and lowest energy ground state to an unstable and high-energy excited state. Since the excited state is unstable, the molecule loses energy by emitting fluorescence and quickly returns from the excited state to the ground state. Fluorescence determination of chlorophyll a content is the use of natural chlorophyll fluorescence characteristics, with a particular wavelength excitation light chlorophyll samples, chlorophyll by stimulating emit red fluorescence, according to the law of lambert - than the ear, in the range of certain concentration, the fluorescent light intensity and the concentration of chlorophyll is proportional relationship, using fluorimeter to issue fluorescence intensity on determination of the sample, can be quantitatively chlorophyll concentration [1].

Fluorescence method has the advantages of fast measurement speed, high sensitivity, good real-time performance and free from interference of other commonly used pigments. It is mostly used in field measurement or online monitoring, especially for eutrophication measurement of water bodies, so as to timely monitor the growth distribution and change status of phytoplankton in water bodies. At present, the determination of chlorophyll a in seawater by fluorimeter is relatively mature. In 2012, Ma Longshan et al. [17] established a derivative synchronous fluorescence detection method for chlorophyll a.This method is simple, fast and does not require complex pretreatment. Thus, the linear range of chlorophyll a is 0.02-125 microns /L, the detection limit is 0.25 microns /L, and the recovery rate is 97.0% ~ 103.8%. In the same year, Dong Dasheng et al. [18] studied the technology and design of chlorophyll a and turbidity sensor. The sensor technology is designed based on the principle of fluorescence induced chlorophyll a detection and the principle of scattering turbidity detection technology, which can measure the relevant parameters of seawater in real time and realize the function of weak optical signal detection required by the design. Now the common seawater chlorophyll a fluorescence sensor on the market mainly includes YSI type sensor developed by the institute of oceanography and instrumentation of Shandong academy of sciences, Alec chlorophyll a sensor from Japan, and YSI sensor from the United States, with resolution up to 0.05 microns /L and accuracy up to 6.72%.

6. Conclusion

With the rapid development of society, the problem of cyanobacteria has become more and more prominent, which has become a key factor threatening the ecological environment of river, sea, lake and reservoir. Chlorophyll a monitoring is an important indicator to evaluate eutrophication level of river, sea, lake and reservoir as well as water ecological health. Its monitoring method has been paid more and more attention by environmental supervision departments and relevant scientific research institutions. Although many monitoring methods have been applied in practice, each monitoring method has some limitations. Therefore, when environmental protection departments carry out routine monitoring or scientific research institutions carry out long-term tracking monitoring, they should choose appropriate methods according to different monitoring objects, or even choose several monitoring methods combined to measure chlorophyll a, so as to provide reliable data support for water environment management and scientific research.

7. Acknowledgements
This work was supported by National Key Research and Development program of China (No. 2016YFC1400800), Key Research and Development Program of Shandong Province (No. 2019GHY112070), the Joint Funds of the National Natural Science Foundation of China (Grant No. U1806202), Youth Foundation of Shandong Academy of Sciences (2018QN0036) and Key Technologies of In-situ Sensor for Oil Pollutants in Seawater (ZR2019BD037).

8. Reference
[1] Liu Jianhua. Study on the influence factors of chlorophyll content in water by fluorescence method [D]. Hebei University of Science and Technology, 2014.
[2] Water and wastewater monitoring and analysis methods (the fourth edition) (M). China Environmental Science Press, 2002.
[3] Zheng Xiaoyan, Discussing about the determination of the chlorophyll a [J]. Energy Environmental Protection,2016, 30 (02): 55-57+34.
[4] Li Huiling, Zhao Feng, Dou Yanxia. An Exploration for the Determination of Chlorophyll a in Water.
[5] Yang Lin. Research Progress in Determination of Phytoplankton Chlorophyll a [J]. Sichuan Environment, 2019, 38 (01): 156-160.
[6] Zhang Jingjing, Liu Dongyan, Sun Huihui, Wang Xiao-feng. Application of UHPLC-MS/MS in determination of the phytoplanktonpigment and size fraction, 2018, 37(02): 158-164.
[7] Mantoura R F C., Llewellyn C. A.. The rapid determination of algal chlorophyll and carotenoid pigments and their breakdown products in natural waters by reverse-phase high-performance liquid chromatography [J]. Anal. chim. Actu, 1983, 151: 197-314.
[8] Wright S W, Jeffrey s w, Mantoura R F,et. Al. Improved HPLC method for the analysis of chlorophulls and carotenoids from marine phytoplankton [J]. Mar.Ecol. Progr. Series, 1991, 77: 183-196.
[9] Nick Welschmeyer. Roy, S., Llewellyn, C., Egeland, E. S., and Johnsen, G. 2011. Phytoplankton Pigments: Characterization, Chemotaxonomy and Applications in Oceanography. Cambridge University Press, Cambridge, UK, 845 pp. [J]. Journal of Phycology, 2013, 49 (1).
[10] Hooker S B, Van Heukelem L, Thomas C S, et al, 2009. The Third Sea WiFS HPLC Analysis Round-Robin Experiment (SeaHARRE-3). NASA Technical Memorandum 2009-215849. Greenbelt: NASA Goddard Space Flight Center.
[11] [11] Alvarez S, Rodriguez F, Riob P, et al, 2013. Chlorophyll c (CS-170) Isolated from Ostreococcus sp. Is [7 -Methoxycarbonyl -8 -vinyl] protochlorophyllide a. Organic Letters, 15 (17): 4430-4433.
[12] Liu Chengliang, Yang Yongpo, Fan Ruiqi. The Application of Hyperspectral Image and Gepgraphic Modeling to Water Dispatching in Yellow River Irrigation Area [J]. Journal of Irrigation and Drainage, 2013, 32 (4): 15-19.
[13] Chen Chuqun, Shi Ping,Mao Qingwen. Study on Modeling Chlorophll Concentration of Surface Coastal Water Using TM Date [J]. Remote Sensing of Environment China, 1996 (3): 168-176.
[14] ANDREO V, DOGLIOTTI A I, TAURO C, et al. Spatio-temporal variations in chlorophyll-a concentration in the patagonic continental shelf: Anexample of satellite time series processing with GRASS GIS temporal modules[C]// IGARSS. 2015.
[15] WATANABE F S, ALCÂNTARA E, RODRIGUES T W P, et al. Estimation of Chlorophyll-a Concentration and the Trophic State of the Barra Bonita Hydroelectric Reservoir Using OLI/Landsat-8 Images [J]. International Journal of Environmental Research & Public Health, 2015, 12 (9): 10 391-10 417.
[16] DALU T, DUBE T, FRONEMAN PW, et al. An assessment of chlorophyll-concentration spatio-temporal variation using Landsat satellite data, in a small tropical reservoir [J]. Geocarto International, 2015, 30 (10): 1 130-1 143.
[17] MA Yong-sha, et al. Determination of Chlorophyll a in Green Algae by Derivative Synchronous Fluorescence Spectrometry [J]. Environmental Monitoring in China, 2012, 28 (2): 58~61.

IOP Conf. Series: Materials Science and Engineering **746** (2020) 012016 doi:10.1088/1757-899X/746/1/012016

[18] DONG Da-sheng, CHENG Qun CHEN Shi-zhe, et al. LI MinDesign of Chlorophyll a and Turbidity Sensor [J]. Instrument Technique and Sensor, 2012 (10): 15 ~ 16.

Subsurface Facet-fisheye Induced Failure of Titanium Alloy under Very High Cycle Fatigue

Wei Li[1,*], Meng Li[1], Xinxin Xing[2] and Ning Gao[3]

[1]School of Mechanical Engineering, Beijing Institute of Technology, Beijing, 100081, China
[2]Research & Development Department, Shipbuilding Power Engineering Institute Co., Ltd, Shanghai, 200129, China
[3]Technical project management, SAIC VOLKSWAGEN, Luopu Rd. 63, Anting, Shanghai, 201805, China
[*]Corresponding author: Tel: +86 010 68918730, Fax: +86 010 68918528, E-mail: lliw@bit.edu.cn

Abstract. Very high cycle fatigue test under pulsating tension was performed to investigate the subsurface failure mechanism of a titanium alloy. A peculiar region, called "facets cluster zone (FCZ)", is located at the center of fisheye. The slip-like patterns on facet surface, coalescence of nanoscale cracks and growth of microcracks in the maximum shear stress plane are observable. Combined with the evaluation of stress intensity factors at crack tips, the subsurface failure processes are summarized as: (a) localization of plastic strain, (b) occurrence of slip lines, (c) nucleation of microcracks, (d) coalescence and growth of microcracks, (f) occurrence of the FCZ, (g) formation of fisheye and (i) unstable crack growth until final fracture.

1. Introduction

Very high cycle fatigue (VHCF) associated with microscale deformation-induced failure of materials has been a topic of growing interest, because fatigue failures in the VHCF regime beyond 10^7 cycles have been widely found in materials that were assumed to be operating at the stresses below conventional fatigue limit [1]. One of the things that we focus in the VHCF regime is the change of failure mode from the surface to the subsurface [2, 3]. Because of this change, some researchers [4, 5] believe that there is no conventional fatigue limit in many metals and alloys such as high-strengthen steels, titanium alloys, high-temperature alloys, irons, etc. According to the *X*-ray analysis, generally internal metallurgical defects such as non-metallic inclusions or pores can be easily the source for cracks to develop in the VHCF regime [2-4]. And, a propagating circular crack shaped like a fisheye can be observable. Sometimes, a fine granular area (FGA) [2] can occur around the inclusion. Some models or methods have been used to explain its formation mechanism [5]. However, for the titanium alloys with non-defect induced failure, the subsurface failure mechanism and related mechanical condition are not yet well understood [6, 7]. In this study, in order to ensure the long-term safety of mechanical components or parts, we therefore investigate the subsurface failure processes with crack nucleation and early growth for a titanium alloy subjected to the VHCF loading.

2. Materials and Method

The investigated material is a TC11 titanium alloy. Its chemical composition (mass percentage) is: 6.3Al, 3.4Mo, 1.68Zr, 0.23Si, 0.10O, 0.05Fe, 0.0006C, 0.0006N and balance Ti. The specimens were first machined into the shape of hourglass with a certain amount of grinding margin, and then heat-treated as: (1) 950 °C × 1 h, air cooled and (2) 530 °C × 6 h, air cooled. The microstructure observed by scanning electron microscope (SEM) consists of about 37% α grains and 63% β grains.

The pulsating tension fatigue test was performed by using an electromagnetic resonant testing machine at the frequency of 100 Hz in open environment at room temperature. The testing cycles were ranged from 10^6 cycles to 10^9 cycles.

3. Results and Discussion

3.1. Subsurface Failure

The *S–N* (stress – number of cycles) data exhibit the continually descending trend without conventional fatigue limit. Based on the observation of SEM and the analysis of energy dispersive *X*-ray spectrometer, failures are all caused by inhomogeneous microstructures. Figure 1(a) shows a typical fracture surface, on which the fisheye, unstable growth zone and momentary fracture zone are all distinguished. A peculiar bright region shown in Figure 1(b) occurs at the center of fisheye, and some flat facets are found within this region. The morphology of facet is given in Figure 1(c). This region is named as "facets cluster zone (FCZ)". From the viewpoint of small crack growth, the maximum stress intensity factors around the facet, FCZ and fisheye, $K_{\text{max-f}}$, $K_{\text{max-FCZ}}$ and $K_{\text{max-FE}}$, are given by [8, 9]

$$K_{\text{max-f or FCZ}} = 0.5\sigma_{\text{max}}\sqrt{\pi\sqrt{area_{\text{f or FCZ}}}} \tag{1}$$

$$K_{\text{max-FE}} = \frac{2}{\pi}\sigma_{\text{max}}\sqrt{\sqrt{\pi area_{\text{FE}}}} \tag{2}$$

Where σ_{max} is the applied maximum stress, $\sqrt{area_{\text{f or FCZ}}}$ is the size of facet or FCZ and $\sqrt{area_{\text{FE}}}$ is the size of fisheye. The values of $K_{\text{max-f}}$ are in the range of 1.17-1.44 MPam$^{1/2}$ and tend to decrease with the increasing of life N_{f}, as showed in Figure 1(d). This means that small crack effect contributes to the fracture of facet. Moreover, the values of $K_{\text{max-FCZ}}$ and $K_{\text{max-FE}}$ are respectively in the limited ranges of 5.17-5.77 MPam$^{1/2}$ and 9.51-11.06 MPam$^{1/2}$, as shown in Figure 1(d). Each of them keeps a constant, regardless of life. Combined with previous studies [2, 10], $K_{\text{max-FCZ}}$ and $K_{\text{max-FE}}$ can be regarded as the threshold values controlling the stable and unstable crack growth, respectively.

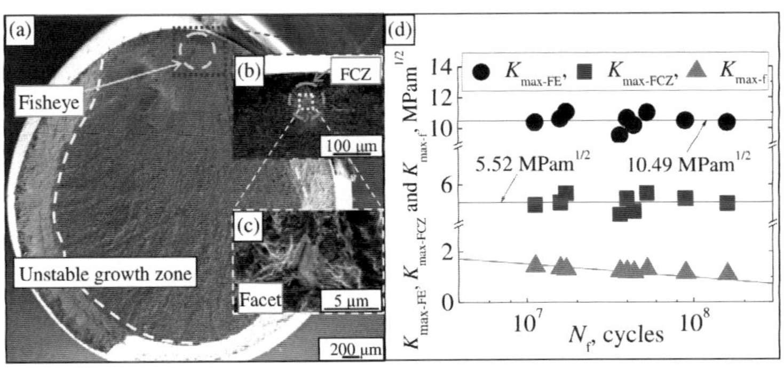

Figure 1. (a) Typical fracture surface (σ_{max}=650 MPa, N_{f}=90,199,000 cycles), (b) FCZ, (c) Facet and (d) $K_{\text{max-FE}}$, $K_{\text{max-FCZ}}$ and $K_{\text{max-f}}$, vs. N_{f}.

3.2. Observation using FIB

The focused ion beam (FIB) was used for site specific milling on fracture surface. For the fracture surface given in Figure 2(a), two sites including the site within FCZ and the site outside FCZ within fisheye are chosen. The milled microscopic holes are shown respectively in Figs. 2(b) and 2(c).

IOP Conf. Series: Materials Science and Engineering **746** (2020) 012017 doi:10.1088/1757-899X/746/1/012017

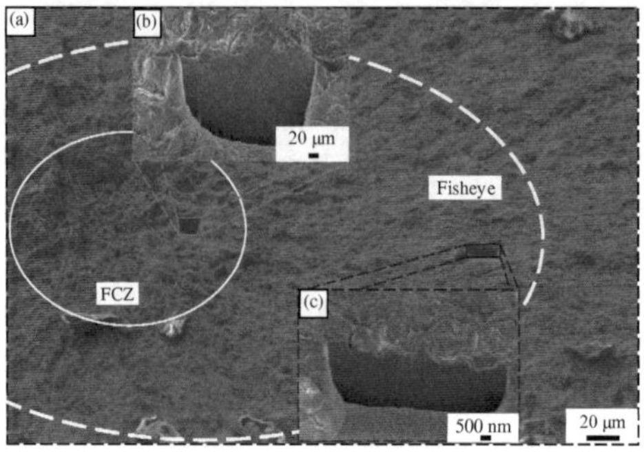

Figure 2. (a) FIB sampling, (b) Milled hole within FCZ and (c) Milled hole within fisheye

Along the base line that is tangent to the edge of facet within the FCZ, the milling direction, i.e. the change of width *W*, is shown Figure 3(a). A number of slip-like patterns on facet surface are found. It is verified that the slip related to dislocation movement plays a role in the formation of facetted crack. With the increasing of *W*, the different morphologies on the inner wall of hole can be observable. When *W*=4 μm, a large grain surrounded by small grains is found in Figure 3(b). The strain discontinuity inevitably occurs on the interface between them. When *W*=8 μm, a few of deformed grains that contact the fracture surface are observed in Figure 3(c). When *W*=10 μm, two embedded microcracks I & II occur in Figure 3(d). It seems that there is no coalescence between them in Figure 3(e). The length of crack II is only about hundreds of nanometers. When *W*=12 μm, a microcrack III shown in Figure 3(f) is inclined with respect to the loading direction. When *W*=13.2 μm, the growth of microcrack III along about 45° is identified in Figure 3(g). This means that the microcracks easily blocked by the grain boundary tend to grow in a plane of maximum shear stress. Furthermore, the coalescence of microcracks I & II is found. For the FIB milling within fisheye, however, besides the propagating secondary cracks shown in Figure 3(h) with *W*=5 μm and non-deformed grains shown in Figure 3(i) with *W*=15 μm, no microcracks can be found below the fracture surface from *W*=0 μm to *W*=15 μm.

IOP Conf. Series: Materials Science and Engineering **746** (2020) 012017 doi:10.1088/1757-899X/746/1/012017

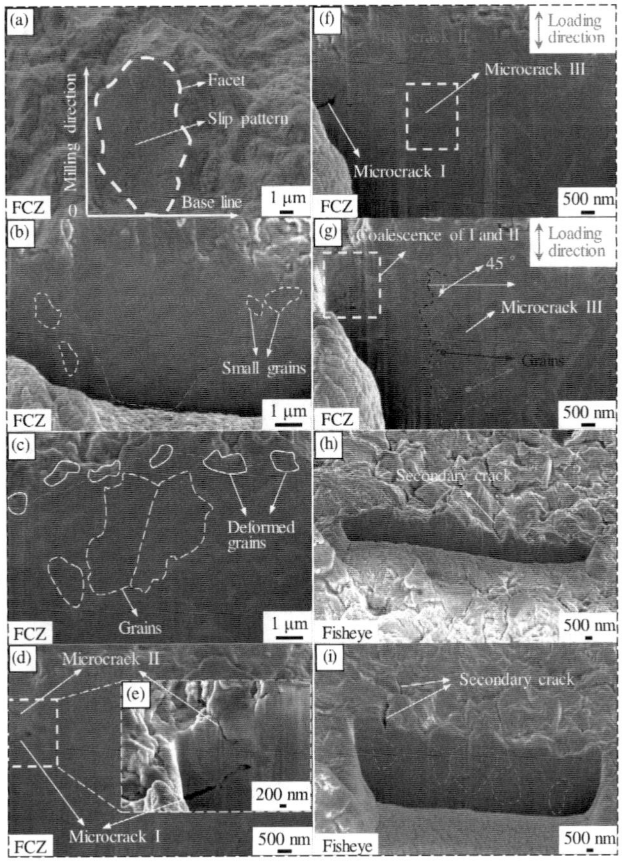

Figure 3. (a) Slip-like patterns and milling direction (W=0), (b) Larger grain and small grains (W=4 μm), (c) Micro-deformation of grains (W=8 μm), (d) Embedded microcracks I & II (W=10 μm), (e) Nanoscale magnification of microcracks I & II, (f) Occurrence of microcrack III (W=12 μm), (g) Coalescence of microcracks I & II, and pattern of microcrack III along 45° (W=13.2 μm); (h) Secondary crack (W=5 μm) and (i) Grains on hole wall (W=15 μm)

3.3. Failure Modeling

Before loading, a number of dislocations can exist in the crystal structure, as shown in Figure 4(a). Under loading, due to the difference of mechanical properties between softer α grains and harder β grains, the large or small stress concentration, i.e. the localization of plastic strain, can occur on the interfaces, as shown in Figure 4(b). On some grains oriented in the maximum shear stress plane, slip lines or bands can preferentially occur, as shown in Figure 4(c). In these stages, K_{max} is still smaller than the threshold value governing local microstructurally small crack, $\Delta K_{th\text{-}L}$. As cycling continuous, more slip lines occur and continue to thicken. Especially at their intrusions, they will become the sources for crack nucleation. As a result, a few microcracks are formed, as shown in Figure 4(d). Some of them will discontinuously grow along the maximum shear stress plane and mutually coalesce. When the crack crosses the entire grain, the facet is formed, shown in Figure 4(e). Meanwhile, K_{max} will exceed $\Delta K_{th\text{-}L}$ governed by the microstructure. When the progressive development of microcracks, followed by the fracture of α-grains, spreads to a given region, and a circular crack, i.e. the FCZ, is formed, as shown in Figure 4(f). Note that K_{max} reaches the threshold value controlling macrocrack growth, $\Delta K_{th\text{-}s}$. Herein, this threshold value is $K_{max\text{-}FCZ}$ =5.52 MPam$^{1/2}$. From then, the mechanics will dominate the growth of crack instead of the microstructure. The formed macrocracks will propagate at a stable rate in Paris region, as shown in Figure 4(g). When the fisheye is fully formed, the subsequent crack will grow in an unstable manner until fracture, and K_{max} is larger than the threshold value controlling unstable growth $\Delta K_{th\text{-}u}$, i.e. $K_{max} \geq K_{max\text{-}FE}$, as shown in Figure 4(h).

IOP Conf. Series: Materials Science and Engineering **746** (2020) 012017 doi:10.1088/1757-899X/746/1/012017

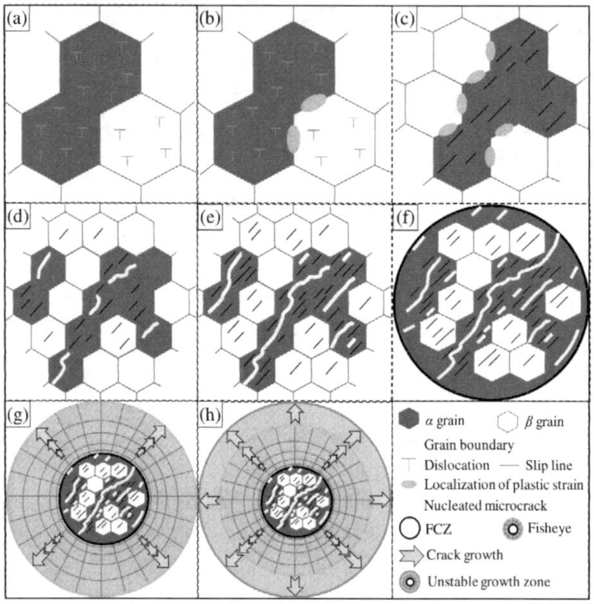

Figure 4. (a) Microstructure with α and β grains ($K_{max} < \Delta K_{th\text{-}L}$), (b) Localization of plastic strain ($K_{max} < \Delta K_{th\text{-}L}$), (c) Occurrence of slip lines ($K_{max} < \Delta K_{th\text{-}L}$), (d) Nucleation of microcracks ($K_{max} > \Delta K_{th\text{-}L}$), (e) Coalescence and growth of microcracks ($K_{max} > \Delta K_{th\text{-}L}$), (f) Formation of FCZ ($\Delta K_{th\text{-}L} < K_{max} \leq \Delta K_{th\text{-}s}$), (g) Formation of fisheye ($\Delta K_{th\text{-}s} < K_{max} \leq \Delta K_{th\text{-}u}$) and (h) Unstable crack growth ($\Delta K_{th\text{-}u} < K_{max}$)

4. Conclusions

In summary, the technique of FIB was used to study the non-defect induced crack nucleation and propagation behavior of titanium alloy in the VHCF regime. The subsurface failure processes are divided as: (a) Localization of plastic strain, (b) Occurrence of slip lines, (c) Nucleation of microcracks, (d) Coalescence and growth of microcracks, (f) Occurrence of the FCZ, (g) Formation of fisheye and (i) Unstable crack growth until fracture.

5. Acknowledgement

This research was supported by National Natural Science Foundation of China (Nos. 51775043 and U1864210).

6. References

[1] Palin-Luc T and Jeddi D 2018 *Int. J. Fatigue* **13** 1545–33.
[2] Sakai T, Oguma N and Morikawa A 2015 *Fatigue Fract. Eng. Mater. Struct.* **38** 1305–14.
[3] Furuya Y 2013 *Mater. Lett.* **112** 139–41.
[4] Bathias C and Paris P C 2010 *Int. J. Fatigue* **12** 894–97.
[5] Hong Y S and Sun C Q 2017 *Theor. Appl. Fract. Mec.* **92** 331–50.
[6] Jha S K and Chandran K S R 2003 *Scripta Mater.* **48** 1207–12.
[7] Bantounas I, Dye D and Lindley T C 2010 *Acta Mater.* **58** 3908–18.
[8] Murakami Y and Endo M 1994 *Int. J. Fatigue* **16** 163–82.
[9] Chapetti M D 2008 *Eng. Fract. Mech.* **75** 1854–63.
[10] Nehila A, Li W, Zhao H Q , Gao N and Xing X X 2018 *Int. J. Fatigue* **6** 112–23.

IOP Conf. Series: Materials Science and Engineering **746** (2020) 012018 doi:10.1088/1757-899X/746/1/012018

Effect of Al Content on Microstructure and Properties of Zn-Cu-Al Alloy

Hongxing Wang[a, b], Yan Zhang[a, b], Changsheng Wang[b], Shuo Cao[b], Wan Bai[b], Canhui Wu[b] and Jiahao Qian[b]

[a]School of Materials Science and Engineering, Nanjing Institute of Technology, 1Hongjing venue, Nanjing 211167, PR China
[b]Key Laboratory of Advanced Structural Materials and Application Technology of Jiangsu Province, 1Hongjing venue, Nanjing 211167, PR China
Email: wanghx@njit.edu.cn

Abstract. Zn-Cu-Al zinc alloy was prepared by metal mould casting. The effects of Al addition on the microstructure, electrical conductivity and wear resistance of Zn-Cu-Al alloy were studied. The results show that the structure of Zn-Cu-Al alloy is composed of Zn solid solution containing Cu and Al (η phase), eutectic phase (α+η) and CuZn5 (ε phase); Al0.4Zn0.6 phase appeared when Al addition content exceeded 2.0wt.%; With the increase of Al addition content, eutectic phase (α+η) content increased and η phase grain refined. Meanwhile, the hardness and electrical conductivity of the alloy increase with the increase of Al addition content. When the addition of Al is 5.0 wt.%, the hardness and electrical conductivity of the alloy reach the maximum, 151HV and 26.4% IACS, respectively. The wear mechanism of the alloy is mainly plastic deformation with a mixture mode of plastic deformation and adhesion.

1. Introduction

With the increasing number of fuel vehicles, a large number of vehicle exhaust emissions have aggravated the degree of environmental pollution. In recent years, new energy electric vehicles have been put into use, effectively reducing the environmental pollution caused by the exhaust emissions of fuel vehicles. With the increasing consumption of new energy electric vehicles, the demand for copper alloy parts related to new energy electric vehicles is also increasing, which increases the price of copper and its alloys and the manufacturing cost of new energy electric vehicles. Therefore, the research and development of low-cost new metal materials, partly replacing copper alloys, can reduce the dependence on copper and its alloys, and also reduce the manufacturing cost of new energy electric vehicles.

Zinc alloy is mainly used in hardware, daily decoration, building materials, automotive parts, coatings and other fields because of their characteristics of low melting point, good casting performance and easy cutting [1, 2]. Zinc alloys mainly consist of zinc-aluminium alloys and zinc-copper alloys. At present, the main focus is on the study of the structure, corrosion and strength of high aluminum zinc alloy and high copper zinc alloy [3-8], while the study of the structure and properties of low copper zinc alloy and low aluminum zinc alloy is relatively less [9, 10], especially the electrical conductivity of zinc alloy.

The addition of titanium and magnesium in zinc can refine the grain size, improve creep resistance and plasticity of the alloy. Magnesium can increase strength, but make alloy brittle [10-13]. In addition, the mechanical properties of zinc alloys are similar to that of brass by means of deformation or heat treatment, which makes zinc alloys hopeful to be substitutes for copper alloys [14, 15]. Al has high conductivity, which is second only to Ag, Cu and Au. In this paper, the effect of Al addition on the microstructure, wear resistance and electrical conductivity of Zn-1.25Cu alloy is studied, which provides experimental basis and theoretical support for the development of the application of zinc alloy.

2. Experimental

Pure Zn (99.9%), pure Cu (99.9%) and pure Al (99.9%) were used as raw materials. The crucible resistance furnace (model: SG2-5-10) is used as the melting equipment, and the alloy is melted in the graphite crucible. After melting, the liquid metal is poured into the Y-shaped metal casting mould to form. The smelting process of zinc alloy is as follows: firstly, pure zinc block is put into the graphite crucible in resistance furnace, at the same time, a certain amount of charcoal particles with diameter of 20 cm are added, and the melting temperature is set at 480℃. When most pure zinc is melted, copper particles with a diameter of 1 cm are added to the zinc solution, and the temperature is adjusted to 600 ℃ for heat preservation. After the copper is completely melted, the temperature rises to 650 ℃, the aluminum block is pressed into with the self-made bell jar and the alloy liquid is rotated and stirred. After the aluminum melts completely, the temperature drops to 550℃. After the slag is drawn, the molten metal is poured into the mould preheated to 150℃. It is cooled to about 100℃ and the parts are taken out. Alloy addition composition (wt.%): Cu is 1.25, Al is 1.25, 1.5, 2.0, 3.0 and 5.0, and Zn is the remainder.

Samples with size of 20 mm × 15 mm × 5 mm were machined by WEDM. The etallographic corrosive formulation of zinc alloy is (mass percentage, %): CrO_3:Na_2SO_4:H_2O=20:1.5:100. OLYMPUS GX51 metallographic microscope was used for metallographic observation and JSM-6380LV scanning electron microscopy (SEM) and energy spectrum analysis (EDS) were used to analyze the microstructure of the alloy. X-ray diffractometer (Bruker D8 Advance) was used to analyze the alloy structure, copper target, tube current is 40 mA, tube voltage is 40 kV, and scanning speed is 5 Degree/min.

The microhardness of the alloy was measured by FM700 microhardness tester. Test conditions: pressure head load 10 N, loading time 15 s, in five different parts of the test, take the average value.

Friction experiments of alloy specimens were carried out using MFT-3000 (Rtec) wear tester. GCr15 ball (ϕ12 mm, quenched state, hardness HRC63) is grinding ball. Test conditions: reciprocating, radius 6 mm, load 30 N, sliding speed 10 $mm.s^{-1}$, time 15 min, ambient temperature room temperature, dry friction.

The resistance test sample is machined into a circular rod of 2.0 mm × 60 mm by WEDM. Before the resistance test, the surface of the rod is polished with sandpaper, and the rod is measured three times in different parts, and the average value is taken. The length of the bar test resistance is fixed at 20 mm, the test position is changed for three times, and the average value is obtained. The digital DC bridge (model QJ84A) is used to measure resistance. The formula for calculating resistivity is as follows:

$$\rho = R \times S/L \tag{1}$$

The electrical conductivity of the alloy can be obtained by formula (1) and formula (2). The formula for calculating electrical conductivity is as follows:

$$K = 0.01724/\rho \times 100\% \tag{2}$$

There is, K, conductivity, IACS; ρ, sample resistivity, $\Omega mm^2/m$; R, sample resistance, Ω; S, Cross Section Area of Sample, mm^2; L, the length of testing the resistance part of the sample, m.

3. Results and Discussion

3.1. Micorstructure

Fig. 1 shows the metallographic structure of Zn-Cu-Al alloys with different additions of Al. It can be seen from Fig. 1 that the structure of Zn-Cu-Al alloy is mainly composed of gray-white dendrite η phase (Zn solid solution containing Cu and Al) and interdendritic black eutectic. With the increase of

Al content, the gray-white dendrites are refined and the volume fraction decreases, while the volume fraction of black eutectic increases, and the η phase becomes rounded. The main reason is that during solidification, on the one hand, Al distributes around primary dendrites, which prevents secondary dendrite growth of solid solution, increases the number of nuclei of primary phase, makes them distribute uniformly, and refines the grain size of primary phase; on the other hand, when the addition content of Al increases, the amount of Al in the melt is also increasing, reaching the composition near eutectic reaction, which leads to the number of eutectic phases in the solidified structure increase significantly.

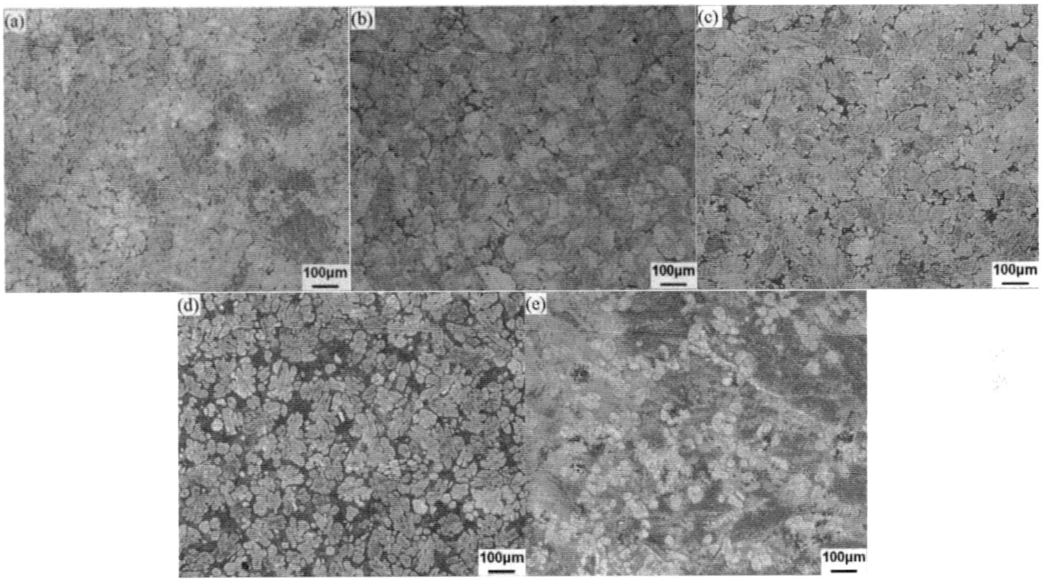

Figure 1. Optical images of Zn-Cu-Al with different addition of Al: (a) 1.25; (b) 1.5; (c) 2.0; (d) 3.0; (e) 4.0

Fig. 2 shows the SEM photos of Zn-Cu-Al alloys with different amounts of Al. It can be seen from Fig. 2 that the eutectic phase between dendrites of η phase in the as-cast structure of Zn-Cu-Al is layer structure, mainly consisting of η phase and α phase (Al-rich phase) [10, 15]. The number of eutectic structure increases significantly with the increase of Al addition. The positions of "1", "2" and "3" in Fig. 2 are analyzed by elemental energy spectrum. The results are shown in Table 1. From Table 1, it can be seen that when the content of Cu is 1.25wt.% and the content of Al is 1.25-5.0wt.%, the content range of elements in the bid "1" in Figure 2 is: Al 1.27-1.82wt.%, Cu 1.75-2.61wt.%, Zn 95.31-96.88wt.%, and the structure of the site 1 is a Zn solid solution phase (η phase)containing solute atoms, Cu and Al; the content range of elements in the bid "2" is: Al 12.24-21.36wt.%, Cu 1.5%. 8-2.08wt% and Zn 77.05 - 85.67wt%. The structure of the site 2 is eutectic structure consisting of η phase and α phase (Al-rich phase).

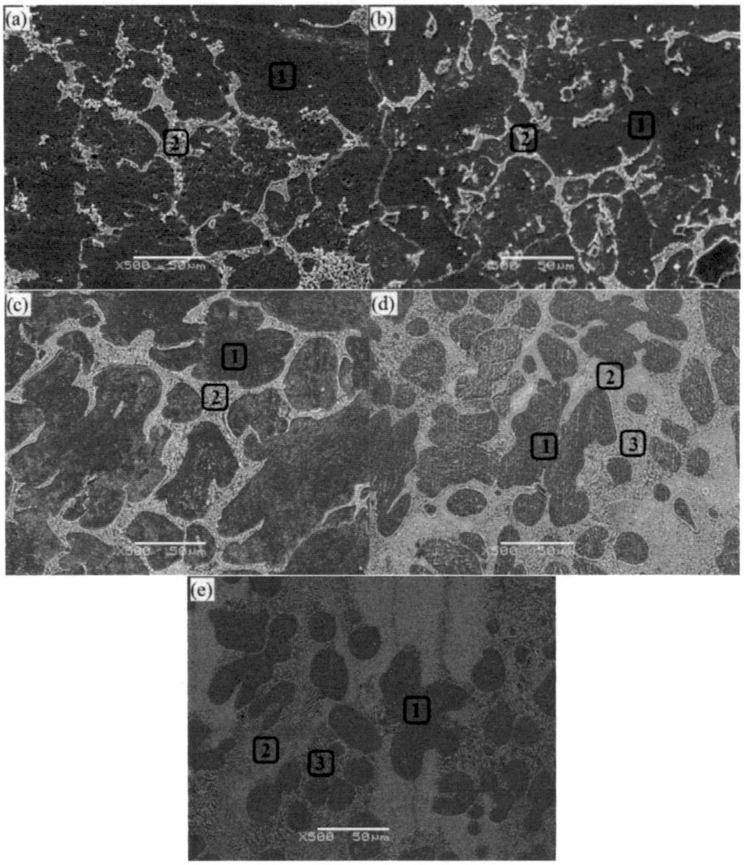

Figure 2. SEM images of Zn-Cu-Al alloy (wt.%): (a) 1.25; (b) 1.5; (c) 2.0; (d) 3.0;(e)5.0

Table 1. Energy spectrum analysis of cast microstructure of Zn-Cu-Al alloy in different locations shown in Fig. 2 (wt.%)

Positions	Elements (wt.%)		
	Cu	Al	Zn
a1	2.87	1.82	95.31
a2	1.76	18.04	80.20
b1	2.61	1.49	95.90
b2	1.71	18.67	79.62
c1	1.85	1.27	96.88
c2	1.78	18.16	80.05
d1	1.75	1.60	96.65
d2	1.71	18.71	79.58
d3	1.91	17.04	81.05
e1	1.68	1.63	96.69
e2	1.70	19.63	78.67
e3	1.74	18.27	79.99

Table. 1 shows the results of elemental energy spectrum analysis at different locations in the structure of Zn-Cu-Al alloy. From the data in Table 2, it can be seen that with the increase of Al addition, the total content of Cu and Al in the marked "1" region decreases basically, from 4.69 wt.% when Al addition is 1.25wt.% to 3.31wt.% when Al addition is 5.0wt.%, indicating that the solid solubility of alloy elements Cu and Al in the zinc matrix is decreasing. The atomic ratio of elements Zn and Al in the "2" region of the table is close to 3:2.

In order to further analyze the effect regularities of Al addition on the structure of Zn-Cu-Al alloy, the area at "1" in Fig. 2 is enlarged, as shown in Fig. 3. From Fig. 3, it can be seen that when Al addition content is 1.25wt.%, fine spherical particles with nano-scale are precipitated and dispersed in η phase; when Al addition content is 1.5wt.%, the size of spherical particles grows and the size difference between particles is not obvious; when Al addition content is 2.0wt.%, some spherical precipitated particles turn into short strips, and the size difference between particles increases obviously; when Al addition content is 3.0 wt.% and 5.0wt.%, some precipitated particles coarsen and the number of precipitated particles increases. The elemental energy spectrum analysis of the precipitated particles in η phase is carried out. As shown in Fig. 3 (f), the results show that the precipitated particles are composed of elements Zn, Cu and Al with atomic contents of 89.38 at.%, 3.45 at.% and 7.18 at.%, respectively.

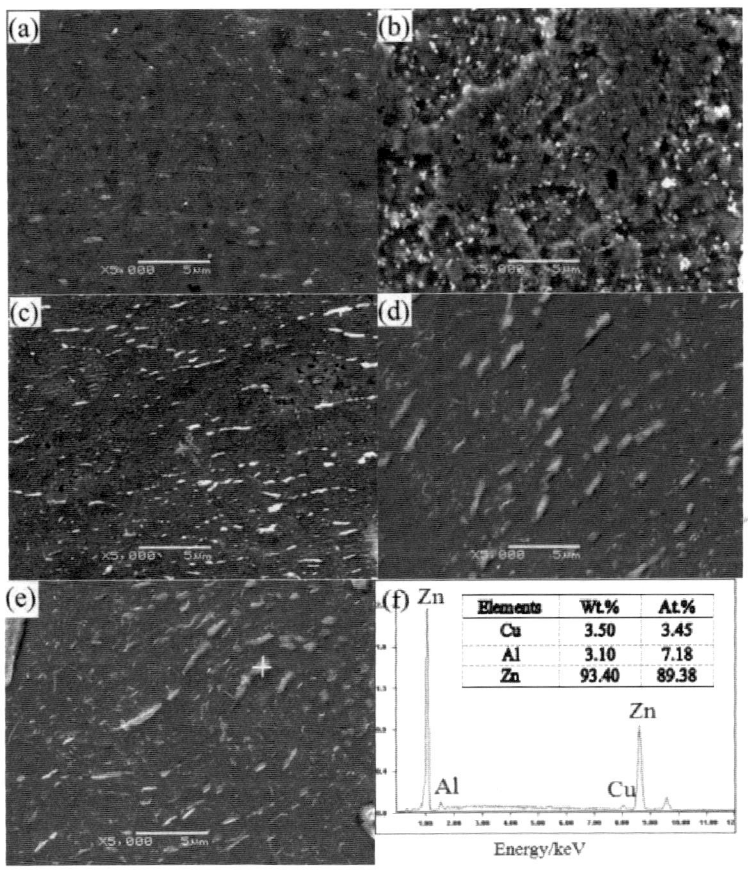

Figure 3. Magnified SEM images of region marked by 1 in Fig. 2: (a) 1.25; (b) 1.5; (c) 2.0; (d) 3.0; (e) 5.0; (f) EDS of plus in Fig. e

3.2. Phase Structure

Fig. 4 shows the XRD spectra of Zn-Cu-Al alloys with different Al contents. It can be seen from Fig. 1 that when the addition amount of Al is 1.25wt.% ~ 2.0wt.%, the Zn-Cu-Al alloy mainly consists of two phases: η phase (Zn) and ε phase (ZnCu5). Al0.4Zn0.6 phase appeared when the addition amount exceeded 3.0wt%. However, the intensity of the diffraction peak is very weak. An important factor affecting the formation of compounds between elements is the electronegativity of elements. The greater the difference of electronegativity between elements, the stronger the possibility of bonding between elements and the easier of the formation of compounds. The electronegativity of Zn, Cu and

IOP Conf. Series: Materials Science and Engineering **746** (2020) 012018 doi:10.1088/1757-899X/746/1/012018

Al is 1.65, 1.90 and 1.50 respectively. The difference of electronegativity between Zn and Al is 0.15, which is much larger than the difference between Zn and Cu (0.4). It is judged from thermodynamic theory that Zn and Cu are easy to form compounds. Previous studies [16-18] have shown that the solid solubility of Cu in Zn is only 0.3wt.% at room temperature in Zn-Cu binary alloy[19] , and increases to 2.7wt.% at 425℃. When the concentration of Cu exceeds 1 wt.%, metastable phase ε (CuZn5) is formed, and the ε phase is generally dispersed in the matrix with fine particles. In this experiment, the addition content of Cu element is 1.25wt%. According to the results of energy spectrum analysis in Table 1, the concentration of copper in different regions of the alloy is more than 1wt%. It can be determined that there is ε phase in the alloy structure. In the XRD diagram of Figure 4, the characteristic peak of ε phase with weak strength appears. However, due to the small size and dispersive distribution of the ε phase, the ε phase can not be clearly observed in the SEM images of the alloy microstructures in Figs. 2 and 3.

According to the Zn-Al phase diagram [19], there are mainly η (Zn) phases, solid solution based on zinc, dense hexagonal structure; α (Al) phase, solid solution based on aluminium, face-centered cubic structure; (α + η) eutectoid phase, solid solution based on aluminium or zinc-aluminium, and the lamellar solid solution η (Zn) based on zinc. Some weaker peaks correspond to the $CuZn_5$ phase, i.e. ε phase. Because the η phase and the ε phase have the same lattice type, hexagonal structures, and the lattice constants are very close, which leads to the coincidence of the two phases on the diffraction peaks at a specific angle. Because of the large volume fraction of η phase, the intensity of the diffraction peak is very high, and its diffraction peak masks the relatively weak peak of the ε phase.

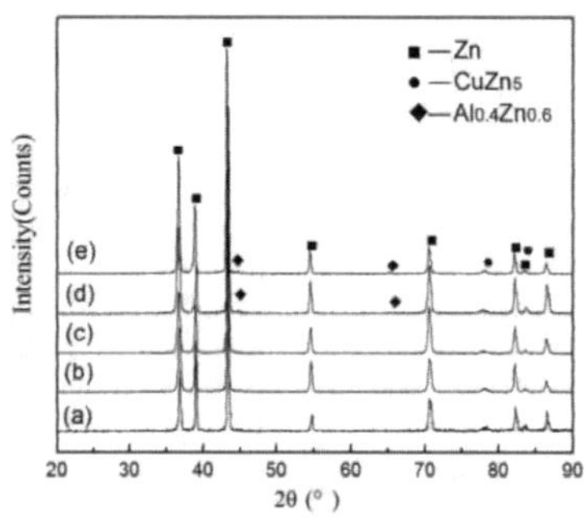

Figure 4. XRD patterns of Zn-Cu-Al alloy with different Al addition content (wt.%): (a) 1.25; (b) 1.5; (c) 2.0; (d) 3.0;(e)5.0

3.3. Electrical Conductivity

Fig. 5 shows the effect of Al addition on the electrical conductivity of Zn-Cu-Al alloy. As can be seen from Fig. 5, the electrical conductivity of the alloy increases with the addition of Al. When the Al addition content in the alloy is 1.25, 1.5, 2.0, 3.0 and 5.0wt.% respectively, the corresponding electrical conductivity is 23.31 IACS, 23.56 IACS, 23.95IACS, 24.61 IACS% and 25.07 IACS%. Comparing to the alloy with 1.25wt.%Al addition content, the conductivity increases by 7.5% when the Al addition content is 5.0wt.%.

Generally, the electrical conductivity of alloys is lower than that of pure metals. This is because the lattice arrangement of pure metal is relatively orderly, and the electron is less hindered in the transmission process. However, after alloying, because of the difference of atomic radius between solute and solvents, the lattice shrinkage or expansion of the matrix will be caused to varying degrees by the formation of solid solution or second phase, resulting in lattice distortion, which will increase the impediment of electronic movement and decrease the conductivity of the alloy. Moreover, the scattering effect of lattice distortion caused by atoms in solid solution on the electrons in the matrix is higher than that caused by reciprocation. The scattering effect caused by the second phase is much stronger and the electrical conductivity is much lower [20, 21]. The atomic radius of Zn is 1.39 Å, that of Cu is 1.28 Å, and that of Al is 1.43 Å. Therefore, in theory, the conductivity of Zn-Cu-Al alloy should decrease with the increase of Al addition content, but in fact, the conductivity of Zn-Cu-Al alloy increases with the increase of Al addition content.

According to the results of microstructures and EDS analysis, when the Al addition content ranges from 1.25% to 5.0wt.%, the total amount of Cu and Al in solid solution in η phase decreases with the increase of Al addition content, which is beneficial to improve the conductivity of the alloy. In addition, with the increase of Al addition content, the grain size of η phase in Zn-Cu-Al alloy decreases, and the area of grain boundary increases. Meanwhile, the particles precipitated from η phase coarsen. Both inhibit the conductivity of the alloy. From the morphology of Figs. 1 and 2, the state of η phase in Zn-Cu-Al alloy changes from continuous flaking at low Al content to discrete distribution at high Al content. On the contrary, the eutectic phase (α+η) which was originally dispersed between η phase crystals joined into pieces, and the number increased significantly. This structural change may be the main reason for the increase of electrical conductivity of the alloy.

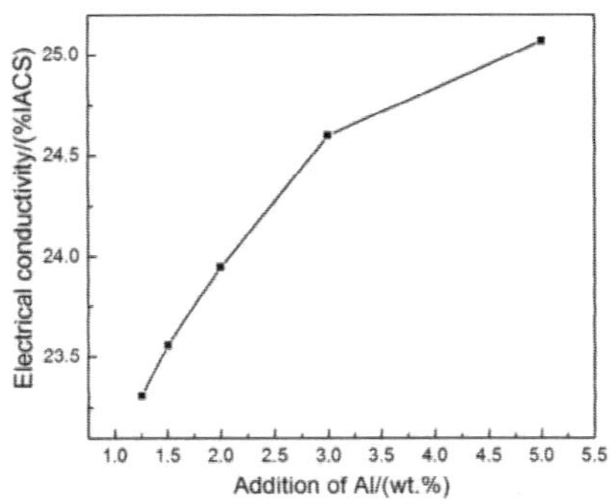

Figure 5. Effects of Al on the electrical conductivity of Zn-Cu-Al alloy

3.4. Microhardness

Fig. 6 shows the effect of Al addition on the microhardness and friction coefficient of Zn-Cu-Al alloy. It can be seen from Fig.6 that the microhardness and friction coefficient of the alloy increase with the increase of Al addition. The hardness of the alloy increases from 99 HV at 1.25 wt.% Al addition to 125 HV at 5.0 wt.% Al addition and the corresponding friction coefficient decreases from 0.55 to 0.45. From the analysis of the microstructure of the alloy, it can be seen that the increase of the hardness of the alloy is mainly attributed to the increase of the eutectic phase amount. In addition, the alloying element Al has inoculation effect. With the increase of content, the refining effect of η phase is obvious [9]. η phase is densely arranged hexagonal crystal structure, lattice constant is a= (2.6655±0.0001) Å, c =(4.9494±0.0003) Å, α phase is face-Centered Cubic with good plasticity, lattice constant a = (4.0474±0.0002)Å, ε phase (CuZn5) belongs to hard and brittle phase with HV150 hardness and is much higher than η phase (HV83.5) [22]. During the plastic deformation of the alloy, the deformation of the ε phase (CuZn5) is not easy to occur. The eutectic phase (α+η) is layered structure and microhardness is 140 HV [23].

3.5. Wear

From the curve of frictional coefficient varying with sliding time in Fig. 6(b), the fluctuation trend of friction coefficient increases with the increase of Al content in the alloy. Especially when Al addition content is 5 wt.%, the fluctuation range of friction coefficient is larger, about 0.07, while the fluctuation range of friction coefficient of other alloys is about 0.02. The fluctuation of friction coefficient reflects the number, size and distribution of hard phases in the alloy. In the friction process, the α phase acts as the main carrier, while the η phase acts as the solid lubricant [24, 25].

IOP Conf. Series: Materials Science and Engineering **746** (2020) 012018 doi:10.1088/1757-899X/746/1/012018

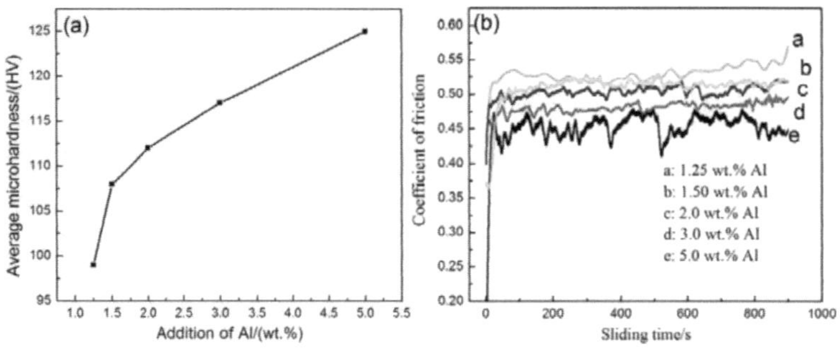

Figure 6. Effects of Al on the microhardness and frictional coefficient of Zn-Cu-Al alloy: (a) micohardness; (b) frictional coefficient

Fig. 7 shows the morphology of wear marks of Zn-Cu-Al with different Al additions with loads of 30N, sliding speed of 10mm/s, sliding time of 15min and dry friction. As can be seen from Fig. 7, there are two characteristic morphologies on the surface of the wear marks of the alloy: plastic deformation and peeling. When the Al addition content is 1.25~2.0wt.%, the friction bearing body is η phase with lower hardness and better plasticity. Under the forward extrusion of the grinding ball, the material in the friction zone is easy to be pushed to edges, resulting in greater plastic deformation. At the same time, under the action of shear stress, the surface material of wear marks is transferred and the appears peeling or delamination. The surface of the wear marks of the alloys shows obvious plastic deformation and local material detachment. The wear mechanism is a mixture of plastic deformation and adhesive wear, and plastic deformation is dominant. When the Al addition content is 3.0wt.% and 5.0wt.%, the friction bearing body is a phase of α+η with high hardness and good plasticity. In the friction process, α phase acts as the main bearing body and the η phase acts as the solid lubricant, the stress state and lubrication state of the alloy under grinding are improved. The surface of alloy wear marks is smooth and the degree of plastic deformation decreases. The wear resistance of the alloy increases with the increase of hardness and strength [26]. The wear mechanism of the alloy is mainly plastic deformation.

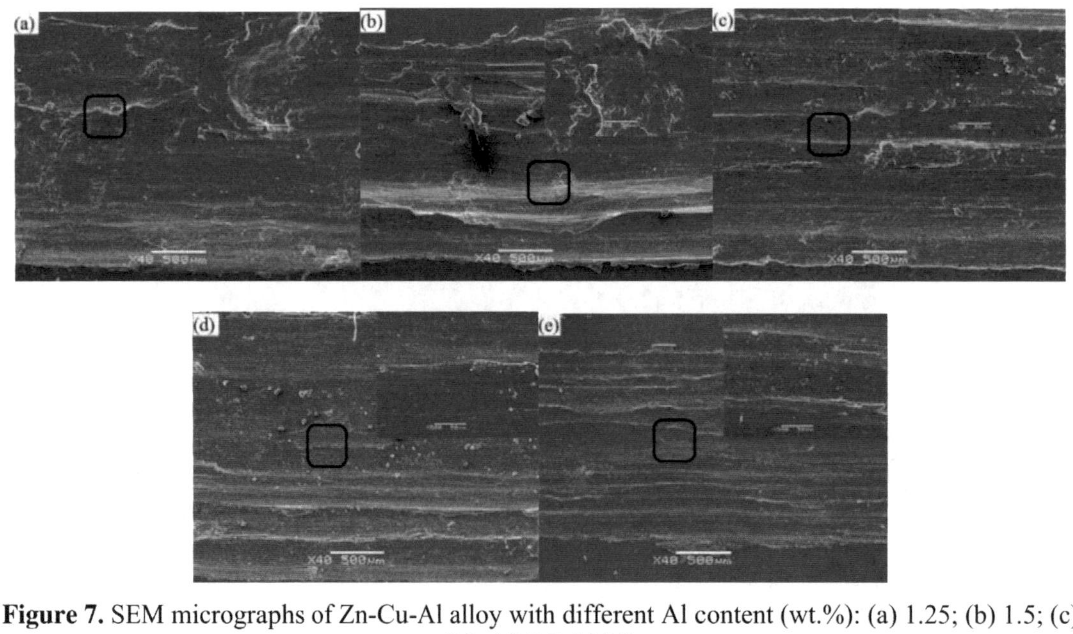

Figure 7. SEM micrographs of Zn-Cu-Al alloy with different Al content (wt.%): (a) 1.25; (b) 1.5; (c) 2.0; (d) 3.0;(e)5.0

4. Conclusions

Zn-Cu-Al zinc alloy with different content of Al was prepared by metal mould casting.

When Al addition content is 1.25wt.%~2.0wt.%, the alloy structure consists of Zn solid solution containing Cu and Al (ηphase), eutectic phase (α+η) and CuZn5 (εphase). Al0.4Zn0.6 phase appeared when Al addition exceeded 2.0wt%.

With the increase of Al content, the eutectic phase content in the alloy structure increases and the η phase grain refines. The microhardness and electrical conductivity of the alloy increase with the increase of Al content. When Al addition content is 5.0wt.%, the hardness and electrical conductivity reach the maximum, 151HV and 26.4% IACS, respectively.

When the Al addition content is less than 3.0wt.%, the wear mechanism of the alloy is a composite mechanism of plastic deformation and adhesive wear, and plastic deformation is dominant. When the Al addition content is less than 3.0wt.%, the wear mechanism of the alloy is mainly plastic deformation.

5. Acknowledgements

The project is financially supported by the Jiangsu Key Laboratory of Advanced Structural Materials and Applied Technology Open Fund (ASMA 201506 and ASMA201706); Jiangsu University Students Innovation and Entrepreneurship Training Program 201711276052X); Science and technology innovation fund for college students of Nanjing Institute of Engineering (TB201902027).

6. References

[1] Savaşkan T, Maleki R A, 2014 *Tribol. T.*, **57** 435.
[2] Tanaka T, Kushibe A, Kohzu M, Takigawa Y, Higashi K, 2008 *Scripta Mater.*, **59** 215.
[3] Yan S Q, Xie J P, Liu Z X, Wang W Y, Wang A Q, Li J W, 2010 *J. Mater. Sci. Technol.*, **26** 648.
[4] Liu T, Si N C, Liu G L, Zhang R, Qi C, 2016 *Trans. Nonferrous Met. Soc. China.*, **26** 1775.
[5] Sánchez-Martínez A, Cruz A, González-Nava M, Suárez M A, 2016 *Mater. Des.*, **108** 494.
[6] Kan T S, Azakl Z, 2008 *Wear*, **264** 920.
[7] Bobi B, Bajat J, Aimovic-pavlovi Z, Bobi I, Jegdi B, 2013 *Trans. Nonferrous Met. Soc. China.*, **23** 931.
[8] Kaya H, Cadirli E, Ulgen A, 2011 *Mater. Des.*, **32** 900.
[9] Liu Z L, Li R Q, Jiang R P, Li X Q, Zhang M X, 2016 *J. Alloys Comp.*, **687** 885.
[10] Wu Z. C., Sandlöbes S., Wu L., Hu W. P., Gottstein G., Korte-Kerzel S., Mater. Sci. Eng. A 651 (2016) 675-687.
[11] Du D, Guan G, Gagnoud A, Fautrelle Y, Ren Z M, Lu X, Wang G H, Dai Y M, Wang Q L, Li X, 2016 *Mater. Charact.*, **111** 31.
[12] Boczkal G, Pałka P, Sułkowski B, Mikułowski B, 2017 *Mater. Sci. Eng. A* **690** 254.
[13] Ares A E, Gassa L M, Schvezov C E, Rosenberger M R, 2012 *Mater. Chem. Phys.*, **136** 394.
[14] Savaşkan T, Hekimoğlu A P, 2014 *Mater. Sci. Eng. A* **603** 52.
[15] Zhang Y M, Yang L J, Zeng X D, Zheng B Z, Song Z L, 2013 *Mater. Des.*, **50** 223.
[16] Prasad B K, 2004 *Mater. Sci. Eng. A*, **367** 63.
[17] Costaa E M, Costab C E, Vecchiaa F D, Ricka C, Scherera M, Santosa C A, Dedavida B A, *J. Alloys Comp.*, **488** 89.
[18] Sharififar M, Mousavi S A A A, 2014 *Mater. Sci. Eng. A*, **594** 118.
[19] Yao C Z, Tay S L, Zhu T P, Shang H F, Gao W, 2015 *J. Alloys Comp.*, **645** 131.
[20] Ying T, Zheng M Y, Li Z T, Qiao X G, Xu S W, *2015 J. Alloys Comp.*, **621** 250.
[21] Zhang D F, Zhao X B, Shi G L, Qi F G, 2011 *Rare Metal Mat. Eng.* **40** 418.
[22] Ma D, Li Y, Ng S C, Jones H, 2000 *Acta Mater.*, **48** 419.
[23] Mahmudi R, Alibabaie S, 2013 *Mater. Sci. Eng. A* **559** 421.
[24] Savaskan T, 2004 *Tribol.Int.*, **37** 45.
[25] Mojaver R, Shahverdi H R, 2010 *Wear*, **268** 605.
[26] Savaşkan T, Maleki R A, Tan H O, 2015 *Tribol. Int.*, **81** 105.

IOP Conf. Series: Materials Science and Engineering **746** (2020) 012019 doi:10.1088/1757-899X/746/1/012019

Synthesis and Characterization of Ultrafine Aluminum Phosphate Powder in CTAB System

Zhifu Wu[1,*], Guiquan Guo[2], Liyun Hu[1] and Hui Chen[1]

[1]School of materials science and engineering, Baise University, 533000
[2]School of Chemistry and Chemical Engineering, Xingtai University, 054001, China
ORCID and E-mail: zhifuwu2013@163.com

Abstract. The phosphoric acid and aluminum hydroxide and CTAB reagents are used as raw materials. The ulrafine aluminum phosphate is prepared by liquid phase method. The sample composition and scanning electron microscope are used to observe the sample. Morphological characteristics were used to analyze the effects of CTAB and temperature of calcination on the morphology of aluminum phosphate powder.

1. Introduction

Aluminum phosphate has the characteristics of high melting point, good fire resistance, good wear resistance and high strength. Aluminum phosphate belongs to polar substance, which shows hydrophilicity, hydrophobicity and poor compatibility. Interfacial separation can be observed in the polymerization reaction with organic polymers [1]. Widely used in fireproof coating, paper industry, textile industry, antifouling agent, flux and other aspects [2]. In the past, there are many methods to prepare aluminum phosphate [3-5], but it is difficult to prepare pure aluminum phosphate, because there are many by-products of direct reaction between aluminum hydroxide and phosphoric acid, a small amount of phosphoric acid reacts with aluminum hydroxide: $H_3PO_4 + Al\ (OH)_3 = AlPO_4 + 3H_2O$; a proper amount of phosphoric acid reacts with aluminum hydroxide: $3H_3PO_4 + 2Al\ (OH)_3 = Al_2(HPO_4)_3 + 6H_2O$.

Excessive phosphoric acid reacts with aluminum hydroxide: $3H_3PO_4 + Al\ (OH)_3 = Al(H_2PO_4)_3 + 3H_2O$; therefore, it is very important to choose the preparation method of partial transportation. Sixteen alkyl three methyl ammonium bromide (CTAB) is a quaternary ammonium cationic surfactant. It can be dissolved in water and produce large amount of foam when it oscillate. It has good compatibility with anionic, nonionic and amphoteric surfactants [6]. It has excellent permeability, flexibility, emulsification, antistatic, biodegradability and bactericidal properties [7]. It is often used in the preparation of nano materials to control the size and morphology of nano particles. CTAB system is also an excellent detergent. In this experiment, CTAB system is used to prepare aluminum phosphate ultrafine powder, which can not only adjust particle size and morphology [8-10], but also increase the anti fouling ability of aluminum phosphate. At the same time, the synthetic route has the advantages of small pollution, high efficiency and low cost, which is a good method worth popularizing.

2. Experimental Part

2.1. Reagents and Instruments

Phosphoric acid (85%), $Al(OH)_3$, and $NH_3{\bullet}H_2O$ were purchased from Tianjin comio Chemical Reagent Co., Ltd. Cetyltrimethylammonium bromide (CTAB) was purchased from Beijing Yili fine chemical Limited company. They were all analytical reagents. SU-5000 Field emission scanning electron microscope (Hitachi, Japan), Panako multifunctional powder X-ray diffractometer (test voltage of 40 kV, current of 20 mA, scanning speed of 0.03°/s, continuous scan, 10° to 80°, Cu Ka (λ = 0.15418 nm)), Instruments include electronic balance (JA5003N, Shanghai Precision Scientific Instrument Co., Ltd.), high-speed centrifuge (TG16K-II, Dongwang Instrument Co., Ltd.), electric heating blast drying oven (HG101-2A, Nanjing Yingxin Experimental Instrument Co., Ltd.).

2.2. Synthesis of Products

After a series of optimization experiments, the concentration and calcination time of CTAB were changed, and the optimized process conditions were obtained. The typical experimental steps are as follows: prepare 40ml of 60% phosphoric acid, transfer it into two 500ml beakers respectively, label a and B, and add 5ml of CTAB with concentration of 0.001mol/l to the solution of beaker B. Weigh two 3.9G aluminum hydroxide, stir it continuously, add A and B beaker respectively, get thick gel, then add 200ml deionized water, add ammonia water to adjust pH value to 5, then put two beaker in 75 degree water bath to keep 2H. After the heat preservation, it can be seen that there are obvious layers in both groups A and B beakers, the upper layer is clear and transparent, and the lower layer is white precipitate. Filter separately and collect white precipitate. The precipitate was placed in a vacuum drying oven and dried at 100 ℃ for 12h. After drying, the white massive solid ball is grinded into powder shape, and then A and B samples are divided into four parts: A1, A2, B1 and B2, and A1 and B1 are tested. A2 and B2 were put into a box resistance furnace and calcined at 800 ℃ for 3h to obtain the product.

3. Results and Discussion

3.1. XRD Analysis

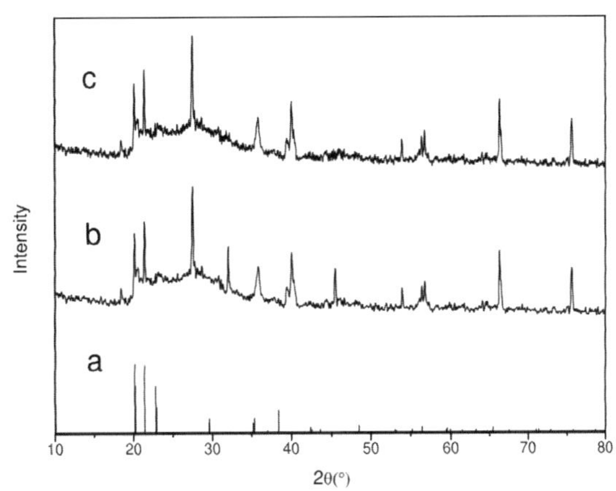

Figure 1. The XRD pattern of the products obtained from the reaction of phosphoric acid and aluminum hydroxide without CTAB (a: aluminum phosphate of PDF standard 50-0303; b: diagram of sample A_1; c: diagram of sample A_2)

Fig. 1 is the XRD pattern of the product obtained from the reaction of phosphoric acid and aluminum hydroxide, FIG. 1a is the diffraction peak diagram of aluminum phosphate PDF standard card 50-0303, FIG. 1b is the XRD diagram of aluminum phosphate produced by the reaction of 60% phosphoric acid and aluminum hydroxide, FIG. 1c is 60% phosphoric acid and aluminum hydroxide reacted at 800 ℃ and baked at high temperature. Compared with FIG. 1a, the diffracted peaks of Fig. 1b and Fig. 1C are close to the diffracted peaks of aluminum phosphate PDF standard card 50-0303, which shows that the main components of sample A2 and A1 are aluminum phosphate, but they all have hetero peaks. Compared with FIG. 1b and Fig. 1C, the hetero peaks are reduced after 800 ℃ high temperature roasting, which proves that high-temperature roasting can remove some impurities.

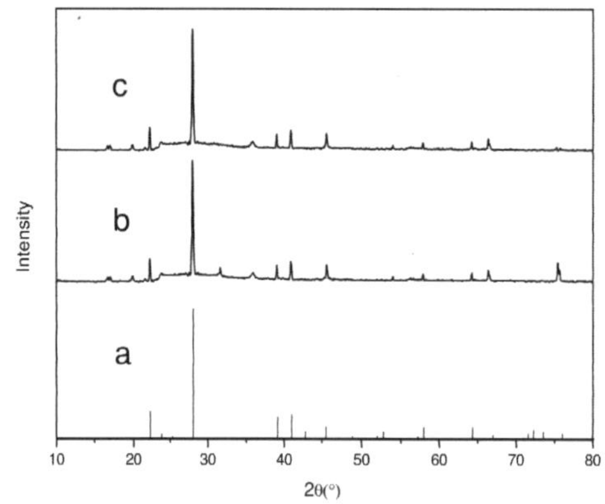

Figure 2. XRD pattern of the products obtained from the reaction of phosphoric acid and aluminum hydroxide after adding CTAB (a: aluminum phosphate of PDF standard 76-0234: b: diagram of sample B_1; c: diagram of sample B_2)

Fig. 2 is the XRD diagram of the product obtained from the reaction of phosphoric acid and aluminum hydroxide after adding CTAB, FIG. 2a is the diffraction peak diagram of aluminum phosphate PDF standard card 76-0234, FIG. 2b is the XRD diagram of aluminum phosphate produced by adding CTAB in reaction of 60% concentration phosphoric acid and aluminum hydroxide, FIG. 2C is the XRD diagram of aluminum phosphate produced by adding CTAB in reaction of 60% concentration phosphoric acid and aluminum hydroxide and calcining at 800 ℃. The diffraction peaks of Fig. 2B and C correspond to the diffraction peaks of No. 76-0234 of aluminum phosphate PDF standard card, but there are miscellaneous peaks, indicating that the main products of sample B2 and B1 are aluminum phosphate, containing impurities. The comparative analysis of Fig. 2B and Fig. 2C shows that the corresponding miscellaneous peaks decrease after high temperature roasting at 800 ℃, which is basically the same as that of No. 76-0234 of aluminum phosphate PDF standard card, further proving that high temperature roasting can remove impurities.

3.2. SEM Analysis

Fig. 3 is the SEM image of the product from the reaction of 60% phosphoric acid with aluminum hydroxide. It can be observed that the size of aluminum phosphate particles in the sample image is large, the average size is about 40 microns, most of them are irregular bodies, and the distribution is relatively disordered.

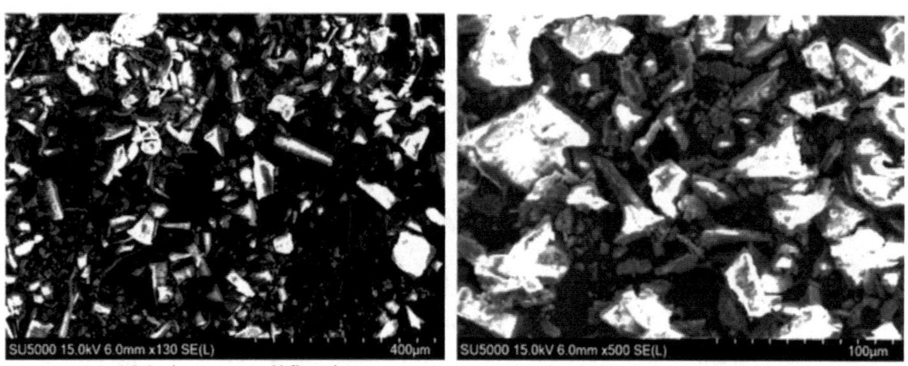

(a) 130 times amplification (b) 500 times amplification

Figure 3. SEM image of the product from the reaction of 60% phosphoric acid with aluminum hydroxide

(a) 600 times amplification (b) 800 times amplification

Figure 4. SEM image of the product obtained after the reaction of 60% phosphoric acid with aluminum hydroxide and calcined at 800 ℃

Fig. 4 is the SEM image of the product obtained after the reaction of 60% phosphoric acid with aluminum hydroxide and calcined at 800 ℃. Compared with Fig. 3, it can be seen that the particle size of the product after calcined at 800 ℃ decreases, the average size is about 25 μ m, and there is a trend of forming regular shape particles. The results show that calcination at high temperature will make aluminum phosphate particles smaller and obtain regular morphology.

(a) 9600 times amplification (b) 2000 times amplification

Figure 5. SEM image of the product obtained by adding CTAB in the reaction of 60% phosphoric acid and aluminum hydroxide

Fig. 5 is the SEM image of the product obtained by adding CTAB in the reaction of 60% phosphoric acid and aluminum hydroxide. Compared with Fig. 3, it can be seen that after adding CTAB, the aluminum phosphate particles become thinner, the average particle size is only about 15 μ m, and the morphology also changes, forming a few regular ellipsoid or square block crystals, which is related to CTAB's ability to adjust the surface morphology of particles.

IOP Conf. Series: Materials Science and Engineering **746** (2020) 012019 doi:10.1088/1757-899X/746/1/012019

(a) 800 times amplification

(b) 2000 times amplification

(c) 10000 times amplification

(d) 10000 times amplification

Figure 6. SEM image of the product obtained from the reaction of 60% phosphoric acid and aluminum hydroxide, adding CTAB, and calcining at 800 ℃

Fig. 6 is the SEM image of the product obtained from the reaction of 60% phosphoric acid and aluminum hydroxide, adding CTAB, and calcining at 800 ℃, compared with Fig. 3, aluminum phosphate particles become smaller, more uniform in size, better distribution, and more regular shape particles. Compared with FIG. 5, the particle size is also smaller, with some regular shapes. However, from Fig. 6, it can be seen that the average particle size after calcination at 800 ℃ is only 1-3 μm, and some aluminum phosphate particles reach nanometer size.

4. Conclusion

The results show that the aluminum phosphate powder prepared by the reaction of 60% phosphoric acid and aluminum hydroxide in 0.001 CTAB system has the smallest particle size; the high temperature calcination at 800 ℃ can make the aluminum phosphate particles smaller and promote the formation of regular morphology, and the high temperature calcination can effectively remove impurities.

5. Acknowledgement

Material Physics and Chemistry Open Project of Guangxi University Key Subject , Baise University Aluminum Matrix Composite Characteristic Research Team Project , Funds for the Construction of Master's Degree Granting Units in Guangxi Zhuang Autonomous Region and Municipal People's Livelihood Science and Technology Security Special Project of Xingtai City in 2018 (2018ZZ18).

6. Reference

[1] Cui S L, Chen N, Huo J C. Research progress in the transformation and control of aluminum phosphate system crystal form, 2012, 09, 21.

[2] Xiong J, Li Y T, Wei C, et al. Effect of aluminum phosphate additive on anti pollution performance of PVDF membrane. Water treatment technology. 2019, 45 (8): 86-88.
[3] Shasha Ma, Hai Lin, Luyi Yang, et al. High thermal stability and low impedance polypropylene separator coated with aluminum phosphate. Electrochimica Acta, 2019, 320.
[4] Huo J C, Lei Y L. Surface hydrophobic modification of aluminum phosphate by solid-phase method. 2012, 09, 11.
[5] Liu Z H, Sun M Y, Zeng Y L, et al. Progress in preparation and application of aluminophosphate molecular sieve , Guangzhou chemistry, 2019, 44 (02).
[6] Harichandran G, Divya, P S. Radha J, et al. Facile and controllable CTAB-assisted sonochemical synthesis of one-dimensional $MnWO_4$ nanorods for supercapacitor application. Journal of Molecular Structure, 2020, 1199.
[7] Galina M. Mamardashvili, Elena Y. Micelles encapsulated Co (III)-tetra (4-sulfophenyl) porphyrin in aqueous CTAB solutions: Micelle formation, imidazole binding and redox Co (III) / Co (II) processes. Journal of Molecular Liquids, 2019, 293.
[8] Chen X Y, Huang Y L, Yuan A Q, Huang Z W, Wei D P, Ma S M. Controllable synthesis of nano zinc phosphate. Inorganic salt industry, 2019, 51 (08): 20-24 + 82.
[9] Ma Y Y, Liu M, Li J, Li X Y, Yang Z Q. Study on the preparation and antibacterial properties of CTAB coated nano gold. Journal of Liaocheng University, 2019, 32 (05): 64-70.
[10] Zhang Q, Ren Y F, Zhao F. Research progress in thermal synthesis of phosphate by ionic liquid. Liaoning chemical industry. 2018, 47 (06), 549-550.

IOP Conf. Series: Materials Science and Engineering **746** (2020) 012020 doi:10.1088/1757-899X/746/1/012020

Accelerating Materials Discovery Based on Generalized Low-dimensional Conformation Performance Relationships

Yuanyuan Qu[1], Xiangying Meng[1*], Zixi Jia[2], Xiaofeng Liu[3], Dongyan Liu[1], Shanshan Li[1] and Fang Bian[1]

[1]College of Sciences; [2]Faculty of Robot Science and Engineering; [3]Information Construction and Network Security Office; [3]College of Information Science and Engineering, Northeastern University, Architecture Building, Room 115, No.3-11, Wenhua Road, Heing District, Shenyang, P. R. China,110819

Email: x_y_meng@mail.neu.edu.cn

Abstract. The construction of conformation-performance relationships (CPRs) is the core issue of efficient material design. Different from traditional CPRs originating from experience summarizations or rational model cognitions, a strategy of generalized low-dimensional conformation-performance relationships (GLD-CPRs) for rapid material discovery has been established in this work by the aid of material information technology, providing us with a new perspective to material gene (MG) and its function on performance. From a pure computing perspective, the strategy has been concreted by a case study of semiconductor bandgap engineering. Based on more than 20,000 items of self-produced semiconductor bandgap and attributes data, a set of CPRs for bandgap regulation towards arbitrary semiconductor of the same kind has be generated using GLD-CPRs strategy. By comparing with the reported studies, the reliability of the strategy is confirmed. Finally, future improvements needed by GLD-CPRs are addressed.

1. Introduction

One principal goal of materials science is to understand conformation-performance relationships (CPRs) of material systems such that specific performances may be achieved via material engineering. For centuries, we have been accustomed to build traditional CPRs based on experience summarizations or rational model cognitions. Now, it is feasible to construct CPRs for producing materials and precisely designing their interiors to address the performance issues with the boom of high-throughput [1] and data processing technologies [2]. In general, a specific substance can be defined by its structure S, composition C, and ingredient I, and CPRs for a given performance is a 3D point-to-point function P (S, C, I) in the conformation-performance space. Since there has been much interest in accelerating the discovery of materials, it is necessary to fix one or two variables in P (S, C, I) to construct point-to-face or point-to-line low-dimensional CPRs (LD-CPRs) to narrow the searching space and thus speed the progress of materials screening.

Although not explicitly proposed, the concept of LD-CPRs has been implied in many successful data-processing-based materials designs. For example, Xue [3] et al determined the optimal composition in $Ni_{50-x-y-z}Ti_{50}Cu_xFe_yPd_z$ low thermal hysteresis shape memory alloys based on the LD-CPR of thermal to composition ΔT(C)|S=R,B19,B19';I=Ni,Ti,Cu,Fe,Pd. Lin [4] et al obtained carbon-capture materials in 2.6 million zeolite-like structures of silica with the LD-CPR of parasitic energy to structure E(S)|I=Si,O; C=1:2. Castelli [5] et al predicted new photoelectrochemical cells with improved light absorption in ABO_2N perovskite oxy-nitrides by mapping the change of bandgap (BG) with respect to A/B site cations BG(IA,B)|S=Perovskite; C=1:1. Different from traditional CPRs pursuing universality, data-processing-based LD-CPRs can be tailored for the specific material system and directly produces materials with merits or demerits for performance screening, making it more attractive in materials discovery.

Another cause for the rise of material data processing technology is the emergence of advanced experimental instruments. With this process, more and more sophisticated intrinsic attributes, namely material genes (MGs), are expected to involve in the material design to affect performances, while the traditional CPRs based on experiences or models are becoming less and less good at describing the role of personalized MGs. The character of multiple attributes of performance leads to a multivariate function of performance to attributes as $P(x_1,\ldots x_i)$, where x_i-set is a collection of various intrinsic attributes that may affect the material performance. Following the LD-CPRs strategy, the generalized low-dimensional CPRs (GLD-CPRs) need to be established to accelerate the process of material oriented design.

To fulfil the goal of GLD-CPRs, three key links, i.e. correlation analysis, weight evaluation and data mining, are necessary to guarantee the rationality and reliability of the strategy. Firstly, it is of great significance to evaluate the correlation between attributes to avoid overlapping representation of material genetic information. We can list arbitrary attributes as original MGs, and then construct nearly independent derived MGs through correlation analysis. Secondly, a weight evaluation mechanism should be established to assess the contribution of MGs to performance. The weight of genes on performance (WGP) is used to judge whether attributes are major or minor MGs. Thirdly, a mature machine learning algorithm ought to be adopted to produce a reliability regression model. The final GLD-CPRs between major MGs and performance in a specific material system will be excavated out based on the regression model, which serves to be material design atlases to guide the subsequent experiments.

The energy bandgap of semiconductor is a fundamental performance restricting light harvesting, conversion and transport technologies, such as photovoltaics [6] and photocatalysts [7]. In this work, the function, efficiency and reliability of GLD-CPRs strategy are embodied in the bandgap design of rutile-TiO_2 binary semiconductors, with a target of screening effective visible light responsive semiconductors. More than twenty thousand items, covering multi-dimension attributes and bandgap, are produced by ourselves and used to train a mature reliable regression model based on data-preprocessing and random forest algorithm [8]. By evaluating WGP, the major MGs that fundamentally influence bandgap are picked out to build GLD-CPRs to disclose new facts and guide bandgap engineering. Our strategy is universal to the target optimization design of materials, also will arouse extensive interest and thinking in other research fields.

2. Details of Algorithms

The implementation of GLD-CPRs is schematically descripted in figure 1. In our own work, we have favoured the analysis of computational data and the validation of predictions through both reported experiments and calculations. Four modules including interface to high-throughput computing database, data pre-processing, regression model training and GLD-CPRs construction are integrated in the routine. In this work, data of attributes and bandgap is uniformly generated from workflows in the high-throughput computing platform AiiDA [9]. An interface to the platform has been developed in our group to automatically download computational data and then push it to the subsequent data preprocessing procedure. To guarantee the reliability of regression model from data layer, data-prerecession is executed to convert data set to training set. By the aid of random forest machine learning algorithm, a mature reliable regression model for bandgap predicting is built. Finally, various constraints for accelerating target materials discovery are manifested through GLD-CPRs. As a result, more than twenty thousand cleaned bandgaps are prepared for the subsequent data training.

3. Results and Discussion

A mature and reliable regression model for bandgap predicting has been established through data processing technique and machine learning algorithm. Then the model is utilized to assess WGP and yield GLD-CPRs. By taking bandgap designs of rutile-TiO_2 binary as a case study, we illustrate the function of the strategy.

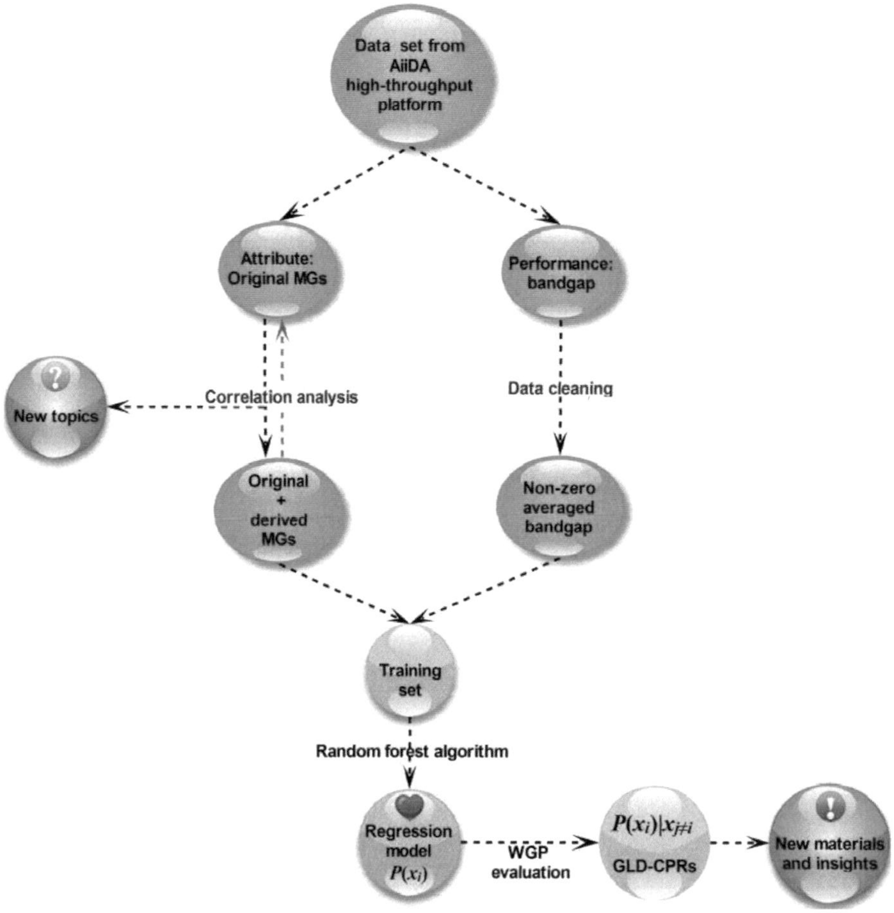

Figure 1. Implement of GLD-CPRs strategy in bandgap design

The big data from AiiDA high-throughput computing platform is divided into two parts: original attributes (MGs) and performance (bandgap). Bandgap cleaning process produces a set of non-zero averaged bandgap. After data pre-processing, a clean training set is obtained and input into the random forest machine learning algorithm for building a reliable regression model. By evaluating the weight of MGs on performance (WGP), major MGs are picked out to establish corresponding GLD-CPRs to suggest new experiments, with the triple goals of model improvement, mechanism recognition and materials discovery.

3.1. General Evaluation of WGP for Binary Semiconductors

In figure 2, there is a clear division between the first five MGs and the rest MGs. The first five MGs, each weights twice the average weight and total weights greater than 60%, are considered to be major MGs affecting bandgap profoundly. The rest 18-diemntion MGs, their respective weight is close to or largely less than the average weight are thus regarded as minor MGs affecting bandgap weakly. As expected in the section of PCC analysis, we found that the weights of some derived MGs have suppressed that of original MGs.

We are more concern about the feedback of WGP and its impact on our physicochemical cognition of materials. First of all, bandgap is found to be a localized electronic property and most sensitive to the nearest interaction distance between cations and anions. This finding is consistent with the understanding of the origin of electronic energy bands in solids, because the electronic bands are formed by the overlap of atomic orbitals and the extent of overlap is determined by the interatomic distance.

In the second place, the deference between ionization energy and affinity energy of ingredients, namely IP_1-EA_2 or IP_2-EA_1, is found to be the major MGs significantly affecting the bandgap. Here WGP evaluation provides a basis for the widely used model of approximating the bandgap of an infinite periodic solid with the atomic energy level difference of a finite ingredient cluster. For the finite system, the first ionization potential (*IP*) and electron affinity energy (*EA*) of ingredients respectively defines their highest occupied atomic orbitals (HOAO) and the lowest unoccupied atomic orbitals (LUAO). WGP informs us that the real bandgap E_g of binary clusters can be approximated by $E_{ga}=IP_i-EA_j$, where subscript *i* and *j* represent different kinds of ingredients.

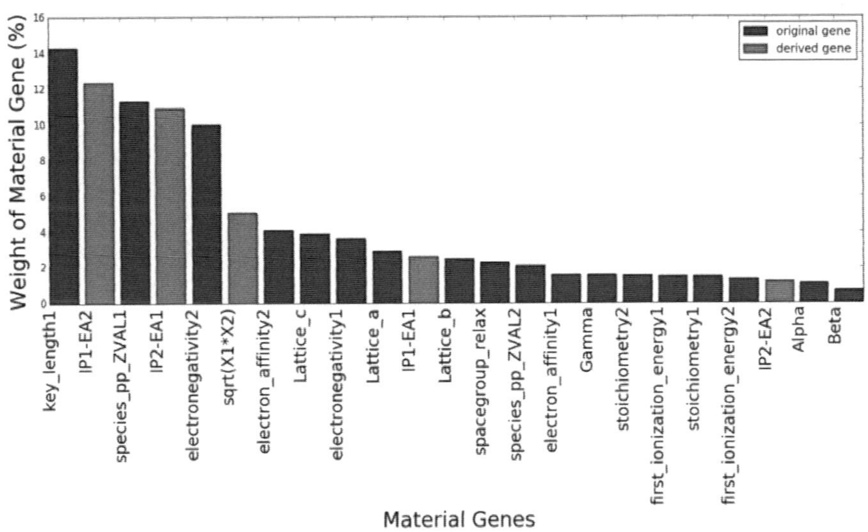

Figure 2. WGP histogram of bandgaps for binary semiconductors

The weights of total 23-dimention MGs on bandgap are sorted according to the size of their contributions. The blue column represents the weight of original MGs and the red one is the weight of derived MGs.

More significantly, WGP shows that bandgap is largely related to the valence electron structure of cation and electronegativity of anion, which suggests an effective way to regulate the bandgap by controlling ingredients. Interestingly, although being the same physical properties, valence electron number of anions and electronegativity of cations are minor-MGs that affect bandgap weakly. Therefore, the influence of ions on the bandgap will vary with their roles, that is, the same element will have different effects on the bandgap with different cationic or anionic identities.

Through above data analysis, the validity and importance of WGP evaluation are verified by checking with the existing knowledge. As long as reasonable MGs and reliable regression model are constructed, WGP can provide us with strong supports for the existing axioms, profound understanding of phenomena and new research topics and directions.

3.2. Construction of Specific GLD-CPRs

Since major-MGs have been distinguished from minor-MGs by WGP evaluation, it is critical to reveal the quantitative relationships between bandgap and major MGs by GLD-CPRs. In our strategy, this process needs to be realized by means of dimensionality reduction, that is, the relation of $P(x_i)|_{x_{j\neq i}=costant}$ need to be clearly expressed based on the bandgap regression model, in which the multivariate function of performance versus attributes $P(x_1,...x_i)$ has been implicated.

Valence electron structure of cation and electronegativity of anion are two major MGs affecting bandgap profoundly according to WGS evaluation. By the aid of GLD-CPRs between bandgap and the two quantities, bandgaps of rutile-TiX_2 and rutile-XO_2 (X represents the specie of cations or anions) are presented in figure 3. Based on the results, we can predict the effect of substitutional doping in rutile-TiO_2.

With the aim of reducing bandgap, there are much more rooms for cation-doping than that for anion doping. For cationic doping, the GLD-CPR suggests that rare-earth elements (from Tb to Lu), *d*-filled transitional metals (Cu, Zn, Ag, Cd) and post-transitional metals (Ga, Ge, As, In, Sn, Sb) are promising dopments for reducing the bandgap of rutile-TiO_2, while alkali, alkaline-earth and *3d*- or *4d*- unfilled transition dopments have a fat chance in reducing the bandgap of rutile-TiO_2. As for anionic doping, the GLD-CPRs prompt that B, C, N, P and S are potential dopments for the bandgap reduction of rutile-TiO_2, but doping halogens (F, Cl, Br, I) in rutile-TiO_2 is not a good choice. These enlightenments from GLD-CPRs are basically coincident with the reported facts [10], indicating that the strategy is generally reasonable and convincing. Further experimental verification and deep-seated explorations should be aroused by these predictions.

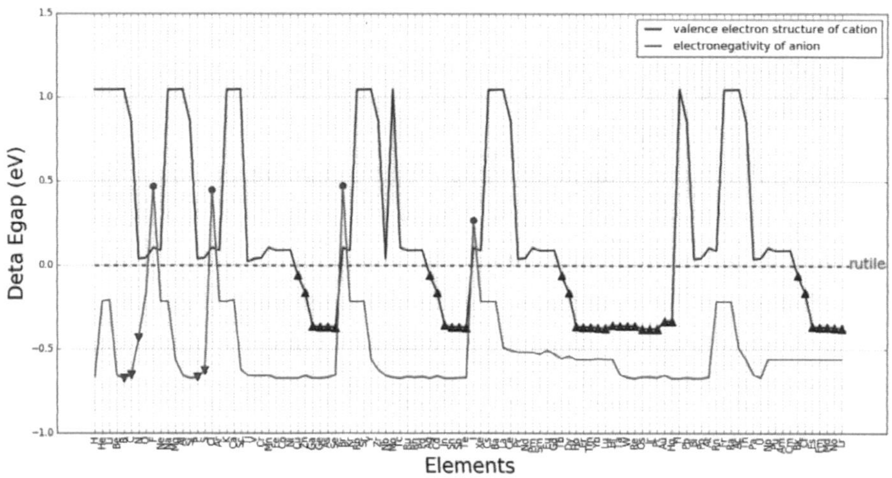

Figure 3. Bandgap variation induced by ingredient change in rutile structure

The bandgap of intrinsic rutile-TiO_2 is set to zero. The bandgap spectrums of XO_2 (blue curve) and TiX_2 (green curve) in the same lattice prototype as rutile structure are predicted based on GLD-CPRs, which provides a guidance for the bandgap regulation of rutile by means of element doping.

4. Future Perspectives

There is no doubt that the prospects of material informatics are promising, but at present the method is far from mature. Aiming at discovering materials with expected performances rapidly and correctly, a complete procedure, including data-generation, data-management and data-utilization, should be established. In these aspects, the reliability of attributes construction, data-cleaning and classification, regression model training and CPRs strategy is directly related to the effectiveness and practicability of material informatics method. In the GLD-CPRs frame, there are two main causes that lead to the uncertainty of the strategy and need to be further refined.

Firstly, attributes (MGs) are not orthogonal to each other as independent variables, so adiabatic approximation is introduced into the model of GLD-CPRs. Adiabatic approximation is reasonable for those weakly correlated MGs, but for strongly correlated MGs, orthogonalization scheme need to be developed to obtain as complete a set of attributes as possible. During the process, the physical analysis of those new emerging derivative genes is challenging but worth looking forward to. In this regard, one of the main tasks of GLD-CPRs strategy is to approach a complete set of MGs, and each element of the set has a clear physical meaning towards specific performance requirements.

Secondly, the reliability of GLD-CPRs relies on the accuracy of regression model, and machine-learning models are only as good as the data used to construct them. Therefore, the model, algorithm and parameter adjustment of machine learning need to be protected at data layer. In addition, the current material design based on machine learning method adopts a "forward" strategy, which means that the performances of unknown materials are predicted by those known materials data, and then the expected target materials are selected by exhaustive method. In the future, it is necessary to further develop "reverse" machine learning technology, that is, by setting the performance requirements, target material systems that meet the criteria are autonomously generated by machine learning method itself. This will be a challenging but significant work.

5. Summary

Data processing technology has shown its great effectiveness in the design of structural and functional materials. At present, the complementary advantages between material data technology and traditional material rational design have emerged. However, material data technology now exists as a "tools" rather than a "field" of material science research. The main reason lies in that it has not formulated a standard data research strategy from the perspective of material design to raise specific scientific issues.

In this work, the strategy of GLD-CPRs for material design has been established based on material informatics method. In the framework of the strategy, performance is a multivariate function of material genes, and constraints imposed on material system are concreted by GLD-CPRs that are used to narrow the searching space and thus speed the materials discovery. With the help of machine learning algorithm, the proposed strategy highlights the construction of GLD-CPRs, adds new depth and width to our understanding of material gene on performance, thus makes materials design more fast, reasonable, and traceable. This strategy offers a universal solution to the efficient material oriented design, as required by the material genome engineering. Facing the challenge of GLD-CPRs, it is hopeful to broaden the research field of materials science.

6. Acknowledgments

This work is financially supported by the National Key Research and Development Program of China 2016YFB0701100, the Fundamental Research Funds for the Central Universities of China (N160208001) and National Natural Science Foundation of China (No. 51525101).

7. References

[1] Curtarolo S, Hart G, Nardelli M, Mingo N, Sanvito S and Levy O 2013 *Nat. Mater.* **12** 191.
[2] Nosengo N 2016 *Nature* **533** 22.
[3] Xue D, Balachandran P, Hogden J, Theiler J, Xue D and Lookman T 2016 *Nat. Commun.* **7**, 11241.
[4] Lin L, Berger A, Martin R, Kim J, Swisher J, Jariwala K, Rycroft C, Bhown A, Deem M, Haranczyk M and Smit B 2012 *Nat. Mater.* **11**, 633.
[5] Castelli I, Olsen T, Datta S, Landis D, Dahl S, Thygesen K and Jacobsen K 2012 *Energy Environ. Sci.* **5**, 5814.
[6] Yu P, Wu J, Liu S, Xiong J, Jagadish C and Wang Z 2016 *Nano Today* **11** 704.
[7] Fujishima A and Honda K 1972 *Nature* **238** 37.
[8] Ho T 1998 *IEEE T. Pattern Anal.* **20** 832.
[9] Pizzi G, Cepellotti A, Sabatini R, Marzari N and Kozinsky B 2016 *Comp. Mate. Sci.* **111**, 218.
[10] Stengl V, Bakardjieva S and Murafa N 2009*Mater. Chem. Phys.* **114** 217.

IOP Conf. Series: Materials Science and Engineering **746** (2020) 012021 doi:10.1088/1757-899X/746/1/012021

Preparation of Latent Curing Agent for Epoxy Resin by Encapsulation Technology

Xiaojun Yang, Liyang Zhao, Qingyang Li and Limin Ma

College of Material Science and Engineering, Beijing University of Technology, Beijing, China

E-mail: yangxj@bjut.edu.cn

Abstract. The microcapsule-type curing agents DDM-PMMA and IZ-PU were prepared by solvent evaporation method and interfacial polymerization method, respectively. The surface morphology of the two microcapsules was characterized by SEM. The coating efficiency and latency were analyzed by DSC analysis of the one-component adhesive composed of the microcapsule-type curing agent and epoxy resin. Two encapsulation methods were compared in terms of hydrophilicity or lipophilicity of the core material, formation time of the microcapsules, coating efficiency and latency of the microcapsules.

1. Introduction

Epoxy resins are widely used as electrical materials, coating materials, adhesives, aerospace materials and composite materials because of their high adhesion strength, excellent chemical resistance and thermal insulation, etc. [1] The curing agents for epoxy resin, such as amines and anhydrides, are commonly used with high curing activity [2]. However, the high activity curing agent has a short storage period and cannot be used in one-component adhesive or prepreg. The latent curing agents with low curing activity at room temperature, such as dicyanamide and diaminodiphenyl sulfone, can be used in one-component epoxy system, but the reaction temperature required for the curing reaction is often too high to be widely used. [3] In comparison, the microencapsulation of the commonly used highly active curing agent is a feasible and economic method. The curing agent is embedded in a wall material so that the curing agent is separated from the epoxy resin. Thus, the curing reaction of epoxy resin can be controlled by the breakage or permeability of the wall material [4, 5].

Since the microencapsulation technology was reported in the 1930s, several methods for preparing microcapsules have been described, including interfacial polymerization, in-situ polymerization, spray drying, and solvent evaporation, etc. The study on microcapsule-type curing agent has attracted much attention. The spray drying process is simple, but the microcapsules obtained are a mixture of the core material and the wall material, so the active ingredients cannot be completely embedded. Therefore, the spray drying method is not suitable for the preparation of microcapsule-type curing agent for epoxy resin. Interfacial polymerization, and solvent evaporation are common methods for preparing microcapsule-type curing agent. In our previous study [6, 7], a microcapsule-type latent curing agent was prepared by solvent evaporation method with diaminodiphenylmethane (DDM) as the core material and PMMA as the wall material. We also prepared a microcapsule-type latent curing agent with imidazole (IZ) as the core material by interfacial polymerization of triethanolamine (TEOA) and diphenylmethane diisocyanate (MDI). In this paper, the two encapsulation processes and microcapsule-type latent curing agents obtained were analyzed.

2. Results and Discussion

2.1. Encapsulation of Hydrophilic or Lipophilic Core Material

The encapsulation process for solvent evaporation method is relatively simple to control, and many studies have been reported [8, 9]. Encapsulation of the lipophilic core material (e.g., DDM) can be carried out by solvent evaporation. DCM is a commonly used volatile solvent for precipitating wall materials. The stability of the emulsion has an important influence on the core/shell structure of the microcapsules. The microcapsules formed in stable emulsions generally have a smooth surface and a regular spherical appearance, indicating that the oil phase droplets remain stable before the solidification of the microcapsules. However, the precipitation of wall materials depends on the evaporation process of volatile solvent in oil-in-water emulsion or water-in-oil-in-water emulsion. The long time for evaporation makes the diffusion of lipophilic DDM from the oil phase to the aqueous phase non-negligible, resulting in a lower core content. Due to the relatively large variety of hydrophilic core material (e.g., amines curing agent), interfacial polymerization techniques using water-in-oil emulsions may be more applicable for encapsulation.

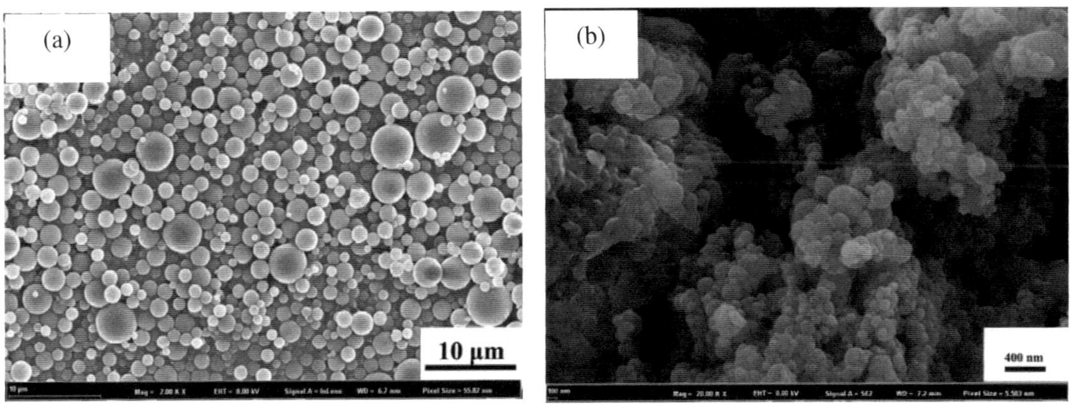

Figure 1. SEM images of DDM-PMMA and IZ-PU microcapsules
(a) DDM-PMMA microcapsules; (b) IZ-PU microcapsules

Imidazole is a commonly used curing agent with high activity. The microencapsulation of imidazole and its derivatives has been reported. Dong et al. [10] and Min et al. [11] prepared the microcapsule of 2-methyl imidazole as the core material by spray drying method. Ham et al. also studied the preparation of microcapsule curing agent for polymer-enclosed imidazole by solvent evaporation [12, 13]. The hydrophilic nature of imidazole makes it more suitable for dispersion in water-in-oil emulsions. Moreover, the interfacial polymerization technique requires relatively short time for the solidification of the microcapsules. However, compared with aqueous phase, the viscous oil phase results in a certain degree of adhesion of the formed microcapsules, but this does not affect its dispersion in epoxy resin. And the residual IZ dispersed on the surface of the adhered microcapsules is difficult to remove with water and ethanol.

Figure 1(a) shows the SEM images of the DDM-PMMA microcapsules. It shows that the DDM-PMMA microcapsules prepared by solvent evaporation method has a smooth surface and a regular spherical appearance. Figure 1(b) shows the SEM images of the IZ-PU microcapsules prepared by interfacial polymerization. It shows that the IZ-PU microcapsules prepared by interfacial polymerization has a relatively small particle size.

2.2. Curing Behavior of the Microcapsule-type Curing Agent for Epoxy Resin

Thermal curing behavior of the DDM-PMMA microcapsules to epoxy resin was examined by DSC under argon atmosphere. The curing exothermic values of DDM-PMMA/EP adhesives with different DDM-PMMA contents were normalized to the values of heat flow per gram of epoxy resin. Figure 2 shows the relationship between the curing exothermic values per gram of epoxy resin and the

DDM-PMMA content. The inflection point of the curve corresponds to the proper DDM-PMMA content of the DDM-PMMA/EP adhesive, indicating complete curing of epoxy resin.

The curing exothermic values of the DDM-PMMA/EP adhesives with the same content of DDM-PMMA microcapsules prepared under different conditions were compared to optimize the preparation parameters. Figure 3 shows the curing exothermic values of the DDM-PMMA/EP adhesives with DDM-PMMA microcapsules as curing agent. The DDM-PMMA microcapsules corresponding to different preparation processes are labeled DPM1, DPM2, DPM3 and DPM4, respectively. The higher the curing exothermic value, the higher the core content of the microcapsule curing agent.

The DSC curves of IZ-PU/EP adhesives are shown in Figure 4. The DSC curve for the original IZ-PU/EP adhesive was marked as 0day, and IZ-PU/EP adhesives after sealed storage for 15days and 30days, were marked as 15d and 30d.

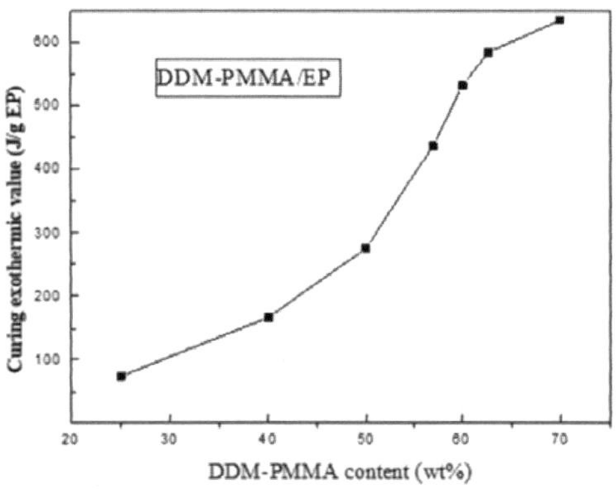

Figure 2. The curing exothermic values of DDM-PMMA/EP adhesive vs. DDM-PMMA content

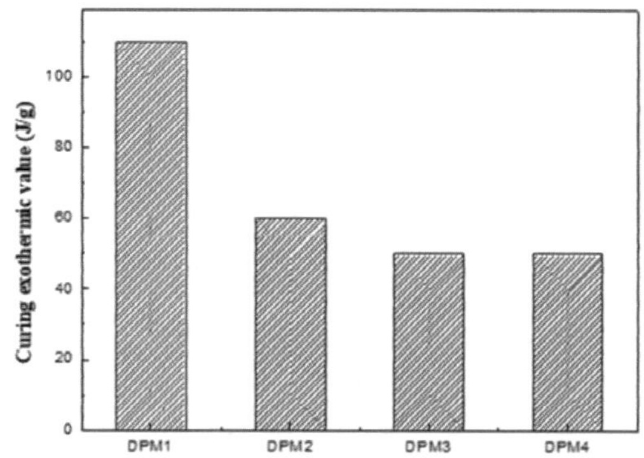

Figure 3. The curing exothermic values of DDM-PMMA/EP adhesive with different DDM-PMMA microcapsules

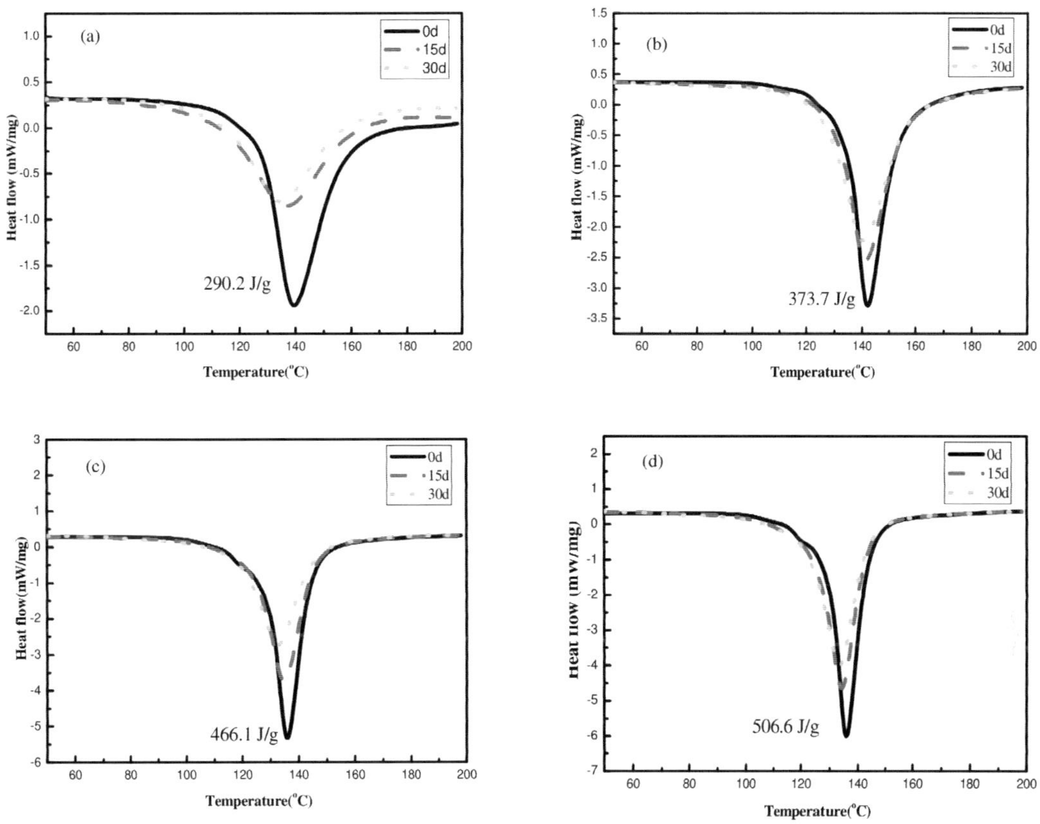

Figure 4. DSC curves of IZ-PU/EP adhesives
(a)-(d) corresponding to the IZ-PU microcapsules of IPM1-IPM4

As shown in Figure 4, the curing exothermic values of 15d and 30d samples are lower than the value of original IZ-PU/EP adhesive. The reduced value is due to the curing reaction of IZ-PU and EP during the storage period. The curing degree of the sample can be estimated by the ratio of the reduced value of the corresponding sample to the original value. The curing degrees of IZ-PU/EP adhesive corresponding to IPM1, IPM2, IPM3 and IPM4 after sealed storage for 15days and 30days are shown in Figure 5.

As discussed above, the residual IZ dispersed on the surface of the adhered microcapsules is difficult to remove by washing. The residual IZ is in direct contact with the epoxy resin, causing a certain degree of curing of the adhesive during storage. However, the one-component adhesive with a curing degree lower than 10% has little effect on the viscosity and applicability of the adhesive.

The original value of curing exotherm and the curing degree of the stored IZ-PU/EP adhesive at room temperature are the main evaluation criteria for evaluating the coating efficiency and latency of IZ-PU microcapsule. As shown in Figure 5, sample IMP2 has a relative low curing degree value after sealed storage at room temperature for 30days, compared with the other three samples. It can be inferred that the shell-core structure of corresponding IZ-PU microcapsule may have the most efficient coating. As the initial curing exothermic value corresponding to sample IMP4 is the highest among the four samples, that is to say, the core content of corresponding IZ-PU microcapsule is higher than the others. Therefore, the coating efficiency and latency of IPM4 might be better than IPM2.

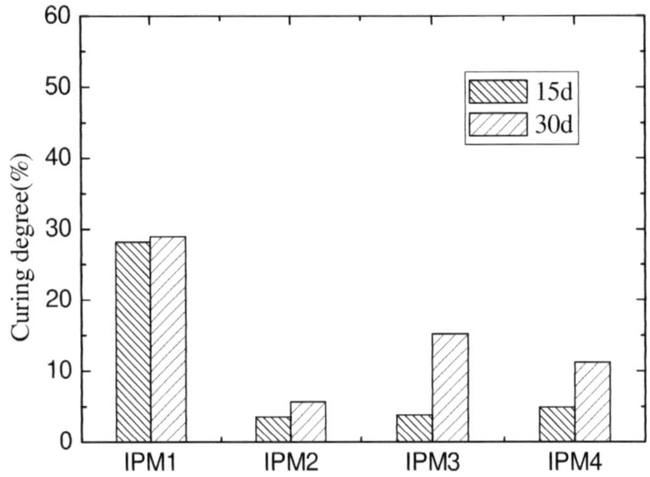

Figure 5. The curing degrees of IZ-PU/EP adhesives after sealed storage for 15days and 30days

3. Conclusions

Solvent evaporation method and interfacial polymerization method were used to prepare the microcapsule-type curing agents DDM-PMMA and IZ-PU, respectively. The application of different encapsulation processes depends on the hydrophilicity or lipophilicity of the core materials. In terms of the latent curing agent suitable for epoxy resin, according to the above results, interfacial polymerization method may be more efficient than solvent evaporation method.

Summarized as follows:

1. Due to the high reactivity of diisocyanate, the interfacial polymerization is rapid, and the shell of the microcapsules is formed quickly, so the diffusion of the core material into the outer phase is much less, compared with solvent evaporation method.

2. Since hydrophilic curing agents for epoxy resin (e.g., amines) are more common than lipophilic curing agents, interfacial polymerization techniques using water-in-oil emulsions may be more applicable for encapsulation.

3. Due to the higher curing efficiency of IZ and the higher core content of IZ-PU microcapsules, the mass fraction of IZ-PU microcapsules required to cure epoxy resin is much lower than that of DDM-PMMA microcapsules. Therefore, the wall material introduced by the microcapsules has a relatively small influence on the mechanical strength of the cured resin.

4. References

[1] Pearce P J, Ennis B C. *J. Appl. Polym. Sci*, 1993, 47 (8): 1401-1409.
[2] Jin F L, Li X, Park S J. *Journal of Industrial & Engineering Chemistry*, 2015,29:1-11.
[3] Xue J Z, Su M K, Zhan Q L. *Advanced Materials Research*, 2011, 150-151 (5): 988-991.
[4] Ahangaran F, Navarchian A H, Hayaty M, et al. *Smart Material Structures*, 2016, 25 (9).
[5] Qi L I, Mishra A K, Kim N H, et al. *Composites Part B Engineering*, 2013, 49 (6): 6-15.
[6] Liyang Zhao, Xiaojun Yang, Qingyang Li, Limin Ma. *J. Appl. Polym. Sci*, 2019, 136 (28), DOI: 10. 1002/APP. 47757.
[7] Xiaojun Yang, Liyang Zhao, Qingyang Li, Limin Ma. *J. Beijing Univ. Technol*, 2020, 46 (2). (online).
[8] Milián Y E, Gutiérrez A, Grágeda M, Ushak S, *Renew. Sustain. Energy Rev*. 2017, 73, 983.
[9] Li M, Rouaud O, Poncelet D, *Int. J. Pharm*. 2008, 363, 26.
[10] Mozaffari S M, Beheshty M H, Mirabedini S M, *Iranian Polymer Journal*, 2016,25:385-394.
[11] Min J S, Jin G K, Shin J S. *J. Appl. Polym. Sci*, 2012, 122 (2): 108-115.

[12] Ham Y R, Lee D H, Kim S H, *et al. Journal of Industrial & Engineering Chemistry*, 2010, 16 (5): 728-733.
[13] Min J S, Shin Y J, Hwang S W, Shin J S, *J. Appl. Polym. Sci.* 2013, 129, 1036.

IOP Conf. Series: Materials Science and Engineering **746** (2020) 012022 doi:10.1088/1757-899X/746/1/012022

Study on Fatigue Limit Test of TCS Stainless Steel

Junlin Ren, Qiang Li, Binjie Wang and Yi Yin

School of Mechanical, Electronic and Control Engineering, Beijing Jiaotong University, Beijing 100044, China
Email: 19126057@bjtu.edu.cn

Abstract. Material fatigue limit plays a vital role in the fatigue life of railway vehicle parts. According to the fatigue test method, the fatigue performance of TCS stainless steel used in railway rolling stock was studied in this paper. The corresponding fatigue limit calculation methods of lifting method and group method were given, and the P-S-N curve of TCS stainless steel under test conditions and fatigue limit under different survival rates were obtained. The results showed that thc fatigue limit of TCS stainless steel was 176.43 MPa under 50% survival rate, 163.21 MPa under 90% survival rate, and 152.44 MPa under 99% survival rate. The results of this paper provide basic data for the application of TCS stainless steel in railway rolling stock.

1. Introduction

The fatigue limit of stainless steel material plays a vital role in the fatigue life of railway vehicle parts. Due to the discreteness of fatigue test data, the relationship between fatigue life and stress level is not a one-to-one single value relationship, but is closely related to survival rate P. A large number of fatigue tests show that the material fatigue limit obeys normal distribution or logarithmic normal distribution when the life is constant. When the stress is constant, the fatigue life obeys logarithmic normal distribution or Weibull distribution.

Before metal materials are used, fatigue limit test should be carried out. Teimourimanesh [1] studied the influence of the stress range lower than the fatigue limit on the fatigue life of materials. Generally, the up-down method is used to test the fatigue limit of materials. Holst [2] proposed the fatigue limit test method based on Bayesian test, which can shorten the time of obtaining the fatigue limit.

Y. Song [3] carried out theoretical and experimental research on fatigue life evaluation of aluminum alloy body of EMU, and proposed three loading test schemes including constant amplitude loading, variable amplitude loading, and random loading. T. Ma [4] studied the fatigue performance of the surface reinforced layer of the axle of high-speed EMU, obtained the fatigue limit of S38C axle steel under the specified confidence and reliability, and drew the P-S-N curve. Q. Wang [5] gave the probabilistic fatigue durability strength of rim and spoke of D1 steel wheel and probability S-N curve containing the ultra-long life. Y. Zhang [6] carried out a study on the fatigue performance and reliability of axle steel with ultra-long life, and proposed two fatigue limit reliability evaluation methods. M. Li [7] analyzed the dynamic stress of dangerous section of EMU axle under random excitation and different line conditions, and carried out axle damage calculation and life evaluation under the impact condition of flat scar of wheels.

J. Zhou [8] studied the rapid prediction of fatigue limit of 20Cr2Ni4A gear steel based on infrared thermal imaging method, and revealed the change rule of damage index of materials during fatigue cyclic loading. S. Guo [9] investigated the method of rapid determination of fatigue limit of metal materials based on temperature distribution characteristics, and obtained the evolution law of heat dissipation rate of AZ31B magnesium alloy under high cycle fatigue load. D. Fu [10] conducted MT transmission fatigue design and test research based on load spectrum, and obtained S-N curve of gear material. W. Xu et al. [11] carried out an experimental study on the fatigue performance of titanium alloy film disk material based on vibration, and provided the fatigue limit of this material sample.

These studies mainly focus on specific materials. One stainless steel co., LTD produced TCS stainless steel material used for railway vehicles under the condition of a certain technology. In order to confirm the fatigue performance of this TCS stainless steel, this paper intends to conduct indoor fatigue test research, which can obtain P-S-N curve of the TCS stainless steel under test conditions and fatigue limit under different survival rates, so as to provide basic data for the application of the TCS stainless steel.

2. Test Method and Sample

2.1. Test Equipment and Loading Conditions
The experiment was completed by Zwick high frequency fatigue testing machine as shown in figure 1. The data collection system is IMC CRONOS Compact data collection system.

Symmetrical cyclic loading was adopted with a loading frequency of 117±1Hz. The temperature of the experiment was 20~25 °C, and the air humidity was 50%~60%.

Figure 1. Sample measurement equipment

2.2. Experimental Sample
According to test requirements and GB/T3075-2008, the shape and size of specimen was designed as figure 2.

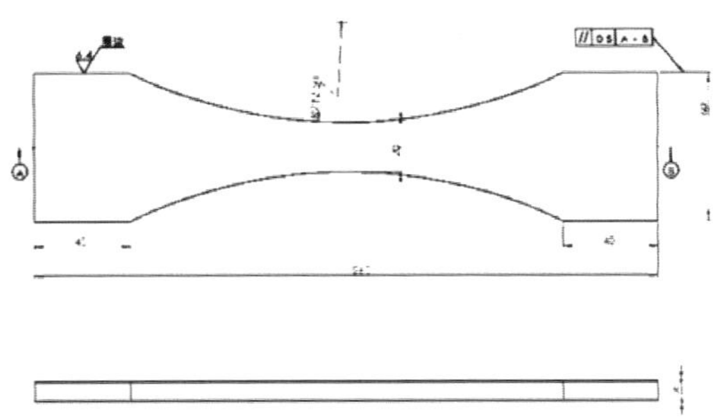

Figure 2. The shape and size of sample (unit: mm)

3. P-S-N Curve Determination Method
P-S-N curve refers to the S-N curve under a given survival rate P and a given confidence. The survival rate represents the failure probability of the sample after a certain number of cycles N under a certain stress level S, and the confidence reflects the credible degree of the event itself.

When measuring P-S-N curve, the combination of group method and up-down method is generally adopted. The low stress level point of the curve (fatigue limit) was determined by up-down method, and the higher stress level of other 3 ~ 4 grades was determined by group method. The number of samples required by the up-down method and the number of samples required by the group method at each stress level depends on the dispersion degree of the test data and the required survival rate and confidence. After statistical analysis, the fatigue limit or safety life of the material under a certain survival rate and confidence can be obtained. The fatigue life and safety life of the materials under the specified survival rate and confidence are drawn on the coordinate paper, and the P-S-N curve with confidence can be obtained by fitting each data point with the curve.

3.1. Up-down Method

The fatigue limit distribution under the specified life cycle 10^7 is measured by the up-down method of small sample. Generally, the stress level of 3~5 grades is taken, and the stress increment is usually chosen as 3%~5% of the predicted fatigue limit. The number of effective samples is no less than 13. A large number of studies show that the fatigue limit obeys normal distribution. It is assumed that n is the total number of pairs of effective samples, m is the stress level of order, n_i is the number of pairs between the stress of order i and i+1, and σ_i the mean value of the stress of order i and i+1. Then the mean value of the fatigue limit $\bar{\sigma}_{-1}$ and its standard deviation s are defined respectively as follow

$$\bar{\sigma}_{-1} = \frac{1}{n}\sum n_i \sigma_i \tag{1}$$

$$s = \sqrt{\frac{\sum n_i (\sigma_i - \bar{\sigma}_{-1})^2}{n-1}} \tag{2}$$

After the minimum paired number and distribution test, the fatigue limit $(\sigma_{-1})_p$ when confidence γ and survival rate P are specified can be obtained from equation (3).

$$(\sigma_{-1})_p = \bar{\sigma}_{-1} + u_p \hat{k} s \tag{3}$$

3.2. Group Method

When the group method is used, the stress level of 3~4 is generally taken. When testing a group of samples under a certain stress level, less samples can be taken when the data dispersion is small, while more samples are needed when the data dispersion is large.

Firstly, the measured fatigue life Ni at this stress level is written in sequence, and the mathematical expectation of the failure rate of the i-th sample is

$$\hat{p}_i = 1 - \frac{1}{n+1} \tag{4}$$

The mean, standard deviation and coefficient of variation of logarithmic fatigue life at this stress level can be determined by equation (5) - (7) respectively.

Mean:

$$\bar{x} = \frac{1}{n}\sum_{i=1}^{n} x_i \tag{5}$$

Standard deviation:

$$s = \sqrt{\frac{\sum_{i=1}^{n} x_i^2 - \frac{1}{n}(\sum_{i}^{n} x_i)^2}{n-1}} \tag{6}$$

Coefficient of variation:

$$C_v = \frac{s}{\bar{x}} \times 100\% \tag{7}$$

It is generally assumed that the fatigue life of the sample obeys lognormal distribution and can be tested on lognormal probability coordinate paper.

Under a certain stress level, the minimum number of samples required for a given confidence can be determined by equation (8).

$$\frac{\delta}{t_r\sqrt{\frac{1}{n} + u_p^2\left(\hat{k}^2 - 1\right)} - \delta u_p \hat{k}} \geq \frac{s}{\bar{x}} \tag{8}$$

In equation (8), δ denotes the relative error limit, n denotes the minimum number of samples required, $\hat{k}$ denotes the standard deviation correction coefficient, t_r denotes the distribution value and is related to confidence γ and number of samples n, is the standard normal deviation related to survival rate. In general, δ takes as 5%, and $\hat{k}$,t_r , u_p can be obtained by referring to the table of fatigue reliability data. Safety logarithmic fatigue life or safety fatigue strength with specified confidence γ and survival rate P can be calculated according to equation (9).

$$\hat{x}_p = \hat{u} + u_p \hat{k} s \tag{9}$$

When the number of samples cannot meet the criterion of equation (8), the unilateral tolerance coefficient k should be used to give the safe fatigue life with confidence γ. The function of the unilateral tolerance coefficient k is to substitute the u_p in equation (8).

$$k = \frac{u_p - u_\gamma\sqrt{\frac{1}{n}\left[1 - \frac{u_\gamma^2}{2(n-1)}\right] + \frac{u_p^2}{2(n-1)}}}{1 - \frac{u_\gamma^2}{2(n-1)}} \tag{10}$$

In equation (10), u_γ is the standard normal deviation related to confidence, which can be obtained by referring to relevant literatures.

3.3. P-S-N Curve Plot

Power function is used to express P-S-N curve equation:

$$\sigma^m N = C \tag{11}$$

Where, are material constants. The logarithm of both sides of this equation is taken, then we get

$$m \lg \sigma + \lg N = \lg C \tag{12}$$

The P-S-N curve is a straight line on the log-log coordinate.

By using the least square method, linear fitting was carried out for the logarithmic safety fatigue life and logarithmic safety fatigue strength under specified reliability and different stress levels.

4. Material Fatigue Performance

4.1. Mechanical Properties of Test Base Material

Table 1 shows the mechanical properties of the tested materials.

Table 1. Mechanical properties of the tested materials

Tensile strength (MPa)	Yield strength (MPa)	Elongation %	Cold bend 180° D=2d	Impact power Akv(J) (Half sample)
≥350	≥480	≥20	Surface perfect	≥34 (-40°C)

4.2. Single-point Test and Data Processing

When the stress of the material is less than 170MPa, no fracture occurs under 1×10^7 cycle load. When the stress of the material is 180MPa and the number of cycles is 2.54×10^6, the material fractures. The fitted line is shown in figure 3. The estimated fatigue limit is 170MPa.

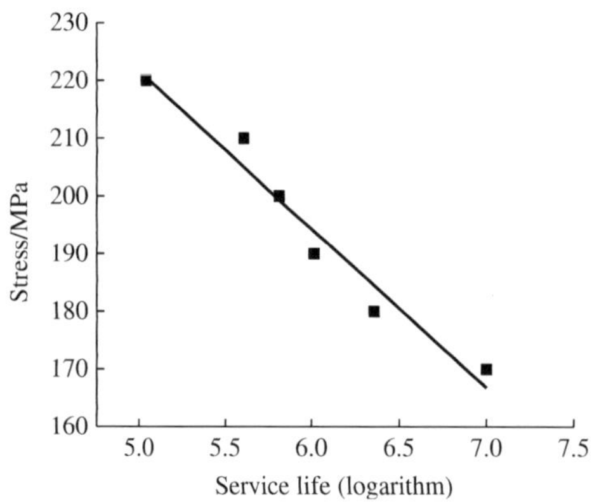

Figure 3. Fitted line of single point method

4.3. Up-down Method Test and Data Processing

It is known that the fatigue limit of the material is about 170MPa. During the test, 190MPa was selected as the stress level for the first specimen test, and the stress level difference should be 10MPa. The fatigue up-down diagram is shown in figure 4. In figure 4, × represents fracture and ○ represents unfractured. The data included 7 pairs, which were used to calculate the fatigue limit, standard deviation and coefficient of variation of the specimen.

$$\bar{\sigma}_{-1} = \frac{1}{n}\sum n_i\sigma_i = 176.43\text{MPa} \tag{13}$$

$$s = \sqrt{\frac{\sum n_i\left(\sigma_i - \bar{\sigma}_{-1}\right)^2}{n-1}} = 9.90 \tag{14}$$

$$v = \frac{s}{\bar{\sigma}_{-1}} = \frac{2.76}{71.57} = 0.0561 \tag{15}$$

Table 2 shows the related coefficients in the process of fatigue test data processing and analysis. It can be seen that when the confidence is 90%, the value of T distribution is relatively stable.

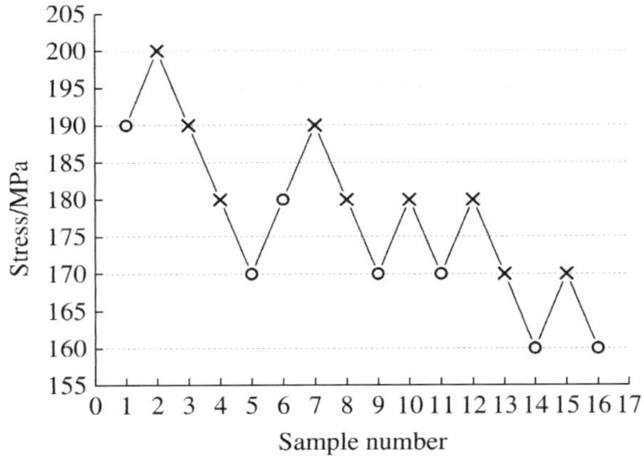

Figure 4. Up-down diagram

The value of k can be obtained by looking up table 2, and when the specified life is 10^7, the conditional fatigue limit is obtained. When confidence is 95% and survival rate is 50%, σ_{-1p}=176.34 MPa. When confidence is 95% and survival rate is 90%, σ_{-1p}=163.21 MPa. When confidence is 95% and survival rate is 99%, σ_{-1p}=152.44 MPa.

Table 2. Mechanical properties of the tested materials

n	6	7	8	9	10	12
Suspicious value test $\left\|(x_i - \bar{x}_j)/s_{xi}\right\|$	1.730	1.800	1.860	1.910	1.960	2.040
Coefficient of variation v tolerance value	0.132	0.139	0.144	0.150	0.154	0.162
Correlation coefficient $r_{\min}$	0.811	0.754	0.707	0.666	0.632	0.576
Standard deviation correction factor $\hat{k}$	1.051	1.042	1.036	1.031	1.028	1.023
When r=90%, p=90%, unilateral tolerance coefficient k	-2.379	-2.244	-2.145	-2.071	-2.012	-1.925
When r=90%, p=99%, unilateral tolerance coefficient k	-4.408	-3.822	-3.658	-3.537	-3.442	-3.301
When confidence is at 90%, the value of T distribution	1.943	1.895	1.860	1.833	1.812	1.782

4.4. Group Method Test and Data Processing

We arrange the fatigue life in descending order and take the corresponding list of logarithms to calculate the corresponding survival rate. Three stress levels of 180MPa, 195MPa and 205Mpa were selected for group test. The data are shown in table 3. The mean value, standard deviation and coefficient of variation of the group method data were calculated respectively and listed in table 4.

Table 3. Mechanical properties of the tested materials

Stress level σ	Number i	Fatigue life N_i (times)	Logarithmic fatigue life $\lg N_i$	Survival rate P_i
205MPa	1	409596	5.612355	0.875
	2	369790	5.567955	0.750
	3	350203	5.544319	0.625
	4	340308	5.531872	0.500
	5	315138	5.498500	0.375
	6	307565	5.487936	0.250
	7	278050	5.444122	0.125
195MPa	1	654878	5.816160	0.875
	2	602486	5.779946	0.750
	3	585411	5.767460	0.625
	4	584614	5.766869	0.500
	5	582456	5.765263	0.375
	6	504567	5.702918	0.250
	7	479596	5.680875	0.125
180MPa	1	2116902	6.325701	0.875
	2	2137565	6.329919	0.750
	3	2233302	6.348947	0.625
	4	2335792	6.368434	0.500
	5	2335835	6.368442	0.375
	6	2589640	6.413239	0.250
	7	2597533	6.414628	0.125

Table 4. Data processing results of group method

Stress level σ (MPa)	Mean $\bar{x}$	Standard deviation s	Coefficient of vibration ν
205	5.5267	0.0514	0.0093
195	5.7542	0.0430	0.0075
190	6.3670	0.0334	0.0052

The P-S-N curve equation and P-S-N curve (figure 5) can be obtained by using the stress value and logarithmic safety life obtained by group method and up-down method mentioned above.

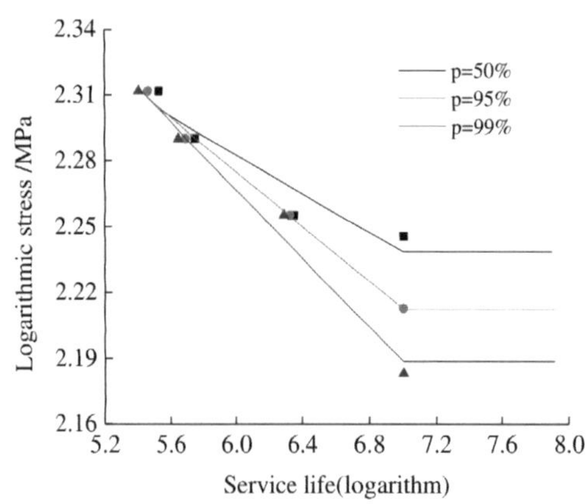

Figure 5. P-S-N curve

IOP Conf. Series: Materials Science and Engineering **746** (2020) 012022 doi:10.1088/1757-899X/746/1/012022

5. Conclusions

According to the fatigue test method, the fatigue performance of TCS stainless steel for railway vehicles was studied, and the P-S-N curve and fatigue limit of TCS stainless steel under the test condition were obtained. The results showed that the fatigue limit of TCS stainless steel was 176.43 MPa under 50% survival rate, 163.21 MPa under 90% survival rate, and 152.44 MPa under 99% survival rate. This data provided basic data for TCS stainless steel applications.

6. Acknowledgments

This work was supported by National Key R & D Program (Grant No. 2016YFB1200506-21) of China and National Natural Science Foundation of China (Grant No. 11790281, 51575036).

7. References

[1] Teimourimanesh S, Nilsson F. Effects of cycles below the fatigue limit on the life of a high strength steel. Journal of Testing and Evaluation. 2002, 25 (1): 79-89.

[2] Holst A. Optimal test planning for high cycle fatigue limit testing. Annals of Operations Research, 2015, 224 (1): 101-110.

[3] Song Y. Evaluation Theory and experimental study of emu carbody fatigue life. Southwest Jiaotong University. 2016.

[4] Ma T. The Study on Fatigue of high-speed EMU axle surface reinforcement layer. Beijing Jiaotong University. 2018.

[5] Wang Q. Experimental study on fatigue and fracture properties of D1 wheel steel. Southwest Jiaotong University. 2011.

[6] Zhang Y. Research on fatigue property and reliability of railway axel steels in giga-cycle regime. Southwest Jiaotong University. 2006.

[7] Li M. Study on fatigue test method of the EMU axles. Southwest Jiaotong University. 2015.

[8] Zhou J. Fast prediction of fatigue limit of 20Cr2Ni4A gear steel based on infrared thermography. Jiangsu University of Science and Technology. 2019.

[9] Guo S. Study on rapid determination for metallic material fatigue limit based on the temperature characteristics. Taiyuan University of Technology. 2012.

[10] Fu D. The manual transmission fatigue design and test study based on road load data and S-N curve. Chongqing University. 2006.

[11] XU W, YANG X., GUO G. Experimental study of titanium alloy film disc fatigue performance of based on vibration fatigue measurement. Journal of experimental mechanics, 2018, 33 (2): 239-244.

IOP Conf. Series: Materials Science and Engineering **746** (2020) 012023 doi:10.1088/1757-899X/746/1/012023

The Effect of Temperature on Properties of CNTs Grown by Chemical Vapor Deposition

Hui Li[1]

[1]Shannxi engineering research center of controllable neutron source, Xijing University, Xian, 710123, China
Email: lzlihui@163.com

Abstract In this article, the carbon nanotubes (CNTs) grown at different temperature by thermal chemical vapor deposition(T-CVD) have been reported, while the effect of growing temperature and catalyst annealing on the quality of CNTs has also been investigated. The longest CNT bundles were grown at the temperature of 730°C by Scanning Electron Microscopy (SEM) images and the better quality of CNT bundles were achieved at higher temperatures of 850°C by Raman spectroscopy.

1. Introduction

CNTs have been widely reported in the gas pollution, water treatment or biomaterial sensing applications due to their small size, relative chemical inertness and large specific surface area. There are also various reports on CNTs in electrical applications because of its extraordinary electrical, mechanical and thermal properties [1-8].

Although many CNTs synthesis methods are available in the literatures [9-12], chemical vapor deposition (CVD) and plasma enhanced chemical vapor deposition (PECVD) are still the most common used methods [13, 14]. However, the CNTs synthesis temperature is very high, normally varies from 650 to 900°C. This high temperature which can cause serious damages on the electronic components in integrated circuit(IC), become a significant barrier to handle and integrate CNT with microelectronic circuits or devices to form a completed system [15-17].

Therefore, a systematic study of the temperature parameter on both the length and quality of CNTs arrays grown by thermal CVD is carried out in this paper. The purpose of the study is to understand how the various synthesis parameters will affect CNTs growth, and determine which synthesis parameters can be used effectively to control the length and quality of the resulting CNTs. It is of great significance in the CNTs growth and silicon based Micro-electro-mechanical systems (MEMS), which will be integrated into through-wafer interconnects and biosensor [18-20].

2. Experimental

A low pressure thermal chemical vapor deposition (LP-TCVD) systems which consists of a 4 inch barrel quartz tube furnace and mass flow controllers was used to synthesis CNTs on Si substrate. 2 μm thick of silicon dioxide was first formed on the Si substrate before an electron beam evaporator was used to deposit a uniform layer of iron (Fe) which acted as the catalyst. After the catalyst pretreatment, C_2H_2 was introduced into the quartz tube to initiate the growth of CNTs. The flow rate of the process gas H_2:Ar:C_2H_2 was fixed at the ratio of 1:4:1, respectively.

The surface morphology of the as grown CNTs on the Si wafer substrate was characterized using scanning electron microscopy (SEM). The root mean square (RMS) roughness of the substrate surface and the mean diameter of the catalyst particles after annealing treatment without CNTs growth were measured. Raman spectroscopy (a 633 nm HeNe laser excitation, Renishaw Raman Microprobe-RM1000) was also used to study and characterize the graphitic ordering of the growing CNTs.

3. Results and Discussion

The CNTs growth was investigated as a function of the CVD growth temperature in the range of 650°C to 900°C. Fig. 1 showed a plot of CNTs length vs. growth temperature, and it could be seen that CNTs length was very temperature dependent, as a slight increased in the temperature range from 700°C to 730°C could increase the CNTs length by about 100 μm. It was also found that the drop in overall vertical CNT length was corresponded to the drop in the CNTs density, which was evident of the crowding effect that determined the overall vertical CNTs length [21].

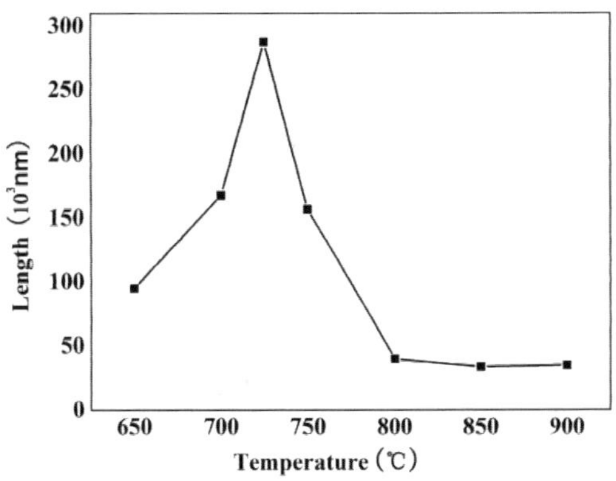

Figure 1. The CNTs length grown at different temperature by chemical vapor deposition

The effect was more noticeable when CNTs synthesized at the temperature of 700°C, 800°C and 900°C compared with Fig. 2. The crowding effect could be explained by the differences in Fe catalyst particle density and the resulting CNTs density because of different temperature. The influence of growth temperature on the morphology of the Fe catalyst particles density could be investigated by simulating the condition of the catalyst layer. To investigate the reason, the prepared substrates with deposited Fe catalyst layer underwent annealing in vacuum with three different temperatures of 700°C, 800°C and 900°C without introducing acetylene gas into initiate CNTs growth. The surface topography of the resulting catalysts surface after annealing was then studied by SEM and RMS roughness.

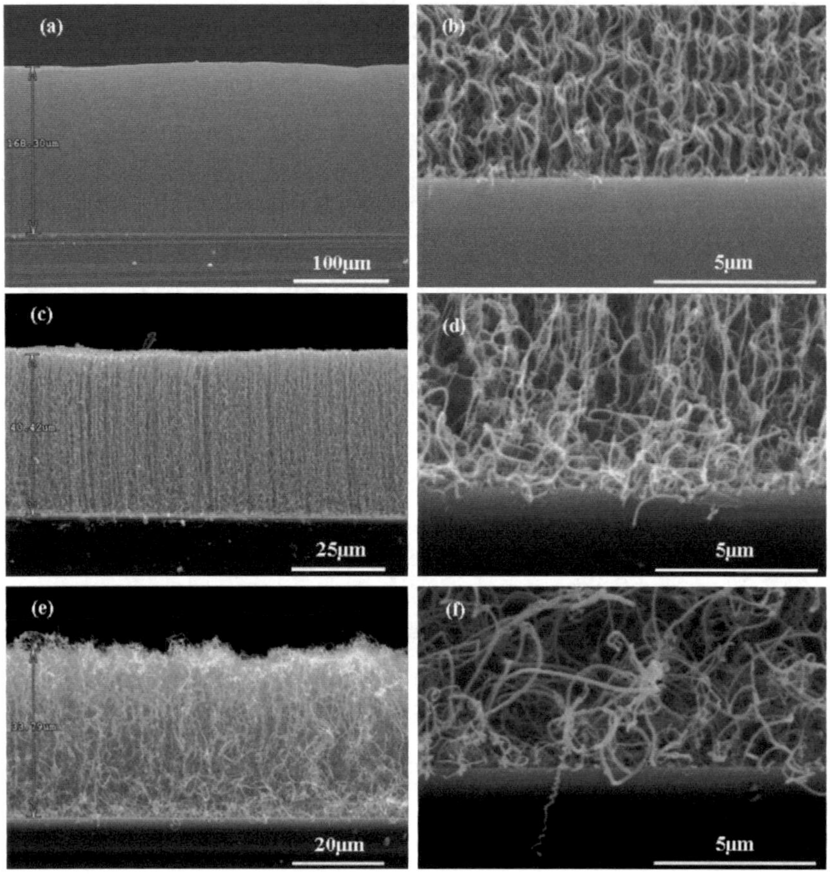

Figure 2. Cross section SEM images of CNTs at different growth temperatures; (a) 168 μm long CNTs grown at 700°C (c) 40 μm long CNTs grown at 800°C, (e) 35 μm long CNTs grown at 900°C; The SEM images taken at the interface between the substrate and CNTs grown at (b) 700°C, (d) 800°C, (f) 900°C

The SEM images (Fig.2) showed that the RMS roughness (Fig.4) and the mean catalyst particles diameter (Fig.3) on the catalyst surface increased with increasing annealing temperature, which was similar to some results made by other papers[21, 22]. This was a result of a coalescence phenomenon, where the mobility of nanoparticles increased on the surface and agglomerated to form larger particles which reducing the surface energy as the result of temperature increasing [22, 23].

Higher temperatures could encourage the formation of bigger Fe catalyst particles which were less uniformly distributed with lower catalyst particles density. At higher temperatures of 800°C and 900°C, the catalyst particles were less dense, thus the resulting CNTs growth was also less dense. The higher rate of reaction and increase in the dissolution, diffusion and precipitation of carbon provided by higher temperatures were not ignored, but this effect was seriously weighed down by the decrease in CNTs density. The low CNTs density allowed more space for the CNTs to grow laterally and in random directions as they suffered from a lack of crowding effect or mechanical leaning between neighboring CNTs. As a result the overall vertical length was much shorter for higher temperatures. Conversely the CNTs synthesized at 700°C formed from dense Fe catalyst particles were denser, and more aligned which explained the long overall vertical length of 168μm. This was similar what was being reported in several papers [15, 20], which investigated the effect of temperature on CNTs growth.

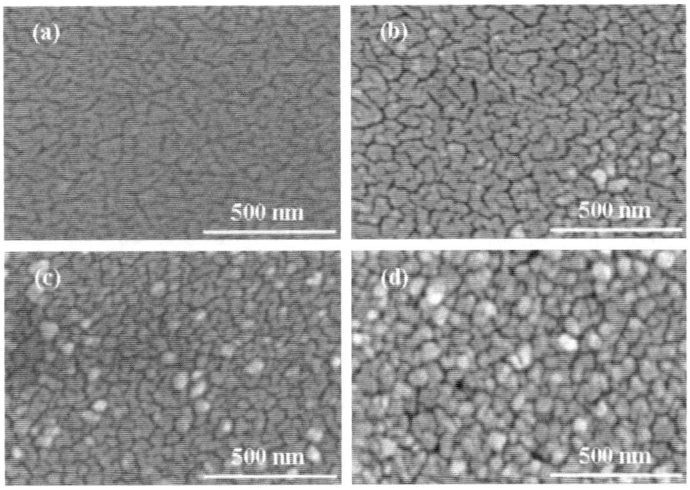

Figure 3. The catalyst surface SEM images (a) before annealing and after annealing pretreatment of at (b) 700°C, (c) 800°C, (d) 900°C.

It was widely accepted that the CNTs diameter was determined by the catalyst particle diameter. However, it was very difficult to directly correlate the CNTs diameter to measured particle diameter because the diameter of the CNTs was so small, and the variation in CNTs diameter could not be verified within the limits of the SEM pictures. It was observed that the distribution of CNTs diameter was wider at 900°C, with some CNTs of significantly bigger diameters (Fig. 3f). It can be stated that these thicker CNTs were formed from the bigger size catalyst particles that coalescence at the higher temperature of 900°C. The larger surface of larger Fe particle permitted more adsorption of carbon atoms, and the diffusion rate increased with the higher temperature and this result in large diameters CNTs [23].

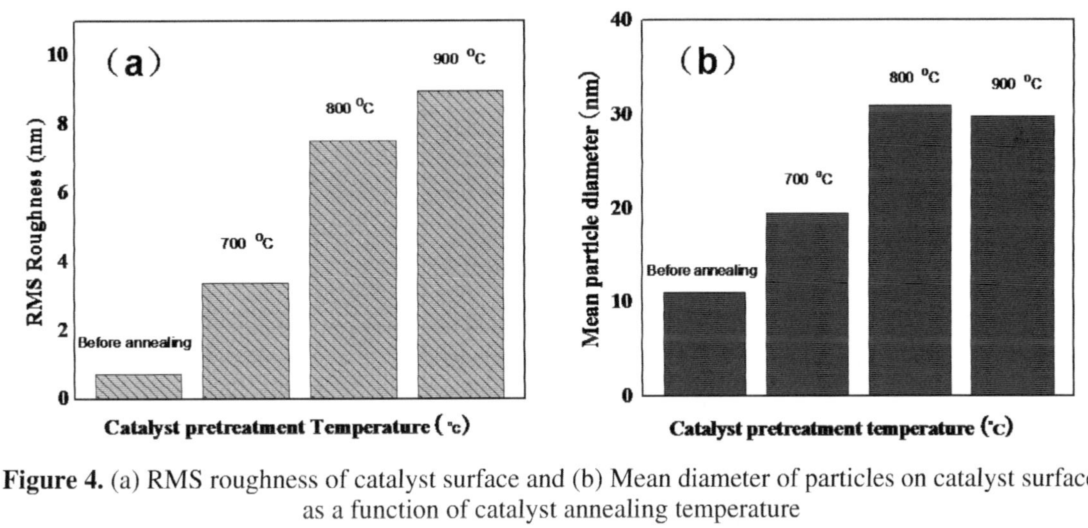

Figure 4. (a) RMS roughness of catalyst surface and (b) Mean diameter of particles on catalyst surface as a function of catalyst annealing temperature

Raman characterization had demonstrated that the crystallinity of graphitic CNTs improved with increasing temperature, and the degree of defectiveness in the CNTs decreased with the growth temperature which was also reported in other papers [15, 17, 21]. Fig.5 showed the relationship of the peak intensity ratio of the G band and the D band as a function of growth temperatures. As the growth temperatures increased, the Raman I_G/I_D ratio reaching a maximum at 850°C showed that highly ordered graphitic tubular structures of the CNTs grown at this temperature.

As the growing CNTs were mainly multi-wall CNT (MWCNT) at lower temperatures, the radial breathing mode of single-wall CNT (SWCNT) started to appear in the lower parts of the spectrum at the highest temperature of 900°C (Fig.5). This point to the presence of SWCNT at higher temperature similar to that reported in some studies [21]. The high temperature at 900°C could have provide the needed activation energy for the nucleation of SWCNT, and the high surface roughness of the catalyst surface after annealing at 900°C (Fig.4) could also help to stabilize the catalyst particle with smaller size and promote the growth of SWCNT [21]. The results show that not only the density and length of CNTs growth by the thermal CVD system could be controlled by adjusting the temperature, but also the type, structure and quality of CNTs is also greatly affected by the temperature. Vertically aligned MWCNT with the longest length were best synthesized at around 730°C while CNTs with better quality were achieved at higher temperatures of 900°C.

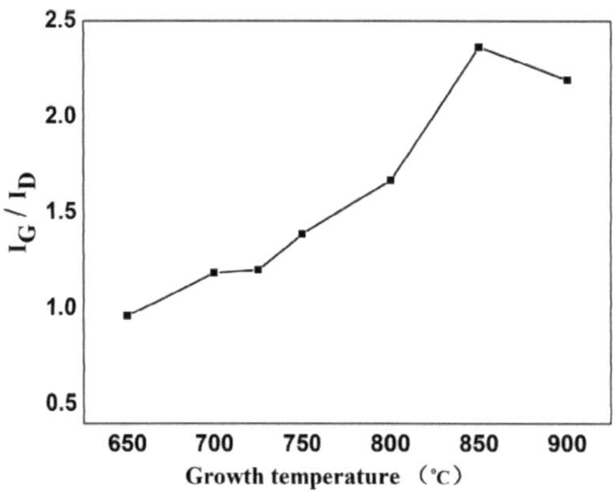

Figure 5. Peak intensity Raman ratio I_G/I_D of CNTs grown at different temperatures

4. Conclusion

In summary, the CNTs grown at different temperature by thermal chemical vapor deposition have been reported, and the surfaces of catalysts annealed at different temperature were investigated. The SEM pictures and the Raman characterization demonstrated that the longest of CNTs bundle were grown at the temperature of 730°C, and the better quality were achieved at higher temperatures of 900°C by chemical vapor deposition.

5. References

[1] Y. Cui, C. M. Lieber, Science. **291** 851-853 (2001).
[2] Q. Li, C. Liu, S. Fan. Nano Lett. **9** 3805-3809 (2009).
[3] P. Gao, H. Li, Q. Zhang, et al, Nanotechnology. **21** 095202 (2010).
[4] L. Ding, S. Liang, T. Pei, et al: Appl. Phys. Lett. **100** 263116 (2012).
[5] G. Q. Han, J. H. Shen, X. X. Ye, et al. Mater. Lett. **181**300-304 (2016).
[6] R. Liu, R. Bai, J.Yi., et al: Journal of Alloys and Compounds. **26** 81-87 (2017).
[7] N. F. Khusnun, A. A. Jalil, S. Triwahyono. et al, Powder Technology. **327** 170-178 (2018).
[8] P. Chen, S. Jing, Y.Chu. et al, Journal of Alloys and Compounds. **730** 42-46 (2018).
[9] M. Zhou, Y. Mai, H. Ling, Mater. Res. Bul. **97** 537-543 (2018).
[10] S. Iijima: Nature. **56** 354 (1991).
[11] D. S. Bethune, R. D. Johnson, J. R. Salem: Nature. **363** 605 (1993).
[12] A. Thess, R. Lee, P. Nikolaev, et al: Science. **273** 483 (1996).
[13] J. H. Hafner, M. J. Bronikowski, B. R. Azamian, et al: Chem. Phys. Lett. **296** 195 (1998).
[14] P. Nikolaev, M. J. Bronikowski, R. K. Bradley, et al: Chem. Phys. Lett. **313** 91 (1999).
[15] C. J. Lee, J. Park, Y. Huh, et al: Chem. Phys. Lett. **343** 33 (2001).
[16] Y.T. Lee, J. Park, Y. S. Choi, et al: J. Phys. Chem. B.**106** 7614 (2002).
[17] O. A. Nerushev, R. E. Morjan, D. I. Ostrovskii, et al: Physica B. **323** 51 (2002).
[18] T. Xu, Z. Wang, J. Miao, et al: Appl. Phys. Lett. **91** 042108 (2007).
[19] T. Xu, J. Miao, H. Li, et al: Nanotechnology. **20** 295303 (2009).

IOP Conf. Series: Materials Science and Engineering **746** (2020) 012023 doi:10.1088/1757-899X/746/1/012023

[20] X. Pan, C. W. Tan, J. Miao, et al: Thin Solid Films. **517** 4905 (2009).
[21] H. Cui, G. Eres, J. Y. Howe, et al: Chem. Phys. Lett. **374** 222 (2003).
[22] E. Terrado, E. Muñoz, W. K. Maser, et al: Diam. Relat. Mater.**16**1082 (2007).
[23] C. Ducati, I. Alexandrou, M. Chhowalla, et al: J. Appl. Phys. **92** 3299 (2002).

Optimal Combined Process for Treating Oily Wastewater Using Membrane Distillation Technology

Xiangxu Cui[1], Yunzhi Ge[1], Ruoting Yu[1], Yan Dong[1,*] and Xiaodong Dai[1]

[1]Shengli College China University of Petroleum, Dongying 257061, China
[*]Corresponding: 24224342@qq.com

Abstract. The membrane distillation technology used in this paper is a new type of wastewater treatment technology, which is now mostly used in seawater desalination and wastewater treatment. The traditional oily wastewater treatment methods have problems such as large energy consumption and low removal rate. It is proposed to use vacuum membrane distillation technology to use a large amount of waste heat in wastewater as the driving force for mass transfer to treat oily wastewater, which not only saves energy, but also effectively counteracts the problem. The solid particles, hydrocarbons and other pollutants contained in the wastewater are intercepted, so that the treated water quality reaches a certain degree of purity and meets the discharge standards. In this paper, the membrane microporous membrane used in the process of membrane distillation is targeted at the problem of membrane pollution Improved to achieve the purpose of reducing costs and increasing membrane life.

1. Introduction
At the current stage, in the treatment of oily wastewater, the traditional separation technology has been relatively developed, but there are still some aspects that cannot be controlled well. The vacuum membrane distillation technology used in the process described in this paper is a new separation technology combining membrane technology and distillation technology [1-3]; the main components of this technology are membrane and vacuum pump, which operate in the membrane distillation process It is also relatively simple, and does not require high environmental conditions, and can operate normally at low temperature and normal pressure. Oily wastewater contains a large amount of solid particles, hydrocarbons, inorganic ions, bacteria and other substances, so the water quality of oily wastewater does not meet the requirements of wastewater. Emission standards are not easy to handle. Traditional processing technologies have problems such as high cost, low removal rate, and large energy consumption. In industrial production, a large amount of waste heat is often generated. Vacuum membrane distillation is used to treat oily wastewater. Its operating temperature is low and can be used. The effective use of waste heat in oily wastewater as a heat source provides a driving force for mass transfer and has important social and economic benefits for the development and utilization of new energy.

2. Principles and Advantages of Vacuum Membrane Distillation Technology

2.1. Principles of Vacuum Membrane Distillation Technology
Vacuum membrane distillation is a membrane separation process that uses a hydrophobic membrane with the vapor pressure difference between the two sides of the membrane as the driving force for mass transfer [4]. The hydrophobic microporous membrane used in the membrane distillation process can separate aqueous solutions of different temperatures on both sides of the membrane. Separately,

one side is connected with the hot solution to be treated, and the other side is evacuated, so under the effect of the steam pressure difference on both sides, the hot side water vapor will enter the vacuum side through the membrane hole to condense, thereby achieving the concentration of the solution And separation [5]; this principle is somewhat similar to the condensation of water vapor into water during ordinary membrane distillation. At the same time, it applies membrane technology, so it is called a membrane distillation process.

2.2. *Advantages of Vacuum Membrane Distillation Technology*
Compared with other wastewater treatment, vacuum membrane distillation technology has the following advantages: (1) the hydrophobic microporous membrane used has good hydrophobicity and non-polarity, which can reduce the interaction between membrane and water; (2) the membrane Corrosion resistance and salt resistance, so it can treat highly salty and corrosive wastewater; (3) Its structure is relatively simple, and the environmental conditions are not harsh; the vacuum on the vacuum side causes the pressure difference on both sides to increase. Compared with other membrane distillation forms, vacuum membrane distillation technology has the advantages of large membrane flux and high processing efficiency.

3. Film Material
The material, pore size, thickness, and porosity of the membrane directly determine the working efficiency of the membrane. Studies have shown that the optimal thickness of the membrane is 30 ~ 60μm, and the pore size of the membrane is mostly between 0.1 ~ 1μm. The membrane is an organic polymer membrane, and its membrane materials mainly include polytetrafluoroethylene (PTFE), polypropylene (PP), polyvinylidene fluoride (PVDF), etc. [8].

Performance analysis of organic film materials

Membrane material	Surface Tension $/10^{-3}N \cdot m^{-1}$	Thermal conductivity/ $W \cdot m^{-1} \cdot k^{-1}$	Thermal stability	Corrosion resistance	Hydrophobic
PTFE	9-20	0.25	Stronger	Stronger	Strong
PVDF	30	0.19	Strong	Strong	General
PP	30	0.17	Stronger	Stronger	Stronger

Comprehensive consideration, this process uses a polyvinylidene fluoride (PVDF) flat film as the base film, and applies surface reduction method to prepare thin film CNTs (carbon nanotubes) with a metal layer on the surface, and prepares a transition metal loaded on only one surface of the polymer film Homogeneous film of additive TiO_2.

4. Application in the Treatment of Oily Wastewater
In the process of oilfield mining and chemical production, a large amount of oily waste water will be generated. The components of oily waste water are more complex and generally contain oil, emulsifiers, fatty acids and other components. Traditional oily wastewater treatment methods such as chemical, mechanical, and thermal methods have problems such as high cost, difficulty in operation, and low efficiency [6]. The vacuum membrane distillation technology is introduced here. According to the surface characteristics of hydrophobic microporous membranes, Optimize the operation mode to achieve high-efficiency separation and realize the treatment of oily wastewater [9]; some studies have shown that membrane distillation technology can effectively separate organic substances such as oil and fat in oily wastewater. First, three polypropylene capillary membranes were selected and the direct contact type was used. Membrane distillation and vacuum membrane distillation treat olive oil industrial wastewater. The permeation flux of direct contact membrane distillation is 6.5L / ($m^2 \cdot h$), and the permeation flux of vacuum membrane distillation is 19L / ($m^2 \cdot h$). The removal rate of pollutants is as high as 99.6% [10]; it can be seen that it is efficient and feasible to use membrane distillation technology to treat oily wastewater.

IOP Conf. Series: Materials Science and Engineering **746** (2020) 012024 doi:10.1088/1757-899X/746/1/012024

5. Membrane Fouling

Membrane fouling is an important factor affecting membrane flux and membrane service life, and is a major problem that affects membrane distillation efficiency and stability for a long time. Fouling is usually formed on the surface of the membrane or inside the membrane pores, which will cause the membrane's permeability to decrease. If it is not handled in time, it will cause damage to the membrane, which indirectly increases the cost of oily wastewater treatment; the main form of membrane pollution for inorganic pollution, organic pollution and microbial pollution.

5.1. Inorganic Pollution

Inorganic pollution generally refers to inorganic fouling. Inorganic fouling is caused by the precipitation of minerals in wastewater. In addition, cement and corrosion products in wastewater can also form inorganic fouling on the membrane. For example, when treating high-salt wastewater, when the soluble ions in the medium exceed the equilibrium solubility product, inorganic salt scale will form. In the process of membrane distillation, the main reason for the formation of inorganic fouling is that with the progress of membrane distillation, the water in the original solution continuously evaporates, so that the solution is in a supersaturated state and the temperature changes continuously. Physical cleaning, chemical cleaning, and operating conditions are carried out in three aspects.

5.2. Organic Pollution

Organic pollutants usually refer to the adsorption and deposition of specific organic substances on the surface of the hydrophobic membrane, and the gel formed on the surface of the membrane. These organic pollutants will not be easily removed without the use of corresponding chemical reagents. In addition, the formation of organic pollutants can also be controlled by physical control methods, heating pretreatment methods, acidic organic liquid washing methods, and directly changing the hydrophilicity of the membrane surface. In the treatment of oily wastewater, in order to prevent petroleum derivatives from Pollutants are formed on the sediment, and the wastewater can be ultra-filtered in advance.

5.3. Microbiological Pollution

The main biological pollution refers to the pollution of microbial metabolites, which is caused by the growth and accumulation of microorganisms on the membrane surface. The high temperature and high salinity of the concentrated liquid during the membrane distillation process are not conducive to the growth of microorganisms, so the application of membrane distillation technology during the treatment of wastewater, there are fewer biological pollutants produced on the membrane, but biological pollutants still exist on the inevitable membrane. The most commonly used fungicides and descaling agents for treating biological pollutants are H_2O_2 and NaClO [11, 12].

6. Specific Introductions to This Optimized Process

6.1. Process Flow Chart

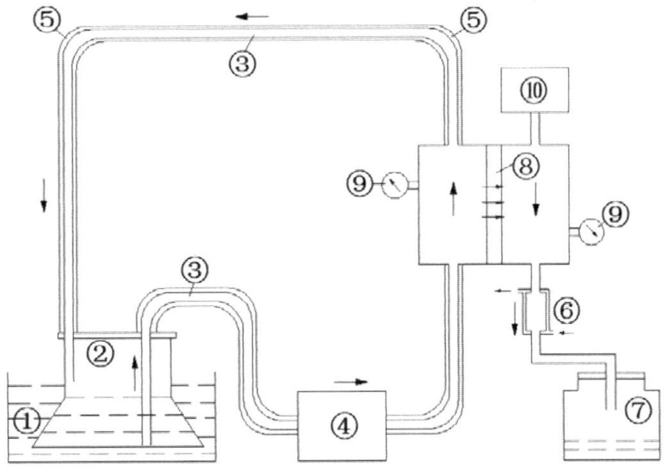

Figure 1. 1. Water bath heating pot; 2. Storage sewage container; 3. Pipe; 4. Peristaltic pump; 5. Insulation layer; 6. Condenser; 7. Purified water collection container; 8. Metal film; 9. Pressure gauge; 10. Vacuum pump

6.2. Main Technologies and Performance Advantages of This Process
This process is mainly applied to the treatment of oily wastewater in oil fields, and is also suitable for oily industrial wastewater with a certain temperature.

6.2.1. Vacuum membrane distillation technology
Vacuum membrane distillation technology (VMD) uses a vacuum pump to maintain a negative pressure state on the permeate side of the membrane, thereby increasing the vapor pressure difference between the two sides of the membrane to improve membrane permeability.

6.2.2. Metal oxide functional composite film
(1) Polyvinylidene fluoride flat film has good abrasion resistance, mechanical and mechanical properties, and has good resistance to pollution and radiation; (2) CNTs (carbon nanotubes) can effectively improve the retention of composite films, as The intermediate transition layer can provide a membrane structure with high porosity and pore density; (3) TiO_2 in the metal oxide functional composite film used in this process has the ability to degrade pollutants, and under certain conditions, catalytic degradation remains Membrane fouling particles on the membrane surface or inside the membrane pores can slow down membrane fouling and enhance the anti-pollution performance of the composite membrane.

6.3. Specific Implementation Plan
(1) First set the raw material liquid storage device and place it in a water bath, in order to keep it at a certain temperature to complete the membrane distillation process; (2) The raw material liquid storage device is directly connected to the peristaltic pump. The peristaltic pump is This experiment provides power to enable the raw material liquid to enter the membrane module smoothly and complete the cycle; (3) Assemble the membrane module, install the metal membrane, one side of the membrane is the feed side, and the other side is the water vapor transmission side; the membrane module After installation, it is connected to the other end of the peristaltic pump; (4) pressure gauges are installed on the feed side and water vapor transmission side of the membrane module, and pressure gauges are installed on the feed side and water vapor transmission side of the membrane module; (5) membrane

The lower part of the water vapor transmission side of the module is connected to the cooling collection device; (6) The upper side of the water vapor transmission side of the membrane module is connected to the vacuum pump; (7) The upper end of the feed side is connected to the return liquid storage device to form a circulation device; (8) Open the peristalsis Pump, the material liquid begins to pass through the membrane module, most of the material liquid will return to the circulation device, a small part of the liquid enters the water vapor transmission side through the membrane hole in the form of water vapor; (9) the water vapor transmission side is connected to the Vacuum pump To maintain a certain pressure difference on both sides, determine the most suitable pressure difference through the pressure gauges on both sides to achieve the most optimal and reasonable treatment effect; (10) Water vapor flows through the serpentine condenser to condense into water and flows into the purification Water collection device; (11) In order to prevent the heat loss of the material liquid during the circulation, the heat insulation layer is covered outside the circulation pipeline; (12) The material liquid will be in the process of the material liquid passing through the membrane module and continuously circulating Gradually becomes a concentrate.

6.4. Application and Prospect of the Process

This process uses a vacuum membrane distillation technology combined process, which uses a large amount of low-grade heat in oily wastewater as the driving force to treat sewage. Among them, waste heat as low-grade thermal energy is often easily ignored. Traditional oily wastewater treatment methods have large energy consumption and costs high, incomplete treatment and other problems, using this combination process, reasonable and effective use of low-grade thermal energy, reducing the use of other energy sources and unnecessary waste, energy saving and environmental protection; improved metal oxide functional composite film will be effective It can slow down and control the pollution of membrane and prolong the service life of membrane. The adoption of this process is not only in line with the concept of contemporary energy saving and environmental protection, but also can greatly reduce the treatment cost of oily wastewater.

7. References

[1] Sun, N.; Li, J. F.; Guo, S. J.; etal. Research status and progress of membrane distillation in wastewater treatment .J. Industrial water and wastewater, 2018, 49 (1): 7-11.

[2] Liu, Y. J.; Wang Y. S.; Han, J. T.; etal. Research and application progress of membrane distillation technology. J. Progress in Chemical Industry, 2018, 37 (10): 3726-3736.

[3] Huan, G. L.; Du, Q. Y.; Wang, W. Research status of membrane distillation technology .J. Journal of Tianjin Polytechnic University, 2009, 28 (4): 12-18.

[4] Yan, J. M. Study on transfer mechanism of membrane distillation and optimization of membrane module .D. Beijing University of Chemical Technology, 2001.

[5] Ding, Z. W. Study on the performance and dynamic process of membrane module for membrane distillation .D. Beijing University of Chemical Technology, 2001.

[6] Ma, Y. C.; Li, Y. J. Application of vacuum membrane distillation in wastewater treatment .J. Chemical Engineering and Equipment, 2018, (08): 334-335.

[7] Duan, M. X.; Wang, J. B.; Wang, J.; etal. Vacuum membrane distillation and membrane absorption and their integrated processes for treating high ammonia nitrogen wastewater .J. Journal of Environmental Engineering, 2015, 9 (2): 659-664.

[8] Shen, J.; Wang, Z. B.; Research progress and development trend of separation membrane materials .J. Guangzhou Chemical Industry. 2018, 46 (22): 17-20.

[9] CARNEVALE M C, GNISCI E, HILA J, et al. Direct contactand vacuum membrane distillation application for the olive mill wastewater treatment .J. Separation & Purification Technology, 2016, 169 (9): 121-127.

[10] ELABBASSI A, GARCAPAYO M C, et al. Concen-tration of olive mill wastewater by membrane distillation for polyphenols recovery .J. Desalination, 2009, 245 (1): 670-674.

[11] Wu, J.; Zhang, C. Y.; Li, J. F.; etal. Experimental study on membrane distillation treatment of biochemical effluent from coking wastewater. J. Industrial Water and Wastewater, 2015, 46 (6): 13-17.

[12] Li, H.; Zuo, Y.; Qin, X. H. Membrane pollution in membrane bioreactors and its control measures. J. Industrial Water and Wastewater, 2011, 42 (1): 1-4.

IOP Conf. Series: Materials Science and Engineering **746** (2020) 012025 doi:10.1088/1757-899X/746/1/012025

Preparation of MAO Film Layer on LY12 Aluminum Alloy Surface in Sodium Hexametaphosphate Solution

Bo Wang[1, a], Qian Liang[2, b], Cheng Li[3, c], Dixing Bai[1, d] and Fengguang Zuo[1, e]

[1]State Grid Shannxi Electric Power Research Institute, Xi'an 710054, Shannxi, China
[2]State Grid Shannxi Electric Power Company, Xi'an 710054, Shannxi, China
[3]State Grid Shaanxi Electric Power Company Hanzhong Power Supply Company, Hanzhong 723000, Shannxi, China
Email: [a]wbgoodman@163.com, [b]951374344@qq.com, [c]179231821@qq.com, [d]2963941563@qq.com, [e]181696114@qq.com

Abstract. Aluminum alloy is widely used in power grids to solve the problem of aluminum excess and copper deficiency. The MAO film is prepared by micro-arc oxidation on LY12 aluminum alloy with phosphate electrolyte. The growth rate of film layer in different treatment time is studied and the variation trend of film in growth and outward growth was carried out. Finally, the morphology and composition of the film were tested by SEM and EDS. The results show that the initial growth of micro-arc oxidation is slow, mainly grow outwards, the medium-term rate is accelerated, the later period is slow, and the middle and late stages are mainly grow inwards. Phosphorus in the electrolyte participates in the film reaction.

1. Introduction

With the development of science and technology, grid security is the fundamental factor for the survival and development of power countries and enterprises [1]. With the continuous development of power grids, the security and reliability of power grids are of great importance. Among them, the safety of equipment caused by the corrosion of metal materials in power grid equipment is more common, and it is increasingly valued by the electric power departments in various regions of the country [2]. Surface treatment on alloys is one of the main measures for anti-corrosion. As a high-strength aluminum alloy, LY12 is widely used in aerospace and civil engineering. In addition, aluminum alloy materials have outstanding advantages in power architecture. Wire and cable "Aluminum instead steel" has strategic significance for solving the excess of electrolytic aluminum and ensuring national resource security [3, 4].

Therefore, in order to strengthen the application of aluminum alloy, the surface modification treatment of aluminum and its alloy materials by micro-arc oxidation technology has the advantages of simple process, high production efficiency and high industrial production. The electrolyte has strong anti-pollution ability and high re-use rate, low environmental pollution, and meet the needs of China's sustainable development strategy [5]. Liu Wenliang et al. [6] studied the MAO of LY12 aluminum alloy in solution systems such as silicate, aluminate and phosphate. It was found that the coating grew faster in silicate and phosphate systems. For the LY12 aluminum alloy, the ceramic layer formed in the phosphate solution is the thickest and fastest. Therefore, in this experiment, a sodium phosphate hexametaphosphate solution was selected as the micro-arc oxidation electrolyte. it is necessary to analyze the micro-arc oxidation process.

IOP Conf. Series: Materials Science and Engineering **746** (2020) 012025 doi:10.1088/1757-899X/746/1/012025

2. Experimental

The LY12 aluminum alloy was selected as the experimental material, and its chemical composition is shown in Table 1. After grinding, the burrs and surface scales and stains generated by the wire cutting are removed, cleaned and dried. The polished sample is drilled at the center of the edge as needed, with a hole diameter of approximately φ3 mm to facilitate suspension with the clamp during the ceramic layer preparation process. The sample is connected to one end of the aluminum wire, the joint is covered with a polymer liquid sealant, and the other end of the aluminum wire is fixed on the power supply bracket, the sample is used as an anode, and the stainless steel piece is used as a cathode. The MAO200H-III micro-arc oxidation equipment developed by the School of Materials Science and Engineering of Xi'an University of Technology is used. The alkaline electrolyte was prepared by using deionized water which is 20 g/L sodium hexametaphosphate solution. The chemical reagents used in the experiments were analytically pure reagents. The whole process is controlled by constant current, manual, current density is 0.5A/dm2, solution temperature is 20 ~ 40 °C, and oxidation time is 2 ~ 120 min. The thickness of the ceramic layer was measured using a TT240 cladding eddy current thickness gauge. The microstructure and chemical composition of the micro-arc oxidation ceramic layer were analyzed by JSM-6700F field emission scanning electron microscope and X-ray energy spectrometer (Oxford INCA).

Table 1. Chemical composition table of LY12 aluminum alloy

chemical composition	Cu	Mg	Mn	Ti	Zn	Fe	Si	Ni	impurities	Al
Content (mass fraction)%	3.8–4.9	1.2–1.8	0.3–0.9	≤0.15	≤0.3	≤0.5	≤0.5	≤0.1	≤0.1	margin

3. Results and Discussion

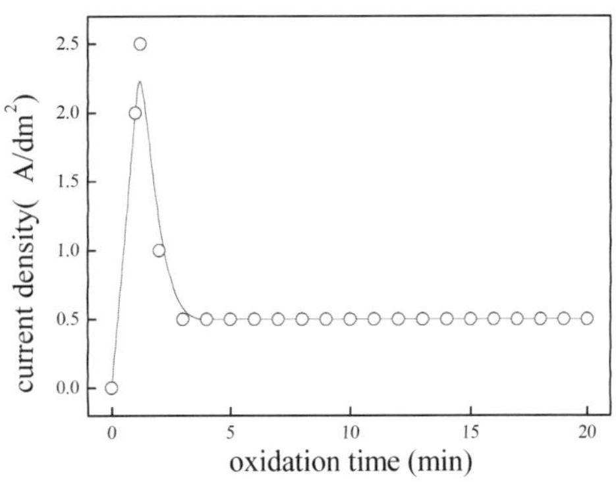

Figure 1. Relationship between current density and oxidation time of LY12 aluminum alloy

Figure 1 is a plot of current density versus oxidation time for micro-arc oxidation of LY12 aluminum alloy in sodium hexametaphosphate solution. During the micro-arc oxidation process, the current density rises first with the oxidation time, and then slowly decreases, during which there is a moment of abrupt change. Before the start of micro-arc oxidation, due to the high voltage preset by the power supply, it is impossible to directly reach such a high voltage, and it can only rise slowly. When the voltage rises above the breakdown voltage, it enters the micro-arc oxidation stage. At this instant, the current density instantaneously increases to a maximum due to the large voltage breakdown the oxide film on the surface of the sample. Then entering the micro-arc oxidation stage, the surface spark of the sample gradually becomes larger and brighter, and the thickness of the formed ceramic film is also thicker and thicker, but the increase in thickness becomes slower and more difficult, The impedance of the formed ceramic film is also getting larger and larger, and at this time, the current density is correspondingly reduced. In the sodium hexametaphosphate solution, when the voltage is kept constant, the film growth rate and the rate of dissolution are substantially the same, the film thickness is substantially not change again, at this time the current density is also stabilized at a certain value.

The variation of the outward and inward growth dimensions and total thickness versus time of the ceramic layer formed during the micro-arc oxidation of LY12 aluminum alloy is shown in Fig. 2(a). (b) is a schematic diagram of the size of the micro-arc oxidation film.

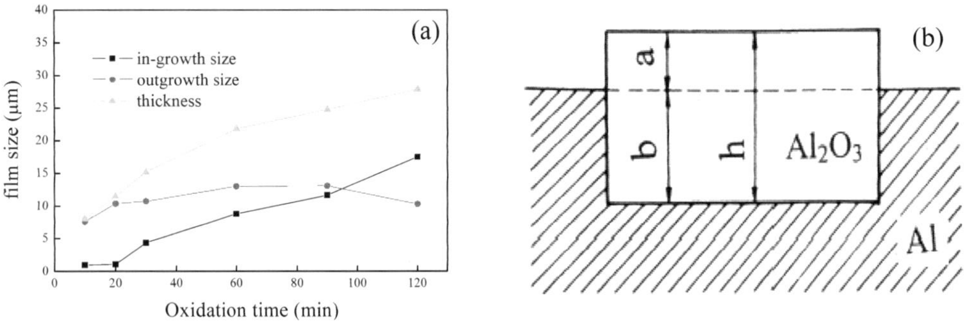

Figure 2. Thickness versus time curve (a) and schematic diagram of the micro-arc oxide film size (b)

It can be seen from Fig. 2(a) that the total thickness of the ceramic layer is close to linear growth. the MAO coating formed on the surface of the sample is thin and not continuous for about 1 min after the start of the micro-arc oxidation stage, As the micro-arc oxidation proceeds, the MAO coating grows inwardly and outwardly on the surface of the sample in such a manner that the density and thickness of the coating are continuously increased. As the thickness of the coating increases, more energy is required to break through the coating, so a bright orange-yellow spark begins to appear frequently on the surface of the sample. This is because of the high spark energy and the larger range of action. The coating growth at the stage was the fastest. When the micro-arc was oxidized for 60 min, the coating quickly increased to about 23 μm, but the coating surface became rougher. As the micro-arc oxidation continues, the greater the thickness of the coating, the more difficult the spark is to break down, so the thickness of the coating increases less and less, and becomes slower and slower. When the micro-arc is oxidized to about 120 minutes, the coating thickness increased to a large spark can not be broken, the growth of the coating is finished, and the thickness of the coating is substantially unchanged.

In the initial stage of micro-arc oxidation, the outward growth rate of the ceramic layer is slightly larger than the growth in the matrix, and then in the middle and late stages, the rate of ingrowth is slowly greater than the rate of outward growth, and the speed difference between the two is getting bigger and bigger. Therefore, in the whole micro-arc oxidation process, the growth of the ceramic film layer is mainly ingrowth. It can also be considered that the growth in the initial stage is mainly based on outward growth, and the growth in the middle and late stages are mainly ingrowth.

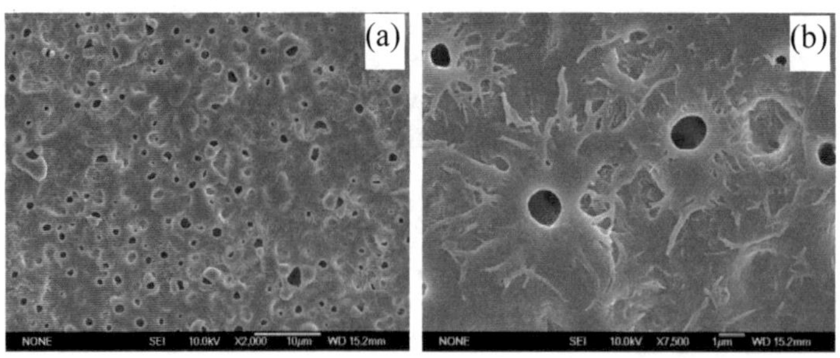

Figure 3. SEM photo of LY12 aluminum alloy micro-arc oxidation ceramic film: (a) 2000x; (b) 7500x

The ceramic layer prepared in the sodium hexametaphosphate electrolyte is fine grayish white, and the microscopic surface morphology of the ceramic coating film in the micro-arc oxidation process is shown in Fig.3. It can be seen from the figure that a large number of micron-sized pores of different sizes are distributed on the surface of the formed ceramic layer, similar to the "volcanic spray" melt, and there are obvious traces of melting and sintering around, This is because, after entering the micro-arc oxidation stage, although the spark exists for a short time, the instantaneous high temperature in the local area around the spark can still melt the anodized film on the surface of the sample and the aluminum alloy substrate under the film, and part of the melt follows many tiny discharge channels are sprayed into the electrolyte and rapidly cooled by the electrolyte, thus forming a porous structure on the surface of the sample similar to a "crater".

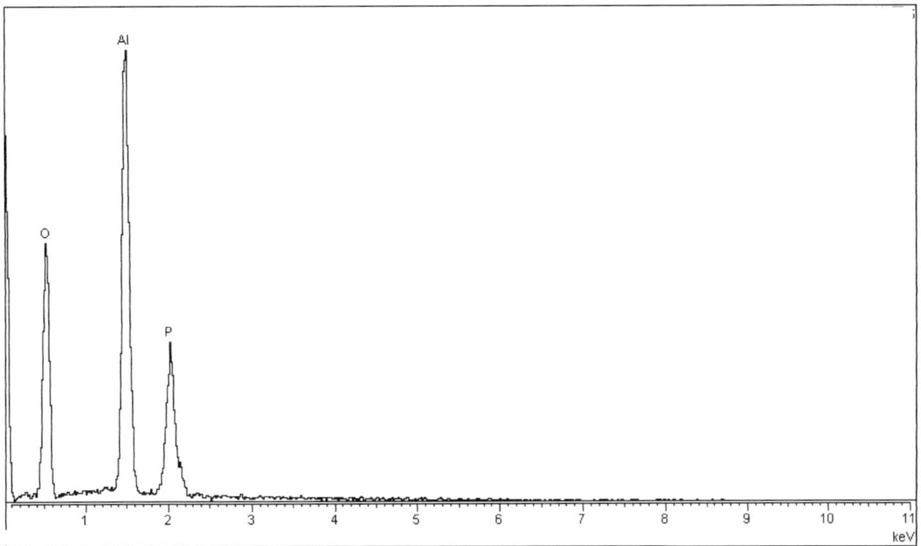

Figure 4. EDS spectrum of micro-arc oxidation ceramic film of LY12 aluminum alloy

Table 2. Chemical composition of micro-arc oxidation ceramic film of LY12 aluminum alloy

element	Weight percentage	Atomic percentage
O K	58.21	70.91
Al K	29.93	21.62
P K	11.86	7.46
total	100.00	

The results of EDS analysis of the obtained ceramic layer are shown in Fig. 4 and Table 2. It can be seen from Figure 4 and Table 2 that the content of O, Al and P is higher and the content of O is the highest. In the initial stage of micro-arc oxidation, anodizing reaction occurs on the surface of the aluminum alloy to passivate the surface of the aluminum alloy, but some aluminum elements are still dissolved in the solution to form a colloid. In the subsequent spark discharge process, since the dissolution function of the electrolyte continues to act on the surface of the aluminum alloy, as the experimental time increases, the more aluminum elements enter the electrolyte, which is the main reason for the loss of aluminum in the aluminum alloy matrix.

4. Summary

(1) The thickness of the micro-arc oxidation ceramic film of LY12 aluminum alloy increases with the increase of oxidation time, and the film thickness is basically linear with the oxidation time. The growth rate is different in different micro-arc oxidation stages, and the initial stage is slow. When the micro-arc oxidation discharge is more intense, the film thickness has the highest growth rate. When

IOP Conf. Series: Materials Science and Engineering **746** (2020) 012025 doi:10.1088/1757-899X/746/1/012025

the micro-arc oxidation is nearing the end, its growth rate gradually decreases, and finally stops growing.

(2) The thickness of the micro-arc oxidation ceramic film is basically linear with the oxidation time. The film thicknesses of inward and outward growth at different times are different. The initial stage of micro-arc oxidation is mainly outward growth, and the growth in the middle and late stages is mainly ingrowth.

(3) The ceramic layer prepared by LY12 aluminum alloy in sodium hexametaphosphate electrolyte is off white, with fine touch feeling, and the microscopic morphology is a porous structure similar to "crater". The elements contained in the ceramic layer are O, Al, and P, and the content of the O element is the highest.

5. References

[1] Bin H, Qi Z. Innovation and application of aluminum alloy support (cable) in photovoltaic power station construction 2018 *World Nonferr. Met.* **10** 209-211.

[2] Yun-xiang C, Qing-zhao N and De-yuan Lin. Research Progress in Corrosion and Service Life Prediction of Metal Materials in Grid Equipment Under Atmospheric Environment 2016 *Mater. Rev.***21** 89-108.

[3] Jv C, Xiao-hong T and Jian-guo T. Application of Aluminum Alloy in Transmission Trusses 2008 *Electric Power Construction*, **29** 26-28.

[4] Bin-bin W, Zhen-yao W and Gong-wang C. Atmospheric Corrosion Behavior of Pre-strained Aluminum Alloys LY12 and LC4 in Salt Lake Environment in Western China 2014 *J. Chin. Soc. Corr. Pro.* **34** 287-293.

[5] Wei-yi M, Zheng-xian L and Ji-hong D. Research Progress of Ceramic Coatings Formed on Aluminum Alloys by Micro-arc Oxidation 2013 *Surf. Technol.* **42** 94-99.

[6] Wen-liang L. Study on Properties of Aluminum Alloy Microarc Oxidation Coating in Different Kinds of Electrolytes 1999 *Plating and Finishing* **21** 9-11.

IOP Conf. Series: Materials Science and Engineering **746** (2020) 012026 doi:10.1088/1757-899X/746/1/012026

Stress Analysis of the Circular Pier Anti-collision Box Made of Sandwich Plate with Stiffening Rib

Wenfang Liu[1] and Chenglin Shan[2,*]

[1]Department of urban construction engineering, Guangzhou City Polytechnic, Guangzhou, China

[2,*](Corresponding Author), School of civil and transportation, South China University of Technology, Guangzhou, China

Email: [1]463375944@qq.com, [2]shanchenglin@163.com

Abstract. The pier anti-collision box made of steel-polyurethane sandwich plate with stiffening ribs can withstand greater impact force. In order to ascertain the influence of stiffening ribs to anti-collision box, the static equivalent method is adopted to numerically calculate the stress and impact depth of the anti-collision box in the conditions of no stiffening rib and the different forms and amount of stiffening ribs. The results show that it can reduce the stress and deformation of anti-collision box to set vertical or horizontal stiffening rib separately, and the effect of the former is much better than the latter, but the decrease degree of vertical stress is different from horizontal stress. The stiffening rib in the box should not be too many; especially horizontal rib and vertical rib should be considered mainly. The anti-collision box with horizontal rib and vertical rib is mainly to reduce collision deformation; however, it can reduce little than the box with only vertical rib in the aspect of stress.

1. Introduction

In recent years, the events of ship impacting bridge have occurred, leading to the destruction of the bridge, resulting in greater social impact. In order to solve this problem, more and more new anti-collision devices appear, the purpose of which is to reduce the collision force of the ship and reduce the energy exchange during the collision, so as to achieve the effect that the bridge, ship and anti-collision device are not damaged [1]. The anti-collision device studied in this paper is a kind of sandwich box made of sandwich plate, which is curved, sheathed on piers or anti-collision piles, and can float up and down with the water level, and has strong adaptability to inland waterway with large water level change [2, 3]. The sandwich plate is added a thick layer of polyurethane core layer between the two layers of steel plates, the research on this aspect abroad early. The sandwich structure has good impact resistance and fatigue performance, and is mainly used in military, ship, bridge reinforcement and other fields [4, 5]. There are few studies in anti-collision bridge pier device made of sandwich plate at home and abroad. The paper [6, 7] conducted the finite element analysis of the collision happened to the bridges by cars or ships. The paper [8] studied the static performance of the box in the case of front impact and slanting impact. The paper [3] studied the change of box energy during the collision course, which is that the box can absorb most of the front impact energy; on the contrary, due to the smooth and rotating surface of the box, the ship is easy to move in the slanting impact, so the most kinetic energy is kept in the ship. The paper [9] simulated the ship impact force using the equivalent static load, which analysed the stress performance of no stiffening rib box by changing the thickness of steel plate and polyurethane core layer.

The results show that when the ship tonnage is large, it is necessary to set up the steel stiffening rib inside the box. In this paper, ANSYS software is used to build the anti-collision system model and calculate numerically repeatedly in the form of equivalent static loads, which was to analyze the influence of different types and different number of stiffening rib on the strength and rigidity of the anti-collision box, so as to properly set up the stiffening rib.

2. Model Building

2.1. Model Size

This paper with a practical project as the background, relying on the waterway level for Ⅳ, according to the design specification of highway bridge [10], it is concluded that the corresponding ship collision force is 550 kN. For example, the diameter of the pier is 1.3 m, and the height is 10 m, the anti-collision box is shown in Figure1. The inner diameter is 1.6 m, the outer diameter is 4.6 m, the vertical section is the ellipse, the long axis is 2.3 m, and the short axis is 1.5 m. The box is the sandwich structure, the thickness of the outer steel plate is 8 mm, the thickness of the inner steel plate is 6 mm, and the intermediate polyurethane thickness is 40 mm, as shown in Figure 2. The box and pier are combined with a circular ring rubber fender, with a thickness of 100 mm and a height of 400 mm.

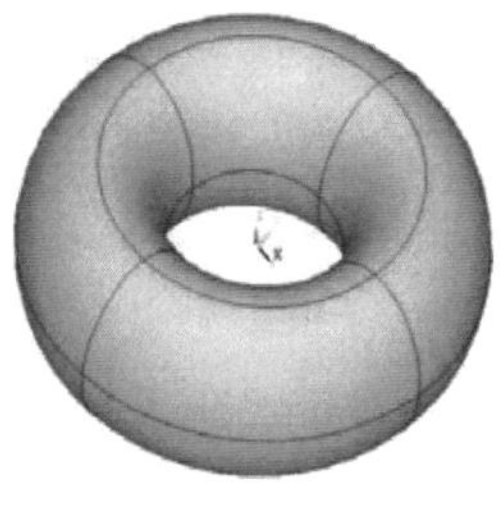

Figure 1. Profile of the anti-collision box made of sandwich

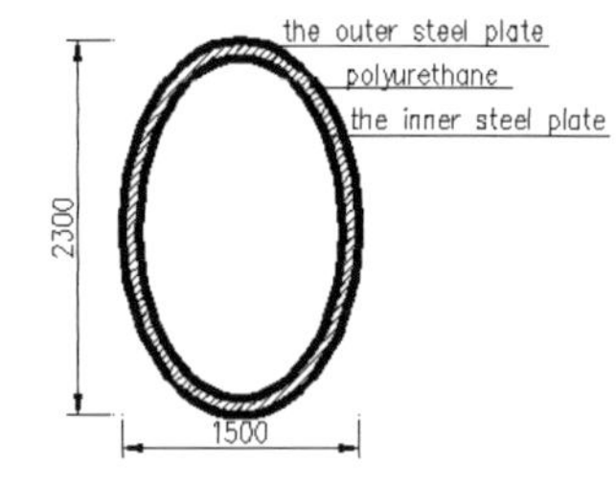

Figure 2. Meshing box element in the horizontal circle

2.2. Material Parameters

The panel and stiffening rib of the anti-collision box are made of Q345 steel, whose elastic modulus is 2.05 x 10^5 MPa, the poisson ratio is 0.3 and the heavy was 78.5 kN/m^3. The hardness of polyurethane is not less than 80HRA, the elastic modulus is 750 MPa, the poisson ratio is 0.48, and the weight is 12 kN/m^3 [11]. The pier adopts C30 concrete, whose elastic modulus is 3 x 10^4 MPa, the poisson ratio is 0.2, the heavy is 25 kN/m^3; The fender adopts the styrene butadiene rubber SBR, the hardness of which is 40 degrees, the density is 940 kg/m^3, the elastic modulus is 6.1 MPa, and the tensile strength is 20.5 MPa.

2.3. Division of Units

The steel - polyurethane sandwich plate is simulated by shell99, which can be used to model the layered shell structure, which can define the material properties of each layer and output the stress and displacement in different directions of each layer. Each node has six degrees of freedom. They are the translation of X, Y and Z directions and the rotation along the X, Y, and Z axes. In this paper, the stress performance of the sandwich panel is studied, so it is reasonable to use shell99 sandwich unit to simulate the stress of the sandwich plate. The pier and fender are simulated with solid45 unit, which has eight nodes and each node has 3 degrees of freedom moving along the X, Y and Z direction.

The box unit is divided into 60 equal parts along the vertical ellipse and 80 equal parts along the horizontal circle. The size of the unit grid is 180 mm in width and 95 mm in height, as shown in Figure 3 and Figure 4. The pier division is divided into 40 equal parts of the loop segment, the radial segment is 20, and the vertical segment is 10.The pier division is divided into 40 equal parts of the loop segment, the radial segment is 20, and the vertical segment is 10. In the middle of the pier, the box section is divided by means of the encryption grid, the loop is 40 equal to the segment, the radial segment is 20, and the vertical segment is 30. For the convenience of modeling, the fender height is 1500 mm, and the grid division of the fender is the same as the middle part of the pier.

The stiffening rib is divided into two kinds, which are circular horizontal and oval vertical stiffening rib. The horizontal stiffening rib is the same as the direction of the latitude, which is divided into 80 equal parts along the circumference. The vertical stiffening rib is the same as the direction of the longitude, which is divided into 60 equal parts along the ellipse direction. The horizontal stiffening rib is arranged in the central height of the box along the circular direction of the inner steel plate in the box, which is coupled with the inner steel plate joint, evenly arranged along the upper and lower sides of the box, and the spacing is the height of a box unit, which is about 100 mm. The vertical stiffening rib is arranged evenly around the circumference of the casing. The arrangement of stiffening rib is shown in Figure 5.

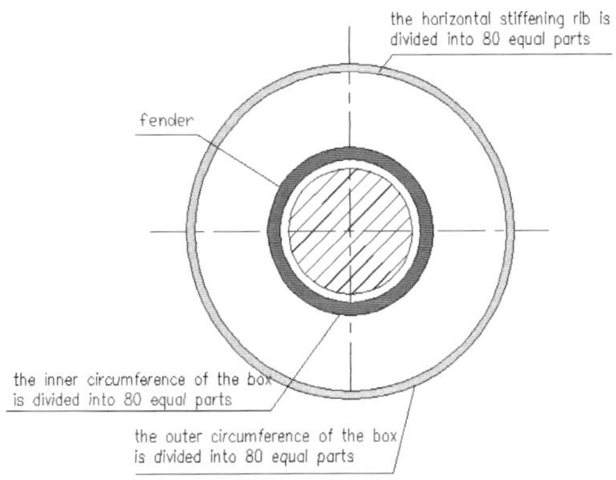

Figure 3. Meshing box element in the horizontal circle direction

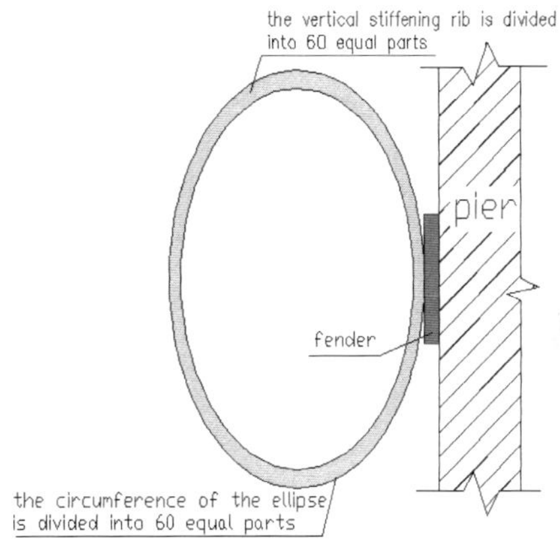

Figure 4. Meshing box element in the vertical ellipse direction

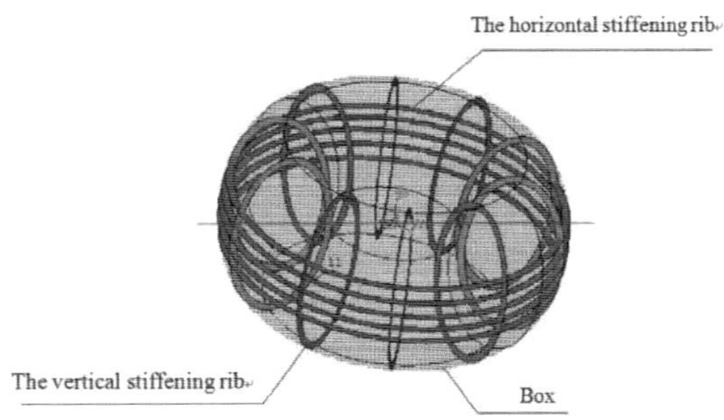

Figure 5. Arrangement of the stiffening ribs

2.4. Boundary Conditions

Because of the large rigidity of the transverse bridge of the pier and bearing platform, the lower part of pier can be assumed to be consolidated, and the upper part is free, without taking account of the upper structure's favorable restraint to the pier top. The fenders are connected to the pier by means of the GLUE command in the ANSYS without relative displacement. The connection between the fender and the box is a connection between the body element and the shell element, which is simulated in a half - cycle coupling manner, that is, the degree of freedom is defined on each pair of coincidence nodes, a coupling set is generated, and the coupling of the coincidence nodes in the model is realized by the CPINTF command in the ANSYS. The boundary conditions of the model are shown in Figure 6.

IOP Conf. Series: Materials Science and Engineering **746** (2020) 012026 doi:10.1088/1757-899X/746/1/012026

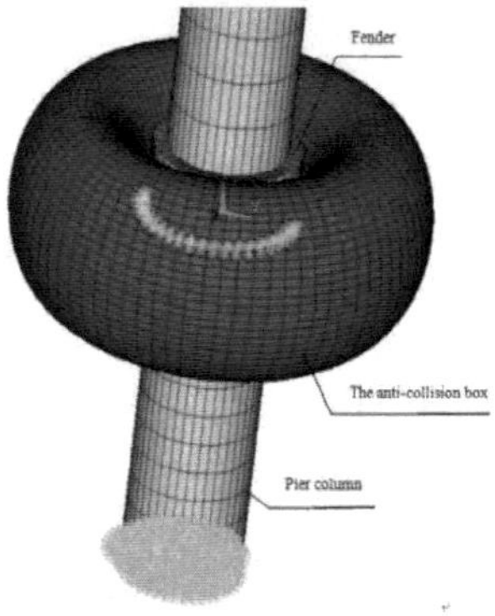

Figure 6. Modal boundary conditions

3. The Force Analysis of Box in Ship's Frontal Collision

The ship's frontal collision is one of the least unfavorable collisions, which is located in the middle of the box height and pointed to the box radial, as shown in Figure 7 and Figure 8. The ship's collision force according to the corresponding technical standard is taken as the local uniformly distributed load acting on the one unit of the collision point. In order to study the influence of stiffening rib on the mechanical properties of this type of anti-collision box, there are two kinds of settings, one is setting circular horizontal stiffeners and elliptical vertical stiffeners ,another one is not setting. The sandwich plate's stress and deformation of the impact point a range of up and down, right and left are calculated to reflect the mechanical performance of the anti-collision sleeve.

Due to the symmetry of the force and structure, the force of the impact point is symmetrical, so the data of the gathering point on one side of the collision point is taken. According to the test, the stress of the sandwich the outer steel plate is the maximum, so in the following analysis, the vertical and lateral stresses and the deformation of the impact direction are only taken from the points of the outer steel plate. Figure 7 shows the stress collection point on the vertical side of the impact point. Within the range of 60° above the impact direction, the length of the elliptic arc is divided into 60 equal parts, with a total of 10 collection points. Figure 8 shows the stress collection point on the horizontal side of the impact point. Within the circumference of 40.5° to the right of the impact direction, the arc length is equally divided into 9, which also has 10 collection points. The following diagram shows that the stress of the collection point is positive, indicating the tensile stress, and the negative value indicates the compressive stress.

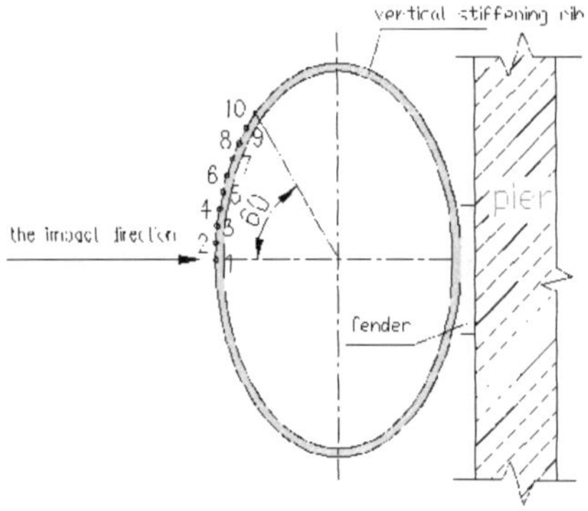

Figure 7. Distribution of the stress collection points of the box in vertical section

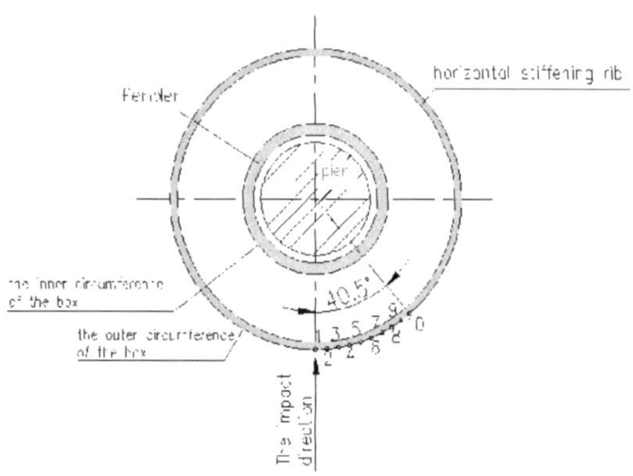

Figure 8. Distribution of the stress collection points of the box in horizontal section

3.1. The Influence of Horizontal Stiffening Rib on the Force of Anti-collision Box

There are two situations that one is setting horizontal stiffening ribs and another one is not setting. The Settings are shown in figure 5. One, three, five and seven horizontal stiffening ribs are set respectively. The thickness of steel plate with stiffening rib is 10 mm and the width is 100 mm.The spacing of the stiffening rib is the height of a box unit, which is about 100 mm. One of the horizontal stiffening rib is located at the impact point, and the vertical stiffening rib is not set.

From Figure 9 to Figure 12, it can be seen that the transverse and vertical stress of the outer plate of the anti-collision box are mainly compressive stress, which sharply decreases after leaving the collision center, then gently, and the stress concentration is prominent.Compared with no stiffening ribs, for the setting, the initial compressive stress of the outer steel plate decreases a lot, but the decrease of the vertical compressive stress is greater than that of the transverse stress. For the same collision position, it is not helpful to reduce the initial compressive stress of the outer layer steel plate by setting multiple or more horizontal stiffening ribs . The horizontal stiffener changes rapidly from the transverse tensile stress to the compressive stress at the impact point, and the less stiffening ribs there are, the larger the undulation is. The depth of impact at the impact point decreases with the increase of the quantity of horizontal stiffeners, but the more stiffening ribs there are, the smaller the reduction is.

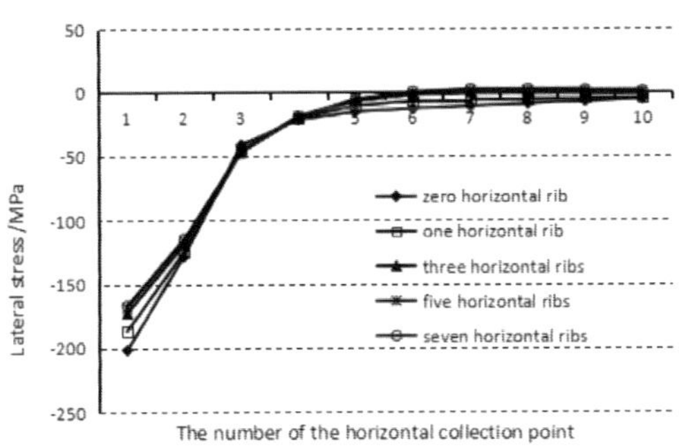

Figure 9. Lateral stress of the collection points on surface of outer steel plate in horizontal direction

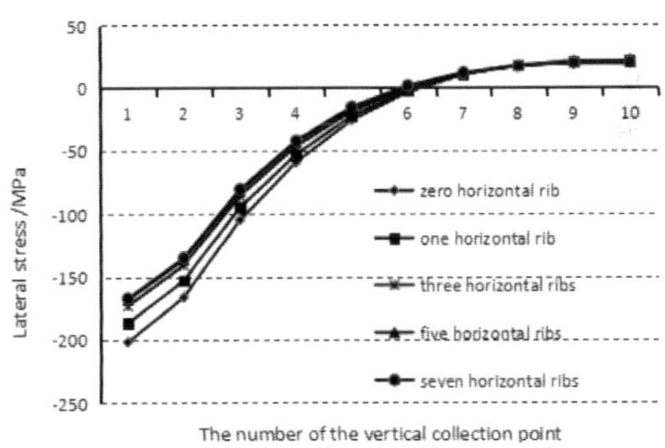

Figure 10. Lateral stress of the collection points on surface of outer steel plate in vertical direction

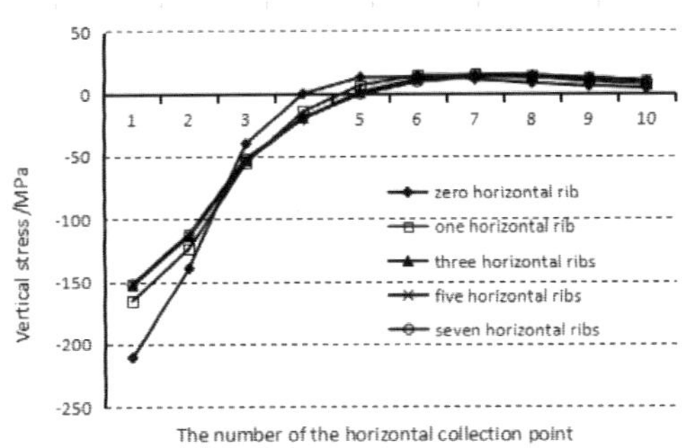

Figure 11. Vertical stress of the collection points on surface of outer steel plate in horizontal direction

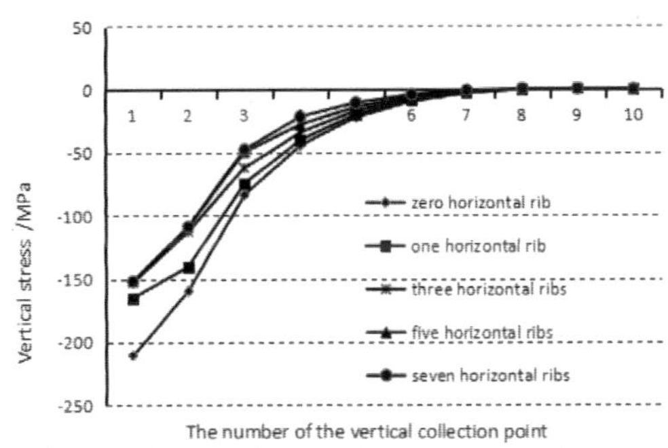

Figure 12. Vertical stress of the collection points on surface of outer steel plate in vertical

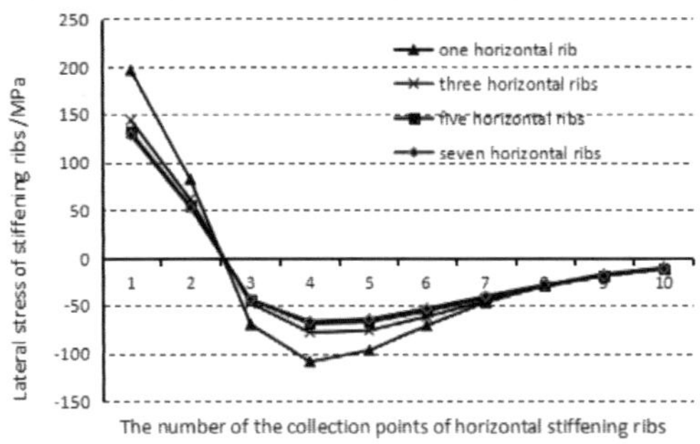

Figure 13. Lateral stress of the collection points on horizontal stiffening ribs

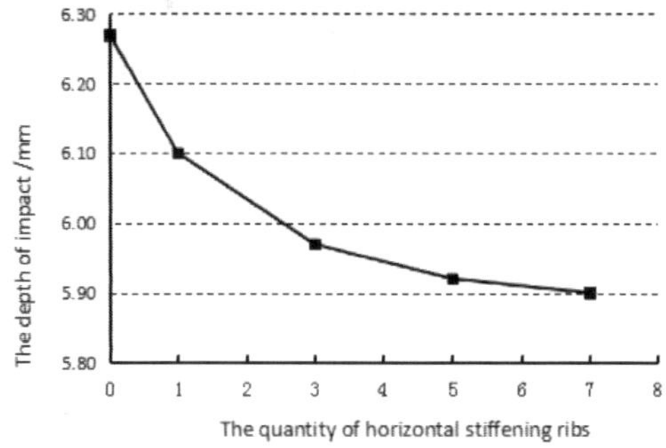

Figure 14. Displacement of the impact point

3.2. The Influence of Vertical Stiffening Rib on the Force of Anti-collision Box

There are also two situations that one is setting vertical stiffening ribs and another one is not setting.The Settings are shown in Figure 5. 10, 20 and 40 vertical stiffening ribs are set respectively, so the interval is 36°, 18° and 9° along the circumference of the box respectively, the thickness of the stiffening rib steel plate is 10 mm, the width is 100 mm, and the stiffening rib is coupled with the corresponding nodes of the inner steel plate. While one of the vertical stiffening ribs is just at the impact point, the horizontal stiffening rib is not set.

From Figure 15 to Figure 18, the transverse and vertical stress of the outer steel plate of the anti-collision box are mainly compressive stress, which sharply decreases after leaving the collision center, then gently, and the stress concentration is prominent. Compared with no stiffening ribs, for the setting, the initial compressive stress of the outer steel plate decreases a lot, the maximum can be decreased by 43.9%.The degree of reduction is much greater than that of setting the horizontal stiffening rib, and the degree of stress reduction in the horizontal direction is greater than that in the vertical direction. It shows that the effect of setting the vertical stiffening rib is better than that of the horizontal stiffening rib. For the same collision position, it is not helpful to reduce the initial compressive stress of the outer layer steel plate by setting multiple or more vertical stiffening ribs. The transverse tensile stress of the vertical stiffening rib at the impact point is rapidly transformed into compressive stress, and the transverse tensile stress of the vertical stiffening rib is slightly larger than that of the horizontal stiffening rib, as shown in Figure 19. The depth of impact at the impact point decreases with the increase of the number of the vertical stiffening rib. However, the decrement becomes smaller when the number is more than one, as shown in Figure 20.

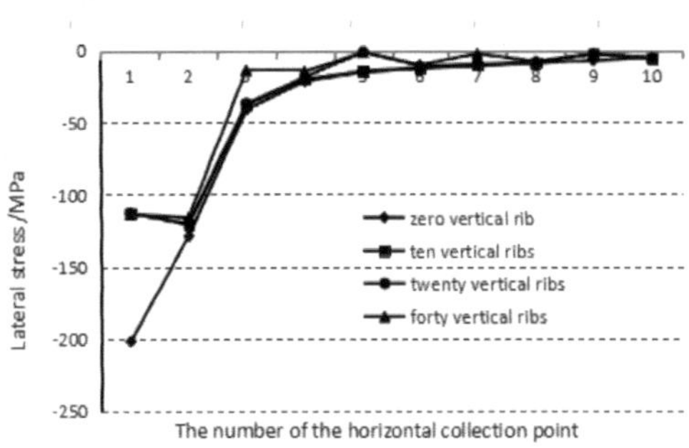

Figure 15. Lateral stress of the collection points on surface of outer steel plate in horizontal direction

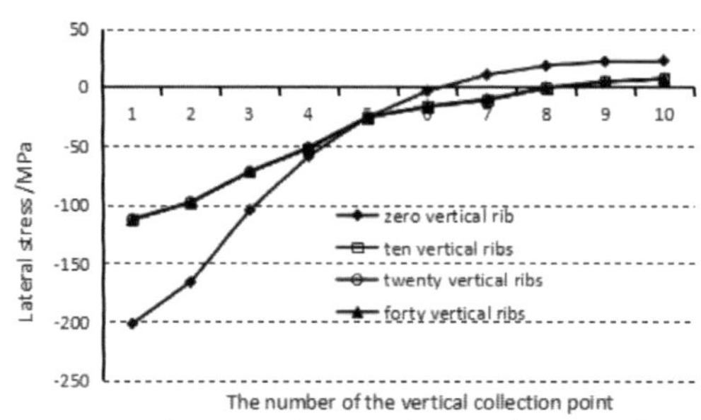

Figure 16. Lateral stress of the collection points on surface of outer steel plate in vertical direction

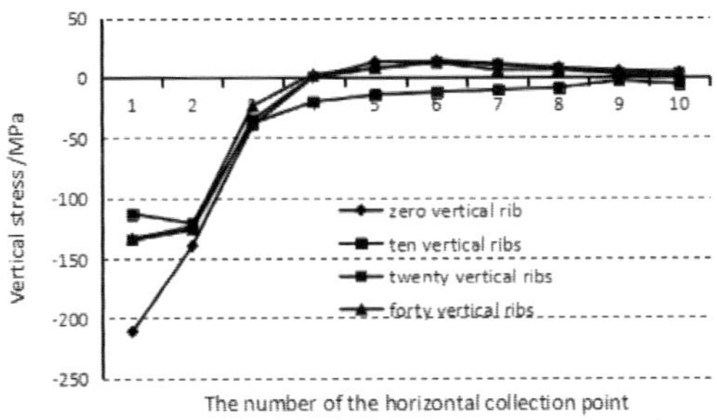

Figure 17. Vertical stress of the collection points on surface of outer steel plate in horizontal direction

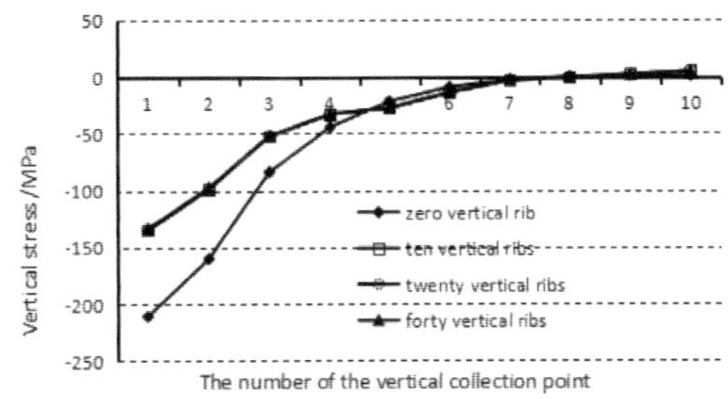

Figure 18. Vertical stress of the collection points on surface of outer steel plate in vertical direction

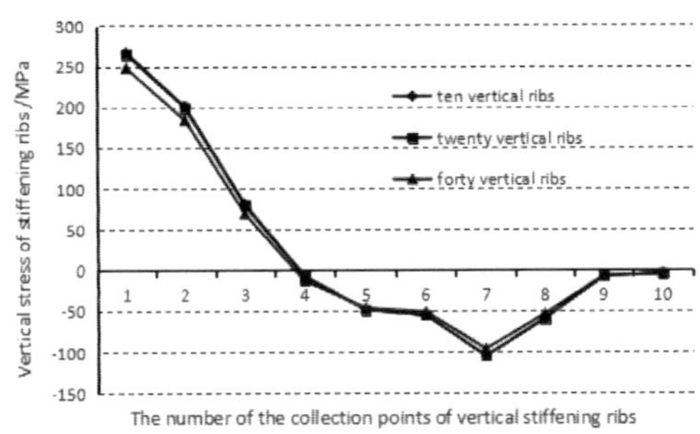

Figure 19. Vertical stress of the collection points on vertical stiffening ribs

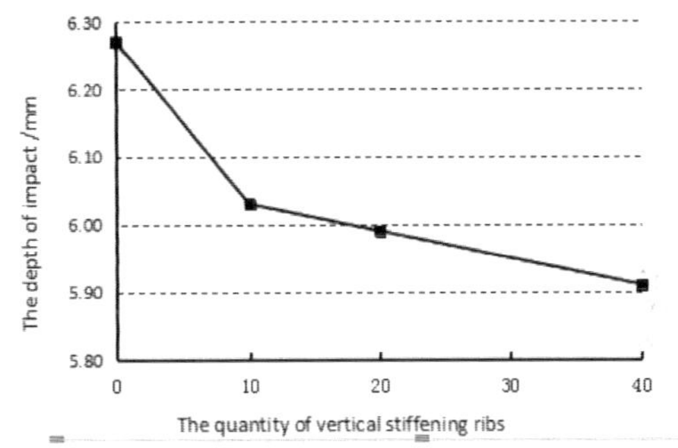

Figure 20. Displacement of the impact point

In the above collision calculation, the collision points are equipped with stiffening rib and the stiffening rib of other positions is not obvious to reduce the force effect of the steel plate. However, the position of the collision point of the ship is random within a certain range of the anti-collision box, and the box is also rotatable, so when the impact force is large, the stiffening rib with a certain density must be installed.Therefore, it is necessary to study the force in the middle of the two vertical stiffening ribs, while other conditions are invariable, and the results of Figure 21 ~ Figure26 are obtained by calculation.In contrast to the previous case, it can be found that the stress of the outer steel plate increases by more than 90% when the load is between the stiffeners, but there's not much difference in the number of stiffening ribs, which indicates that the encryption of the stiffening rib does not effectively reduce the force of the outer plate. But the force of the stiffening rib is much smaller than that of the previous one, as shown in Figure 25. The impact depth increases slightly and decreases linearly with the increase of the stiffener number, as shown in Figure 26.

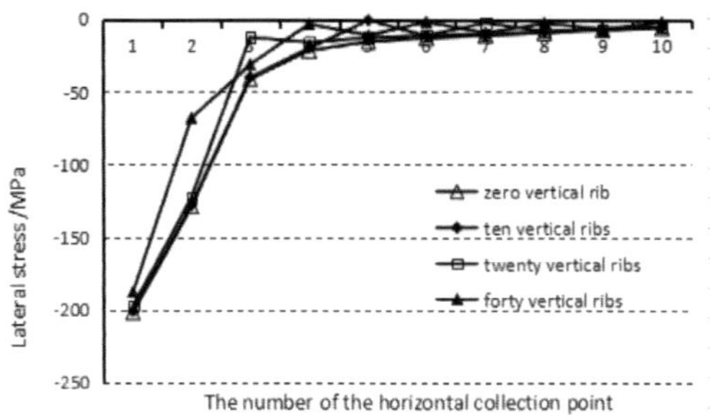

Figure 21. Lateral stress of the collection points on surface of outer steel plate in horizontal direction

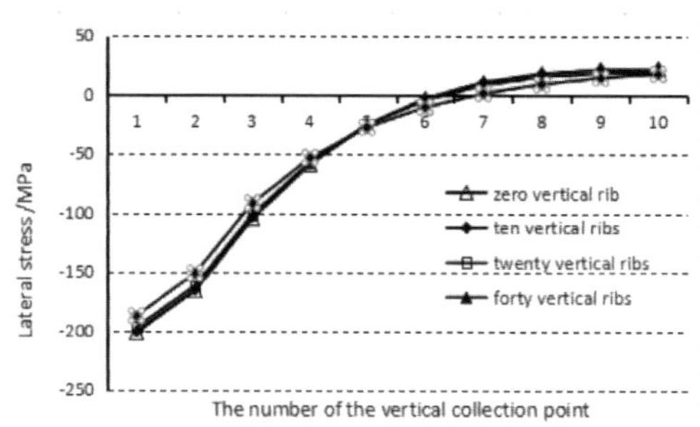

Figure 22. Lateral stress of the collection points on surface of outer steel plate in vertical direction

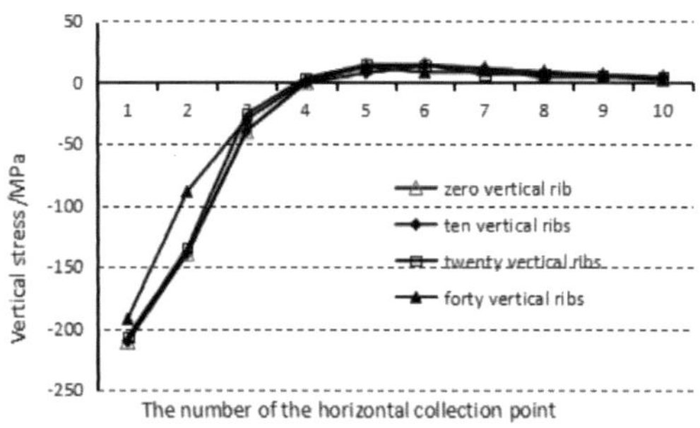

Figure 23. Vertical stress of the collection points on surface of outer steel plate in horizontal direction

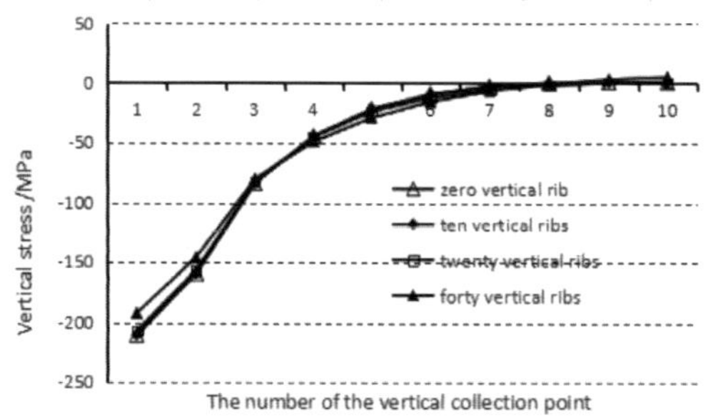

Figure 24. Vertical stress of the collection points on surface of outer steel plate in vertical direction

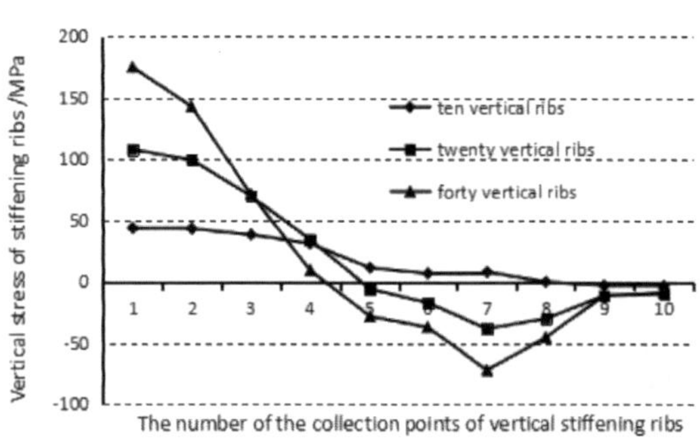

Figure 25. Vertical stress of the collection points on vertical stiffening ribs

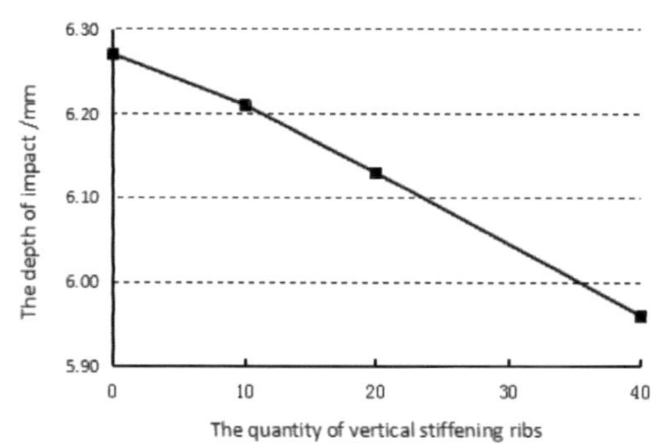

Figure 26. Displacement of the impact point

3.3. The Influence of Horizontal and Vertical Stiffening Rib on the Force of Anti-collision Box

The above is the situation of setting the horizontal stiffening rib and the vertical stiffening rib separately.When setting the horizontal stiffening rib and vertical stiffening rib at the same time in the box, then the influence on the force of the box is analyzed. At the same time, the number of the 5 horizontal stiffening rib remains unchanged, and the size of the stiffened steel plate also remains unchanged.In this case, the stress analysis is done by changing the spacing between the vertical stiffening ribs.The vertical and horizontal stiffeners are installed at the impact point.

From Figure 27 to Figure 30, it can be seen that the transverse and vertical stresses of the outer steel plate of Anti-collision box are mainly compressive stress, which sharply decreases after leaving the collision zone, and the stress concentration is prominent. For the same impact position, setting 20 or more vertical stiffeners is of little help to reduce the initial compressive stress of the outer layer steel plate. When the horizontal and vertical stiffening ribs are set at the same time, the effect is relatively ideal, which can greatly reduce the stress level of the inner and outer steel plate, but the distance between stiffening ribs is not as dense as possible, because of an optimal combination problem of horizontal and vertical stiffening ribs. The vertical stiffening ribs changes rapidly from the transverse tensile stress to the compressive stress at the impact point. The displacement at the impact point decreases with the increase of the stiffening rib number, but not linearly.

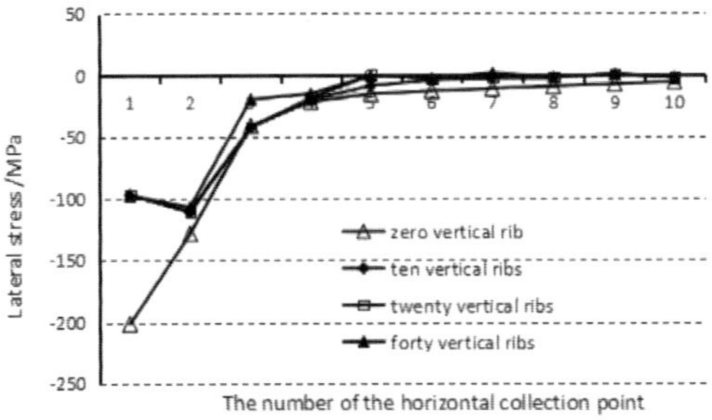

Figure 27. Lateral stress of the collection points on surface of outer steel plate in horizontal direction

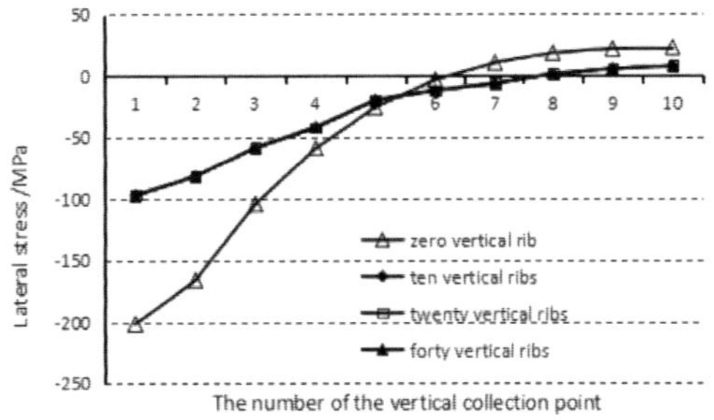

Figure 28. Lateral stress of the collection points on surface of outer steel plate in vertical direction

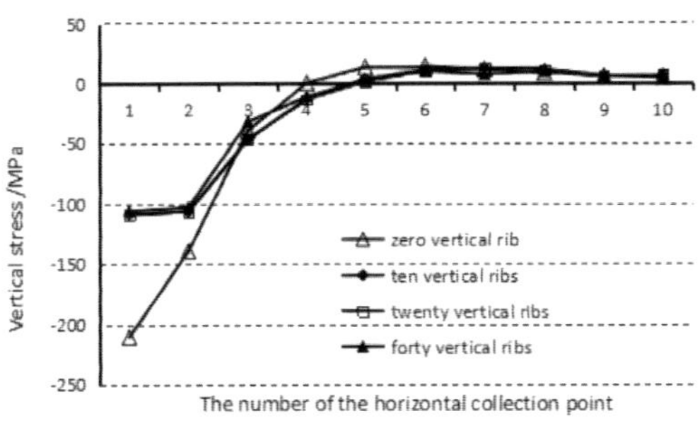

Figure 29. Vertical stress of the collection points on surface of outer steel plate in horizontal direction

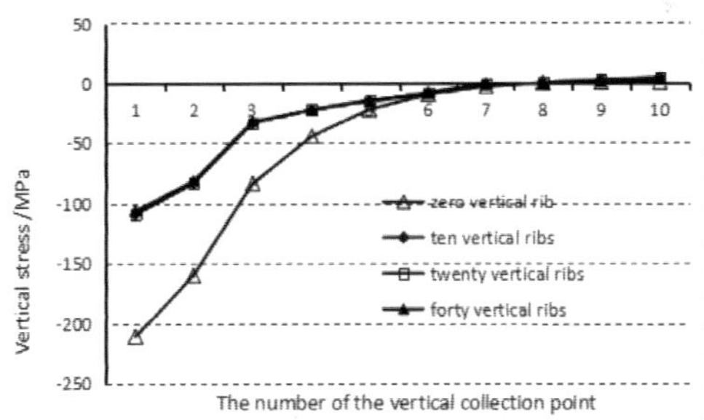

Figure 30. Vertical stress of the collection points on surface of outer steel plate in vertical direction

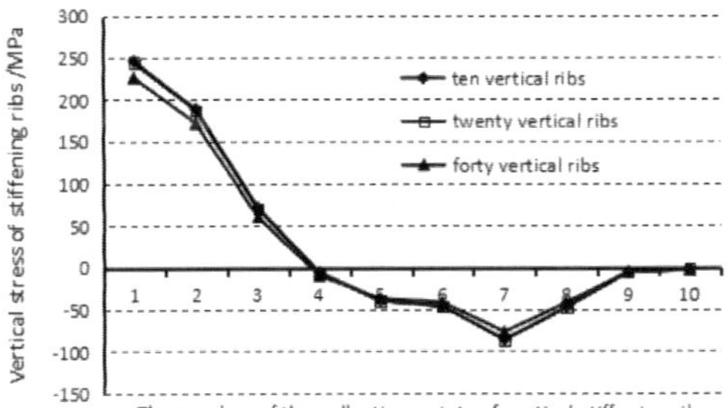

Figure 31. Vertical stress of the collection points on vertical stiffening ribs

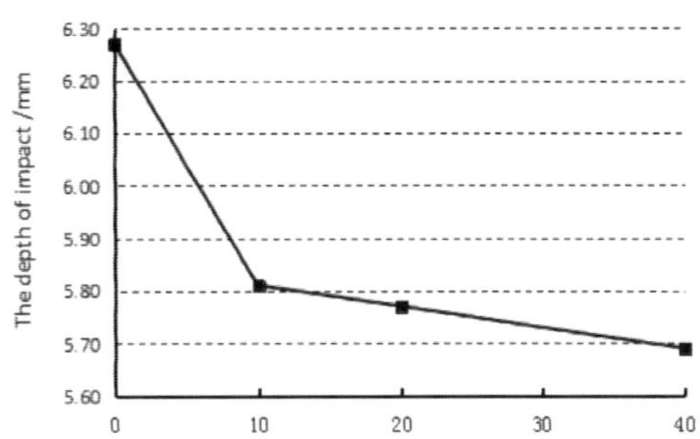

Figure 32. Displacement of the impact point

3.4. Force Condition of the Pier

Figure 33 shows that the stress nephogram of the collision surface of pier and the maximum tensile stress at the bottom of pier is 10.6 MPa, when 5 horizontal stiffening ribs and 40 vertical stiffening ribs are installed in the box. In fact, in the static calculation, the number of stiffening ribs in the anti-collision box changes, the stiffening ribs and strength of the collision box also change, which only affects the force of the box itself, and does not affect the force of the pier. Only in the dynamic calculation, the stiffness change or rotation of the anti-collision box will prolong the impact time and change the direction of impact, so as to reflect the impact force change on the bridge pier.

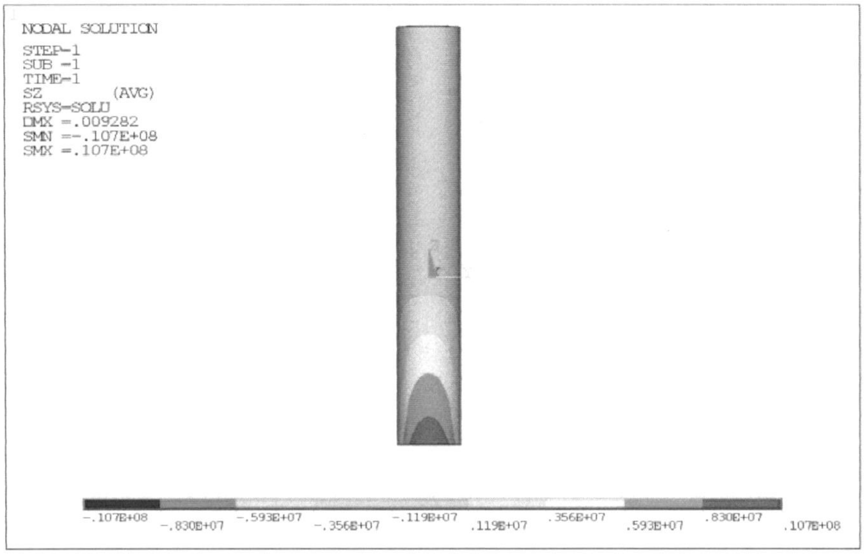

Figure 33. Stress cloud of the pier

4. Conclusion

(1) The horizontal stiffening rib can reduce the stress of the sandwich plate, and the stress reduction in the vertical direction is greater than that in the horizontal direction. The numbers of horizontal stiffening rib have little difference in reducing the initial stress of the sandwich plate.

(2) The vertical stiffener can greatly reduce the stress of the sandwich plate, and the stress reduction in the horizontal direction is greater than that in the vertical direction. The effect is better than that of the horizontal stiffening rib, but the stress of the stiffening rib itself will increase. The numbers of vertical stiffening rib have little difference in reducing the initial stress of the sandwich plate.

(3) The stiffening rib should be mainly considered in the installation of stiffening box, and the density should be maintained. When the horizontal and vertical stiffeners are set up at the same time, compared with only the vertical stiffening rib, the stress of the sandwich plate decreases less, but the impact depth decreases slightly.

(4) From the analysis, it can be seen that the number of stiffening rib is too much to improve the force of anti-collision box, on the contrary, which will increase the cost and weight. But considering the randomness of the collision location, it must also set a certain density stiffening rib. Therefore, the horizontal stiffening rib should be set according to odd symmetry in the middle of the box, the number of 5 is appropriate, the spacing is determined by the size of the box. The vertical stiffening rib should be set along the circumference of the box, the number depends on the spacing of about 350 mm, in order to form the best combination of optimization.

5. References

[1] Guoyu Chen and Lili Wang. 2008. *M. China Railway Publishing House.* Ship hiting bridge and its defense. Beijing, 2008: 20-28.

[2] Chenglin Shan, Xuejun Xu, Qingwen Zhong, et al. 2015. *P. China, Invention patent.* A whole curved surface anti-collision floating box for bridge pier. ZL 201210499404. 0. 2015, 6.
[3] Xuejun Xu and Chenglin Shan. 2015. *J. Journal of Hunan University (Natural Sciences).* Impact analysis of the bridge pier anti-collision floating box sets made by sandwich structure with curved–shaped, 2015, 42 (3): 106-111.
[4] Kennedy S J and Kennedy D J L. 2006. *C. Proceedings of the 5th International Conference on Composite Construction in Steel and Concrete V, ASCE.* A True innovation: Steel plates with a structural elastomer core. 2006: 114-126.
[5] Farmer I. 2006. *J. Structural Engineer.* Sandwich plate system for new build and repair. 2006, 84 (5): 22-26.
[6] Amin H Almasri, Rajai Z Alrousan and Al-harith Manasrah. 2015. *J. KSCE Journal of Civil Engineering.* Finite element analysis of a 2-span pedestrian bridge collapse due to trucks collision. 2015, 19 (6): 1845–1851.
[7] Li Tian and Fei Huang. *2014. J. Transactions of Tianjin University.* Numerical simulation for progressive collapse of continuous girder bridge subjected to ship impact. 2014, 20 (4): 250-256.
[8] Chenglin Shan and Xuejun Xu. 2014. *J. China Journal of Highway and Transport.* Analysis of mechanical properties of pier anti-collision device by sandwich plate with overall curved face. 2014, 27 (10): 46-54.
[9] Shuchao Zhao. 2013. *D. South China University of Technology.* Research on mechanical behavior for composite structure of anti-collision bridge pier. 2013.
[10] People's Republic of China Ministry of Transportation. 2004. *S. China Communication* Press General code for design of highway bridge and culverts. Beijing, 2004.
[11] Yuhua Yi and Chaofeng Shi. 2007. *J. Marine Technology.* The composite steel-polyurethane-steel plate and its manufacturing method. 2007, 280 (6): 36-38.

IOP Conf. Series: Materials Science and Engineering **746** (2020) 012027 doi:10.1088/1757-899X/746/1/012027

Flexible Wire Based on Silver Nanowires and Elastomer

Jianhui Zhou[1, 2], Jianguo Cao[1, *] and Chunfu Cheng[1]

[1]School of Mechanical Engineering, University of Science and Technology Beijing, Beijing, China

[2]Qian'an College, North China University of Science and Technology, Tangshan, China

[*]Email: geocao@ustb.edu.cn

Abstract. In order to solve the problem that the wire is difficult to have both elasticity and conductivity, a novel method for producing flexible wires based on silver nanowires (AgNWs) and elastomer has been proposed. This type of flexible wire is fabricated using a pre-stretching process and a glass sheet transfer process. Stretching test shows that after 50 times 0-30%-0 cycles of tensile test, the resistance growth rate relative to initial resistance is only 510%. The flexible wire can be widely used in flexible electronic devices.

1. Introduction

Flexible wires are the key components in flexible electronic devices such as intelligent manufacturing, pension service, disability assistance and so on. The flexible wires required to be both elastic and conductive. In recent years, the research of flexible wires has become a hotspot [1, 2].

At present, research on flexible wires is progressing rapidly after decades of development. There are three main methods for making flexible wires: mechanical buckling, multidimensional metal wires, composite material of conductive material and elastomer. Suo et al. deposited a gold membrane on a pre-stretched ultra-thin elastomer to make wires [3]. Stretching rate of the wire can reach 100%, but it is easy to wear. M. Jablonski et al. make curved metal wires buried into elastomer to improve the flexibility of the wire, but it is only suitable for single direction stretching [4]. The composite material of conductive material and elastomer has attracted intensive attention of researchers due to its good flexibility and conductivity. Feng Xu [5] and Haopeng Wang [6, 7] et al. produced stretchable conductors by the way of embedding AgNWs in the surface of PDMS. And the conductive composites are conductive and flexible. Based on the work of predecessors, this paper explores new methods to further improve the flexibility of wires.

Here, starting from the selection of materials and the new manufacturing process, the new method for manufacturing flexible wires was explored. The novel AgNWs/PDMS composites were selected, and a pre-stretching process and a glass sheet transfer process were used to produce flexible wires. The flexibility and elasticity of this new type of wires are improved greatly.

2. Experimental Detalls

2.1. Materials

The flexible wire is made by the method of mixing conductive fillers into elastomer. The conductive composite material not only maintains the excellent conductivity of conductive fillers, but also has excellent flexibility of elastomer. Therefore, it is an ideal material for making flexible wires.

The conductive properties of this conductive composites are determined by the conductive filler mainly. Here, AgNWs are selected for their excellent conductivity, good mechanical flexibility and well transparency. Currently, AgNWs have demonstrated application value in solar cells, touch screen, tactile sensor and other areas [8]. Though experiments, it's found that the size, concentration, and solvent of AgNWs all have effects on the flexible wire. In order to ensure the conductivity and flexibility of the flexible wire, the size and concentration of AgNWs should be with a diameter of 90 nm, a length of 2 ~ 20 μm and a concentration of 10 mg/mL. In addition, the water solvent is selected to obtain a uniform thickness during the production process of flexible wires.

Elastomer mainly affects the flexibility of conductive composite material. Here, well-known PDMS is chosen for its outstanding physical and chemical properties. PDMS film has excellent flexibility and transparency, so it is an extremely suitable material for making flexible wires as substrates. In this paper, the type of PDMS is Dow Corning Sylgard 184.

Therefore, the flexibility of the wire is guaranteed from the perspective of materials.

2.2. Method

The flexible wire is fabricated by using a pre-stretching process and a glass piece transfer process. The specific steps are as follows:

First, a PDMS film needs to be prepared in advance. The preparation method comprises the following steps: mixing the main agent and the binder of the liquid PDMS in a certain ratio; stir the composite uniformly and pour it on a glass plate; allow the liquid PDMS to naturally flow to the required thickness; then remove the bubbles inside the liquid PDMS; at last, heat and cure the PDMS film.

Second, cut the prepared PDMS film into a strip with a width of 5mm and a length of 100 mm. Stretch the strip PDMS film and fix it on a glass plate.

Third, drop AgNWs solution on a glass piece with a width of 3 mm and a length of 60 mm. Wait for the AgNWs solution dried leaving AgNWs on the glass piece, then repeat the dropping process three times.

Forth, make the side of glass piece coated with AgNWs adhered to the pre-stretched PDMS film by thin liquid PDMS. Remove bubbles inside the middle between AgNWs and liquid PDMS, then heat and cure the entire structure.

Finally, remove the glass piece, then the flexible wire is obtained as shown in Figure 1.

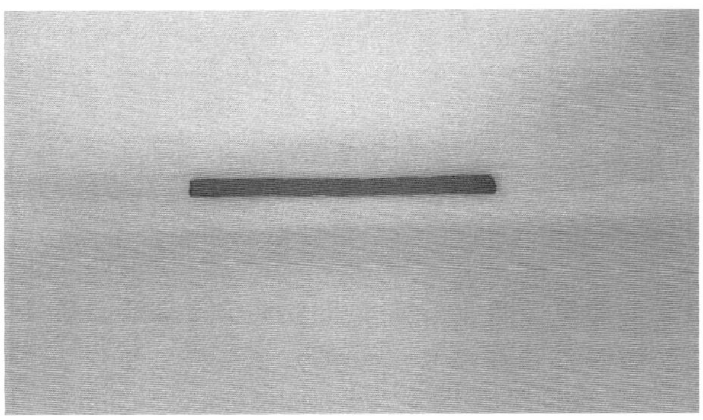

Figure 1. Physical photo of the flexible wire

3. Results and Discussion

In order to test the flexibility of the flexible wire, a stretching test is designed using a tensile tester provided by Shanghai Hualong Testing Instrument Co., Ltd. The flexible wire was fixed in the middle of the clamps of the tensile tester and was repeatedly stretched to 130% at a constant speed during the stretching test. The stretching speed would have effects on the conductive stability of the flexible wire. Here, the stretching speed selected in this experiment is 3 mm/min. The method of stretching test is to make fifty times 0-30%-0 stretching cycles on the flexible wire.

IOP Conf. Series: Materials Science and Engineering **746** (2020) 012027 doi:10.1088/1757-899X/746/1/012027

Figure 2. Stretching test

Figure 3 shows the relationship between resistance and stretching cycles of the flexible wire. The initial resistance of the flexible wire is 3.2Ω, and the resistance value increases slightly with the number of stretching increases. At the state of fifty times stretching to 130%, the resistance of the flexible wire is just 19.5Ω, and the resistance growth rate relative to initial resistance is only 510%. The stretching test proved that the flexible wire produced by a pre-stretching process and a glass sheet transfer process has good conductivity and elasticity.

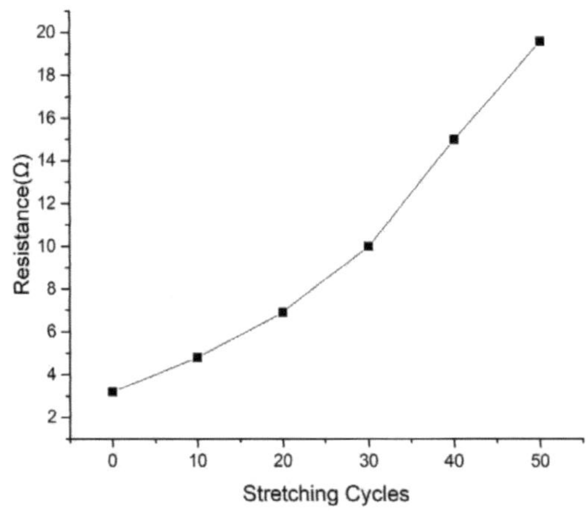

Figure 3. The relationship between the resistance and the stretching cycles

Figure 4 shows the SEM image of the flexible wire after repeated stretching cycles. It is found that, the surface of the flexible wire has no obvious cracks. Therefore, good electrical conductivity of the flexible wire is maintained after repeated stretching.

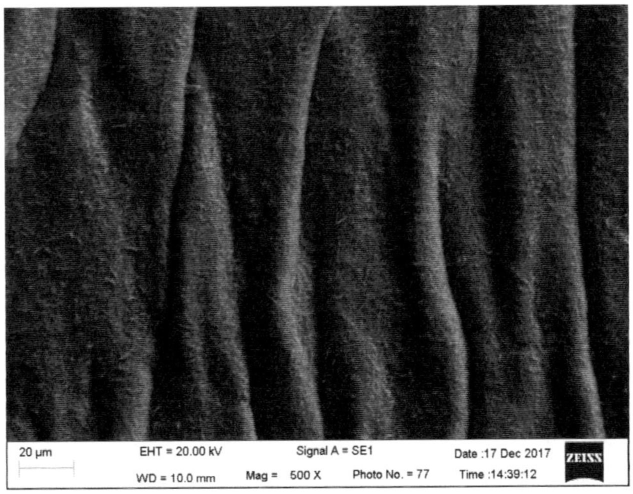

Figure 4. SEM image of the flexible wire

4. Conclusion

In summary, the novel flexible wire based on AgNWs/PDMS composites produced by a pre-stretching process and a glass sheet transfer process exhibits excellent flexibility and elasticity. Therefore the flexible wire has potential applications in flexible devices such as bionic robots, medical instruments, health monitoring equipment, and so on.

5. Acknowledgments

The authors would like to acknowledge the financial support provided by Innovation Method Fund of China (2016IM010300) and the Fundamental Research Funds for the Central Universities (FRF-GF-18-010B).

6. References

[1] Wang Guobiao, CHEN Diansheng, CHEN Kewei, et al. The current research status and development strategy on biomimetic robot [J]. Journal of Mechanical Engineering, 2015, 51 (13): 27-44.

[2] CAO Jianguo, ZHOU Jianhui, MIAO Cunxiao, et al. Research progress and development strategy on tactile sensors for e-skin [J]. Journal of Harbin Institute of Technology, 2017, 49 (1): 1-13.

[3] S. P. Lacour, J. Jones, S. Wagner, T. Li, Z. Suo. Stretchable interconnects for elastic electronic surfaces. P IEEE. 93 (8): 1459-1467, 2005.

[4] I. Chtioui, F. Bossuyt, M. de Kok, J. Vanfleteren, M. H. Bedoui. Arbitrarily Shaped Rigid and Smart Objects Using Stretchable Interconnections. IEEE T COMP PACK MAN. 6: 533-544, 2016.

[5] F. Xu, Y. Zhu. Highly conductive and stretchable silver nanowire conductors. Adv. Mater. 24 (37): 5117-5122, 2012.

[6] Wang H P, Zhou D B, Cao J G. Development of a skin-like tactile sensor array for curved surface [J]. IEEE Sensors Journal, 2014, 14 (1): 55-61.

[7] Wang H P, Zhou D B, Cao J G. Development of a stretchable conductor array with embedded metal nanowires [J]. IEEE Transactions on Nanotechnology, 2013, 12 (4): 561-565.

IOP Conf. Series: Materials Science and Engineering **746** (2020) 012028 doi:10.1088/1757-899X/746/1/012028

Effects of Shot Peening on Fatigue Life of Ground Component of 7075-T6 Aluminum Alloy

Xiuquan Cheng[1], Junhao Zhang[2], Xiangfei Sheng[2] and Qinxiang Xia[2,*]

[1]School of Aircraft Maintenance Engineering, Guangzhou Civil Aviation College, Guangzhou 510403, China
[2]School of Mechanical and Automotive Engineering, South China University of Technology, Guangzhou 510640, China
[*]Corresponding author: Qinxiang Xia, Email: meqxxia@scut.edu.cn

Abstract. The aircraft corroded components can continue to be used after removing the corroded portion by grinding and strengthening the ground area by shot peening. The finite element analysis model of the ground specimen with complex shape surface was established based on ABAQUS software. The residual stress distributions under three conditions of without shot peening, ground surface shot peening and three surfaces shot peening were analysed. On this basis, the cyclic alternating load was introduced into the shot peening model, and the fatigue analysis was carried out based on MSC. Fatigue software. The result shows that the position of the maximum tensile stress moves from the surface to the subsurface of the ground specimen after shot peening; shot peening conditions will affect the location of the crack sources, the propagation direction of the crack; with the increasing of the numbers of peened surface, the location of stress danger point shifts from ground surface to side surface or sub-surface, and the crack propagation velocity decreases. All of these increases the fatigue life. The experimental results conform well to the simulation one.

1. Introduction

Aircraft structural components are easily corroded due to the working environment. Some slight corroded components can continue to be used after removing the corroded material by grinding and then strengthening the ground area by shot peening before the final surface anti-corrosion treatment. Residual stress field is induced in the surface layer of the ground component after shot peening. The fatigue life and the corrosion resistance of the strengthened component improve substantially under the influence of residual compressive stress [1, 2].

In the past decades, fatigue life improvement by shot peening is a hot research topic [3, 4]. Y.K. Gao [5] investigated the fatigue properties of 7050-T7451 aluminum alloy after shot peening, and the results show that the fatigue life of the shot peened component improves significantly compared with the machined one. Y.K. Gao et al. [6] analyzed the influence of shot peening on small crack growth and fatigue life of 7475-T7351 aluminum alloy. Their results show that cracks initiate at the second phase particles of the alloy and the growth rate of fatigue crack reduces greatly after shot peening. G.H. Majzoobi et al. [7] focus on the fretting fatigue life of 7075 aluminum alloy after shot peening, and the results show that the low cycle fatigue life increased by 300% after shot peening.

In this paper, the 7075-T6 aluminum alloy ground specimen was taken as the researching object. The stress response of shot peened ground specimen under the cyclic alternating load was investigated. The comparison between the simulation and experimental results of fatigue life and the position of the most dangerous area or crack sources of the ground specimen was carried out.

2. Establishment of the Shot Peening Macro-scale Finite Element Analysis (FEA) Model

In practice, the corrosion penetrating through the width direction of the structural components occurs frequently, as shown in Figure 1. The specimen with corrosion is firstly disassembled and then ground the corroded area.

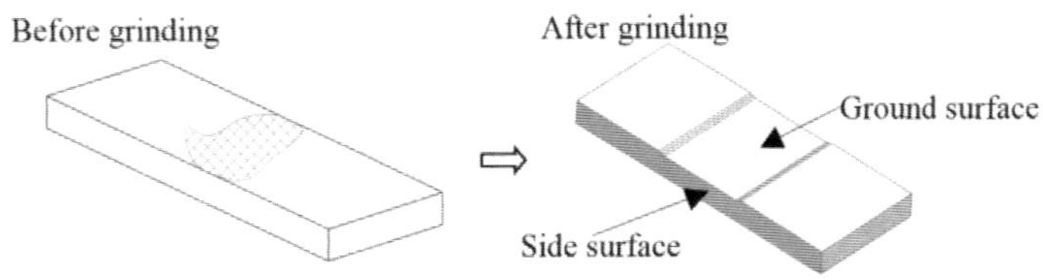

Figure 1. Ground structure component

Afazov et al. [8] proposed an element mapping (EM) method, by which the residual stress was mapped from the micro-scale FEA model to the macro-scale one. Li et al. [9] analysed the influence of the residual stress and on the crack propagation based on the EM method. In Figure 2, the residual stress was obtained by the fixed peening angle (as shown in Figure 3) through the micro-scale FEA model established by ABAQUS software.

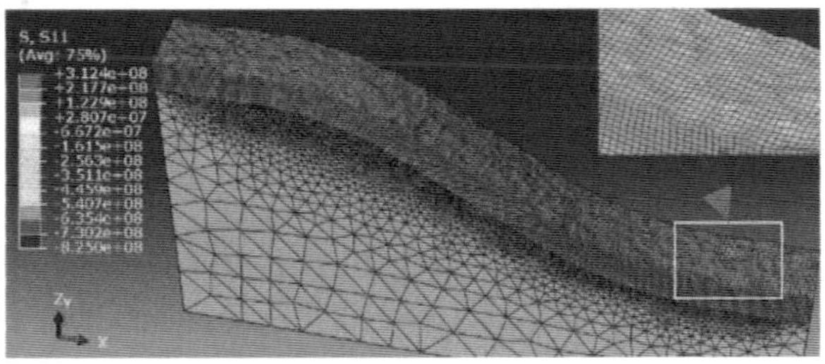

Figure 2. Distribution of residual stress

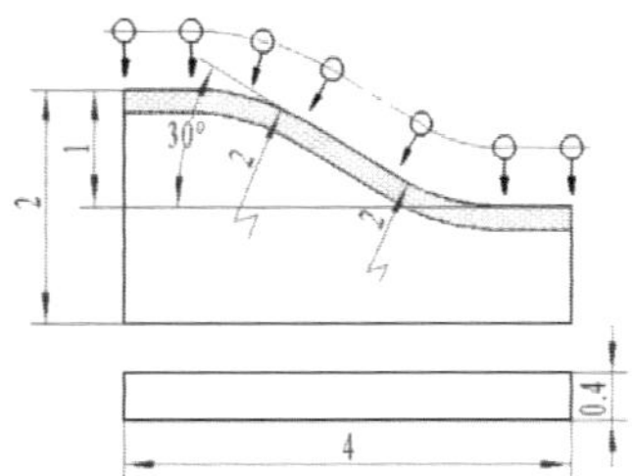

Figure 3. Micro-scale FEA model (Units: mm)

The macro-scale FEA model was shown in Figure 4. The depth of the ground portion h_d=2 mm, the concave and convex fillet radius r_d=2 mm, and the inclination angle of oblique plane a_d=30°, which are the same as the micro-scale one; while the length, width, and height of the specimen is 30 mm, 2.5 mm and 10 mm. The model was used to obtained the residual stresses under three different shot peening conditions (condition-1: without shot peening; condition-2: top surface shot peening; condition-3: three surfaces shot peening).

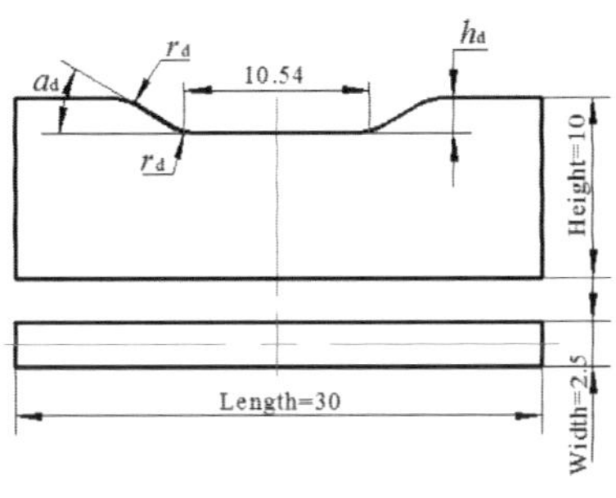

Figure 4. The macro-scale FEA model (mm)

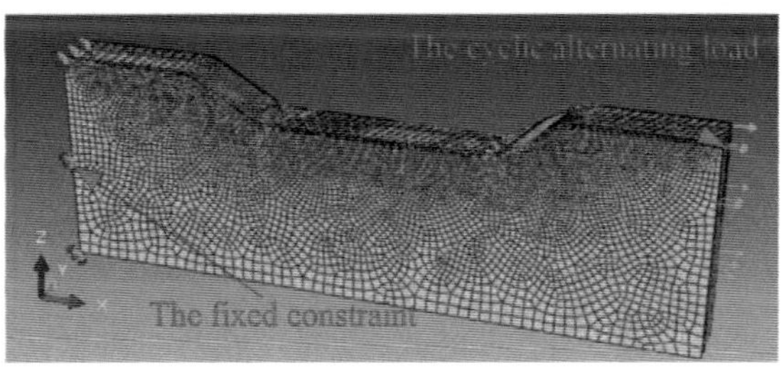

Figure 5. Macro-scale FEA model with load and residual stresses

Figure 5 shows the macro-scale FEA model based on the size of Figure 4. The establishment of the model includes two steps. The first one is to obtain the residual stress field of the micro-scale FEA model after shot peening and map it to the macro-scale FEA model. The second one is to set the boundary conditions and loads of the macro-scale FEA model. In the macro-scale FEA model, the left end face is fixed constraint, and the right end face is applied with cyclic alternating load in X-axis direction (maximum stress is 120 MPa, stress ratio is 0.1, frequency is 100 Hz). Finally, the obtained stresses were used to analyze the fatigue crack growth of the ground specimens based on software MSC. Fatigue [10].

To verify the reliability of the residual stress obtained by the micro-scale FEA model, X-ray diffraction method was used to measure the residual stress on the specimen surface. The results show that the maximum deviation between the simulated and measured values is 5.15%. This means that the residual stress field obtained by the model is reliable.

3. Numerical Simulation Analyses

3.1. Stress Distributions of the Shot Peened Specimens under the Cyclic Alternating Load

Figure 6 shows the simulation results of the peened specimen with cyclic alternating load. The results show that the maximum stress on the ground surface $\sigma_{max\text{-}surf}$ locates near the centre point of the concave surface due to the stress concentration; the maximum stress on the side surface $\sigma_{max\text{-}side}$ locates at the point *B*, which is in the centreline of the side surface and near the upper surface.

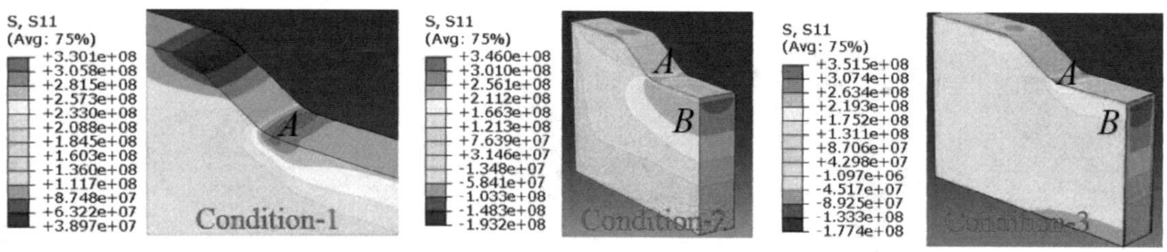

Figure 6. Simulation results of the ground specimen with cyclic alternating load

Table 1 shows the maximum stress $\sigma_{max\text{-}surf}$ at the concave surface point *A*, the maximum stress $\sigma_{max\text{-}side}$ at the side surface point *B*, the maximum stress σ_{max} in the specimen. Li et al. [11] proposed the concepts of surface fatigue limit (SFL) and interior fatigue limit (IFL), and investigated the relationship of the critical stress causing the mesoscopic-yield between the surface material and the interior material. The results show that the ratio of the interior maximum stress to the surface one is approximately $\sqrt{2}$. It indicates that the fatigue crack source is more difficult to initiate from the interior than from the surface of specimens.

Table 1. Stress characteristics under three different shot peening conditions

	$\sigma_{max\text{-}surf}$ (MPa)	$\sigma_{max\text{-}side}$(MPa)	σ_{max}(MPa)	$\sigma_{max}/\sigma_{max\text{-}surf}$	$\sigma_{max}/\sigma_{max\text{-}side}$
Condition-1	330.1	272.3	330.1	1.000	1.212
Condition-2	212.9	260.1	346.0	1.625	1.330
Condition-3	238.5	167.3	351.5	1.625	2.101

The ratio $\sigma_{max}/\sigma_{max\text{-}surf}$ and $\sigma_{max}/\sigma_{max\text{-}side}$ listed in Table 1 shows that:

1) Under condition-1, both the value of $\sigma_{max}/\sigma_{max\text{-}surf}$ and $\sigma_{max}/\sigma_{max\text{-}side}$ are smaller than $\sqrt{2}$ and $\sigma_{max\text{-}surf}$ is greater than $\sigma_{max\text{-}side}$. It indicates that the fatigue crack source is located on surface near point A.

2) Under condition-2, the value of $\sigma_{max}/\sigma_{max\text{-}surf}$ is bigger than $\sqrt{2}$, and the $\sigma_{max}/\sigma_{max\text{-}side}$ is still smaller than $\sqrt{2}$. The ground surface is strengthened by shot peening. However, the two side surfaces become the new dangerous areas, the fatigue crack source is located on side surface near point *B*.

3) Under condition-3, both the value of $\sigma_{max}/\sigma_{max\text{-}surf}$ and $\sigma_{max}/\sigma_{max\text{-}side}$ are bigger than $\sqrt{2}$. It indicates that the whole ground specimen is strengthened by shot peening, and the position of fatigue crack source moves to the sub-surface of specimen. The fatigue property of the ground specimen

improves effectively. Meanwhile $\sigma_{max\text{-}surf}$ is greater than $\sigma_{max\text{-}side}$ under condition-3. The stress dangerous point of fatigue crack source is located on the sub-surface near the point *A*.

3.2. Effect of Shot Peening on the Fatigue Life of the Specimens

Figure 7 shows the simulation results of fatigue crack propagation under the three different shot peening conditions. The results show that when the crack length propagated to 5mm, the specimen almost fatigue fracture. The propagating lives were 8.29×10^4 cycles, 1.53×10^5 cycles, and 1.29×10^6 cycles. When the crack length is below 1 mm, the crack propagation velocity is significantly affected by the surface stress level of the specimen, while the crack propagation velocity increases rapidly after the crack length exceeds 1 mm. From condition-1 to condition-3, the initial crack propagation velocity gradually decreases due to the gradual increase of the surface residual compressive stress, which resulted in a gradual increase in the total fatigue life.

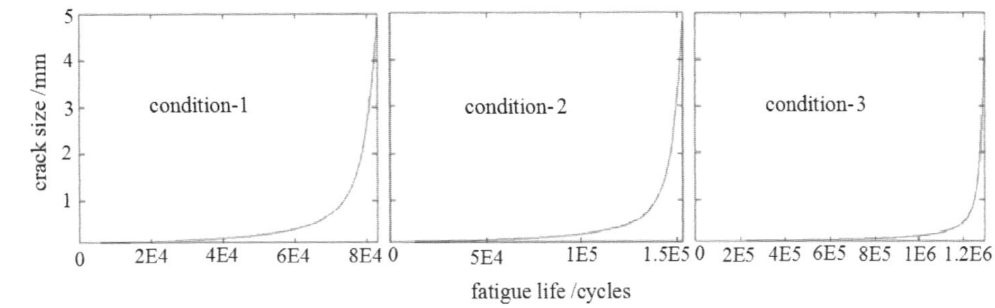

Figure 7. Fatigue crack growth under three different shot peening conditions

4. Experimental Research on the Effect of Shot Peening on the Fatigue Life

4.1. Experimental Conditions

The shape and size of the fatigue specimen is shown in Figure 8, all of the specimens were machined along with the rolling direction of material, where the shaded areas of the specimens being shot peened under condition-2 (only ground surface A_1 was shot peened) and condition-3 (ground surface A_1, two side surfaces A_2 and A_3 were shot peened). Four specimens were tested repeatedly for each condition and the average value of cycles was regard as the fatigue life of each condition. The definition of three shot peening conditions, the process parameters of shot peening, the cyclic alternating load and the main geometry sizes of specimen are the same as that in simulation. All of the fatigue tests were carried out according to the standard HB 5287-96 [12] by using MAG-100 KN high frequency fatigue testing machine.

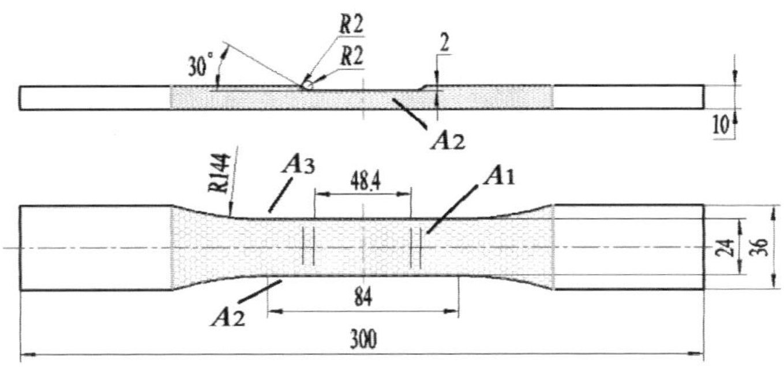

Figure 8. Shape and size of the fatigue specimens (mm)

4.2. Results and analysis

4.2.1. Position of fatigue fracture

Figure 9 shows the fatigue fracture of specimens. It shows that the fracture under condition-1 and -3 located near point A. As analyzed above, the dangerous area of the fatigue specimen located on the concave surface (near point A) under condition-1, and in sub-surface near point A under condition-3. Therefore, both of the two positions of fractures locate near the concave surface. For the specimen under condition-2, the dangerous area locates in the center-line of the side surface near the ground surface (near point B). Therefore, the fracture located near the center of the specimen.

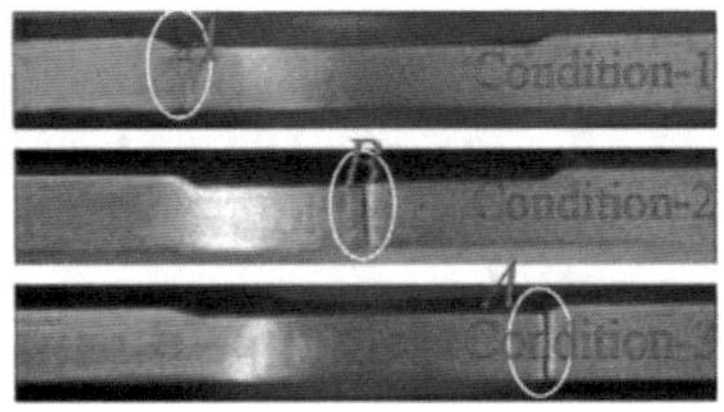

Figure 9. Fracture location of fatigue specimens

4.2.2. Initiating position and propagating direction of crack

Figure 10 shows the fatigue fracture morphology. The small circles mean the position of the fatigue crack sources and the arrows indicate the cracks propagating direction. The results show that: 1) For the specimen under condition-1, multiple fatigue crack sources occur and located on the concave arc of the ground surface. The cracks propagate mainly along the thickness direction of the specimen; 2) For the specimen under condition-2, only one fatigue crack source occurs and located on the side surface of the specimen, the crack propagates mainly along the width direction of the specimen; 3) For the specimen under condition-3, also only one fatigue crack source occurs and located in the sub-surface of concave arc below point A, the crack propagates mainly along the thickness direction of the specimen.

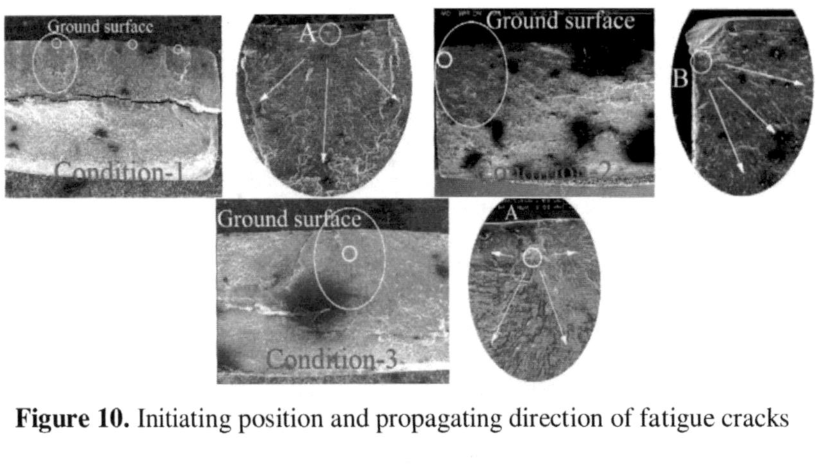

Figure 10. Initiating position and propagating direction of fatigue cracks

4.3. Comparison between Simulation and Experimental Results

Under the experimental and simulation conditions of this article, the average fatigue life obtained by experimental under three different shot peening conditions were 8×10^4 cycles, 1.52×10^5 cycles, and 1.50×10^6 cycles, and that obtained by the simulation were 8.29×10^4 cycles, 1.53×10^5 cycles, and 1.29×10^6 cycles. The fatigue life coincides well each other.

5. Conclusions

The main conclusions are as follows:

1) The stress dangerous points of fatigue cracks for the ground specimen without shot peening is located on the ground surface, while that of the specimen with ground surface or three surfaces shot peening is located on the side surface or the sub-surface.

2) Shot peening can reduce the number of specimen crack sources. For the specimen without shot peening, there are multiple fatigue crack sources; for the ground specimen with shot peening, there is only one fatigue crack source.

3) The fatigue life improves greatly by shot peening. The fatigue life of the ground specimen with ground surface and three surfaces shot peening are 1.84 times and 15.56 times of that without shot peening, respectively.

6. Acknowledgments

The authors are grateful for the financial support provided by National Natural Science Foundation of China (NSFC) (No. 51575117 and 61179051).

7. References

[1] Fathallah R, Laamouri A, Sidhom H, Braham C 2004 High cycle fatigue behaviour prediction of shot-peened parts *Int J Fatigue* 26 (10) pp 1053-1067.

[2] Curtis S, De los Rios ER, Rodopoulos CA, Levers A 2003 Analysis of the effects of controlled shot peening on fatigue damage of high strength aluminum alloys *Int J Fatigue* 25 (1) pp 59-66.

[3] Bagherifard S, Guagliano M 2012 Fatigue behavior of a law-alloy steel with nanostructured surface obtained by severe shot peening *Eng Fract Mech* 81 (10) pp 56-68.

[4] Terres MA, Laalai N, Sidhom H 2012 Effect of nitriding and shot-peening on the fatigue behavior of 42CrMo4 steel: Expermental analysis and predictive approach *Mater Des* 35 (7) pp 741-748.

[5] Gao YK 2011 Improvement of fatigue property in 7050-T7451 aluminum alloy by laser peening and shot peening *Mater Sci Eng* A 528 (10-11) pp 3823-3828.

[6] Gao YK, Wu XR 2011 Experimental investigation and fatigue life prediction for 7475-T7351 aluminum alloy with and without shot peening-induced residual stresses *Acta Mater* 59 (9) pp 3737-3747.

[7] Majzoobi GH, Azadikhah K, Nemati J 2009 The effects of deep rolling and shot peening on fretting fatigue resistance of aluminum-7075-T6 *Mater Sci Eng A* 516 (1-2) pp 235-247.

[8] Afazov SM, Becker AA, Hyde TH 2012 Mathematical modeling and implementation of residual stress mapping from microscale to macroscale finite element models *J Manuf Sci Eng* 134 (2) pp 1-10.

[9] Li YZ, Wang Q, Zhang ZP, He J 2011 Exploring fruther fatigue crack closure in residual stress field through numerical simulation *J Northwestern Polytechnical Univ* 29 (1) pp 97-102.

[10] Zhang J, Gu X, Zhu L, Sun AH 2010 Numerical simulation of fatigue life of 7050 aluminum alloy processed by laser shock processing *Chin J Lasers* 37 (12) pp 3192-3195.

[11] Li JK, Li HT, Yao M 1990 Tensile residual stress field induced by shot peening and internal fatigue limit of materials *Acta Aeronaut Astronaut Sin.* A11 pp 369-377.

[12] Novovic D, Dewes R C, Aspinwall D K 2004 The effect of machined topography and integrity on fatigue life *Int. J. Mach. Tools Manuf.* 44 (2-3) pp 125-134.

IOP Conf. Series: Materials Science and Engineering **746** (2020) 012029 doi:10.1088/1757-899X/746/1/012029

Diamond Scintillation and Quality Evaluation

Zhongkui Zhang[1], Ying Guo[2, *], Qiutong Liu[2] and Yijiang Shen[2]

[1]School of Information Engineering, China University of Geosciences (Beijing) pp Beijing, China

[2]School of Gemmology China University of Geosciences (Beijing) pp Beijing, China

[*]Corresponding author: guoying@cugb.edu.cn

Abstract. The quality of a diamond is not only affected by the 4Cs but also affected by fire and scintillation. The relationship between the color, clarity, fluorescence, and scintillation of a diamond is obtained by associating its scintillation area and gemological characteristics. Take images of 18 round bright cut diamonds in total, for each to rotate 30° to take one and get 12 images by rotating a circle. Rotate the images to the same angle, cut out the square image of 1000*1000 pixels, then calculate the color difference of the image of adjacent angles, and then judge whether scintillation occurs at each point to get the scintillation area of the diamond. The results showed that the lighter the color, the better the clarity and the weaker the fluorescence, then the larger the scintillation area and the more obvious the scintillation. This indicates that the scintillation area has a certain reference value for diamond quality evaluation.

1. Introduction

The static properties of diamonds have been studied in depth. With the help of weighted light returns (WLR) model Hemphill et al (1998) made comprehensive analysis of multiple crown angle, pavilion angle and table size combinations as well as ray trace the 3D model of diamond, found that different combinations can produce the same results, and gave the advice to improve the brilliance in diamond cutting. Zhang et al (2006) proposed that only diamonds that meet the requirements of precise cutting grinding ratio and high symmetry can produce the pattern of eight arrows and eight hearts. Liu et al (2014) on the one hand, collected images of colorless synthetic cubic zirconia which was similar to diamonds in optical properties; on the other hand, input cutting parameters of samples, establish models, and simulate the optical effect pattern of samples. The comparison between the two verified the feasibility of studying lightness through diamond models. What's more, Moses et al (2004) believed that the quality evaluation of diamonds should not only unify the observation environment, consider personal preferences and understand the market tendency, but also comprehensively take the interaction of various facet of diamonds integrated into account instead of being limited to a few attributes. Reinitz et al (2001) emphasized that every facet matters in a round brilliant diamond. Therefore, Caspi et al (1997) pointed out that diamond dealers had significantly improved the processing and output level of rough stones, despite the short history of using computer automated analysis, cutting and polishing. Not only in the field of diamond processing, but the diamond rating has also already entered the stage of software automation. Wang et al (2011) pointed out that diamond cutting analysis can be completed automatically by the instrument.

In addition, Yu et al (2016) calculated the total reflection of the plane model of the incident light of the vertical section of the diamond and found that the diamond has the maximum light return value and a smaller dark area ratio when pavilion angle within the range of 39.0° ~ 41.0°. When the size of the pavilion angle is fixed, there is a unique crown angle to enable the diamond to achieve the best fire effect. Reinitz et al (2001) analyzed the relationship between diamond fire and various parameters with the help of a 3D mathematical model and pointed out that in some cases, subtle changes in various attributes would significantly affect the final fire and lightness effect. In addition to computer modeling, Fourier Transform Infrared Spectroscopy (FTIR) has also been widely used in the classification of diamonds, the identification of synthetic diamonds, and researches on colored diamonds. King et al (2005) discussed the genesis of yellow diamonds and further divided them into 5 groups according to their absorption spectra. Shigley et al (2015) further studied the absorption spectra of diamonds of various colors and found that diamonds with different absorption spectra and different types can present similar colors. Yan discussed the feasibility of distinguishing CVD synthetic diamonds from natural diamonds through the fluorescence characteristics of DiamondView diamonds. Since the synthetic diamond has already reached the level of gems, Breeding et al (2009) still suggested sending the uncertain type of diamond to the gemology laboratory for further testing even after he had put forward the methods of the absorption spectrum, gem microscope, electric conductivity and so on. In other aspects, studies on the setting of diamond selection lamps showed that it is more appropriate to sort cut diamonds above VG level when the incidence angle of diamond lamp is between 20° and 25°.

However, further research is needed to determine the specific relationship between the dynamic characteristics of a diamond, such as its scintillation as well as clarity, lightness, and fluorescence. Scintillation refers to the light produced by the diamond as seen by the observer when the diamond, light source or observer moves. In order to better study the relationship between diamond scintillation and quality, we use the color difference to quantitatively analyze the scintillation phenomenon in diamond pictures, and the relationship between gemological characteristics and scintillation of diamonds is studied by quantifying these factors.

2. Sample Overview and Image Processing

2.1. Experimental Samples

We used a total of 18 round brilliant cut diamonds, the sizes of which are ranged from 0.300 ct to 0.387 ct, the color grading: D ~ H; the cutting, polishing, and symmetry grading: VG and above. And the cutting parameters: the table size is 58% ± 3%; the total depth is 61.4% ± 3.1%; the pavilion depth is 43% ± 1%; the crown angle range is 32.5° ~ 37.5°. As shown in table 1.

Table 1. The parameters of the diamond used in the experiment

Number	Weight/ct	Color	Clarity	Fluorescent	Total depth/%	Table size/%	Crown angle/°	Pavilion depth/%	Cut	Polish	Symmetry	Scintillation area
D-3-51	0.300	D	VVS1	FNT	61.7	56	36.0	42.0	VG	VG	VG	4.40%
D-3-52	0.302	D	VVS2	NON	62.2	59	37.0	42.5	VG	EX	VG	2.73%
D-3-53	0.313	D	VS1	NON	61.5	56	34.5	43.5	EX	VG	EX	4.37%
D-5-10	0.387	E	VS1	MED	62.6	55	34.5	44.0	VG	EX	EX	0.42%
D-5-14	0.300	F	VS1	NON	64.5	56	37.0	43.0	VG	VG	EX	0.26%
D-5-15	0.344	F	VS2	FNT	59.7	60	32.5	43.5	EX	EX	EX	0.43%
D-5-16	0.316	G	VVS1	NON	61.2	58	34.5	43.5	EX	EX	EX	0.32%
D-5-17	0.331	G	VS2	MED	61.3	55	34.0	43.0	EX	EX	VG	0.34%
D-5-18	0.302	G	SI1	NON	63.8	56	37.5	43.0	VG	VG	EX	0.56%
D-5-19	0.381	G	SI2	FNT	62.6	58	35.5	43.5	EX	EX	EX	0.25%
D-5-20	0.345	H	VVS1	MED	62.7	55	35.0	43.5	VG	EX	EX	0.53%
D-5-4	0.300	F	SI1	NON	60.8	56	33.0	43.5	EX	EX	VG	1.17%
D-5-5	0.318	G	IF	NON	62.3	56	35.0	43.0	EX	EX	EX	0.64%
D-5-7	0.306	D	SI1	NON	60.1	59	33.5	43.0	EX	VG	EX	1.23%
D-5-8	0.302	D	VS2	NON	62.6	55	33.0	44.0	EX	EX	VG	0.53%
D-5-9	0.303	D	SI2	NON	58.3	61	32.5	43.0	VG	EX	VG	1.25%
KS1	0.306	E	IF	NON	61.1	57	34.5	43.5	EX	VG	EX	1.81%
KS2	0.305	E	VVS1	NON	62.3	57	35.0	43.5	EX	VG	EX	2.87%

2.2. Photography and Preprocessing

Under the condition of indoor natural light scattering, control the shooting background to be pure black, take an image of each diamond every 30° rotation, and get 12 images by rotating once. Then remove the edge of the image and rotate it to the same angle to get 12 figures of the diamond. As shown in figure 1.

Figure 1. Diamond rotates 360°

2.3. Calculate the Colorimetry Component of the Image

Colorimetry studies the change and relation of color. In CIE 1976 colorimetry space, color has L, a and b components to quantify color. The judgment of scintillation between two pictures can refer to the color difference, lightness difference and hue difference of two pictures.

Take images of a diamond at 240° and 270° (9th and 10th of figure 1) as an example. Apply applycform function in MATLAB to calculate these two images to get the results of color difference, lightness difference, and hue difference. Five pictures (a) ~ (e) in figure 2 represent rotating the diamond 240° and 270° respectively, as well as the hue difference, color difference and lightness difference between them. In figure (a), (b), (d) and (e), two flashing areas are marked with gray and black respectively at the same position, named as the upper and the lower region. As can be seen from the figure (a) and (b) pp the upper region shows a significant scintillation, while only the left part of the lower region shows scintillation. The performance of the color difference in the upper region is very similar to the lightness difference, but the noise suppression in the lower region is obviously better than the lightness difference. Although the lightness difference can show the change of lightness in the upper region, there is too much noise information in the lower region, that is, the lightness difference cannot well filter out the noise interference. As for the hue difference, it shows no information in two regions, so we take no account of it. Considering comprehensively, the color difference is selected as the criterion of scintillation.

As can be seen from the color difference diagram, there are other areas with obvious color difference changes which haven't been mentioned, corresponding to the scintillation areas in figure (a) and (b).

Figure 2. Images and colorimetry component of diamonds

2.4. Scintillation Area

In CIELAB colorimetry space, the unit of color difference is NBS. As can be seen from the figure a and b, a distinct scintillation occurs in the upper region; in the lower region, the left half shows a more obvious scintillation, while the right half shows only a slight change in lightness, which should not be judged as scintillation. In the upper region of the color difference diagram, the color difference value of this region is shown in the diagram is about 10 NBS to above 90 NBS. There are some color differences in the left half of the lower region. Some regions about 60 NBS show obvious scintillation, while other regions on the right (below 40 NBS) shows an only slight change in lightness and should not be judged as scintillation. Therefore, take 60% of the maximum color difference as the threshold. When the color difference value exceeds 60% of the maximum value, the point determines that scintillation has occurred. Although this standard will sacrifice the effective value of some scintillation areas, it can filter out the influence of the non-flashing part.

Calculate the color difference of each pixel of the two images with 30° adjacent of the diamond, and determine the scintillation successively. The scintillation area S of the diamond is defined as N_s divided by the total number of pixel points N:

IOP Conf. Series: Materials Science and Engineering **746** (2020) 012029 doi:10.1088/1757-899X/746/1/012029

$$S \equiv \frac{N_s}{N} \tag{1}$$

The calculation above was performed on all adjacent images of 12 images with 360°, and the arithmetic mean value was taken. The result was expressed as a percentage to obtain the scintillation area of a diamond. After all the diamonds are photographed, processed and computed, each diamond corresponds to a scintillation area. The result is given in table 1.

3. Effect of Gemological Properties of Diamond on Scintillation Area

3.1. Quantification of Diamond Index

Three attributes of clarity, color, and fluorescence of diamond are selected to quantify each one. The advantages and disadvantages of each attribute are represented by numbers, with 0 representing the best index and 8 representing the worst index. For example, the "clarity" attribute is represented by 0 to 8 from FL to P1. Color grade D, E, F, G, and H are represented by 0, 2, 4, 6 and 8 respectively. Fluorescence NON is 0, FNT is 4 and MED is 8, as given in table 2. In this way, the gemological characteristics of diamonds are quantified into the Numbers 0~8 for uniform comparison.

Table 2. Diamond index grading quantification

Index	Clarity	Color	Fluorescence
0	IF	D	NON
1	VVS1	-	-
2	VVS2	E	-
3	VS1	-	-
4	VS2	F	FNT
5	SI1	-	-
6	SI2	G	-
7	SI3	-	-
8	P1	H	MED

3.2. Plot Results

Take the scintillation area as the independent variable and the above three indexes as dependent variables to make scatter plots respectively, and observe the correlation between each attribute and the scintillation area. In this way, scatter plots of the three properties and their regression curves are obtained, as shown in figure 3. With the increase of scintillation area, all three properties have a decreasing tendency, that is, the lighter the color of the diamond, the better the clarity, the weaker the fluorescence, then the larger the scintillation area, the more obvious the scintillation.

4. Conclusion

After calculating the scintillation area through the color difference of adjacent angle pictures of diamonds, it is found that:

The higher the clarity, the lighter the color, the weaker the fluorescence of the diamond, then the larger the scintillation area, the stronger the scintillation.

Take a picture of the diamond every 30°, rotate it to the same angle, and calculate the color difference between two adjacent pictures, and then find the area of the scintillation. The lightness difference will magnify subtle changes in lightness and interfere with real scintillation areas. The hue difference, which performance is the worst, cannot reflect scintillation, so choose color difference as the criterion of scintillation.

The area of scintillation is related to the gemological characteristics of a diamond. The clarity, color, and fluorescence are quantified as 0 ~ 8, and it is found that the scintillation area of a diamond is positively correlated with its quality and can be used as one of the reference standards to measure the quality of a diamond.

The influence of clarity, color, fluorescence and other attributes on scintillation area conforms to the GIA and GB standards for diamond quality identification, which also verifies the rationality that the weaker the fluorescence, the higher the value.

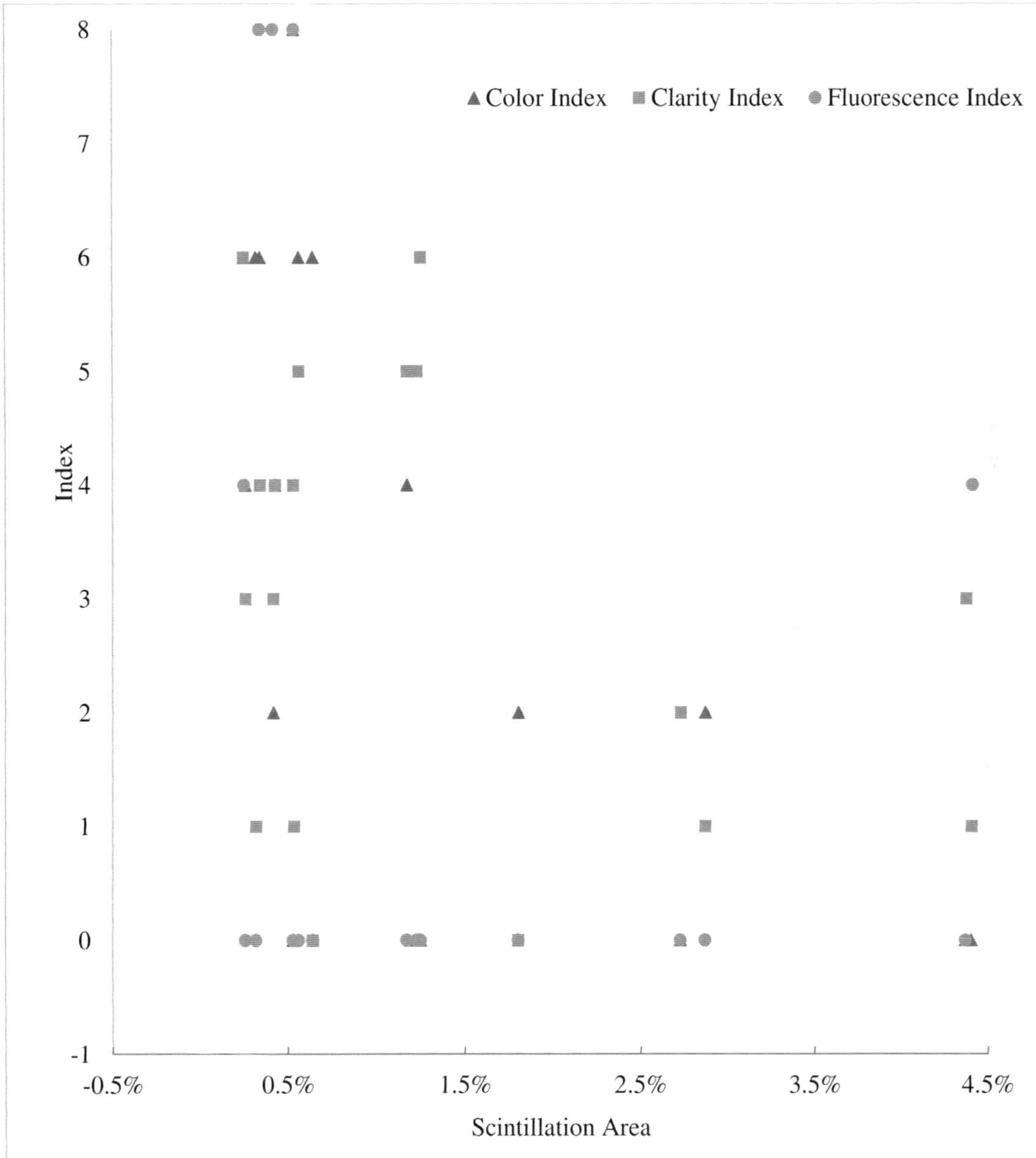

Figure 3. The relationship between scintillation area and gemological properties

5. Acknowledgments

Thanks for the diamond samples and test site provided by the laboratory of School of Gem Studies. Thanks to Mr. Guo Ying, teachers of the institute and Liu Fukang senior for their careful guidance.

IOP Conf. Series: Materials Science and Engineering **746** (2020) 012029 doi:10.1088/1757-899X/746/1/012029

6. References

[1] Hemphill T S, Reinitz I M, Johnson M L and Shigley J E 1998 Modeling the appearance of the round brilliant cut diamond: An analysis of brilliance *Gems & Gemology* 34 (3) pp 158-183.

[2] ZHANG Y T and HE X M 2006 Cutting Condition of Arrows and Hearts Effect of Diamond *Journal of Gems & Gemmology* 8 (1) pp 33-35.

[3] LIU P P, YUAN X Q, SHI B and WANG Y J 2014 Study on Measuring Brilliance of Round Brilliant Cut Cubic Zirconia under Ring Light *Journal of Gems & Gemmology* 16 (1) pp 62-69.

[4] Moses T M, Johnson M L, Green B, Blodgett T, Cino K, Geurts R H, Gilbertson A M, Hemphill T S, King J M, Kornylak L, Reinitz I M and Shigley J E 2004 A Foundation for Grading the Overall Cut Quality of Round Brilliant Cut Diamonds *Gems & Gemology* 40 (3) pp 202-228.

[5] Reinitz I M, Johnson M L, Hemphill T S, Gilbertson A M, Geurts R H, Green B D and Shigley J E 2001 Modeling the appearance of the round brilliant cut diamond: An analysis of fire, and more about brilliance *Gems & Gemology* 37 (3) pp 174-197.

[6] Caspi A 1997 Modern diamond cutting and polishing *Gems & Gemology* 33 (2) pp 102-121.

[7] CHAI Y, CHEN Z and CHANG F 2003 New Development in Diamond Cut *Journal of Gems & Gemmology* 5 (1) pp 14-16.

[8] SHI B and YUAN X Q 2007 Study on Emulation Model and Principle of Optical Effect of Round Brilliant Cut Diamond *Journal of Gems & Gemmology* 9 (2) pp 8-12.

[9] WANG Y J, SHI B and YUAN X Q 2011 Measurement Precision of DC2000 Diamond Proportion Analyzer *Journal of Gems & Gemmology* 13 (1) pp 47-51.

[10] YU X Y, RUAN Q F, YU Y M, SONG L, YANG Y and ZHANG S 2016 Investigation on cutting ratio for round brilliant cut diamond *Journal of Guilin University of Technology* 36 (3) pp 435-443.

[11] King J M, Shigley J E, Gelb T H, Guhin S S, Hall M and Wang W 2005 CHARACTERIZATION AND GRADING OF NATURAL-COLOR YELLOW DIAMONDS *Gems & Gemology* 41 (2).

[12] YAN J, SHAO H P, HUANG X B, YAN X J, YU S Y and HU X C 2017 Study on the spectral and fluorescent characteristics of natural colorless and near-colorless diamonds with or without N3 *CHINA MEASUREMENT & TEST* 43 (9) pp 53-58.

[13] Shigley J E and Breeding C M 2015 VISIBLE ABSORPTION SPECTRA OF COLORED DIAMONDS *Gems & Gemology* 51 (1) 41-43.

[14] Wang W, Moses T, Linares R C, Shigley J E, Hall M and Butler J E 2003 Gem-quality synthetic diamonds grown by a chemical vapor deposition (CVD) method *Gems & Gemology* 39 (4) pp 268-83.

[15] Breeding, C M and Shigley, J E 2009 The "type" classification system of diamonds and its importance in gemology *Gems & Gemology* 45 (2) pp 96-111.

[16] WU Q Q and YUAN X Q 2017 Diamond Cut Characteristic under Diamond Lamp by Simulation and its Application *Journal of Gems & Gemmology* 19 (2) pp 57-63.

[17] LIU P P, YUAN X Q, ANG N and SHI B 2016 A Method for Quantitative Evaluation of Scintillation in Round Brilliant Diamond *Journal of Gems & Gemmology* 18 (3) pp 9-17.

IOP Conf. Series: Materials Science and Engineering **746** (2020) 012030 doi:10.1088/1757-899X/746/1/012030

Dynamic Response of Biodynamic Model Seated on a Crashworthy Seat with Composite Absorber

Na Li and Yi Fu

Shanghai Aircraft Design and Reseach Institute
Email: lina18@comac.cc, Fuyi2015@sjtu.edu.cn

Abstract. This paper presents an energy absorber consists of an inward folding of composite tube. There is no composite debris overflow after the composites collapsing and the debris will fill in the inner of tube to increase the energy absorption. The absorber maintains integrate during crushing and then could be used in a crashworthy structure. Impact tests were conducted to explore the energy absorption performance. To investigate the application of the absorber to the shock-resistance of the seat in the helicopter, a five-degree-of-freedom nonlinear biodynamic model corresponding to 50th-percentile male occupant was setup. Simulations on the responses of the biodynamic with and without absorber were conducted, and the results present effect of the absorber on shock-resistance.

1. Introduction

Metal materials have been widely used in modern commercial aircraft fuselages. The energy generated by impact damage typically absorbed by the plastic deformation of the metal material, but the energy absorption of composite structure is relatively low if the structure was broken suddenly. The energy-absorption composite materials has attracted the attention of researchers in the automotive and aerospace industries. Extensive testing on various types of tubular structure has shown that composite materials can offer extremely high values of specific energy absorption (SEA) [1] [2]. For example, in a detailed review of energy-absorption in composite structures, Jacob et al. [3] determined that only 0.66 kg of a high-performance thermoplastic matrix composite is required to absorb the energy of a 1000 kg car travelling at 15.5 m/s (35 mph). Values for the SEA of widely-used composites, such as carbon fibre reinforced epoxy, generally ranges in 50–80 kJ/kg [3] [4]. So it is essential that the composite materials should be widely used and developed potential of SEA more thoroughly.

Further bumper structure design concept is to provide the bottom plate structure with foam [5] or honeycomb [6] [7]. In order to achieve the energy absorption, it is necessary to set up additional energy-absorbing device while setting up such a complex structure. The crush behaviors of thin-walled hollow square and circular tubes [8] with chamfer failure triggers were compared against those with steeple failure triggers. Results showed that s steeple failure trigger for square tube was more effective than the chamfer failure trigger at maintaining a higher sustained crush load, but the opposite effect was observed for circular tubes.

The weight-specific energy absorption (SEA) of crush devices of composite materials could be superior to metallic absorbers. In the past, the characteristics of various geometrical shapes [9], fibre architectures or trigger mechanisms of the energy absorber have been investigated extensively [10] [11] . Some works shown that circular tubes can achieve high SEA values, which is much higher than any other cross-sections investigated, e.g., [9] [12] [13]. As for the trigger mechanisms, Heimbs [14] presents a solution by cutting composite tubes into strips through a special joint under axial crash, and composite materials will produce relatively complex form of damage to absorbing energy, but the teared strips will expand outside cap in the process of implementation, spreading around and orbiting around cap. Deepak Siromani [15] conducted some experimental study on the effect of failure trigger mechanisms on the energy absorption capability of CFRP tubes under axial compression, but the load condition reveals instability because the composite tube would be split instead of being crushed. The energy absorbing structure should be completely crushed, so there is a larger space to improve the designing.

Crushing load-induced injury mitigating is an important issue in helicopter seat design. Crushing energy-absorbing crew seats have greatly enhanced helicopter crash survivability. In order to predict the biodynamic response as accurately as possible for high-amplitude vertical, frontal, rear, and side impacts, several biodynamic models representing seated human subjects have been developed. Liu *et al.* [16] modelled a seated occupant consisting of four main lumped body parts: pelvis, upper torso, viscera, and head, which were developed based on nonlinear mechanical. Patil *et al.* [17] proposed a modified 7-DOF biodynamic model, which consisted of seven mass segments with the pelvis, abdomen, diaphragm, thorax, torso, back, and head,. These parts were formulated as nonlinear mechanical models comprising masses interconnected with springs with stiffness and dampers with viscous damping constants. Singh *et al.* [18]also developed a modified model in predicting peak magnitude, overall shape, and duration of the biodynamic transient response, with minimal phase shift. And the classic model of Liu [16] is adopted in the calculation, using the date from the test on the energy absorber. And the experimental occupant response date from a full-scale crash testing of the Sikorsky advanced composite airframe program helicopter [19]. The predicted load was used to verify the superiority of the absorber.

This work presented a shock absorber with composite material, which can be used as structural parts in the normal working state. The composite tube is subjected to the axial forced only rather than bent, so the crush process is stable. A 5-DOF nonlinear biodynamic model with seat was used to demonstrate the effect of the absorber during vertical crash.

2. Design the Absorber

The absorber comprises the crush cap, flat pressing cap, the cutter and positioning sleeve (Fig. 1). The cutter is installed in the crush cap, the lower end of which is connected to the flange in the crush cap and the upper end is connected to the sleeve. The sleeve is placed in the crush cap closely contacting the inner wall, while the lower surface contacts the cutter. The crush cap, sleeve and composite tube are connected by pins. Device without cutter can also be used to absorb energy in the form of delamination by the inward-folding of composite tube. The configuration of the absorber is shown in Fig.1.

The composite tube in the absorber services as the main bearing member to absorb the impacting energy. The steel sleeve ensures the functionality of composite strut straight towards to the upper face of the cutter. The steel cutter includes variable cross-section holes (Fig 2), reinforcing ring and guide rounded corners; The variable cross-section hole described are arranged along the circumference, forming the cutting edge blade upward, at the same time of the smaller cross section will squeeze the cut strips to absorb energy further. The cutting blades are formed by the intersection of variable cross-section hole, whose blade upward, making it convenient to cut the composite pipe moved axially.

The cutting blades were reinforced by the reinforcing ring connects with the inner face of all variable cross-section role. The guiding-corner makes the composite material torn easier to pass through the reinforcing ring when the strips turn over, which impel composite tube damaged to store in the inner cavity of tube.

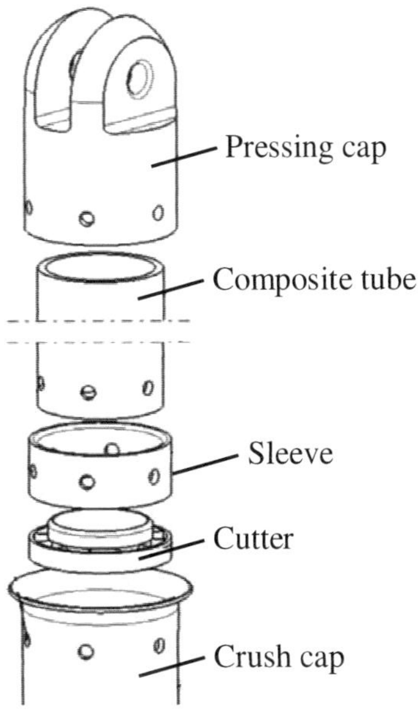

Figure 1. Configuration of the composite absorber

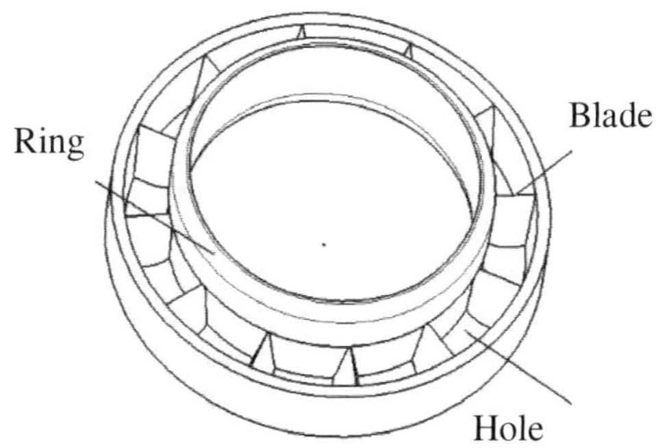

Figure 2. The cutter

The crush cap at the bottom of absorber makes the composite tube fold inward and connects external fixed structure. The cutter is fixed on the flange in the cap whose upper structure is the sleeve. When the composite tube is split by the cutter, the composite strips will reach arc at the bottom of crush cap. So the strips will flip back into the inside of composite tube, the tube will be filled with debris of composite after crush, and the bearing capacity gets a further improvement.

On the other end of the composite tube, the pressing cap connects with the external structure by the lugs. A groove inside the flat pressure cap has been fitted for flatly pressing the on the end surface of the composite materials. There are pin holes on the outer wall of the cap to fix the composite tube, so that the absorber can bear a certain degree of pull force.

Two working patterns are shown in Fig.3. When the device is subjected to axial impact, the composite tube presses the crush cap. After the failure of pins, composite tube will axially move down towards the cutter and will be cut into strips. Delamination may also appear because the strips will be squeezed through the variable diameter hole in the cutter. Meanwhile, the guide arc surface at the bottom of crush cap will promote the composite strips bending towards the inside of the composite tube. The teared strips reverse movement will be more easily under the guidance of the fillet on cutter, so the damaged composite would fill into the cavity of the tube, which makes full use of the space. When the remaining length equals the length of the rolling-over part, debris of composite reaches the flat pressing cap and be gradually compacted to absorb energy furtherly.

Another working pattern for the energy absorber as showed in Fig.3 (b). The cutter in the crush cap is removed, and the inner diameter of the cap is the same as the outer diameter of the composite tube. So the composite tube directly contacts with the inner wall of the crush cap. After the failure of pins, the composite strut moves down axially towards to the cap when absorber under the axial impact. Then the composite tube moves axially until contacting the guide arc surface at the bottom of crush cap, and the tube moves reversal in the direction into the internal cavity of tube. Delamination and fiber failure mainly occurs in this turning process. The whole debris of composite packs into the cavity after the destruction. Delamination absorbs most of the impacting energy in the process, together with the bending of the tube wall, friction and the fiber fracture.

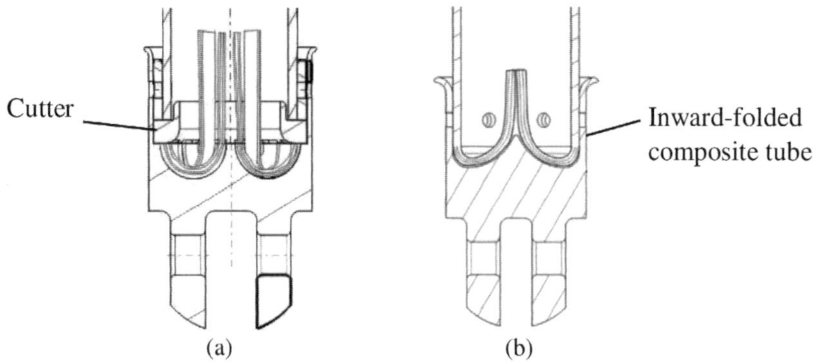

Figure 3. Working patterns of the energy absorber: (a)with cutter and (b) without cutter

3. Experiments

3.1. Experimental Setup

The composite tube was made from a carbon fibre/epoxy prepreg laminate with 50% uni-direction tape in 0° (axial) and 90° direction. To avoid corrosion problems with the cap supports, an outside layer of braided carbon fibres was used. All the external diameter of composite tube is 30 mm, the thickness is 1.5 mm. The tube length was 120 mm. The composite tube has the outer diameter of 30 mm with 8 unidirectional composites layers.

Axial impact crushing tests were performed by using a drop hammer testing system (Fig. 4) to investigate energy absorption characteristics. The drop hammer with the mass of 62.9 kg was lifted by the pulley to a height of 3 m and then released through the trip gear to reach a velocity of 5.4 m/s. The specimens were fixed at the platform. A force sensor was mounted on the base to measure the impact force. An optical sensor was used to measure the velocity of the hammer before crushing the specimens.

Figure 4. Impact test equipment

3.2. Experimental Results

The curve of force time history with the debris filling the tube is shown in Fig.5, where there is an ascending process due to the debris of composite filling the inside of tube at the end of curve. If the cutter was adopted, the energy absorption performance is more efficient under the dynamic (Fig.6) condition. Not only the cutting by the blades, but also the squeezing of strips of composite when passing through the apertures between the blades, which contributes the energy consumption. The delamination was made more thoroughly combining with reversal radius under the crush cap.

IOP Conf. Series: Materials Science and Engineering **746** (2020) 012030 doi:10.1088/1757-899X/746/1/012030

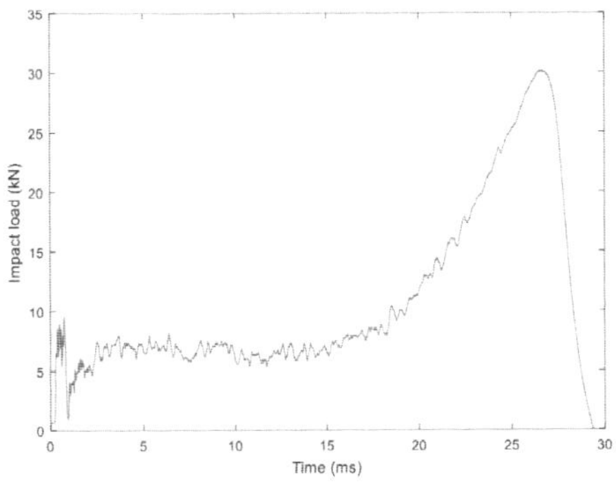

Figure 5. Force-time curve without cutter

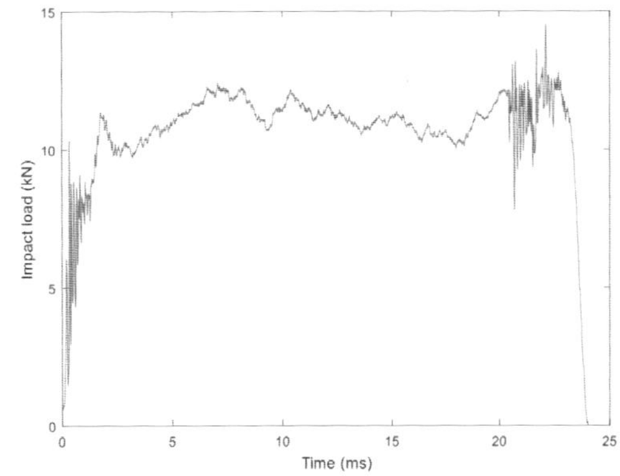

Figure 6. Force-time curve with cutter

4. Simulation of Helicopter Crew Seat

4.1. Mathematical Model

A nonlinear 5-DOF biodynamic lumped parameter model representing a 50th-percentile male exposed to high-speed vertical impacts was employed, as shown in Fig. 7. It is assumed that the human is seated in a perfect upright position and that 29% of the body weight is supported by the feet. The occupant body comprises four parts: pelvis, upper torso, viscera, and head, represented by mass M_i, stiffness K_i, and damping C_i, where i =2, 3, 4, and 5, respectively. These masses were connected via nonlinear springs and dampers. The lumbar spine was represented as a stiff nonlinear spring and a damper connecting the chest to the pelvis. The occupant was assumed to undergo pure vertical displacement (i.e., z direction only), and the motion in the forward direction and sideways was not considered.

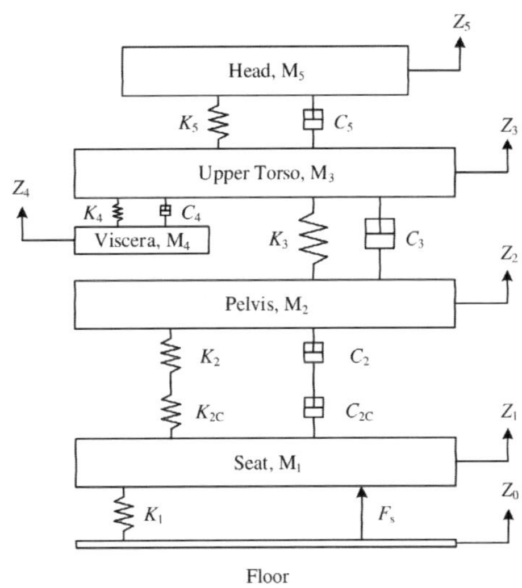

Figure 7. 5-DOF biodynamic model of a seated occupant [18]

The motion of this system is governed by the following equations [18]:

$$M_1\ddot{z}_1 = -K_1(z_1 - z_0) + K_{2t}(z_2 - z_1) + C_{2t}(\dot{Z}_2 - \dot{Z}_1) - F_S \tag{1}$$

$$M_2\ddot{z}_2 = -K_2(z_2 - z_1) + C_2(\dot{Z}_2 - \dot{Z}_1) + K_3(z_3 - z_2) + C_3(\dot{Z}_3 - \dot{Z}_2) \tag{2}$$

$$M_3\ddot{z}_3 = -K_3(z_3 - z_2) + C_3(\dot{z}_3 - \dot{z}_2) - K_4(z_3 - z_4) - C_4(\dot{z}_3 - \dot{z}_4) + K_5(z_5 - z_3) + C_5(\dot{z}_5 - \dot{z}_3) \quad (3)$$

$$M_4\ddot{z}_4 = K_4(z_3 - z_4) + C_4(\dot{z}_3 - \dot{z}_4) \quad (4)$$

$$M_5\ddot{z}_5 = -K_5(z_5 - z_3) - C_5(\dot{z}_5 - \dot{z}_3) \quad (5)$$

Converting these equations to matrix form gives:

$$\begin{bmatrix} M_1 & \cdots & 0 \\ \vdots & \ddots & \vdots \\ 0 & \cdots & M_5 \end{bmatrix} \begin{bmatrix} \ddot{z}_1 \\ \ddot{z}_2 \\ \ddot{z}_3 \\ \ddot{z}_4 \\ \ddot{z}_5 \end{bmatrix} + \begin{bmatrix} -K_1 & K_1+K_2 & -K_2 & 0 & 0 & 0 \\ 0 & -K_2 & K_2+K_3 & -K_3 & 0 & 0 \\ 0 & 0 & -K_3 & K_3+K_4+K_5 & -K_4 & -K_5 \\ 0 & 0 & 0 & -K_4 & K_4 & 0 \\ 0 & 0 & 0 & -K_5 & 0 & -K_5 \end{bmatrix} \begin{bmatrix} z_0 \\ z_1 \\ z_2 \\ z_3 \\ z_4 \\ z_5 \end{bmatrix} +$$
$$\begin{bmatrix} C_2 & -C_2 & 0 & 0 & 0 \\ -C_2 & C_3+C_2 & -C_3 & 0 & 0 \\ 0 & -C_3 & C_3+C_4+C_5 & -C_4 & -C_5 \\ 0 & 0 & -C_4 & C_4 & 0 \\ 0 & 0 & -C_5 & 0 & C_5 \end{bmatrix} \begin{bmatrix} \dot{z}_1 \\ \dot{z}_2 \\ \dot{z}_3 \\ \dot{z}_4 \\ \dot{z}_5 \end{bmatrix} + \begin{bmatrix} F_s \\ 0 \\ 0 \\ 0 \\ 0 \end{bmatrix} = 0 \quad (6)$$

where

$$K_{2t} = \frac{K_2 K_{2c}}{K_2 + K_{2c}} \quad (7)$$

and

$$C_{2t} = \frac{C_2 C_{2c}}{C_2 + C_{2c}} \quad (8)$$

In the Eq. (1), z_0 is the displacement of the floor. The initial conditions for this problem are $z_i = 0$ and $\dot{z}_i = -v_0$, where v_0 is the initial vertical landing velocity of the helicopter. However, it is a fair assumption that the spinal loads are represented by the nonlinear spring and damper connecting the upper torso and pelvis. All the lumped masses are assumed to be descending at the same velocity before the impact, $v_0 = 11.58$ m/s [19].

The stiffness of the upper torso is also nonlinear:

$$K_3 = \begin{cases} 3780 & (Z_2 - Z_3) < 0 \\ 3.78e3 + 1.09e7(Z_2 - Z_3) - 2.69e7(Z_2 - Z_3)^2, & 0 \le (Z_2 - Z_3) \le 0.04 \\ 77043, & (Z_2 - Z_3) > 0.04 \end{cases} \quad (9)$$

The initial conditions given as

$$z_i(0) = 0; \dot{z}_i(0) = -v_0; \ \forall\, i = [0-5] \quad (10)$$

The damping C_i is given by

$$C_i = 2\zeta_i\sqrt{M_i K_i} \quad i = 2,3,4,5 \quad (11)$$

For K_2 and K_3 are nonlinear functions, C_2 and C_3 are also nonlinear. The parameters of the systematic seat model used for this study are specified in Table 1 [16].

Table 1. Parameters of the systematic seat model

Quantity	Symbol	Value	Units
Mass of seat	M_1	11.5	kg
Mass of pelvis	M_2	29	kg
Mass of upper torso	M_3	21.8	kg
Mass of viscera	M_4	6.8	kg
Mass of head	M_5	5.5	kg
Stiffness of coil spring	$K1$	25500	kN/m
Stiffness of soft seat cushion	K_{2c}	37.7	kN/m
Stiffness of viscera	$K4$	2.84	kN/m
Stiffness of head	$K5$	202.3	kN/m
Cushion damping	C_{2c}	159	N.s/m
Pelvis damping	ζ_2	0.25	—
Torso damping	ζ_3	0.11	—
Viscera damping	ζ_4	0.5	—
Head damping	ζ_5	0.1	—

4.2. Results and Discussion

From the figures 5 and 6, the impacting force for composite remains stable for a considerable length of time. So that, in order to simplify the calculation, the reaction force of the energy absorber is considered as a constant. The stroke is assumed to be 0.8 m, and then F_s was optimized as 5000 N.

Figure.8 to figure 11 show the comparison of acceleration of each part of the biodynamic model, with and without absorber. The decrease of acceleration on pelvis and viscera is significant. There is merely change on the acceleration of head and upper torso. So additional control method should be applied for the seat.

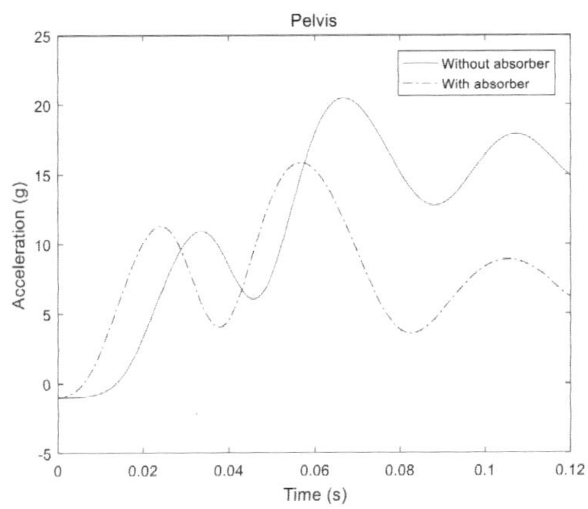

Figure 8. Acceleration of pelvis

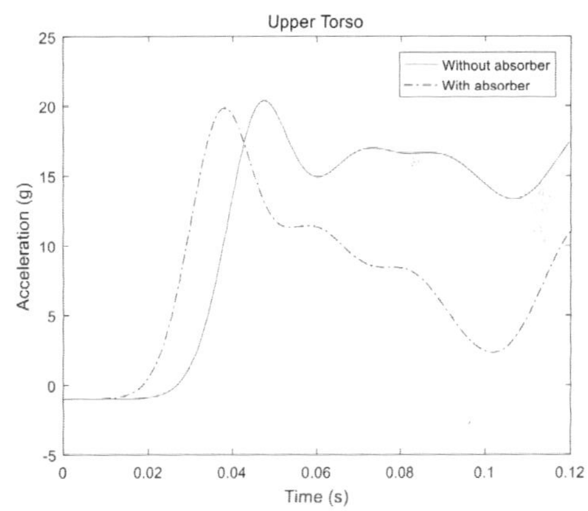

Figure 9. Acceleration of upper torso

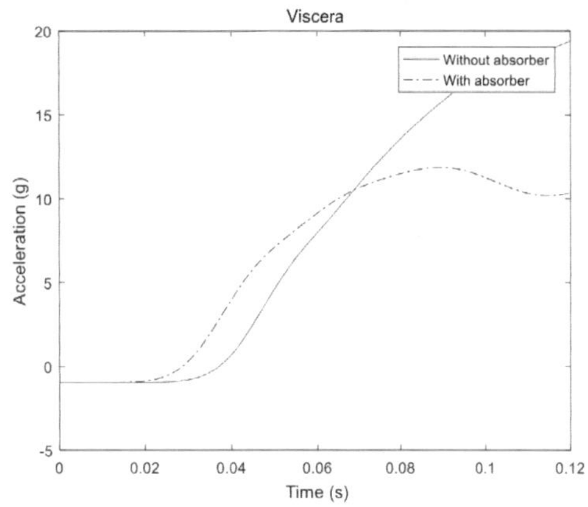

Figure 10. Acceleration of viscera

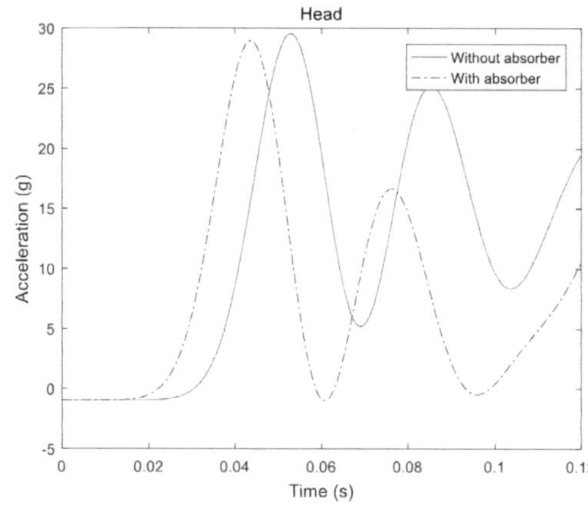

Figure 11. Acceleration of head

5. Conclusions

In the study, an energy absorber based on composite tube was proposed, the design concept and the detailed structure of the absorber had been presented. The corresponding experiments were conducted to explore the performance of the absorber in the energy absorption. This absorber had been used in the nonlinear 5-DOF biodynamic lumped parameter model to test the absorber in practical application. The following conclusions were drawn:

(1) The crushing-energy absorber contains high energy absorption efficiency, it is very suitable for lightweight crashworthiness design in terms of decreasing the initial peak load, and the SEA of absorber still maintains at a high level.

(2)The simulation results show that the absorber affect the acceleration of pelvis and viscera significantly. But the acceleration of head and upper torso decreased a little.

6. References

[1] Chiu CH, Lu CK, Wu CM. Crushing characteristics of 3-D braided composite square tubes. J Compos Mater 1997; 31: 2309–27.

[2] Jacob GC, Fellers JF, Simunovic S, Starbuck JM. Energy absorption in polymer composites for automotive crashworthiness. J Compos Mater 2002; 36: 813–49.

[3] Brachos, V.; Douglas, C.D.: Energy absorption characteristics of hybrid composite structures. 27th International SAMPE Technical Conference, Albuquerque, NM, pp. 421–435, 1995.

[4] Mamalis, A.G.; Robinson, M.; Manolakos, D.E.; Demosthenous, G.A.; Ioannidis, M.B.; Carruthers, J.J.: Crashworthy capability of composite material structures. Composite Structures, 37 (2), pp. 109–134, 1997.

[5] Fasanella, E.L.; Jackson, K.E.; Sparks, C.E.; Sareen, A.K.: Water impact test and simulation of a composite energy absorbing fuselage section. American Helicopter Society 59th Annual Forum, Phoenix, AZ, 2003.

[6] Meng, F.X.; Zhou, Q.; Yang, J.L.: Improvement of crashworthiness behaviour for simplified structural models of aircraft fuselage. International Journal of Crashworthiness, 14 (1), pp. 83–97, 2009.

[7] Fasanella, E.L.; Jackson, K. E.; Kellas, S.: Soft soil impact testing and simulation of aerospace structures. 10th International LS-DYNA Users Conference, Dearborn, MI, 2008.

[8] Palanivelu S, Paepegem W, Degrieck J, Kakogiannis D, Ackeren J, Hemelrijck D, et al. Experimental study on the axial crushing behavior of pultruded composite tubes. Polym Test 2010; 29 (2):224–34.

[9] Palanivelu S, Paepegem W, Degrieck J, Kakogiannis D, Ackeren J, Hemelrijck D, et al. Crushing and energy absorption performance of different geometrical shapes of small-scale

glass/polyester composite tubes under quasi-static loading conditions. Compos Struct 2011; 93 (2): 992–1007.
[10] Pein, M.; Krause, D.; Heimbs, S.; Middendorf, P.: Innovative energy absorbing concept for aircraft cabin interior. International Workshop on Aircraft System Technologies (AST2007), Hamburg, Germany, 375–384, 2007.
[11] Garner; D.M.; Adams, D.: Test methods for composites crashworthiness: A review. Journal of Advanced Materials, 40(4), pp. 5–26, 2008.
[12] Meredith J, Ebsworth R, Coles SR, Wood BM, Kirwan K. Natural fibre composite energy absorption structures. J Compos Sci Technol 2012; 72 (2): 211–7.
[13] Taher ST, Zahari R, Ataollahi A, Mustapha F, Basri S. A double-cell foam-filled composite block for efficient energy absorption under axial compression. Compos Struct 2011; 89 (3): 399–407.
[14] Heimbs, S., F. Strobl, P. Middendorf, & Guimard, J. M. (2010). Composite Crash Absorber for Aircraft Fuselage Applications. International Conference on Structures under Shock and Impact (Vol.113).
[15] Siromani, Deepak, et al. "An experimental study on the effect of failure trigger mechanisms on the energy absorption capability of CFRP tubes under axial compression." Composites Part A 64.21 (2014): 25-35.
[16] Liu, X. X., Shi, J., Li, G., Le, X., Zhao, B., Yue, M., Liu, J., Bai, G., and Ke,W., "Biodynamic Response and Injury Estimation of Ship Personnel to Ship Shock Motion Induced by Underwater Explosion," Proceedings of the 69th Shock and Vibration Symposium, Vol. 18, Shock and Vibration Information Analysis Center, Richmond,VA, 1998, pp. 1–18.
[17] Patil, M. K., Palanichamy, M. S., and Ghista, D. N., "Dynamic Response of Human Body Seated on a Tractor and Effectiveness of Suspension Systems," Soc. of Automobile Engineers Paper 770932, Warrendale, PA, 1977, pp. 755–792.
[18] Singh H J, Wereley N M. Biodynamic Model of a Seated Occupant Exposed to Intense Impacts[J]. AIAA Journal, 2015, 53(2):426-435.
[19] Jackson, K. E., Fasanella, E. L., Boitnott, R., McEntire, J., and Lewis, A., "Occupant Responses in a Full-Scale Crash Test of the Sikorsky ACAP Helicopter," Journal of the American Helicopter Society,Vol. 49, No. 2, 2004, pp. 127–139. doi:10.4050/JAHS.49.127.

IOP Conf. Series: Materials Science and Engineering **746** (2020) 012031 doi:10.1088/1757-899X/746/1/012031

Study on the Performance of Vertical Axis Wind Turbine with Two Sets of Blades

Yixiang Peng, Yonghui Guo, Kui Wan and Ye Gao

Collage of Aerospace and Civil Engineering, Harbin Engineering University, Harbin, China

Email: peng0624@vip.qq.com

Abstract. This paper investigated the impact of added auxiliary blades on the performance of vertical axis wind turbine through adjusting blade spacing and blade pitch angle by means of experimentation. The results show that a critical value of blade spacing affects the performance of turbine, and with the decrease of blade spacing, the output power coefficient of the turbine decreases but the automatic startup performance increases when under low tip speed ratio. In addition, the output power of pitch angle β was the largest when it was 3°, which increased by 3.9% compared with that of 0°. The power of pitch angle β was relatively close in the range of -3°~5°.

1. Introduction

At present, the wind turbines which are generally installed on the shoal near the shore are difficult to make full use of offshore wind energy. Single-layer wind turbine has low power generation efficiency, high failure rate and small application range. Single-layer horizontal axis wind generator cannot meet this special environment [1].

Wind turbine can be divided into horizontal axis wind turbine (HAWT) and vertical axis wind turbine (VAWT) according to the direction of its rotation axis and airflow direction [2]. In the past nearly a century, through the continuous efforts of scholars, the theory of wind turbines has been constantly improved, from the initial windmill model to Bates theory to obtain the theoretical maximum efficiency [3]; From the improvement of the model of the horizontal axis wind turbine to the study of the distribution of air distribution in computational fluid dynamics, the field of wind power has developed very fully [4].

The main reason for the development of horizontal axis wind power generation devices at home and abroad in the early years is that the wind energy utilization rate is higher than that of vertical axis wind turbines, and the automatic-startup performance is obviously better [5], so it has been widely concerned and studied. At present, the research theory of horizontal axis wind turbine is rich [6]. After a lot of research and experiments, excellent blade profile and mature variable pitch technology have been obtained, which has high power generation efficiency and mature market commercialization [7]. But many shortcomings is not horizontal axis wind turbine with the deepening of the research and completely eliminate [8], horizontal axis due to the structure of the horizontal axis itself under the condition of the turbulent flow of power generation efficiency is very low, at the same time since the launch of horizontal axial turbine in order to optimize performance and output performance, design of leaf size and speed usually larger, caused by noise pollution, and its size limits its use of the environment [9]. In recent years, due to the vertical axis wind turbines with a small structure, easy installation, low noise and able to take advantage of the obvious advantages of turbulent wind, able to adapt to the special environmental requirements [10], the vertical axis wind turbine research gradually rise, still belongs to the early stage, the direction of the need to conquer mainly for how to improve the utilization rate of wind energy and increase since the start of the wind turbine performance [11].

Vertical axis wind turbine according to the structure of blade and the wind forces can be divided into different lift type vertical shaft type vertical axial wind turbine. Since the start of the wind power generator capacity is relatively good, able to start under low wind speed, tour of requirements is low, easy to install, but its low utilization rate of wind power, efficiency is generally lower than 20%, thus seriously restrict the development of resistance type wind turbine. Lift type wind turbine are more widely used, the main advantage is that the wind energy utilization rate is higher, after optimization of lift type vertical axis wind turbines have been able to exceed the single-layer horizontal axis wind turbine, the single-layer vertical shaft lift type wind power generator has two shortcomings: since the launch of instability and poor performance.

Taking h-type vertical shaft turbines as the research object, this paper compares the output performance and self-starting performance of double-layer blade turbines with that of ingle-layer single-layer blade turbines. The output power and starting torque of wind turbines with different structures can be obtained by changing the size of turbines and blade. It will provide some help for the future research and development of vertical shaft turbines with different structures.

2. Experimantal Consideration

2.1. Laboratories and Wind Tunnels

In this paper, the number of blade layers, the distance between double-layer blades, and the pitch angle of blades are taken as variables to conduct experimental research. Data such as wind speed, wind turbine rotational speed, and output power of the wind turbine under different working conditions are recorded by experimental data acquisition equipment, and the results of different groups of experiments are compared and analyzed.

The wind tunnel is a self-built small wind tunnel, which is composed of an axial blower in the front section, a fairing in the middle section and a ventilation duct at the back end. The schematic diagram of the model is shown in figure 1.

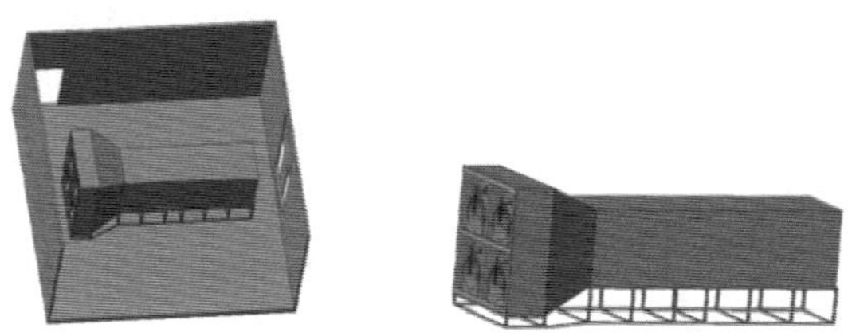

Figure 1. Schematic model of wind tunnel

The basic size of the laboratory is 11.2 m * 9.3 m * 3.9 m, and the overall size of the wind tunnel is 8.22 m * 2.55 m * 2.55 m. The controlled wind speed of the turbine is 0 ~ 10.5 m/s, and its actual wind speed is 0 ~ 9.3 m/s. Four axial flow blowers are placed on the support at the front of the wind tunnel. The maximum diameter of the single blower is 790 mm, the maximum height is 1008 mm, the thickness is 550 mm, and the maximum weight is 60 kg. Fairing and grilles are installed in the middle of the wind tunnel to ensure a relatively stable flow field around the wind turbine, and the maximum wind speed within the fairing is 9.3 m/s. In order to make the flow field of the wind tunnel as stable as possible under the condition that the wind speed of the turbine is sufficient, it is required to withstand the maximum wind speed without excessive noise in the design. The side opening of the wind tunnel facilitates the change of turbine structure and placement of turbines. The tail opening is used to pull out the generator wiring and connect it to the output power test device. The wake flow of the wind tunnel flows out from the outlet to ensure that the airflow field at the outlet is the same as that at the far field. The rest of the whole structure adopts bracket and skin structure, the physical picture of the side of the wind tunnel is shown in figure 2.

IOP Conf. Series: Materials Science and Engineering **746** (2020) 012031 doi:10.1088/1757-899X/746/1/012031

Figure 2. General picture of wind tunnel (a). Axial blower material

Figure 2. General picture of wind tunnel (b). Real side view of wind tunnel

2.2. Vertical Axis Wind Turbine

The vertical axis wind turbine is mainly composed of a square base at the bottom, a power generating rotor at the bottom, a rotating axis in the middle, a symmetrical crossbar at the top and bottom, it absorbs wind energy and converts it into kinetic energy through blades. The blades rotate for generating electricity from the rotor. The upper end is connected with the rotary shaft disc, which has a diameter of 8 cm and the diameter of the rotary shaft is 3.6 cm. The base of the turbine is placed at the bottom of the lower end, and the base leads to the wiring. Due to the small size and strong structural stability of the device, the thickness of the rotary shaft has a great impact on the aerodynamic performance of the wind turbine, so the finer shaft is selected. The physical drawing of power generation rotor is shown in figure 3.

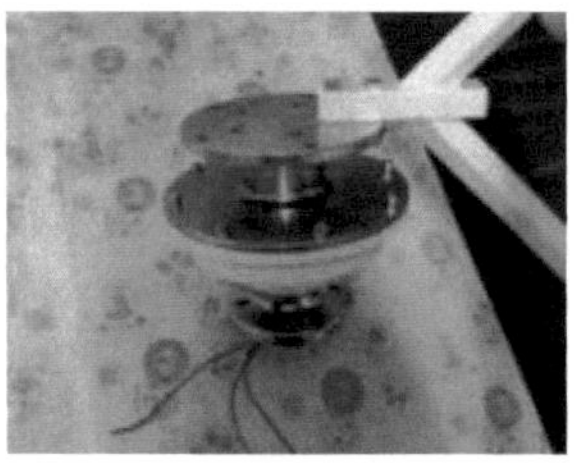

Figure 3. Physical picture of generator rotor

The upper part of the wind turbine is equipped with symmetrical upper and lower crossbar, which can be used with a length of 57 cm. The outer side of the crossbar is connected with the outer blade through the pitch angle changing device, that is, the rotation radius of the outer blade is 0.57 m. The length of the outer blade is 0.55 m, while the inner blade is drilled in the middle of the upper and lower connecting rod according to the required rotation radius, and its length is 0.35 m. The farthest pin hole is 0.36 m away from the rotation center, and holes are drilled every 3 cm to change the blade spacing experiment. The physical drawing of the wind turbine is shown in figure 4.

Figure 4. Physical picture of wind turbine

IOP Conf. Series: Materials Science and Engineering **746** (2020) 012031 doi:10.1088/1757-899X/746/1/012031

Inside section of up and down bars is connected to the rotor disc, a total of three sets spaces between 120°, then install variable pitch device of rail lateral, the outer part of the pitch changing device is connected with the blade pin, and the inner part is connected with the crossbar with a single degree of freedom, and the rotation angle can be changed by -15° ~ 15°, variable pitch angle device model diagram as shown in figure 5.

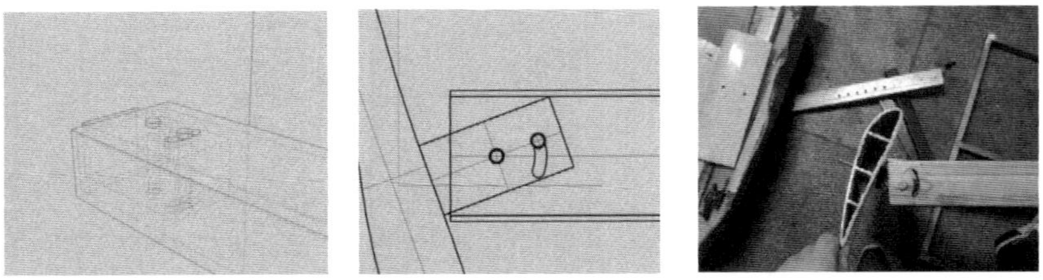

Figure 5. Variable pitch angle model and variable pitch angle object

2.3. Measuring Device

The device for changing wind speed in the experiment is shown in figure 6 (a). The range of wind speed adjustment is 0-10 m/s. During the experiment, four frequency converters are simultaneously adjusted to ensure stable air supply. In this experiment, the output power coefficient of wind turbine is measured by a power meter, which is shown in figure 6 (b).

Figure 6. The ignited lithium (a). Speed control system

Figure 6. The ignited lithium (b). wattmeter

3. Results and Discussion

Based on the basic requirements of scientific experiments in repetitive experiments, this section uses the above devices and measurement methods to conduct repeated experiments on the output performance of double-layer blades in the laboratory environment for two consecutive times. Take the output performance and auto-startup performance experiments of single-blade turbine as example, the pitch angle $\beta = 0°$changes the output power coefficient of wind speed measurement under different wind speed, and wind turbine power output change with wind speed curve as shown in figure 7.

IOP Conf. Series: Materials Science and Engineering **746** (2020) 012031 doi:10.1088/1757-899X/746/1/012031

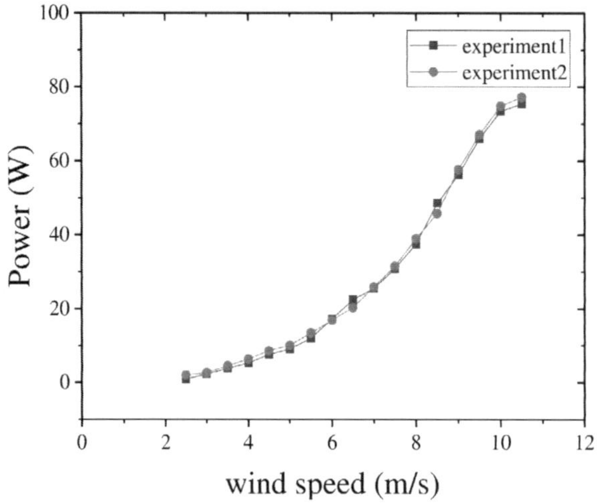

Figure 7. The output power varies with the wind speed

The figure above shows that the output power of the single-blade turbine is basically the same after two experiments under the same rotation radius, with an average error of less than 5%. At the same time of measuring the output power, a repetitive experiment of automatic-startup performance was carried out, and the minimum starting wind speed of two consecutive times in this condition was obtained, and the automatic-startup wind speed was almost identical. As mentioned above, the performance test of vertical axis wind turbine carried out by this experimental device has well repeatability.

The automatic-startup performance of turbines with different blade spacing was tested. Double-layer wind turbines with inner/outer blade rotation radius of 0.42 m/0.48 m, 0.39 m/0.51 m, 0.36 m/0.54 m, 0.33 m/0.57 m and single-layer turbines with r = 0.45 m (blade spacing of 0 m) were measured respectively, as is shown in figure 8.

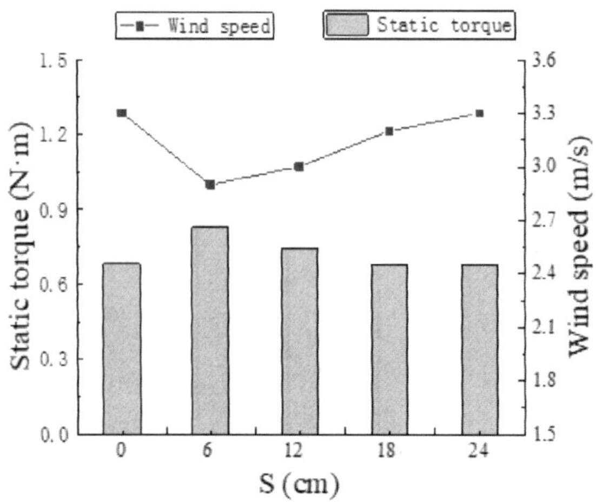

Figure 8. Experimental results of automatic-startup performance of wind turbines with different blade spacing

The experimental results show that the automatic-startup wind speed of the single-blade turbine is basically the same as that of the double-layer turbine with larger blade spacing S. The minimum starting wind speed is 3.3 m/s. With the decrease of blade spacing S, the starting wind speed gradually decreases, reaching a minimum of 2.9 m/s.

Experiments were conducted on the output power of turbines with different blade spacing. Double-layer turbines with inner/outer blade rotation radius of 0.42 m/0.48 m, 0.39 m/0.51 m, 0.36 m/0.54 m,

0.33 m/0.57 m and single-layer turbines with r = 0.45 m (blade spacing of 0 m) were measured at different wind speeds. The output power of turbines with different sizes at different wind speeds was shown in figure 9.

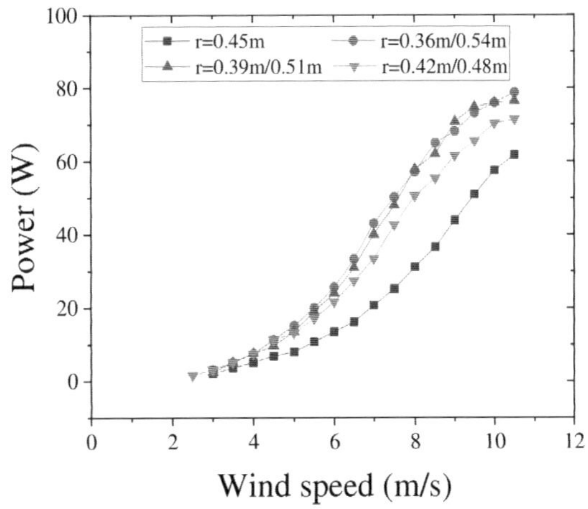

Figure 9. Measurement results of output power of wind turbines with different blade spacing

Experimental results show that the blade pitch S = 12 cm and the experimental results showed that the output power of the two groups of experiments with blade spacing S = 12 cm and S = 18 cm was basically the same. When S = 6 cm, the power decreases significantly, and the power of any double-blade turbine at different wind speeds is higher than that of the single-blade turbine, but it is not reflected in the fast attenuation of the output performance of the double-blade turbine at high velocity ratio. During the observation and experiment, the turbine rotation speed increased with the increase of wind speed, and the rotation speed remained basically unchanged after the wind speed reached 8 m/s. In the case of the turbine with the highest wind speed of 10.5 m/s, the tip speed ratio $\lambda \approx 1.5$.

The output power of turbine with different pitch angle β is measured. The double-layer turbine with blades in the range of pitch angle β = -15° ~ 15° is measured at different wind speeds. When the pitch angle β < -10° or β > 10°, the experiment finds that the output power has decreased significantly. Therefore, the experimental data of seven groups of turbines with different pitch angles in the range of -10° ~ 10° are given here, the different pitch angle under different wind speed wind turbine power output as shown in figure 10.

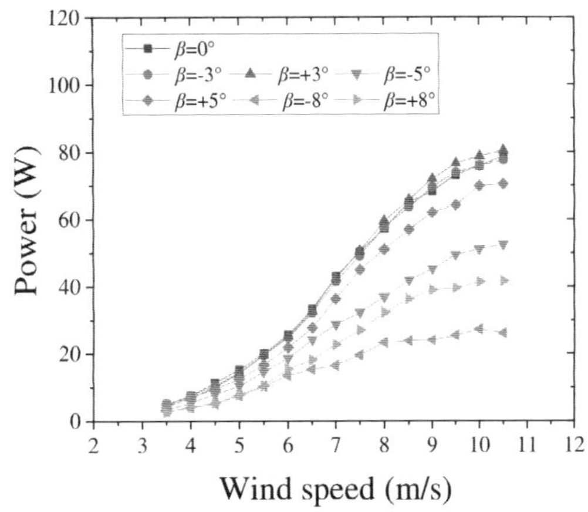

Figure 10. Measurement results of turbine output power with different pitch angle

The output power of several groups of turbine with different pitch angle is compared. Between -3° and 3°, the output power of the turbine is very close. After the wind speed is more than 7 m/s, when the pitch angle is 3°, the output power is gradually greater than that of β = 0 ° and β = -3°.In the wind speed between 3.5 m/s and 10.5 m/s, the output power of pitch angle β = 3° increased by 3.9% compared with that of β = 0°, and the power of pitch angle β = -3° was basically the same as that of β = 0°.Comparing the output power of positive pitch angle and negative pitch angle, the output power of positive pitch angle is obviously larger than that of negative pitch angle. When the pitch angle is between -3° and 5°, the output power is close. When the pitch angle changes more than this range, the power decreases rapidly

4. Summary and Conclusion

In this paper, the single and double layer turbine with different sizes is tested by changing blade spacing and blade pitch angle. In the process of experiment, repeatability is verified to reduce the contingency.

According to the effect of blade spacing on performance, the output performance and automatic-startup performance were tested. When the blade spacing was S>12 cm, the output power was basically equal. With the decrease of blade spacing, the output power decreased. In terms of automatic-startup performance, the automatic-startup performance increases with the decrease of blade spacing.

The influence of pitch angle on output performance was tested. In the case of turbines with experimental size, the output power of pitch angle β was the largest when it was -3°, which increased by 3.9% compared with that of 0°. The power of pitch angle β was relatively close in the range of -3°~5°.

5. Acknowledgments

The work described in this paper was supported by the Natural Science Foundation of China (Grant numbers: NSFC11572095).

6. References

[1] Tescione G, Ferreira C J S and Van G J W. 2016 Analysis of a Free Vortex Wake Model for the Study of the Rotor and Near Wake Flow of a Vertical Axis Wind Turbine *J. Renewable Energy* 87(1): 552-563.

[2] Rezaeiha A, Kalkman I and Blocken B 2017 CFD simulation of a vertical axis wind turbine operating at a moderate tip speed ratio: Guidelines for minimum domain size and azimuthal increment *J. Renewable Energy* 107: 373-385.

[3] Ying Wang, Xiaojing Sun and Xiaohua Dong 2016 Numerical investigation on aerodynamic performance of a novel vertical axis wind turbine with adaptive blades *J. Energy Conversion and Management* 108: 275-286.

[4] Zhonghe Han, Peng An, Shufeng Zhang and Yalei Jia 2017 Aerodynamic performance of wind turbine with double blade type airfoil *J. Renewable Energy Resources* 35(10): 1523-1529.

[5] Chen Zhang 2017 The research on the key technology of horizontal axis wind power generation with concentrated wind *D. Northeast Forestry University.*

[6] Elkhoury M, Kiwata T and Aoun E. 2015 Experimental and numerical investigation of a three-dimensional vertical-axis wind turbine with variable-pitch *J. Journal of Wind Engineering and Industrial Aerodynamics* 139: 111-123.

[7] Firdaus R, Kiwata T and Kono T. et al 2015 Numerical and experimental studies of a small vertical-axis wind turbine with variable-pitch straight blades *J. Journal of Fluid Science and Technology* 10(1): JFST0001-JFST0001.

[8] Rezaeiha A, Kalkman I and Blocken B 2017 Effect of pitch angle on power performance and aerodynamics of a vertical axis wind turbine *J. Applied Energy* 197: 132-150.

[9] Qiuping Yang and Deke Xi 2017 Study on the influence of blade installation angle on aerodynamic performance of H-type vertical axis wind turbine *J. Acta Energiae Solaris Sinica* 38(09): 2544-2551.

IOP Conf. Series: Materials Science and Engineering **746** (2020) 012031 doi:10.1088/1757-899X/746/1/012031

[10] Siyue Qian, Zhenzhou Zhao, Chen Tian, Ruixin Wang, Guanyu Zeng and Yuan Zheng 2018 Numerical simulation of new variable-pitch strategy for H-type wind turbine *J. Journal of Drainage and Irrigation Machinery Engineering* 36(03): 230-236.
[11] Lijun Zhang, Dongchen Ma, Xinhui Zhao, Yuxia Mi, Song Zhang, Hanxiang Wang and Hao Jiang 2018 Real-time variable-pitch control laws of H-type vertical axis wind turbine under dynamic stall *J. Journal of Central South University (Science and Technology)* 49(10): 2610-2617.

IOP Conf. Series: Materials Science and Engineering **746** (2020) 012032 doi:10.1088/1757-899X/746/1/012032

Relation between Dynamic Characteristics and Parameters of a Two-degree-of-freedom Vibro-impact System with Multi-elastic Constraints

Fengwei Yin [1, 2] and Guanwei Luo [1, 2]

[1]School of Mechatronic Engineering, Lanzhou Jiaotong University, Lanzhou, China
[2]Key Laboratory of System Dynamics and Reliability of Rail Transport Equipment of Gansu Province, Lanzhou, 730070, China
Email: 365518280@qq.com

Abstract. A mechanical model of mechanical vibration system with multi-elastic constraints was founded. Based on the multiple parameter and multiple objective cooperative simulation calculation, the three-dimensional parameter space was constructed by the key parameters such as constrained stiffness, excitating frequency and clearance threshold. The large data onto two-dimensional parameter bifurcation diagrams reflecting the correlation between dynamic properties and parameters of the model are obtained. The mode types, existing regions and bifurcation characteristics of periodic impact vibration of the system on the parameter plane of frequency and clearance are described with the change of constrained stiffness. Similarly, the three-dimensional parameter spaced consisting of mass ratio, or damping ratio and excitating frequency and clearance threshold are analyzed. The results show that the high mass ratio excites rich and wonderful periodic vibration, and different damping ratio determines, the bifurcation characteristics and the number of chattering impacts groups in the small frequency domain. The mode types and transition rules of the fundamental periodic impact vibration in small frequency domain and subharmonic impact vibration in high frequency domain induced by different parameters of the system are revealed.

1. Introduction

Nonlinear vibration of mechanical structure and system is a complex problem in engineering practice. Therefore, to study the mechanism and characteristics of various non-smooth bifurcation phenomena of vibration systems with multi-elastic constraints, so as to describe the dynamic properties of mechanical systems more accurately, it is very important to design and manufacture mechanical systems and structures with excellent safety and performance. By means of qualitative analysis, numerical calculation and experiments, the periodic motion and bifurcation characteristics [1-3], grazing singularity [4-6] and chattering-impact motions systems [7-10] with clearances and constraints have been studied by scholars at home and abroad.

On the basis of Wagg's research, the non-smooth bifurcation problem of a two-degree-of-freedom vibro-impact system with multi-elastic stops is systematically researched by utilizing multiple parameter and multiple objective cooperative simulation calculation. The vibration system with rigid constraints studied by Wagg is only a special case of the vibration system with elastic constraints.

IOP Conf. Series: Materials Science and Engineering **746** (2020) 012032 doi:10.1088/1757-899X/746/1/012032

2. Mechanical Model

2.1. Mechanical Model

A two-degree-of-freedom vibration system with multi-elastic stops is expressed in Figure 1. Two masses M_1 and M_2 are linked by linear spring damper components K_1 and C_1, and mass M_2 is fastened to the base by a linear spring damper components K_2 and C_2. The harmonic exciting force $P_i \sin(\Omega T+\tau)$ loads on the corresponding mass M_i (i=1,2), among them, P_i, Ω and τ represent the amplitude, frequency and phase angle individually. The elastic stops on the same stiffness (K_0) were placed on the right stop of mass M_1 and on the symmetric stops of mass M_2, the range of constrained stiffness was $K_0 \in (0,\infty)$. For low amplitude, the system shows free vibration without impact. However, as the vibration amplitude of the system increases sufficiently large, when the displacement of the masses is equal to the clearance threshold of corresponding constraints, the soft impact of the masses occurs to the corresponding constraints [11-13].

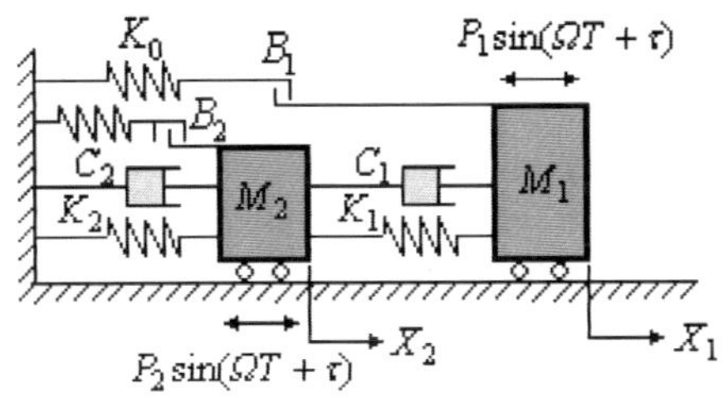

Figure 1. Mechanical model

2.2. Defined Dimensionless Parameters, Variables and Time:

$$m=\frac{M_2}{M_2+M_1},k=\frac{K_2}{K_2+K_1},k_0=\frac{K_0}{K_0+K_1},c=\frac{C_2}{C_2+C_1},\ \delta_i=\frac{B_iK_1}{P_1+P_2},f=\frac{P_2}{P_2+P_1},x_i=\frac{X_iK_1}{P_1+P_2},t=T\sqrt{\frac{K_1}{M_1}}. \tag{1}$$

2.3. Dimensionless Differential Equation of Motion

$$\begin{bmatrix}1 & 0\\ 0 & \dfrac{m}{1-m}\end{bmatrix}\begin{Bmatrix}\ddot{x}_1\\ \ddot{x}_2\end{Bmatrix}+\begin{bmatrix}2\zeta & -2\zeta\\ -2\zeta & \dfrac{2\zeta}{1-c}\end{bmatrix}\begin{Bmatrix}\dot{x}_1\\ \dot{x}_2\end{Bmatrix}+\begin{bmatrix}1 & -1\\ -1 & \dfrac{1}{1-k}\end{bmatrix}\begin{Bmatrix}x_1\\ x_2\end{Bmatrix}+\begin{Bmatrix}f_1(x_1)\\ f_2(x_2)\end{Bmatrix}=\begin{Bmatrix}1-f\\ f\end{Bmatrix}\sin(\omega t+\tau) \tag{2}$$

In which $f_1(x_1)$ and $f_2(x_2)$ were expressed by

$$f_1(x_1)=\begin{cases}\dfrac{k_0}{1-k_0}(x_1-\delta_1)\ , & x_1>\delta_1\ ,\\ 0\ , & x_1\le\delta_1,\end{cases}\qquad f_2(x_2)=\begin{cases}\dfrac{k_0}{1-k_0}(x_2-\delta_2)\ , & x_2>\delta_2\ ,\\ 0\ , & |x_2|\le\delta_2,\\ \dfrac{k_0}{1-k_0}(x_2+\delta_2)\ , & x_2<-\delta_2.\end{cases} \tag{3}$$

Equation (1) allowed us to make a conclusion that the maximum range of partial dimensionless parameters can be identified as $m\in(0,1)$, $k\in(0,1)$, $k_0\in(0,1)$, $c\in(0,1)$. In the above parameter domain, the vibration system with multi-elastic constraints in Figure 1 showed a variety of periodic impact vibration. Symbol q/n was used to describe the periodic impact vibration of a unilateral constrained mass M_1, where n was the ratio of the vibration period $T_n=2n\pi/\omega$ to the periods of the excitating force $T_0=2\pi/\omega$ (n = 1, 2, 3...), q means the number of impacts of mass M_1 at its right elastic constraints during the vibration period (q = 0, 1, 2, 3...). Meanwhile, the symbol n-p-q was used to represent the periodic impact vibration characteristics of the mass M_2 with symmetric constraints, and p and q were used to represent the impact times of the mass M_2 with left and right constraints, respectively. More generally, in order to find the values of n, p and q, defined four kinds of Poincaré section.

$$\sigma_{\mathrm{II}} = \{(x_1, \dot{x}_1, x_2, \dot{x}_2, t)R^4 \times T \mid x_1 = x_{1\min}, \mathrm{mod}(t = 2\pi/\omega)\}\ ;\ \sigma_{1r}^1 = \{(x_1, \dot{x}_1, x_2, \dot{x}_2, t) \in R^4 \times T \mid x_1 = \delta_{1r}, \dot{x}_1 > 0\}\ ;$$

$$\sigma_{1l}^2 = \{(x_1, \dot{x}_1, x_2, \dot{x}_2, t) \in R^4 \times T \mid x_2 = -\delta_{2l}, \dot{x}_2 < 0\}\ ; \quad \sigma_{1r}^2 = \{(x_1, \dot{x}_1, x_2, \dot{x}_2, t) \in R^4 \times T \mid x_2 = \delta_{2r}, \dot{x}_2 > 0\}\ . \quad (4)$$

For periodic impact vibration, the fixed points of Poincaré section σ_{II} , σ_{1l}^k and σ_{1r}^k means the periodic number n of the mass corresponding to the k constraints and the impact number p and q of the left and right constraints, respectively. The equation of the impact Poincaré mapping σ_{1r}^1 of the system was expressed as

$$X^{(i+1)} = f(X^{(i)},\ \upsilon)\ , \quad (5)$$

Where $X^{(i)}=(\tau^{(i)}, \dot{x}_{1+}^{(i)}, x_2^{(i)}, \dot{x}_{2+}^{(i)})^{\mathrm{T}}$, $X^{(i+1)} = (\tau^{(i+1)}, \dot{x}_{1+}^{(i+1)}, x_2^{(i+1)}, \dot{x}_{2+}^{(i+1)})^{\mathrm{T}}$, $X \in R^4$, υ are parameters, $\upsilon \in R^8$.

3. Results and Discussion

3.1. Effect of Constraint Stiffness Ratio on System Dynamic Properties

Considering a two-degree-of-freedom forced vibration system with multi-elastic constraints, the interaction between the dynamic properties of the system and its parameter constrained stiffness ratio k_0 is revealed. Firstly, the effects of constraints (constrained stiffness k_0, clearance threshold δ) and excitating frequency ω of the dynamic characteristics of the system are studied. Selection of dimensionless parameters: $m = 0.5$, $k = 0.5$, $c = 0.5$, $\zeta = 0.1$. The range of constrained stiffness ratio $k_0 \in (0,1)$, the values are discretized at equal intervals. Figure 2 representatively shows the mode types and occurrence regions of the periodic vibration of the system in the (ω,δ) -parameter plane in Figure 1 corresponding to three sets of k_0. When the constrained stiffness is taken as a value $k_0 \in (0,0.5)$, the system masses M_1 and M_2 exhibit simple and conventional periodic impact vibration in the parameter domain of figure 2 (a). With the increase in k_0, the dynamic characteristics of the system show complexity and diversity, see figure 2 (b). As the frequency ω decreases in the small clearance δ region, a series of grazing contacts occur to between each mass and its corresponding constraints, resulting in the transition of q/1 vibration to (q+1)/1 vibration via Grazing bifurcation. As a result, the number of impacts p (or q) of each mass in the system increases step by step during the fundamental periodic. In the small frequency ω and clearance δ region of the lower left corner of the parametric plane, the mass M_2 shows the different types of fundamental periodic motion groups, while the mass M_1 clearly shows tongue-shaped non-hysteretic transition region along the boundary of adjacent fundamental periodic impact vibration. The fundamental periodic motion groups (q/1 or 1-p-q) of the system exhibit a chattering-impact motions characteristic when the number of impacts of the system masses to become large enough under the condition of minimal frequency ω . When $k_0 > 0.95$, with the increase of frequency ω , the sequence of periodic doubling bifurcation may be interrupted by grazing contact between a periodic vibration trajectory and elastic constraints, and the system shows complex dynamic characteristics. Based on the above analysis, the constrained stiffness $k_0 = 0.95$ is taken as the reference parameter, as shown in Figure 2 (c), and The effects of other parameters of the model on the mode types and bifurcation characteristics of periodic vibration are further analyzed. In the biparametric bifurcation diagram, 60% of the unlabeled black regions are chaotic.

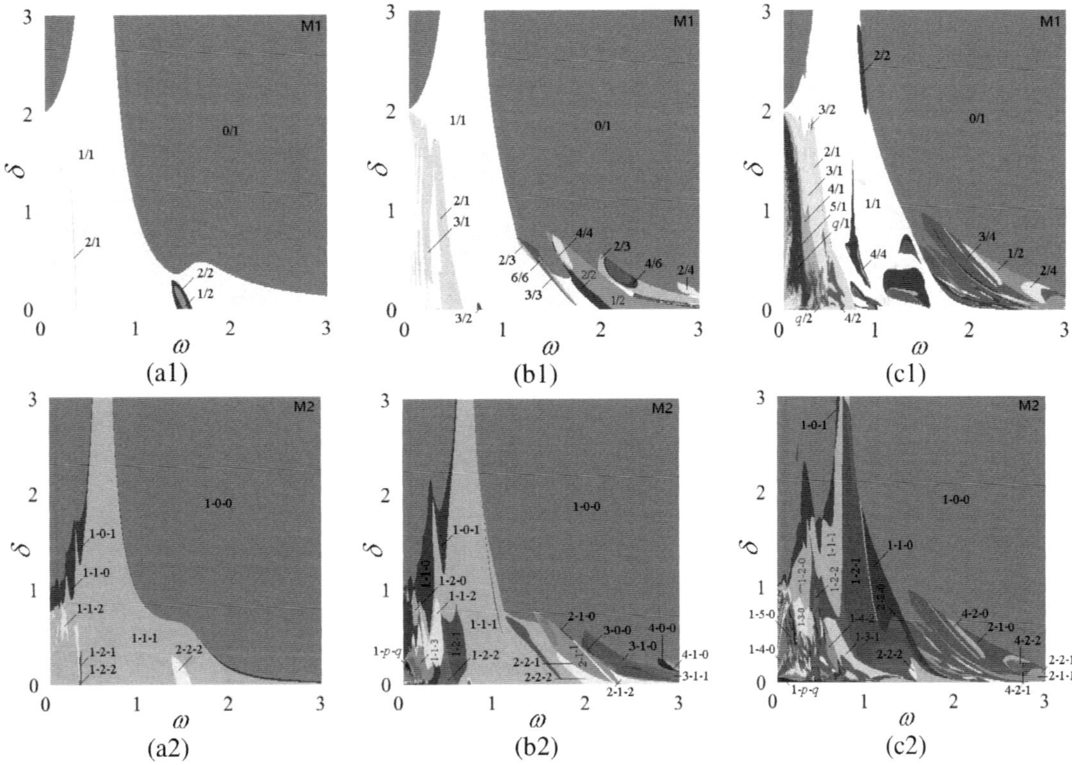

Figure 2. Under different constraint stiffness ratios, the system presents dynamic characteristics in the (ω,δ) -parameter plane: (a1) M_1: $k_0 = 0.25$; (b1) M_1: $k_0 = 0.75$; (c1) M_1: $k_0 = 0.95$; (a2) M_2: $k_0 = 0.25$; (b2) M_2: $k_0 = 0.75$; (c2) M_2: $k_0 = 0.95$

3.2. Effect of Mass Ratio on System Dynamic Properties

The range of mass distribution ratio is $m \in (0,1)$, and the values are discretized at equal intervals. The damping distribution ratio c of the system parameters is calculated by the same method. In Figure 3 are plotted the dimensionless reference parameters: $k_0 = 0.95$, $k = 0.5$, $c = 0.5$, $\zeta = 0.1$, the typical three groups of m values correspond to the mode types and bifurcation characteristics of periodic vibration on the (ω,δ) - parameter plane. When the mass ratio m is small, the equivalent mass mainly concentrates on the mass M_1. In the low clearance δ and frequency ω region of the lower left corner of the parametric plane, the mass M_1 presents the fundamental periodic vibration groups, and tongue-shape non-hysteretic transition regions are formed along the boundary of two adjacent fundamental periodic vibrations (q/1 and (q+1)/1). There are different types of fundamental periodic vibration groups (1-p-q) in the mass M_2, and the number of impacts on the left stop p is more than that on the right stop q, namely $p > q$. In the high frequency ω and small clearance δ region, As the frequency ω decreases, mass M_1 exhibits subharmonic periodic vibration of 1/2, 2/2 and 2/3, or doubling bifurcation or grazing bifurcation, resulting in subharmonic periodic vibration of 2/4, 3/4, 3/3 and 6/6. Mass M_2 induces periodic doubling bifurcation or grazing bifurcation, subharmonic periodic vibrations of 3-1-0, 2-1-1, 2-2-1, 2-3-1, 2-3-0, 2-2-0 and 2-1-0 are shown. When the mass ratio m is large, the equivalent mass focuses on the mass M_2, and the system exhibits abundant dynamic characteristics. The parameter domain of (q/1 ($q \geq 2$)) the fundamental motion groups of mass M_1 rapidly extends to the large clearance threshold δ. However, (1-p-q) the different types of fundamental motion groups of mass M_2 are significantly reduced and the existing regions are obviously contracted with the direction of frequency ω reduction, see Figures 3 (b) and 3 (c). In the clearance δ and frequency ω region of the upper right corner of the parametric plane, the system displays (0/1 or 1-0-0) non-impact vibration. The results show that with the increase m of system parameters, the fundamental motion groups of mass M_1 exhibit diversity, and the different types of fundamental motion groups of mass M_2 exhibit fierce competition in the parameter domain. The larger m of system parameters, the more complex the dynamic characteristics of the system.

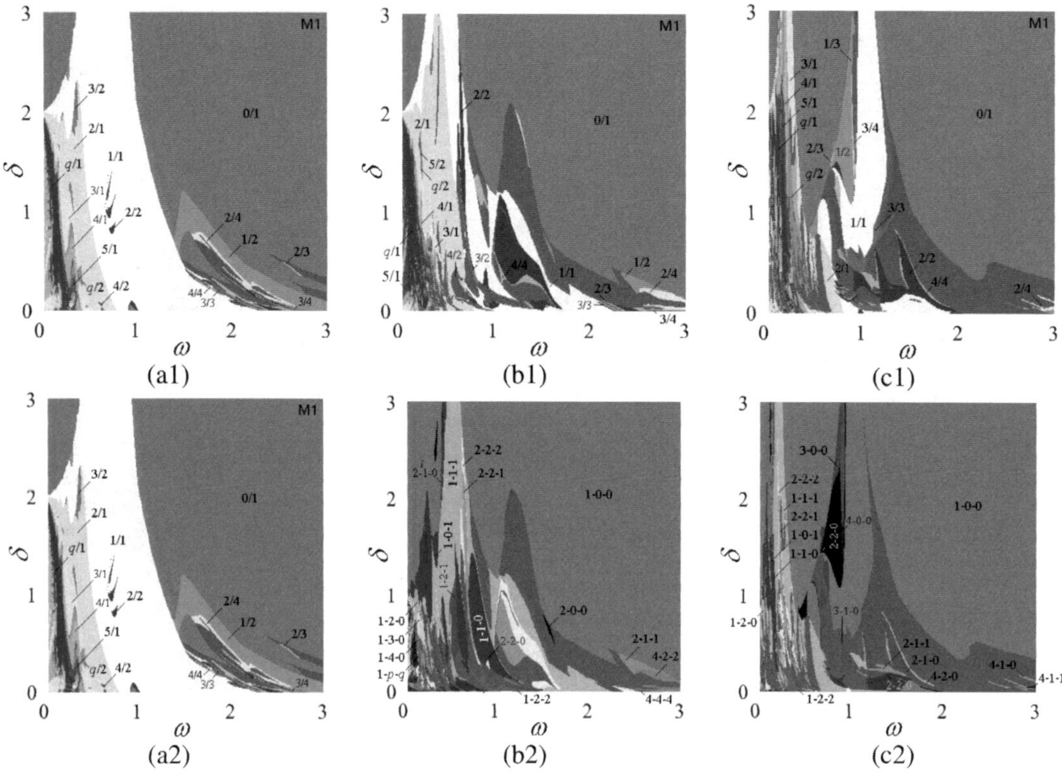

Figure 3. Under different mass ratios, the system presents dynamic characteristics in the (ω,δ) - parameter plane: (a1) M_1: $m = 0.25$; (b1) M_1: $m = 0.75$; (c1) M_1: $m = 0.95$; (a2) M_2: $m = 0.25$; (b2) M_2: $m = 0.75$; (c2) M_2: $m = 0.95$

3.3. Effect of Damping Ratio on System Dynamic Properties

The typical three groups of damping distribution ratios c correspond to the mode types and occurrence areas of periodic impact vibration on the (ω,δ) - parameter plane, as seen in figure 4. The results are compared with those under reference parameters (I), see figure 2 (c). When $c < 0.5$, the mode types and bifurcation characteristics of mass M_1 and M_2 show little change, and they all show complex bifurcation characteristics in the low frequency ω and small clearance threshold δ region. When c increased, the total area of (q/1 ($q \geq 2$)) the fundamental motion groups of the low frequency region of the mass M_1 have almost no change, but the number of impacts (q) decreases gradually. At the same time, the tongue-shaped non-hysteretic transition regions of the adjacent fundamental periodic impact vibration are obviously reduced, and even disappeared, and the existence area of subharmonic periodic vibrations in the high frequency ω region of mass M_1 is also reduced. The mode types and existing areas of the different types of fundamental motion groups of mass M_2 are obviously reduced. When the value of c is larger, the peak values of the corresponding fundamental periodic vibration regions and the subharmonic periodic vibration regions of the two masses decrease with the increase of parameter c, see figures 4 (b) and 4 (c). The results show that the change of damping ratio c in the system parameters has a great influence on the mode types and bifurcation characteristics of the fundamental motion groups corresponding to the masses M_1 and M_2, and the dynamic characteristics of masses M_1 and M_2 show similar evolution regularity.

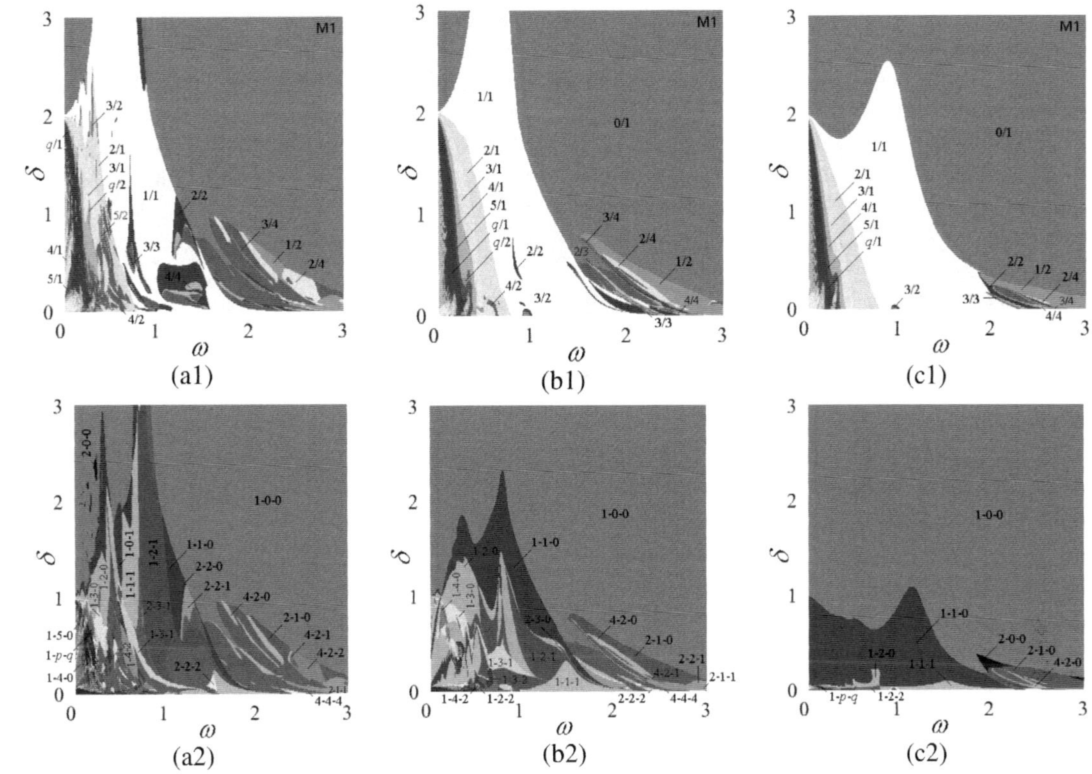

Figure 4. Under different damping ratios, the system presents dynamic characteristics in the (ω,δ) -parameter plane: (a1) M_1: c = 0.25; (b1) M_1: c = 0.75; (c1) M_1: c = 0.95; (a2) M_2: c = 0.25; (b2) M_2: c = 0.75; (c2) M_2: c = 0.95

4. Conclusion

In summary, the constrained stiffness ratio k_0 determines the number of fundamental motion groups of the system. Under the condition of large mass ratio m, the dynamic characteristics of masses M_1 and M_2 show complexity and diversity. The lower the damping ratio c is, the more significant the non-smooth characteristic of the system is.

The numerical results can reveal the essential relationship between the dynamic characteristics of mechanical vibration system with multiple-elastic constraints and its parameters and the matching law. Mechanical systems with different functions can determine the reasonable matching range of their parameters, and realize the collaborative optimization of dynamic characteristics and functional objectives of the system.

5. Acknowledgments

The author heartily acknowledges the support by National Natural Science Foundation (11672121, 11862011), Innovation and Entrepreneurship Talents Training Project of Lanzhou city (2014-RC-33) and Science and Technology Planning Project of Gansu (18YF1WA059).

6. References

[1] A.B. Nordmark. Non-periodic motion caused by grazing incidence in an impact oscillator [J]. Journal of Sound and Vibration, 1991, 145 (2): 279–297.

[2] G.W. Luo, X.F. Zhu, Y.Q. Shi. Dynamics of a two-degree-of freedom periodically-forced system with a rigid stop: Diversity and evolution of periodic-impact motions [J]. Journal of Sound and Vibration, 2015, 334:338-362.

[3] G.W. Luo, X.H. Lv, Y.Q. Shi. Vibro-impact dynamics of a two-degree-of freedom periodically-forced system with a clearance: Diversity and parameter matching of periodic-impact motions [J]. International Journal of Non-Linear Mechanics, 2014, 65:173-195.

[4] H.Y. Hu. Detection of grazing orbits and incident bifurcations of a forced continuous, piecewise-linear oscillator [J]. Journal of Sound and Vibration, 1994, 187(3): 485–493.

[5] G.L. Wen, S. Yin, H.D. Xu, et al. Analysis of grazing bifurcation from periodic motion to quasi-periodic motion in impact-damper systems [J]. Chaos, Solitons & Fractals, 2016, 83:112-118.

[6] S. Yin, J.C. Ji, G.L. Wen, et al. Use of degeneration to stabilize near grazing periodic motion in impact oscillators [J]. Communications in Nonlinear Science and Numerical Simulation, 2018:S1007570418301801-.

[7] D.J. Wagg, S.R. Bishop. Dynamics of a two degree of freedom vibro-impact system with multiple motion limiting constraints [J]. International Journal of Bifurcation and Chaos, 2004, 14(01):119-140.

[8] D.J. Wagg. Rising phenomena and the multi-sliding bifurcation in a two-degree of freedom impact oscillator [J]. Chaos, Solitons and Fractals, 2004, 22(3):541-548.

[9] D.J. Wagg. Multiple Non-Smooth Events in Multi-Degree-of-Freedom Vibro-Impact Systems [J]. Nonlinear Dynamics, 2006, 43(1-2):137-148.

[10] D.J. Wagg. Periodic sticking motion in a two-degree-of-freedom impact oscillator [J]. International Journal of Non-Linear Mechanics, 2005, 40(8):1076-1087.

[11] Y. Ma, J. Ing, S. Banerjee, et al. The nature of the normal form map for soft impacting systems [J]. International Journal of Non-Linear Mechanics, 2008, 43 (6): 504–513.

[12] Soumya Kundu, Soumitro Banerjee, James Ing, et al. Singularities in soft- impacting systems [J]. Physica D, 2012, 241: 553–565.

[13] B. Blazejczyk-Okolewska, K. Czolczynski, T. Kapitaniak. Hard versus soft impacts in oscillatory systems modelling [J]. Communications in Nonlinear Science and Numerical Simulation, 2010, 15(5):1358-1367.

IOP Conf. Series: Materials Science and Engineering **746** (2020) 012033 doi:10.1088/1757-899X/746/1/012033

Research on the Path Following Control of Intelligent Vehicle with Method of Added Planning Layer

Chuanyin Tang[1], Yafeng Zhao[1], Jiahui Hou[1], Simin Chen[2], Quansheng Hu[1] and Jue Gong[1]

[1]School of Mechanical Engineering &Automation, Northeastern University, Shenyang 110819, China
[2]School of Science, Northeastern University, Shenyang 110819, China
Email: monsigor@aliyun.com

Abstract. Aimed at intelligent vehicles operating in complex driving environment, the paper is described on the basis that the reference trajectory has been determined. According to the vehicle running state and obstacle information, the vehicle's running track can be adjusted in real time to solve the path planning problem and according to the re-planned reference track, the following control is carried out in the paper to avoid obstacles. Firstly, a point mass model is established which ignores the car body size. Secondly, on the premise that the size of obstacles is smaller than the car body size, according to the obstacle information obtained by the sensor and the established point mass model, the obstacle avoidance penalty function is established, then the objective function of the path planner is established. Finally, two working conditions: single obstacle and two obstacles, are set up to simulate and verify the designed obstacle avoidance trajectory following system. By observing the distance between the actual track and obstacles under different simulated speeds, varied range of front wheel angle and centre of mass side deflection angle, different range of angular velocity of lateral swing and lateral acceleration to verify whether the vehicles can successfully avoid obstacles with different speed and numbers of obstacles in order to maintain good driving stability in the process. Simulation results show that, under the control of the path following system added to the planning layer, intelligent vehicles in a complex driving environment can autonomously plan their paths, avoid obstacles, follow the path, and ensure good driving stability meanwhile.

1. Introduction

The key technologies of intelligent vehicle [1-5] mainly include the following three parts: (1) Technology of environmental awareness: Intelligent vehicles identify the surroundings through on-board cameras, radars and other electronic devices and obtain such information. Environmental awareness technology is a very important research topic in the field of intelligent vehicle research. Nowadays, sensors and electronic devices commonly used in cars are mainly included as: on-board cameras, GPS positioning system, million-meter-wave radar and so on. (2) Technology of decision-making: In the driving process of intelligent vehicles, the driving operation behaviour of intelligent vehicles can be completed by processing the road and surrounding environment information obtained by the system. At present, the commonly used decision-making technologies mainly include PID control, optimal control, fuzzy control, robust control and so on. (3) Technology of path planning: When the intelligent vehicle encounters obstacles or moves laterally, it needs to plan an optimal feasible path for it by using pre-designed rules or strategies. Path planning plays an important role in motion control technology. (4) Technology of motion control: The planned driving path is used to

IOP Conf. Series: Materials Science and Engineering **746** (2020) 012033 doi:10.1088/1757-899X/746/1/012033

determine the driving behaviour of the car through effective control strategies or methods. It plays a decisive role in judging the performance of the driving system of intelligent vehicles.

When the intelligent car driving on the road, due to the complexity and changeability of the driving environment, the following control of a given reference trajectory has some limitations. For example, when there are obstacles in the given reference path, intelligent vehicles cannot well follow the path, therefore, the trajectory needs to be re-planned to avoid obstacles and then the reference trajectory needs to be followed again. According to the principle of model predictive control, a new control system combining path planning and path following is formed by combining the vehicle's current driving state and obstacle information. Among them, path planning belongs to the upper controller, its main function is to adjust the trajectory of intelligent vehicles in real time according to the vehicle running state, obstacles and reference track information, and solve the problem of path planning. Path following belongs to the lower controller, its main function is to follow and control the path according to the planned path, so as to avoid obstacles. The work of this part of the path planning is based on the reference trajectory has been determined, belongs to the local planning.

2. Path Following Control System with Added Planning Layer

The path following control system added to the planning layer is shown in figure 1. According to the model predictive control principle, the path planner processes the received obstacle information, reference track information and vehicle status information, then output the new trajectory information. The path following controller receives the path information generated by the path planner, combined with the current vehicle status, the front wheel Angle can be output, so as to avoid obstacles and continue to go back to the original reference track.

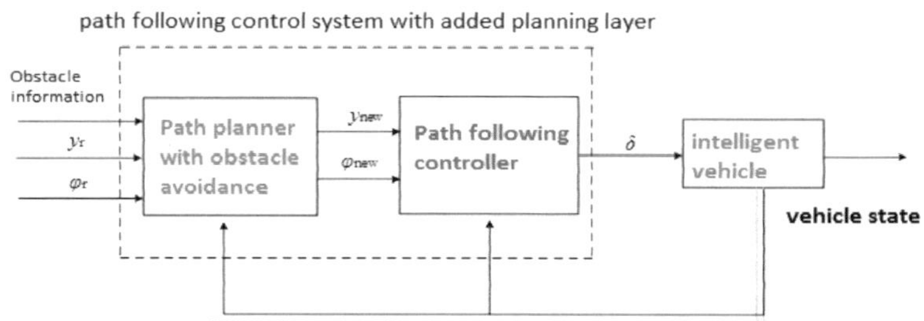

Figure 1. Path following control system with the added planning layer

3. Design of the Path Planning Controller

The path planner is designed in the part. The main task of the path planner is to solve the trajectory that can meet constraints and avoid obstacles according to the obtained obstacle information, track information and the current vehicle state, the planned local trajectory should be as close as possible to the reference trajectory. Therefore, the objective function should include the deviation between the planned trajectory and the reference trajectory, the control quantity and the obstacle avoidance function.

3.1. Model of Point Mass

Due to the large computation amount of path planning algorithm itself, therefore, a lower precision model is needed to reduce the computational burden. The lower precision model is adopted in the path planning layer and the higher model is adopted in the path following control layer, not only the control performance is good, but also the calculation speed has been greatly improved. Therefore, the point mass model that ignores the body size is adopted in the path planning layer, as shown in figure 2.

The model is represented as follows:

$$\begin{cases} \ddot{y} = a_y \\ \ddot{x} = 0 \\ \dot{\varphi} = \dfrac{a_y}{\dot{x}} \\ \dot{Y} = \dot{x}\sin\varphi + \dot{y}\cos\varphi \\ \dot{X} = \dot{x}\cos\varphi - \dot{y}\sin\varphi \end{cases} \tag{1}$$

The vehicle's dynamic constraints are:

$$\left|a_y\right| \le \mu g \tag{2}$$

Equation (1) can be written as follows:

$$\dot{\xi}(t) = f(\xi(t), u(t)) \tag{3}$$

The state amount is defined as: $\xi = [\dot{y} \quad \dot{x} \quad \varphi \quad Y \quad X]^T$, the control amount is defined as $u = \delta$.

3.2. Objective Function of Path Planner

In the process of path following, intelligent vehicles acquire and process obstacle information through sensors, the information from the sensors is scattered, this chapter focuses on obstacle avoidance trajectory planning and path following control, think less about the size of the obstacles. Therefore, in order to prevent the vehicle from passing through the obstacles, it is assumed that the size of the obstacles is smaller than the body size, as is shown in figure 3.

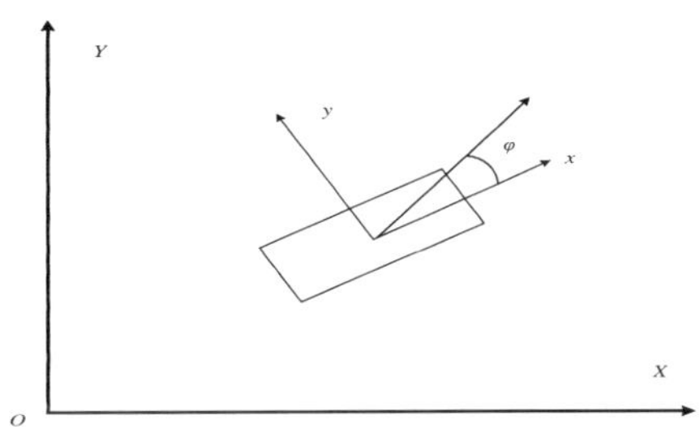

Figure 2. Model of point mass

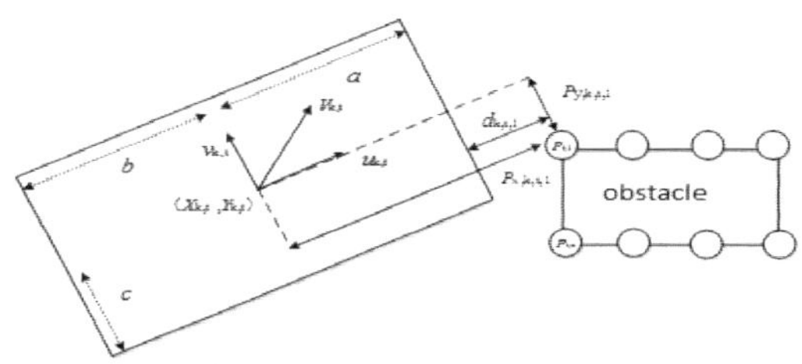

Figure 3. Penalty function for obstacle avoidance

The obstacle avoidance penalty function mainly depends on the distance between the front of the vehicle and the obstacle, at time t, sensors on intelligent vehicles can obtain information about obstacles, the centre of mass coordinates of the vehicle are ($X_{k,t}$, $Y_{k,t}$), the position of the obstacle can be represented by n discrete points, where the position coordinate of the point j is : $p_{t,j}$=($P_{X,t,j}$, $P_{Y,t,j}$), j=1,2,3…n, convert these points to the body coordinate system, the conversion relationship is:

$$P_{x,k,t} = (p_{X,t,j} - X_{k,t})\cos\varphi_{k,t} + (p_{Y,t,j} - Y_{k,t})\sin\varphi_{k,t}$$

$$P_{y,k,t} = (p_{Y,t,j} - Y_{k,t})\cos\varphi_{k,t} - (p_{X,t,j} - X_{k,t})\sin\varphi_{k,t}$$

At time t, the obstacle avoidance penalty function can be expressed as:

$$J_{obs} = \frac{K_{obs} V_{k,t}}{d_{\min_{k,t}} + \varepsilon}$$

Among them, K_{obs} is collision weight function. When the weight coefficient increases, the result of path planning becomes conservative and the deviation between the planned trajectory and the reference trajectory also increases, $V_{k,t}=u_{k,t}^2+v_{k,t}^2$, $d_{\min\ k,t}$ represents the minimum distance from the vehicle to all obstacle points, among them, the $d_{k,t,j}$ is :

IOP Conf. Series: Materials Science and Engineering **746** (2020) 012033 doi:10.1088/1757-899X/746/1/012033

$$d_{k,t,j} = \begin{cases} P_{x,k,j} - a & P_{y,k,j} \in [-c,c] \cap P_{x,k,j} > a \\ 0 & P_{y,k,j} \in [-c,c] \cap P_{x,k,j} \in [-b,a] \\ M & other \end{cases}$$

In conclusion, the objective function of the path planner can be expressed as:

$$\min_{\Delta U} \sum_{i=1}^{N_p} \| y(k+i) - y_r(k+i) \|^2{}_Q + \sum_{i=0}^{N_{c-1}} \| u(k+i) \|^2{}_R + J_{obs}$$

Among them, the first term represents the deviation between the planned trajectory and the reference trajectory, reflecting the control requirements for the deviation; the second item reflects the control requirements of the input quantity; the third item represents the function of obstacle avoidance.

4. Simulation Verification to Path Following System with Added Planning Layer

By designing two working conditions -- single obstacle and two obstacles to simulate and verify the designed obstacle avoidance trajectory following system, the reference path selection is shown in figure 4. The simulation speed is set as v=10m/s, v=15m/s, v=20m/s, road adhesion coefficient is μ=0.8, the parameters of the controller are shown in table 1.

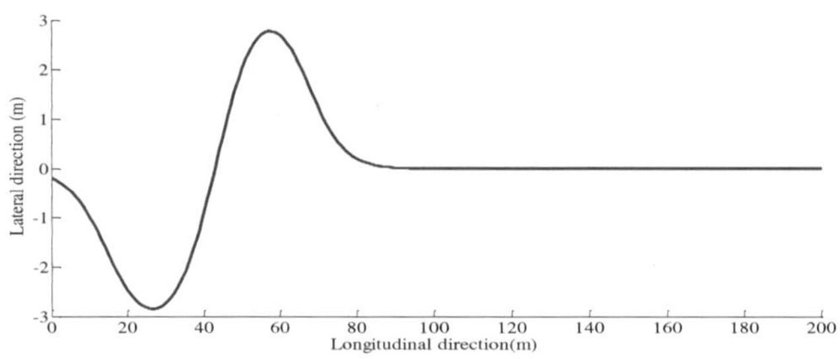

Figure 4. The Reference track

Table 1. Parameters of controller

name of parameter	Path planning controller	Path following controller
Predicted time domain N_p	15	20
Controlled time domain N_c	2	10
Sampling time *T*/s	0.05	0.01

4.1. Single Obstacle Condition

The position of the obstacle is (-2.5, 15), the size is 4m*1m, the simulation results of intelligent vehicles with different speeds is shown from figure 5 to figure 7. The red rectangles represent obstacles, figure 5 shows the trajectory of intelligent vehicles when they avoid obstacles at the speed of 10m/s, figure 6 is the track diagram of the intelligent vehicle when it avoids obstacles at the speed of 15m/s, figure 7 is the trajectory map of intelligent vehicle when it avoids obstacles at a speed of 20m/s.

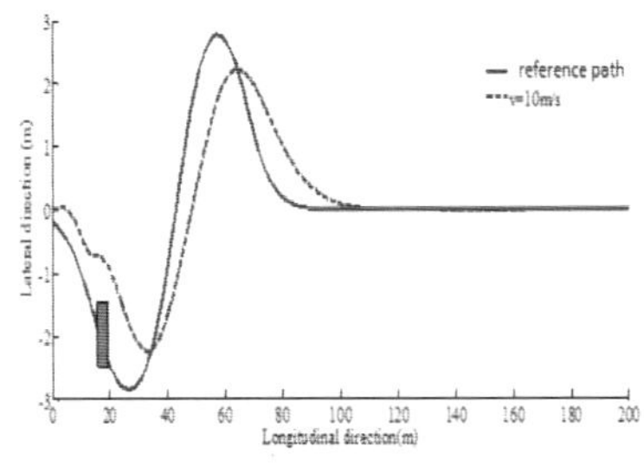

Figure 5. Simulation result of a single obstacle with speed v=10m/s

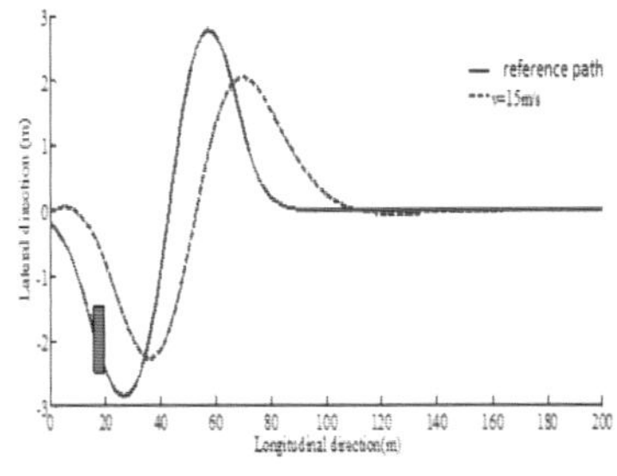

Figure 6. Simulation result of a single obstacle with speed v=15m/s

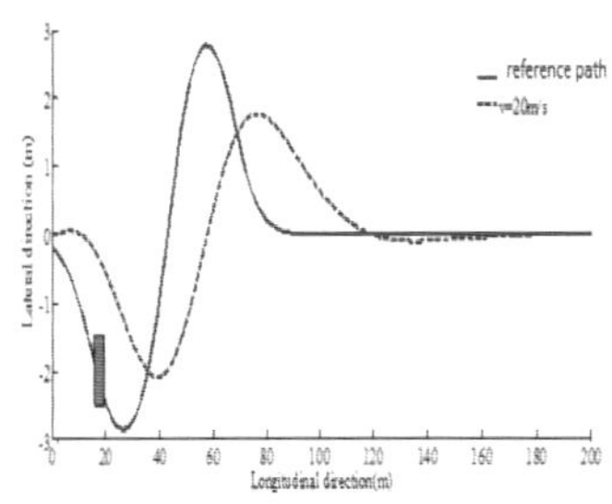

Figure 7. Simulation result of a single obstacle with speed v=20m/s

The distance between the actual track and the obstacle also increases, as seen from the figure 5 to figure 7, with the increase of speed. Vehicles began to avoid obstacles and ensure safety. And the curvature of the vehicle's trajectory becomes smaller, the higher the speed the earlier the vehicle to avoid obstacles, which is consistent with practical driving habits. Finally, the vehicle can return to the reference track after completing the obstacle avoidance. The intelligent vehicle with the controller can successfully complete autonomous path planning, avoid obstacles and follow the path with different speeds.

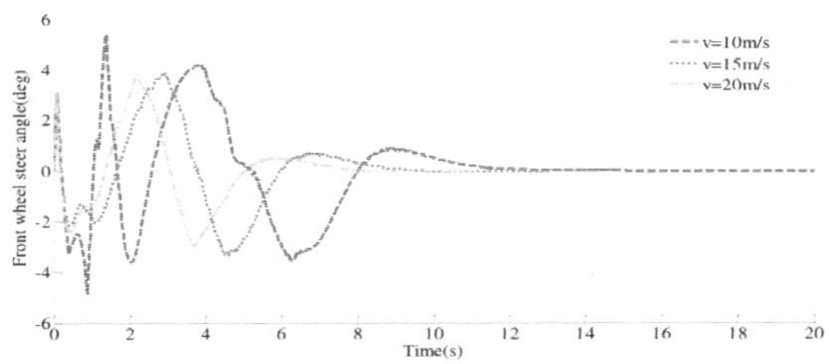

Figure 8. Simulation results of the front wheel steer angle at different speeds

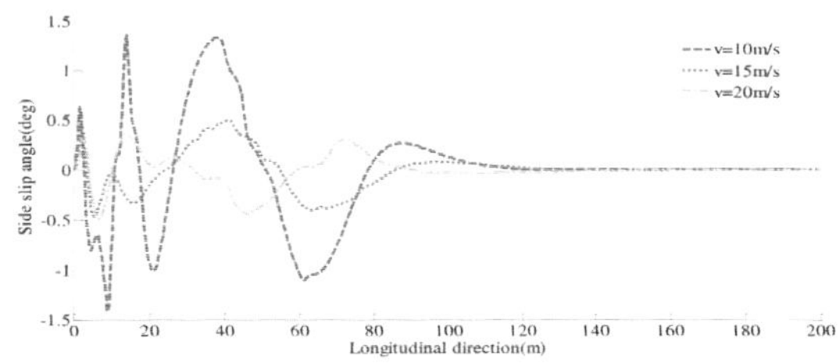

Figure 9. Simulation results of side slip angle at different speeds

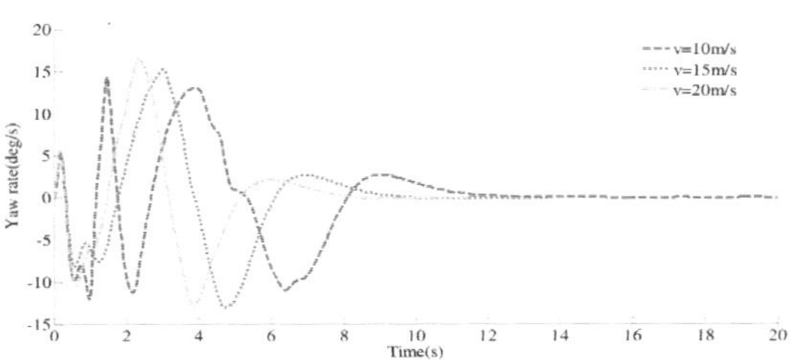

Figure 10. Simulation results of yaw rate at different speeds

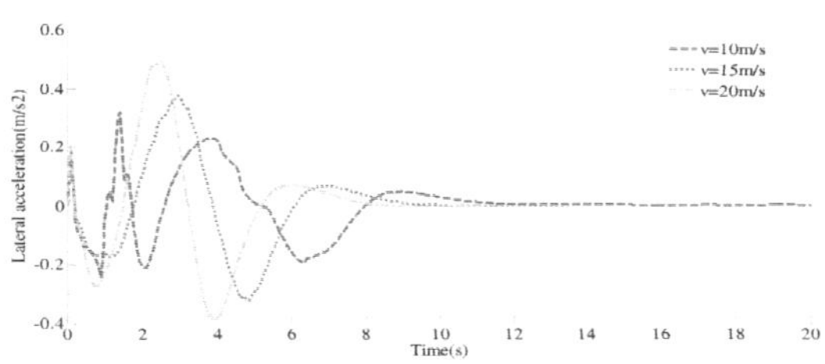

Figure 11. Simulation results of lateral acceleration at different speeds

The overall variation range of the front wheel angle is from -5deg to 5deg, which is always within the constraint range, and the variation is stable, which can ensure the smooth implementation of the actuator, according to figure 8. The lateral deviation angle of the centre of mass and the lateral angular velocity of swing are both important parameters to measure the stability of a vehicle. From figure 9 and figure 10, the centroid side-slip Angle under different speed of the overall change in the range of - 1.4 deg ~ 1.4 deg, far below the scope of centre of mass of side-slip Angle constraint set, and stable, the change of the transverse swing angular velocity amplitude as the speed increases, but the overall range for - 14 deg/s ~ 17 deg/s, it indicates that the vehicle has no dangerous situation such as side-slip or rollover in the process of driving. As can be seen from figure 11, the range of lateral acceleration is-$0.4m/s^2$~$0.5m/s^2$, which is far below the constraint range, it indicates that the vehicle has good driving stability.

4.2. Condition with Two Obstacle

The positions of the two obstacles with different sizes are (-2.5, 15), (0,140), the dimensions are 4m*1m and 4m*0.5m respectively. The simulation results of intelligent vehicles with different speeds is shown from figure 12 to figure 14. The red rectangles representing obstacles, among them, figure 10.1 shows the trajectory of intelligent vehicles when they avoid obstacles at the speed of 10m/s, figure 10.2 is the track diagram of the intelligent vehicle when it avoids obstacles at the speed of 15m/s, figure 10.3 is the trajectory map of intelligent vehicle when it avoids obstacles at a speed of 20m/s.

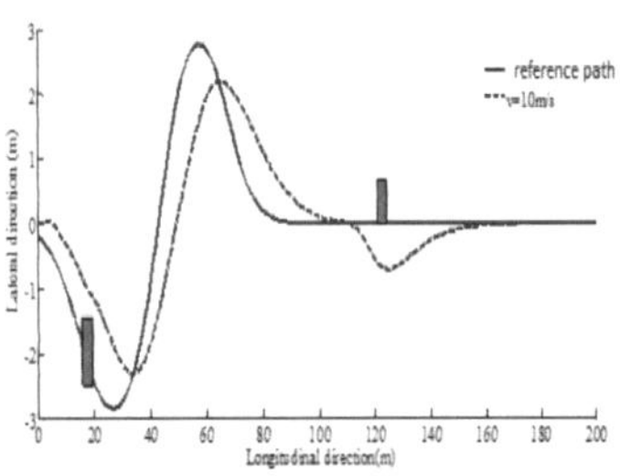

Figure 12. Simulation result of a single obstacle with speed v=10m/s

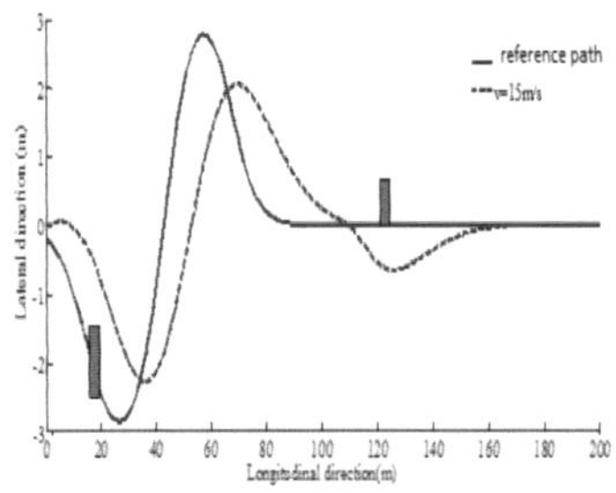

Figure 13. Simulation result of a single obstacle with speed v=15m/s

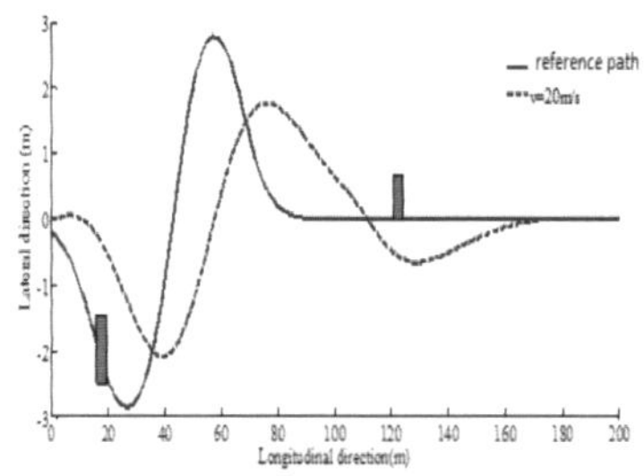

Figure 14. Simulation result of a single obstacle with speed v=20m/s

The distance between the actual track and the obstacle also increases, as seen from the figure 12 to figure 14, with the increase of speed. Vehicles began to avoid obstacles and ensure safety. And the curvature of the vehicle's trajectory becomes smaller, this is because the higher the speed, the earlier the vehicle to avoid obstacles, which is consistent with practical driving habits, finally, the vehicle can return to the reference track after completing the obstacle avoidance. Therefore, at different speeds, intelligent vehicles can successfully complete autonomous path planning, avoid obstacles and follow the path.

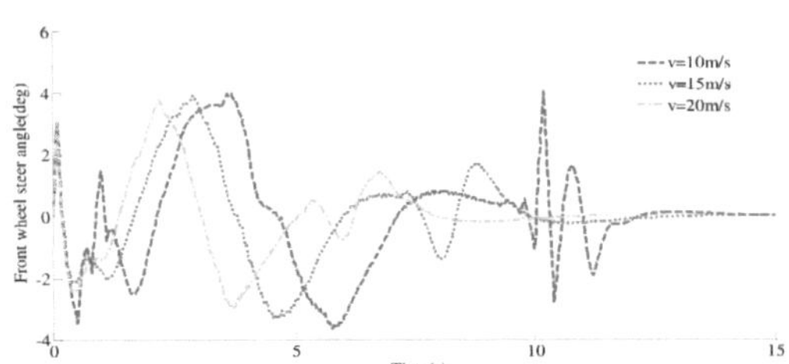

Figure 15. Simulation results of the front wheel steer angle at different speeds

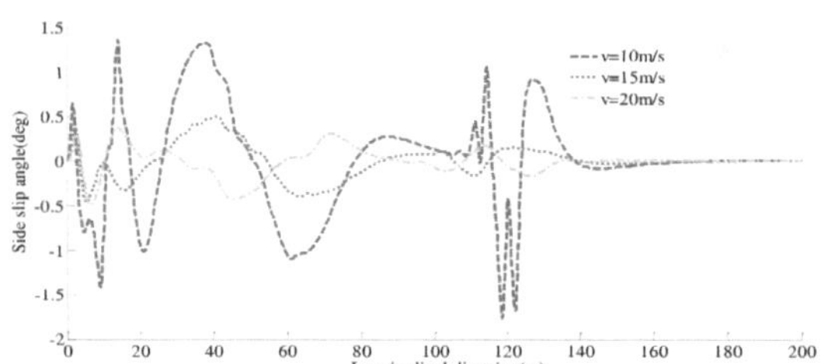

Figure 16. Simulation results of side slip angle at different speeds

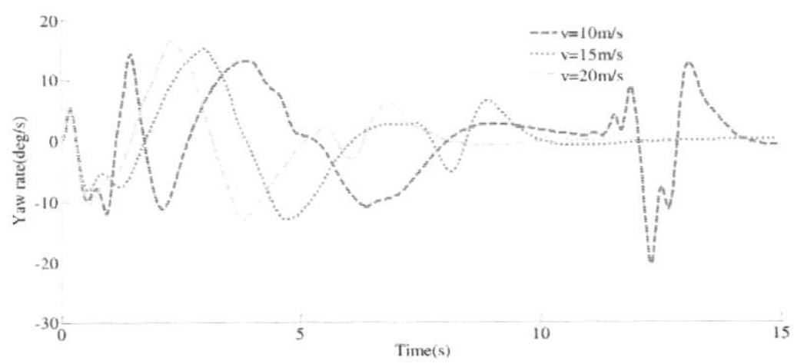

Figure 17. Simulation results of yaw rate at different speeds

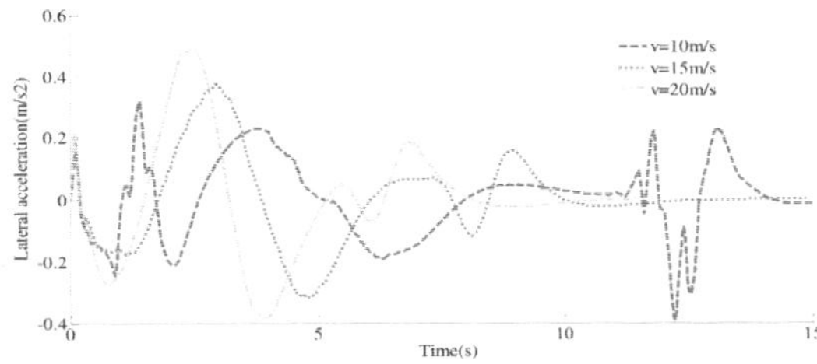

Figure 18. Simulation results of lateral acceleration at different speeds

The overall variation range of the front wheel angle is from -4deg to 4deg, which is always within the constraint range, and the variation is stable, which can ensure the smooth implementation of the actuator, according to figure 15. The lateral deviation Angle of the centre of mass and the lateral angular velocity of swing are both important parameters to measure the stability of a vehicle. From figure 16 and figure 17, the centroid side-slip Angle under different speed of the overall change in the range of -1.7 deg ~ 1.5 deg, far below the scope of centre of mass of side-slip Angle constraint set, and stable, the change of the transverse swing angular velocity amplitude as the speed increases, but the overall range for - 20 deg/s ~ 16 deg/s, it indicates that the vehicle has no dangerous situation such as side-slip or rollover in the process of driving. As can be seen from figure 18, the range of lateral acceleration is-0.4m/s2~0.5m/s2, which is far below the constraint range, it indicates that the vehicle has good driving stability.

5. Conclusion

When the intelligent car driving on the road, due to the complexity and changeability of the driving environment, the following control of a given reference trajectory has some limitations. According to the vehicle running state and obstacle information, the vehicle's running track can be adjusted in real time, to solve the path planning problem, and according to the re-planned reference track, the following control is carried out to avoid obstacles in the paper. Firstly, a point mass model is established which ignores the car body size. Secondly, on the premise that the size of obstacles is smaller than the car body size, according to the obstacle information obtained by the sensor and the established point mass model, the obstacle avoidance penalty function is established, then the objective function of the path planner is established. Finally, two working conditions -- single obstacle and two obstacles -- are set up to simulate and verify the designed obstacle avoidance trajectory following system.

6. Acknowledgments

Thanks for the found of NSFC (51505071) and the support of Northeastern University.

7. References

[1] He Z Z 1992 Architecture of intelligent mobile robot. *Robot*, vol 14 pp 55-58.
[2] Kritayakirana K, Gerdes J C 2012 Using the centre of percussion to design a steering controller for an autonomous race car. *Vehicle System Dynamics*, vol 50 pp 33-51.
[3] Ziegler J, Bender P, Schreiber M , et al 2014 Making Bertha Drive-A*n Autonomous Journey on a Historic Route* . IEEE Intelligent Transportation Systems Magazine, vol 6 pp 8-20.
[4] Mu D Ling Q, Jiewang R, Jun M 2014 Advanced technology and development of driverless vehicle. *Agricultural Equipment and Vehicle Engineering*, vol 52 pp 30-33.
[5] Youchun X, Rongben W, Bing L, et al 2001 Overview of intelligent vehicles in the world. *Automotive Engineering*, 2001, vol 23 pp 289-295.
[6] Heng Z, Guanghan L, Xiaohua Z, et al 2019 Adaptive driving force distribution control for multi-axle special hybrid vehicles .*China Mechanical Engineering*, vol 30 pp 528-534.

IOP Conf. Series: Materials Science and Engineering **746** (2020) 012034 doi:10.1088/1757-899X/746/1/012034

Design of the Transmission Device of the Some Type of the Manipulator

Yan Ma[1,*], Jingxi Xu[2] and Xingqiao Jiang[1]

[1]Shenyang Academy of Instrumentation Science, Shenyang 110043, China
[2]Hangzhou Forward Gear Box Group co. LTD., Hangzhou, Zhejiang 311203, China
[*]Corresponding author

Abstract. The transmission part of a manipulator is presented in the paper. The part of manipulator transmission device is analysed and the part of manipulator transmission device is designed. Design and calculation of suction hand is represented. The driving motor is checked, the calculation of the equivalent moment of inertia of the transmission system is gained, the motor moment frequency characteristic is presented, and the starting torque of the stepping motor is determined. And the schematic diagram of the overall structure and the schematic diagram of the motor connection plate is presented. The theoretical support and design reference for the design of the transmission device of the manipulator is presented in the paper.

1. Introduction

Manipulator is a new device developed in the process of automation and mechanization production. In industrial production, manipulator has a very wide range of use. The basic function is to grab an work piece from a specified location and transport it to another specified location. The mechanical arm replaces manual labour, with high operation precision, improving product quality and production efficiency [1]. In the mechanical industry, it can be used for the assembly of parts, the loading and unloading of processing work piece, handling. The assembly manipulator takes the rigid high arm as the main body, it may carry on each kind of operation automatically according to the signal which sends from the outside. It can save huge work piece conveying device, compact structure and strong adaptability. When the work piece changes, the flexible production system is easy to change, which is beneficial for the enterprise to continuously update the marketable variety, improve product quality, and better meet the needs of market competition. At present, the level of industrial robot technology and its engineering application in China is still far from that of foreign countries, and the application scale and industrialization level are relatively low. The research and development of manipulator has an important impact on the improvement of China's automatic production level, the research and design of manipulator is very meaningful [2-4].

2. Design of the Air Suction Hand

2.1. Structure of the Air Suction Hand

The suction cup is used to form negative pressure to hold the work piece. The suction hand is suitable for carrying thin pieces, such as thin iron sheet, silicon steel sheet, plate, paper, thin-walled fragile glassware, curved shell, etc., especially glassware and non-metal sheet, the adsorption effect is more obvious. Compared with the gripper hand, the air-suction hand has a simple structure, light weight and uniform surface adsorption force distribution, but the suction surface should be smooth, without holes and oil. According to the method of forming negative pressure (or vacuum), air suction hand can be divided into vacuum type, air negative pressure type and squeeze exhaust type suction cup. The suction hand used in the design is shown in figure 1.

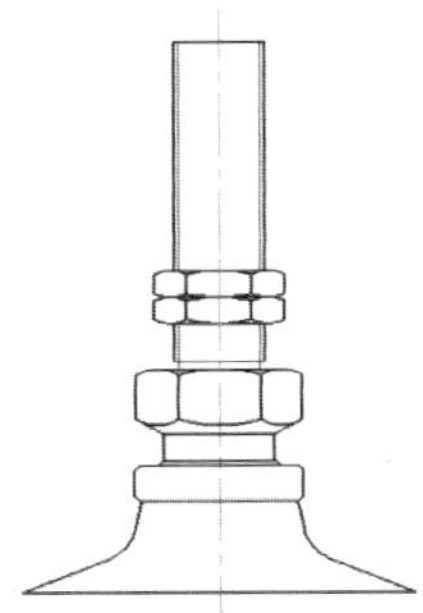

Figure 1. The Schematic diagram of the suction hand

2.2. Calculation of the Suction Hand

Proposed operating conditions: load W=10N.

The suction force of the vacuum chuck during the start-up process is:

$$P = \frac{S}{K_1 K_2 K_3} \Delta p$$

S - suction cup area (mm2), $S = \frac{\pi}{4} D^2$, D is suction cup diameter;

K1 - the safety factor of the suction cup when the work piece is started, K1=1.2 ~ 2;

K2 - coefficient of the operating condition, generally K2=1 ~ 3;

K3- coefficient of azimuth. When the sucker is vertically adsorbed, K1=f^{-1}, f is the friction coefficient. When the rubber sucker is adsorbed, *f*=0.5 ~ 0.8. When the sucker is absorbed horizontally, take K3=1;

Δp – the differential pressure of the suction cup from inside to the outside (MPa).

For vacuum suction cup, the suction cup adsorption work piece should be guaranteed to be absorbed at the start. During the transfer process, the vacuum degree in the suction cup increases with time, and the adsorption work piece becomes more and more reliable.

Micro vacuum pump KVP04, vacuum degree -40kpa, suction cup ZPT32UN-A8, diameter 32mm, nitrile butadiene rubber material; Take 1.2 for K1 and take 1 for K2, take 0.8 for f, and take 0.04 MPa for Δp .

P= (PI /4... 32 ^2)/ (1.2... 1... 0.8)

Then P= 21.45n, and the suction cup meets the requirement.

3. The Check of the Drive Motor

3.1. Calculation of the Equivalent Moment of Inertia of the Transmission System [5, 6]

(1) Inertia of motor rotor J_D.

The inertia of motor rotor from the stepper motor product manual is represented as:

$$J_D = 0.8\text{kg} \cdot \text{cm}^2$$

(2) The translation of the moment of inertia of the ball screw J_S.

Consider the screw as a homogeneous cylinder.

$$J_S = \frac{1}{2} m R^2$$

$$m = \frac{1}{4} \pi {d_1}^2 L \rho_{\text{steel}}$$

Where: d1- outside diameter of ball screw thread

L- the length of the screw

$$m = \frac{1}{4}\pi \times 15.1^2 \times 0.888 \times 7.85 \times 10^3 = 1.25\text{kg}$$

$$J_S = \frac{1}{2} \times 1.25 \times \left(\frac{1}{2} \times 1.51\right)^2 = 0.36\text{kg/cm}^2$$

From the size and performance parameters of the inner circulation ball screw pair series in the mechanical design manual, d_1=15.1 mm.

The full length L of the lead screw is 888 mm.

(3) The reduction of the inertia of the forearm movement J_G.

The working table is a moving part, whose mass is converted to the inertia of horizontal movement of the ball screw axis, which can be calculated according to the following formula:

$$J_G = \left(\frac{V}{2\pi n}\right)^2 m_\Sigma$$

Where: v - moving speed of table, cm/min;

n - screw speed, RPM;

m_Σ - moving parts mass, kg;

$$J_G = \left(\frac{240}{2\pi \cdot 600}\right)^2 \cdot 9.57 = 0.04\text{kg/cm}^2$$

The moment of inertia of the coupling:

$$J_L = 0.085\text{kg/cm}^2$$

The equivalent moment of inertia of the system

$$\text{J} = J_D + J_S + J_G + J_L = 0.8 + 0.36 + 0.04 + 0.085 = 1.285\text{kg/cm}^2$$

3.2. Check for the Moment Frequency Characteristics of the Motor

The maximum static torque of the stepping motor M_{jmax}refers to the positioning torque of the motor, and $M_{\text{jmax}} = 2.4\text{N} \cdot \text{m}$.

The nominal starting torque of the stepping motor isM_{mq}; the relation of the maximum static torque M_{jmax} is: $M_{\text{mq}} = \lambda M_{\text{jmax}}$. The value for λ can be found from the product specification of the motor.

$$M_{\text{mq}} = 0.707 \times 2.4 = 1.70\text{N} \cdot \text{m}$$

The starting torque of the stepper motor can be calculated as follows:

$$M_{\text{kq}} = M_{\text{ka}} + M_{\text{kf}} + M_0$$

Where: M_{kq}- the no-load starting torque;

M_{ka} - the acceleration torque of motor shaft from static to maximum fast forward speed during no-load start;

M_{kf}-the friction torque on motor shaft when no load is applied;

M_0 -the additional friction torque;

The calculation of no-load starting torque $M_{kq}$is presented as follows:

(1) Acceleration torque M_{ka}

The acceleration torque M_{ka} can be calculated according to the following formula:

$$M_{\text{ka}} = J\varepsilon = J\frac{2\pi n_{\text{max}}}{60t} \cdot 10^{-2}$$

$$n_{\text{max}} = \frac{v_{\text{max}}\theta}{\delta_{\text{p}}}$$

Where: J - equivalent moment of inertia of the system;

ε -Epsilon and loss is the maximum angular acceleration of the motor;

$n_{\max}$-motor speed corresponding to the maximum feeding speed of moving parts;
$v_{\max}$-maximum feeding speed of moving parts;
θ-the step moment Angle of the stepper motor;
δ_{p} -impulse equivalent of the stepping motor;

$$n_{\max}=600\mathrm{rpm}$$

(2) No-load friction torqueM_{kf}
The no-load friction torque M_{kf} can be calculated according to the following equation

$$M_{\mathrm{ka}} = 1.285 \cdot \frac{2\pi \cdot 600}{60 \cdot 0.2} \cdot 10^{-2} = 4.04\mathrm{N} \cdot \mathrm{cm}$$

Where: L- maximum stroke of ball screw, cm
η – the total efficiency of the transmission system, eta = 0.9

$$M_{\mathrm{kf}} = \frac{4.96 \cdot 60}{2\pi \cdot 0.9} = 52.63\mathrm{N} \cdot \mathrm{cm}$$

(3) Additional friction torque M_0
Additional friction torque M_0 can be calculated according to the following equation:

$$M_0 = \frac{F_p L}{2\pi\eta}$$

$$M_0 = \frac{2.29 \cdot 60}{2\pi \cdot 0.9} = 24.30\mathrm{N} \cdot \mathrm{cm}$$

Thus, the starting torque required by the stepper motor can be obtained as follows:

$$M_{\mathrm{kq}} = M_{\mathrm{ka}} + M_{\mathrm{kf}} + M_0 = 4.04 + 52.63 + 24.30 = 80.97\mathrm{N} \cdot \mathrm{cm}$$

$$M_{\mathrm{kq}} = 0.81\mathrm{N} \cdot \mathrm{m} < M_{\mathrm{mq}} = 1.70\mathrm{N} \cdot \mathrm{m}$$

The starting torque required by the stepper motor is less than the nominal torque of the stepper motor, and the stepper motor meets the requirements.

The schematic diagram of overall structure and motor connection plate, the schematic diagram of the base and the attachment plate of the suction, are shown from figure 2 to figure 5.

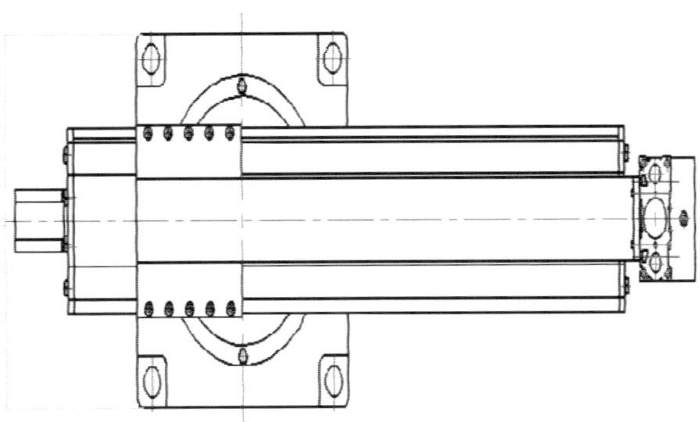

Figure 2. The schematic diagram of the overall structure

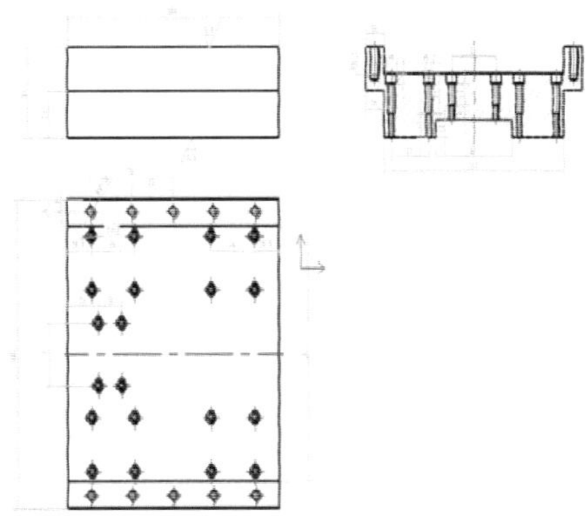

Figure 3. The schematic diagram of the motor connection plate

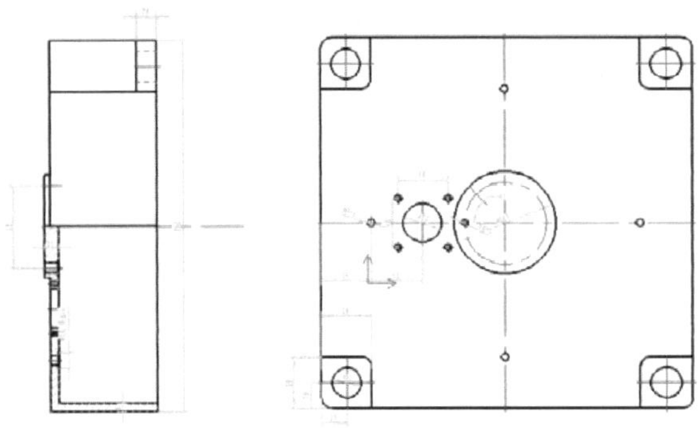

Figure 4. The schematic diagram of the **base**

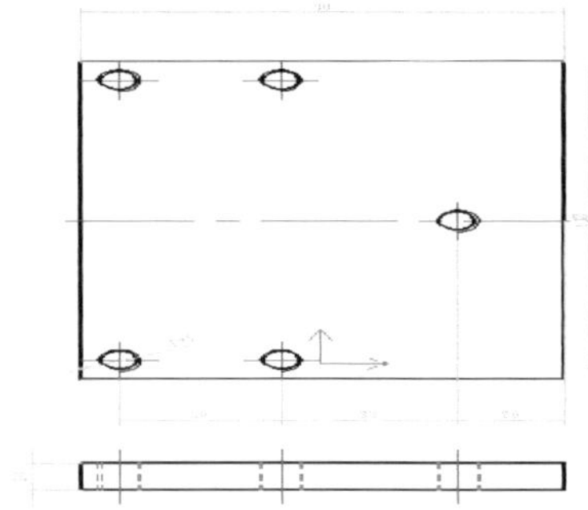

Figure 5. Attachment plate of the suction

4. Conclusion
The part of manipulator transmission device is analysed and the part of manipulator transmission device is designed. Design and calculation of suction hand. The driving motor is checked, the calculation of the equivalent moment of inertia of the transmission system is gained, the motor moment frequency characteristic is presented, and the starting torque of the stepping motor is determined. And the schematic diagram of the overall structure and the schematic diagram of the motor connection plate is presented. In the field of machinery, generation after generation of engineers constantly optimize the product structure and data in the actual production. The experience of predecessors is very worthy of reference, but continuous innovation is needed. Better products can be obtained in this way.

5. Acknowledgments
Thanks for the support of Shenyang Academy of Instrumentation Science and Northeastern University.

6. References
[1] Gao R H 2014 *Industrial Manipulator Design*. Hefei: hefei university of technology press.
[2] Guo T Y, An D 2017 *Robot system design and application*. Beijing: chemical industry press.
[3] Giberti h, Cinquemani S,Ambrosetti S. R 2013 2dofparallel kinematic manipulator-A multidisciplinary test case in mechatronics . *Mechatronics* vol 23 pp 949-959.
[4] Pellicciari M, Berselli G, Leali f, et al. 2013 A method for reducing the energy consumption of pick-and-place industrial robots. *Mechatronics*, vol 23 pp 326-334.
[5] SUN Z L, Ma X G, Huang Q B, Yan Y To 2008 *Mechanical Design*. Beijing: science press.
[6] Gong Y P, Tian W L, Zhang W H, Huang Q B 2008 *Course Design of Mechanical Design*. Beijing: science press.

IOP Conf. Series: Materials Science and Engineering **746** (2020) 012035 doi:10.1088/1757-899X/746/1/012035

Numerical Simulation on Cooling Process of Phase Change Cool Storage Device for Rail Vehicle Air Conditioning System

Hongyu Wang[1], Bing Han[1], Mengmeng Li[2, 3], Xiao Zeng[2] and Suna Wang[2]

[1]CRRC SIFANG Institute Co, Ltd, Qingdao 266031, Shandong, China
[2]School of Energy Science and Engineering, Central South University, Changsha 410083, Hunan, China
Email: 2960944083@qq.com

Abstract. In order to face emergency situation of the rail vehicle air conditioning system, a cool storage device was designed. The device was made up of phase change paraffin and heat exchanger. For the purpose of optimum system design, the mathematic model of phase change and heat transfer in process of cool discharging was established, numerical simulation was finished. Discharging rate of cool and cooling capacity of the device was analyzed. According to the calculation results, the wind speed and air temperature had the greatest influence on the cooling effect, while the thermal conductivity had little effect on it.

1. Introduction

The rapid development of rail vehicles such as high-speed railway, motor train and subway has met people's growing transportation needs, but how to make full-closed carriages meet passenger's minimum requirements for comfort within a certain period of time when emergency situations occur is the biggest problem. A phase change cold storage device, which has the advantages of high energy storage density and approximate isothermal storage and cooling process, is invented in this paper to adjust the temperature of the full-closed carriages.

Ye Hong et al. [1] found the effective thermal conductivity of the phase change material on the tube side has a significant impact on the performance of the heat exchanger. An experimental study on the shell-and-tube phase change heat exchanger with paraffin as heat storage material is carried out by Anita Trp [2]. Zanxia Wu et al. [3] found that the high temperature phase change material meet the requirements of Submarine cold storage air conditioning. After detailed researched, solid-solid phase change paraffin is selected as the energy storage material for air conditioning of railway vehicles, which not only can be repeatedly melted and solidified without supercooling but also remains solid state.

2. Cooling Process Simulation Calculation of Cold Storage Device

Numerical simulation method is used to simulate complex phase-change process.

2.1. Geometric Model Simplification

For the structure of air duct and the condition of wind resistance, Circular tube bundles and flat tube bundles is placed in the air supply duct of train air conditioning system as two kinds of cold storage device's structure.

The circular tube heat exchange unit is simplified into a 2D model of the section, as shown in Figure 1. Among them, L1 is the width direction spacing of each heat exchange tube in the heat exchanger, L2 is the height direction spacing of the heat exchange tube, and d is the outer diameter of

IOP Conf. Series: Materials Science and Engineering **746** (2020) 012035 doi:10.1088/1757-899X/746/1/012035

heat exchange tube.

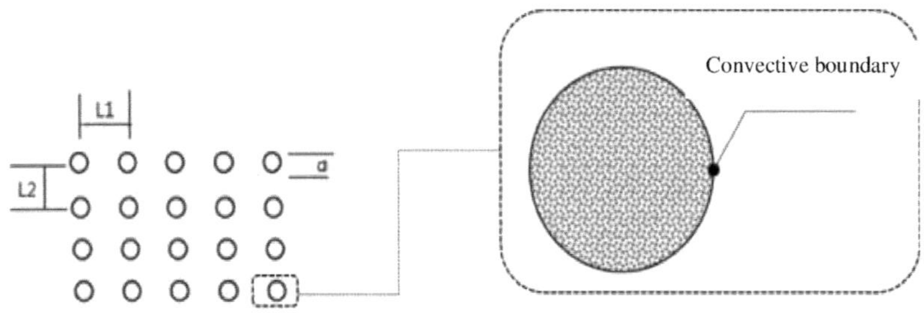

Figure 1. Simplified schematic diagram of simulation model for heat transfer unit of circular tube

Similarly, only a small section of the cross section in flat tube heat exchanger unit is intercepted as the research object, as shown in Figure 2. Among them, L is the width direction spacing of each heat exchange flat tube, H is the height of the heat exchange flat tube, and D is the thickness of heat exchange flat tube.

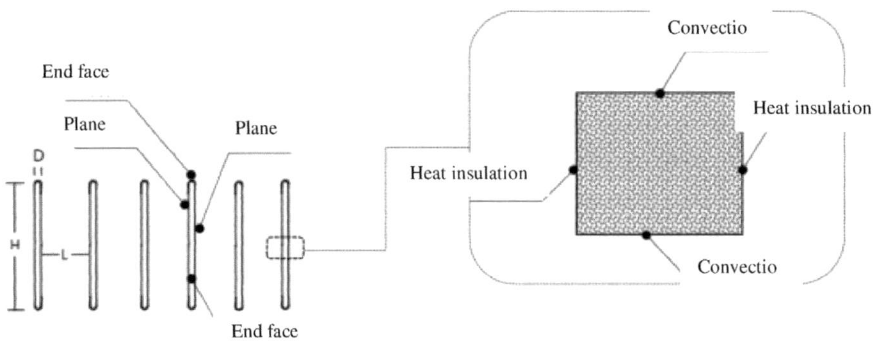

Figure 2. Simplified schematic diagram of simulation model for flat tube heat transfer unit

2.2. Physical Model

The calculation object is circular tube and flat tube bundles (filled with phase change materials) in Figure 1, 2. The pipe is made of aluminium alloy; the phase change material is solid-solid phase change paraffin. And its phase change temperature is 20°C. Other physical properties are listed in the following table:

Table 1. Physical properties of materials

Material name	density /kg·m^{-3}	Thermal conductivity /W·(m·K)$^{-1}$	Specific heat capacity /J·(kg·K)$^{-1}$	Kinematic viscosity /m^2·s^{-1}
Air	1.28	0.0026	1005	1.59×10^{-5}
phase change material	0.88	2.38	3.22	solid
aluminium alloy	2700	170	880	solid

The following simplifications and assumptions are made for the heat transfer of the calculated object:

1) The main heat transfer direction of circular tube is radial; the main heat transfer direction of flat

tube is perpendicular to the wall.

2) The air velocity and temperature distribution between the voids of each single tube in the heat exchanger unit are uniform.

3) The heat exchange process of each single tube in the heat exchanger unit is completely the same.

4) The wall thickness is 0.5 mm, so the heat transfer in the pipe wall is simplified to heat transfer resistance.

2.3. Mathematical Model

In summary, the simulation calculation of the unit type cold storage heat exchanger cooling process is simplified to the simulation calculation of the single tube cooling process.

1) Control equation:

$$\rho c\frac{\partial T}{\partial t}=\nabla\bullet(\lambda\nabla T)+S \tag{1}$$

In this formula, ρis the phase change material density, T is the instantaneous temperature of phase change material, λis the thermal conductivity of phase change material, S is the phase change latent heat of phase change material.

2) Initial conditions: The initial temperature of phase change material and pipe was 17°C, and the air supply temperature of duct was 35°C.

3) Boundary conditions: The convective heat transfer boundary conditions on side wall of pipe are as follows:

$$\lambda\frac{\partial T}{\partial n}=h(T-T_a) \tag{2}$$

In this formula, λ is the thermal conductivity of phase change material, T is the instantaneous temperature of phase change material, Ta is the ambient temperature (air supply temperature of duct), h is the convective heat transfer coefficient of tube wall [4].

4) Source term processing:

How to deal with the source term is the core problem and difficulty of phase change heat transfer. The equivalent heat capacity method is used to deal with it in this simulation calculation, that is, the real-time latent heat of phase change in the phase change temperature range is superimposed on the original specific heat capacity of the phase change material according to the phase change curve obtained from DSC test, which enables the equivalent heat capacity method to reflect actual change of phase change's latent heat in the phase change temperature range in real time. As shown in the following figure:

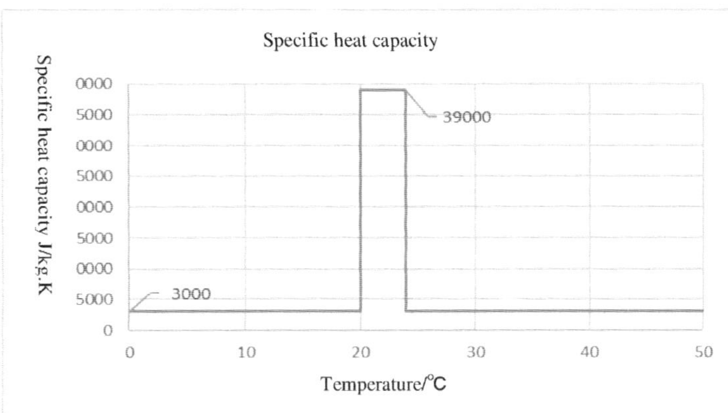

Figure 3. Specific heat capacity of cold storage material in phase-change temperature region (19°C - 24°C)

IOP Conf. Series: Materials Science and Engineering **746** (2020) 012035 doi:10.1088/1757-899X/746/1/012035

3. Calculation Results and Analysis

3.1. Influence of Wind Speed Change on Cooling Time

The influence of air velocity outside the tube on heat transfer of circular tube and flat tube is ultimately reflected in the influence of air flow on convective heat transfer, and the size of air velocity directly determines the size of convective heat transfer coefficient, The relationship between convective heat transfer coefficient and wind speed obtained by theoretical calculation of the air longitudinally passing through the tube bundle is shown in the following table [5]:

Table 2. Convective heat transfer coefficients for different wind speeds and pipe sizes / $W \cdot m^{-2} \cdot K^{-1}$

Pipe size /mm	Outer diameter of tube:10	Flat tube: 100 × 5	Flat tube: 100 × 10	Flat tube: 100 × 15
5m/s	20	19	19	19
7m/s	26	25	25	25
9m/s	31	31	31	31

For example, when the flat tube thickness is 5 mm, the initial temperature is 17°C, the ambient temperature is 35°C, and the thermal conductivity of the cold storage material is 2.38 W/m K, the heat flux density and the cumulative cooling capacity per unit tube length change with time at different wind speeds as shown in Figures 4 and 5:

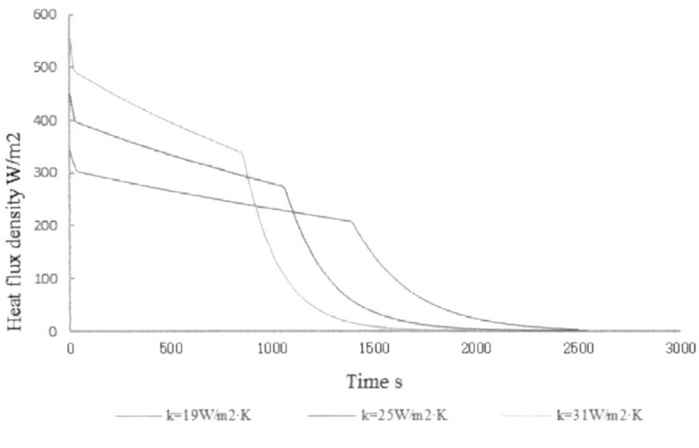

Figure 4. Variation curves of heat flux density on tube wall under different wind velocities

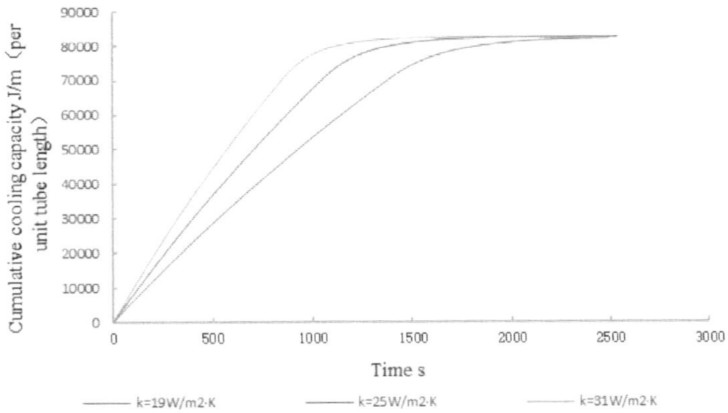

Figure 5. Variation curve of cooling capacity per unit tube length under different wind velocities

Under the same working conditions, the heat transfer unit of circular tube with diameter of 10 mm is also simulated, simulation results is listed:

Table 3. Cooling time of 5 mm pipe thick flat pipe and 10 mm pipe diameter circular pipe at different wind speeds

5 mm thick flat pipe				10 mm diameter circular pipe			
wind speed/ $m \cdot s^{-1}$	100% cooling time/s	Decreased cooling time /%	Cooling capacity per unit length of pipe /$J \cdot m^{-1}$	wind speed /$m \cdot s^{-1}$	100% cooling time /s	Decreased cooling time /%	Cooling capacity per unit length of pipe /$J \cdot m^{-1}$
5	1395			5	1504		
7	1062	23.87	82520	7	1163	22.67	13138
9	858	19.21		9	979	15.82	

3.2. Influence of Ambient Temperature Change on Cooling Time

For example, when the flat tube thickness is 5 mm, the initial temperature is 17°C, the wind speed is 9 m/s and the thermal conductivity of the cold storage material is 2.38 W/m K, the heat flux density and the cumulative cooling capacity per unit tube length under different ambient temperatures are shown in Figures 6 and 7 respectively:

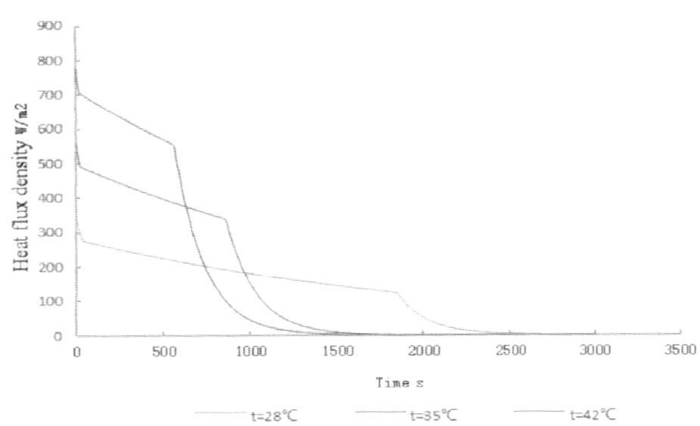

Figure 6. Variation curves of heat flux density on tube wall at different environmental temperatures

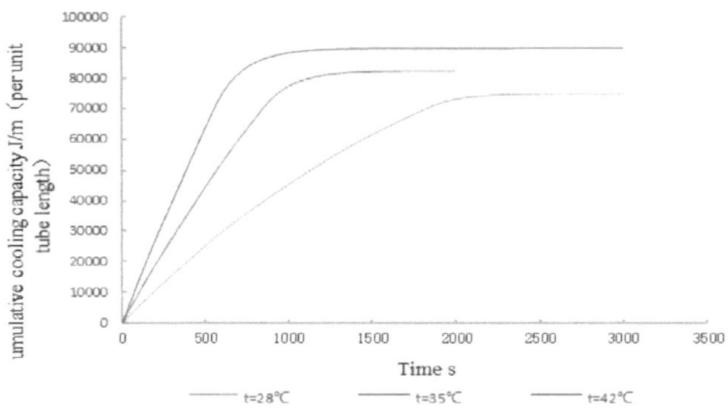

Figure 7. Variation curve of cooling capacity per unit tube length at different environmental temperatures

Under the same working conditions, the heat transfer unit of circular tube with diameter of 10 mm is also simulated, simulation results is listed:

Table 4. Cooling time of 5 mm pipe thick flat pipe and 10mm pipe diameter circular pipe at different ambient temperatures

5 mm thick flat pipe				10 mm diameter circular pipe			
ambient temperatures /°C	100% cooling time/s	Decreased cooling time /%	Cooling capacity per unit length of pipe /J·m^{-1}	ambient temperatures /°C	100% cooling time /s	Decreased cooling time /%	Cooling capacity per unit length of pipe /J·m^{-1}
28	1848		75096	28	2098		11939
35	858	53.57	82520	35	979	53.34	13138
42	562	34.50	89899	42	645	34.12	14363

3.3. Effect of Thermal Conductivity on Cooling Time

A simulation is calculated under the conditions of initial temperature 17°C, ambient tempe-rature 35°C, wind speed 9 m/s but different thermal conductivity to investigate the influence of thermal conductivity of cold storage material on its cooling time. The simulation results of heat transfer units of 5 mm thick flat tube and 10 mm diameter circular tube are listed in the following table:

Table 5. Cooling time of 5 mm pipe thick flat pipe and 10mm pipe diameter circular pipe at different thermal conductivity

5 mm thick flat pipe				10 mm diameter circular pipe			
thermal conductivity /W·(m·K)$^{-1}$	100% cooling time/s	Decreased cooling time /%	Cooling capacity per unit length of pipe /J·m^{-1}	thermal conductivity /W·(m·K)$^{-1}$	100% cooling time/s	Decreased cooling time /%	Cooling capacity per unit length of pipe /J·m^{-1}
1.4	865			1.4	996		
2.38	858	0.72	82520	2.38	979	1.71	13138
4	854	0.47		4	971	0.82	

3.4. *Influence of Tube Thickness (Tube Diameter) on Cooling Time*

A simulation is calculated under the working conditions that 17°C initial temperature, 35°C ambient temperature, 2.38 W/m·K thermal conductivity coefficient of the cold storage material, 9 m/s wind speed, but different tube thickness (tube diameter) to investigate the effect of tube thick- ness (tube diameter) on the cooling time of the cold storage device, and simulation results is listed:

Table 6. Cooling time of 5mm pipe thick flat tube and 10mm tube diameter circular tube at different tube thickness (tube diameter)

Flat tube				circular tube			
tube thickness /mm	100% cooling time/s	Decreased cooling time /%	Cooling capacity per unit length of pipe /J·m^{-1}	tube diameter /mm	100% cooling time/s	Decreased cooling time /%	Cooling capacity per unit length of pipe /J·m^{-1}
5	858	56.18	82520	5	508	44.09	3147
10	1958	36.53	185381	7	732	25.23	6561
15	3085		286314	10	979		13138

4. Conclusions and Recommendations

The following conclusions can be drawn:

1) The most important factors affecting the cooling time are wind speed (convective heat transfer coefficient) and ambient temperature, the slightly weaker factor is tube thickness (tube diameter). In addition, the change of ambient temperature has a great influence on the cumulative cooling capacity.

2) Thermal conductivity has little effect on cooling time.

3) The following principles should be followed when selecting materials and designing the structure: phase change latent heat should be taken into account while thermal conductivity should not be taken into adequate account; flat tubes are recommended for comprehensive consideration;

The structure, size and air supply temperature of heat exchanger should be reasonably selected.

5. References

[1] Hong Ye, Xiao Hao, Danpeng Cheng,et al 2008 Numerical investigation on heat storage and heat storage and heat exchange of the shell-and-tube phase change heat exchanger *Journal of Solar Energy* **29 (12)** 1499-1503.

[2] Trp A 2005 An experimental and numerical investigation of heat transfer during technical grade paraffin melting and solidification in a shell-and-tube latent thermal energy storage unit *Solar Energy* **79 (6)** 648-660.

[3] Zanxia Wu, Gang Wu, Qianghong Zeng, et al 2008 Numerical simulation of cold storage materials used in submarine air conditioning *Fluid machinery* **36 (5)** 75-78.

[4] Shuying Wu,Nan Wang, Dongsheng Zhu, et al 2012 Study on thermal conductivity of nano-copper/paraffin composite phase change heat storage materials *New chemical materials* **40 (5)** 104-106.

[5] Rosnaud. Handbook of Basic Heat Transfer Science [M]. Science Press, 1992.

[6] Yudong Liu, 2005 *Preparation and thermophysical properties of nanocomposite low temperature phase change cold storage materials* (Chongqing: Chongqing University).

[7] Aigang Pan, Junbiao Wang, Xianjie Zhang 2014 Phase change heat transfer numerical analysis based on equivalent heat capacity method and enthalpy method *Computer simulation* **31 (2)** 315-319.

IOP Conf. Series: Materials Science and Engineering **746** (2020) 012036 doi:10.1088/1757-899X/746/1/012036

Design and Experimental Study of Wire Saw Machine for Cutting Oil Pipeline

Liwen Cao and Shuaitong Miao

School of Mechanical and Electrical Engineering, Heilongjiang University, Harbin, Heilongjiang Province, China
Email: caoliwen68@163.com

Abstract. According to the structural characteristics of the oil pipeline, repair cutting quality requirements, designed a special cutting wire saw machine. The structural characteristics of the oil pipeline and the variation of thc area of the steel pipe section and the slit width of the oil pipeline are analyzed. The structural form, working principle and technical parameters of the hydraulic power system of the wire saw machine are determined. The wire sawing machine cut the oil pipeline and tested it, and established a mathematical model of the cutting force, two correspondences were obtained, one of which is the beading cutting force and wear rate of the wire saw machine, and the other is the feed speed and cutting speed. Engineering practice has proved that the use of wire saws to cut oil pipelines is an effective cutting method, both cutting quality and cutting efficiency are superior to other cutting methods.

1. Introduction

The oil pipeline is the main production facility for oil and gas field development, production and oil and gas transportation [1]. With the rapid and efficient development of the petroleum industry, the newly developed oil and gas fields have increased year by year. Due to various reasons, pipeline damage accidents have occurred successively, and the damage forms are various [2, 3]. Once there is a problem with the oil pipeline, one is a huge loss, and the other is that repair work cannot be delayed. The oil pipeline repair should adopt reasonable maintenance methods and necessary maintenance equipment, and repair according to the maintenance process.

When the oil pipeline is repaired or constructed, the common cutting methods include flame cutting, plasma arc cutting, high pressure water jet cutting, explosive cutting, and mechanical cutting. Various cutting methods have their own adaptation conditions and scope, and the processing objects and working conditions have certain limitations [4, 5]. According to the structural characteristics of the oil pipeline, the wire saw machine developed can ensure high-efficiency cutting operation under the premise of cutting section quality, and the application prospect is very impressive.

2. Structural Characteristics of Oil Pipelines

Please follow these instructions as carefully as possible so all articles within a conference have the same style to the title page. This paragraph follows a section title so it should not be indented.

2.1. Structure of Oil Pipeline

The cross area of seabed single oil and gas pipeline see figure 1, and it's thickness of every layers see table 1.

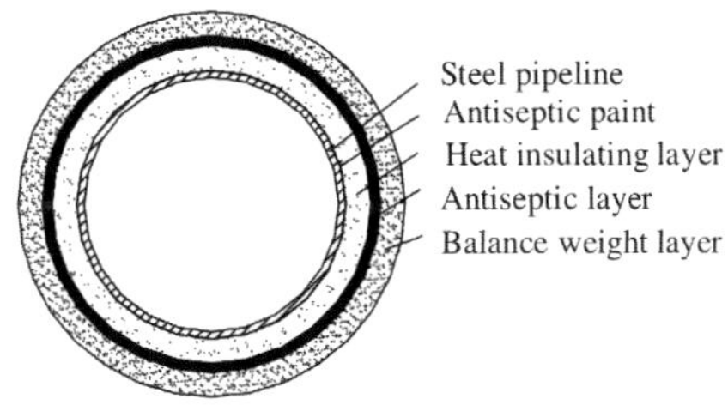

Figure 1. Structure of single oil and gas pipeline

Table 1. Parameter of single pipeline.

External diam	Balance layer	Polyethylene layer	Heat insulating layer	Thickness of pipeline
522 mm	47 mm	9.5 mm	40 mm	12 mm

2.2. Analysis of the Area and the Width of the Pipe Section of the Pipeline during the Cutting of the Oil Pipeline

When the oil pipeline is cut, the steel pipe part is difficult to cut than other parts of the pipeline. After analyzing the cutting area and the slit width of the steel pipe part of the cutting pipeline, the cutting process parameters can be adjusted reasonably.

2.2.1. Area analysis of steel pipe part of oil pipeline

Establish a Cartesian coordinate system as shown in Figure 2.

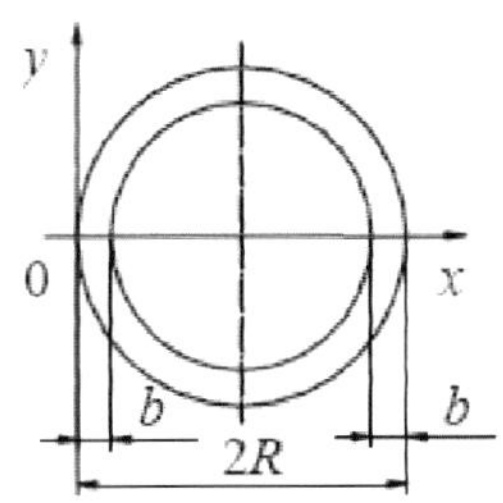

Figure 2. Rectangular coordinate system of steel pipeline

The outer circle equation of the pipe section obtained from the coordinate system is

$$(R-x)^2+y^2=R^2 \quad (0\le x\le 2R) \tag{1}$$

Where *R*— the radius of the pipe section of the pipe (m).
When the y takes a positive value,

$$y=\sqrt{R^2-(R-x)^2} \tag{2}$$

The inner whole equation of the pipe section obtained from the coordinate system is

$$(R-x)^2+y^2=(R-b)^2 \quad (b\le \mathrm{x}\le 2R\text{-}b) \tag{3}$$

Where *b*— the wall thickness of the pipe section of the pipe (m).
When the y takes a positive value,

$$y=\sqrt{(R-b)^2-(R-x)^2} \tag{4}$$

The area of the outer circle of the steel pipe of the beaded rope cutting pipe is

$$S_d = 2\int_0^x \sqrt{R^2 - (R-x)^2}\,dx \quad (0 \le x \le 2R) \tag{5}$$

The area of the inner hole of the steel pipe part of the beaded rope cutting pipe is

$$S_D = 2\int_b^x \sqrt{(R-b)^2 - (R-x)^2}\,dx \quad (b \le x \le 2R\text{-}b) \tag{6}$$

The area of the steel pipe part of the beaded rope cutting pipe is
When $0 \le x \le b$

$$S = S_d = 2\int_0^x \sqrt{R^2 - (R-x)^2}\,dx \tag{7}$$

When $b \le x \le 2R\text{-}b$

$$S = S_d - S_D = 2\int_0^x \sqrt{R^2 - (R-x)^2}\,dx - 2\int_b^x \sqrt{(R-b)^2 - (R-x)^2}\,dx \tag{8}$$

When $2R\text{-}b \le x \le 2R$

$$S = S_d - S_D = 2\int_0^x \sqrt{R^2 - (R-x)^2}\,dx - (R-b)^2\pi \tag{9}$$

2.2.2. Analysis of section width of steel pipe during oil pipeline cutting
The beaded rope cuts the pipe section of the pipe, and the chord length of the outer circle is

$$L_d = 2\sqrt{R^2 - (R-x)^2} \quad (0 \le x \le 2R) \tag{10}$$

The chord length of the inner hole of the beaded rope cutting pipe is

$$L_D = 2\sqrt{(R-b)^2 - (R-x)^2} \quad (b \le x \le 2R\text{-}b) \tag{11}$$

The width of the slit of the steel pipe portion of the beaded rope cutting pipe is
When $0 \le x \le b$

$$L = L_d = 2\sqrt{R^2 - (R-x)^2} \tag{12}$$

When $b \le x \le 2R\text{-}b$

$$L = L_d - L_D = 2\sqrt{R^2 - (R-x)^2} - 2\sqrt{(R-b)^2 - (R-x)^2} \tag{13}$$

When $2R\text{-}b \le x \le 2R$

$$L = L_d = 2\sqrt{R^2 - (R-x)^2} \tag{14}$$

The corresponding relationship between the feed amount of the bead rope and the slit width of the pipe is shown in Figure 3.

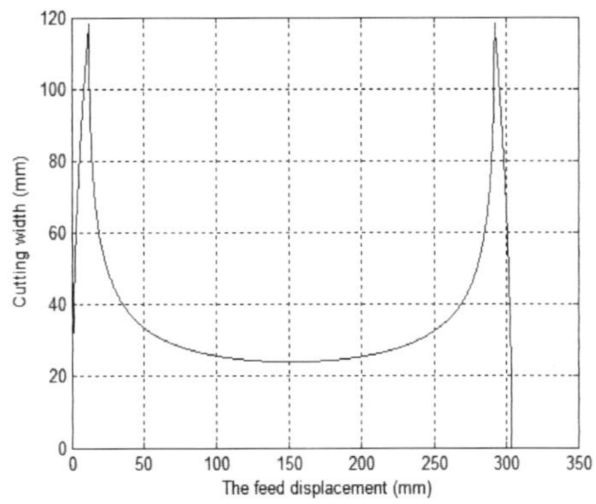

Figure 3. Relation between feeding motion and cut width of diamond string bead wires

3. Diamond Wire Saw

The diamond wire saw is specially designed for cutting oil pipeline equipment. The working performance of the wire saw machine is directly related to the stability of the cutting operation process, the quality of the pipe cutting section and the cutting efficiency.

3.1. Structural Design of Wire Saw Machine

The three-dimensional modeling of the wire saw machine developed for cutting the oil pipeline is shown in Figure 4. The wire sawing machine is composed of a main motion system (driving device), a feeding system (feeding device), a tensioning device, a clamping device, a guiding device, a saw blade frame, a cutting frame, a control system, a hydraulic power system and the like [6].The developed diamond wire saw machine is shown in Figure 5.

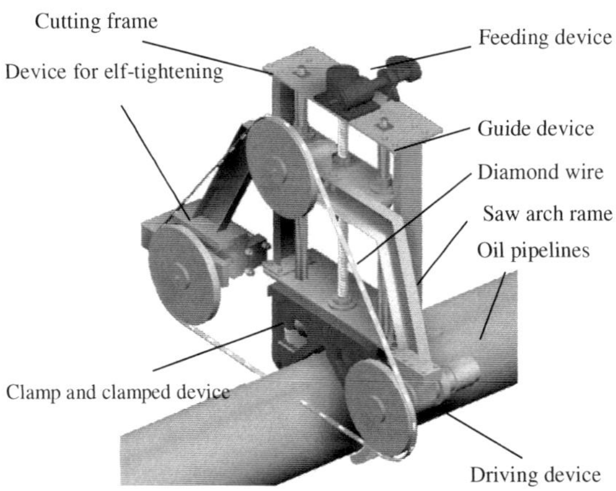

Figure 4. 3D modelling of wire saw

Figure 5. Diamond wire saw

The main part of the wire saw, the clamping device shown in Figure 6, the chain clamp device is a fast clamping method, using V-shaped block positioning, the clamping chain is wrapped around the oil pipeline, the wire saw machine The cutting frame is fixed at the designated position of the pipe to be cut.

The tensioning device is shown in Figure 7. The tensioning device is used to ensure the initial tension of the bead rope and the stable tension during the processing, and to ensure the convenience of assembling and disassembling the bead rope. The tensioning method is a screw transmission method, and the guide wheel assembly is mounted on the nut assembly. When the screw is rotated, the nut drives the guide wheel to move left and right to complete the tension and relaxation of the bead rope.

IOP Conf. Series: Materials Science and Engineering **746** (2020) 012036 doi:10.1088/1757-899X/746/1/012036

Figure 6. Clamp device

Figure 7. Device for self-tightening

The driving device is shown in Figure 8. The main movement of the wire saw machine is the circular motion of the bead rope. The beaded rope is supported along the driving wheel and the guiding wheel to form a closed loop, which is driven by the driving wheel to realize the circular motion. The driving wheel adopts a hydraulic driving mode, and the hydraulic motor directly drives the driving axle to rotate.

Figure 8. Driving device

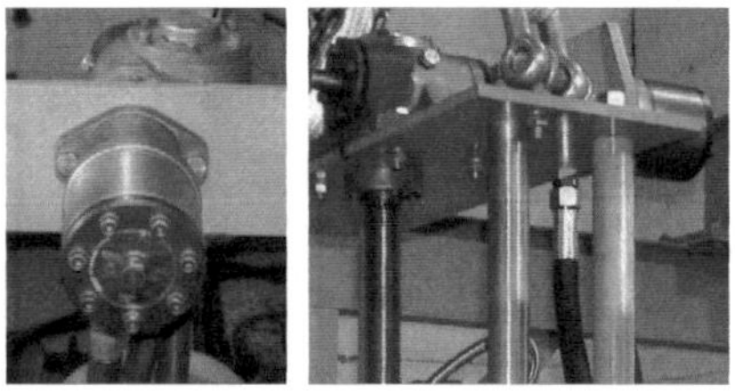

Figure 9. Feeding device

The feeding device is shown in Figure 9. In the feeding device, the worm wheel screw lift is driven by the cyclical hydraulic motor, and the screw rotation of the elevator drives the saw blade frame to move up and down to realize the radial feeding and rapid retracting movement of the cutting member.

3.2. The Working Principle of the Wire Saw

The basic working principle of the wire sawing machine cutting pipeline: firstly, the wire sawing machine is fixed in the oil pipeline to be cut by the clamping device; the tensioning device is adjusted to maintain a certain working tension of the bead rope; then the driving device and the feeding device Start working and cut the oil pipeline.

3.3. Wire Saw Machine Hydraulic Power System

3.3.1. The composition of the hydraulic power system of the wire saw machine

The wire saw machine hydraulic power system includes a hydraulic source and a hydraulic control oil circuit system. The hydraulic source system includes a hydraulic source oil circuit system and an electrical control system. The hydraulic control oil circuit system includes a hydraulic control valve, a hydraulic manifold and a connecting line.

3.3.2. Hydraulic source oil system

The hydraulic oil source is shown in Figure 10 and is the power source for the wire saw machine. It converts electrical energy into the hydraulic energy required for a wire saw. The hydraulic system power supply can provide a pressure of 20 MPa and a maximum flow of 90 L/min.

The schematic diagram of the hydraulic source oil circuit system is shown in Figure 11. The system mainly includes fuel tank, three-phase AC motor, variable pump, electromagnetic relief valve, and remote pressure regulating valve, oil pressure indicator, and oil filter and buzzer alarm composition.

Figure 10. Hydraulic source

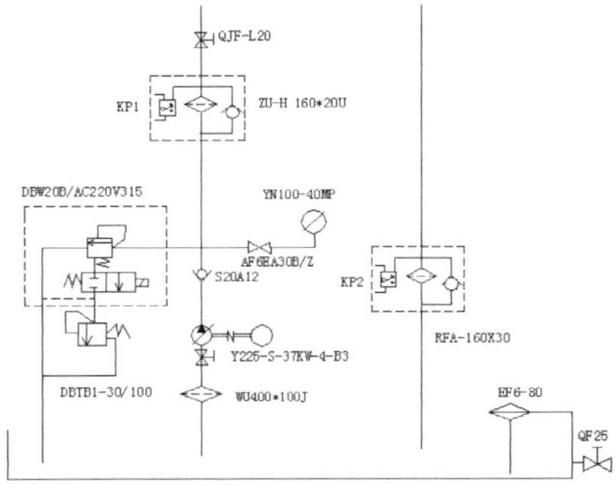

Figure 11. Principle diagram of oil piping system of hydraulic source

3.3.3. Hydraulic control oil system

The control regulating valves in the hydraulic control oil circuit system are connected to the hydraulic manifold by a plate connection. The wire saw machine control oil circuit system is shown in Figure 12, including a feed hydraulic motor control oil circuit and a drive hydraulic motor control oil circuit. The physical diagram of the control valve and hydraulic manifold is shown in Figure 13.

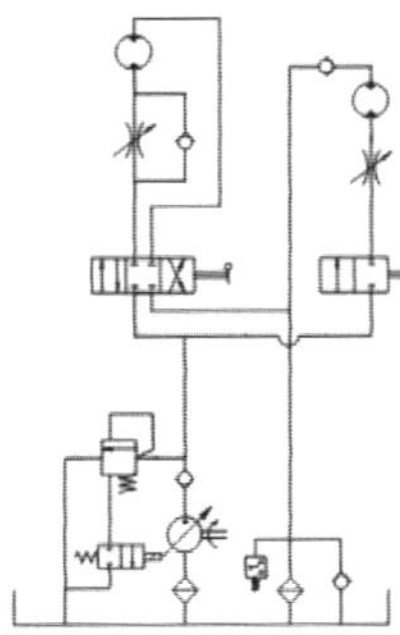

Figure 12. Oil piping system of the diamond wire saw

Figure 13. Hydraulic control valve and integrated lump of hydraulic system

4. Wire Sawing Machine Cutting Oil Pipeline Test

In order to ensure the cutting precision and surface processing quality, and also consider the cutting efficiency, it is necessary to select reasonable cutting process parameters during the cutting process. The cutting process parameters of different materials and different positions of the cutting pipeline need to be determined by experiments.

4.1. Mathematical Model Establishment of Cutting Force of Wire Sawing Machine

The power function form of the beaded rope cutting force indicates that the mathematical model of the bead rope cutting force is as follows:

$$F_t = k v_q^{\alpha} v_j^{\beta} b^{\gamma} \tag{15}$$

$$F_n = k' v_q^{\varphi} v_j^{\lambda} b^{\mu} \tag{16}$$

Where F_t — tangential cutting force (N); F_n — radial cutting force (N); b — the slit width (m) of the cut work piece; v_j — radial feed rate (m/s); v_q — tangential feed rate (m/s).

The regression orthogonal design method was used to analyze the test data of the wire sawing machine and the mathematical model of the tangential and radial cutting force of the beaded rope cutting pipeline was established.

$$F_t = 39.7854 v_q^{0.3097} v_j^{0.4188} b^{0.1489} \tag{17}$$

$$F_n = 23.2081 v_q^{0.5242} v_j^{0.544} b^{0.1923} \tag{18}$$

4.2. Relationship between Cutting Parameters and Cutting Force

4.2.1. Relationship between radial cutting force and feed rate

During the test, the position of the cross section of the steel pipe with three different slit widths (24 mm, 54 mm, 120 mm) was cut, the cutting speed was taken as v_q=13m/s. The relationship between the radial cutting force and the feed speed of the bead wire saw is shown in Figure 14 when cutting the positions of the steel pipes of three different slit width sections. It can be seen from the figure that as the feed speed of the bead rope increases, the radial cutting force acting on the bead rope also increases, and the radial cutting force when the cutting section has a large slit width is significantly larger than the radial direction of the cutting section slit width.

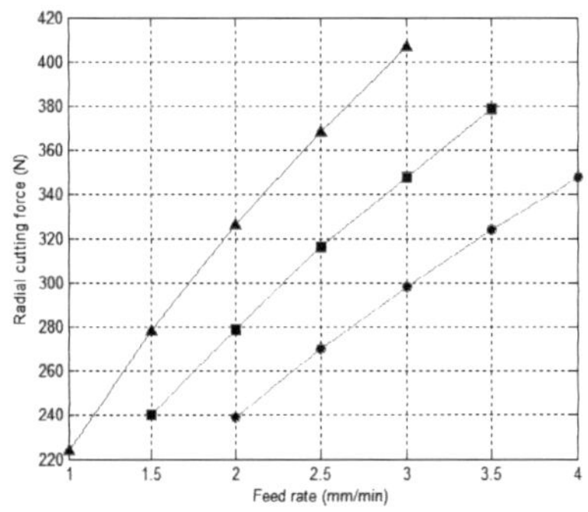

Figure 14. Relation between radial cutting force and feed rate

4.2.2. Relationship between tangential cutting force and feed rate

Figure 15 shows the relationship between the tangential cutting force acting on the beaded rope and the feed rate with respect to the varied cross-sectional width of steel tubes. The feed speed of the bead rope increases, the tangential cutting force acting on the bead rope also increases, and the tangential cutting force is larger than the cutting section slit width when the cutting section slit width is large.

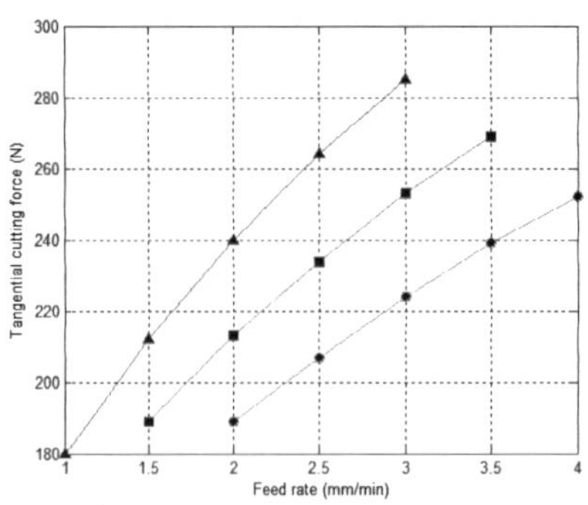

Figure 15. Relation between tangential cutting force and feed rate

4.2.3. Relationship between radial cutting force and cutting speed

During the test, the feed rate was taken as v_f=2.5 mm/min, and the initial tension of the beaded rope was F=1400N. The relationship between the radial cutting force and the cutting speed of the bead wire saw when cutting the position of three different slit width steel pipes is shown in Figure 16. The cutting speed of the bead rope increases, the radial cutting force acting on the bead rope also increases, and the radial cutting force is larger than the cutting section slit width when the cutting section slit width is large.

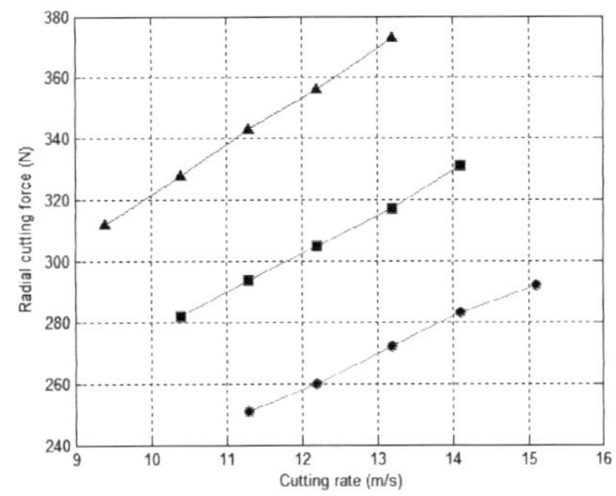

Figure 16. Relation between radial cutting force and cutting rate

4.2.4. Relationship between tangential cutting force and cutting speed

The relationship between the tangential cutting force acting on the beaded rope and the cutting speed when the feed rate of v_f=2.5mm/min and the initial tension of the bead rope of F=1400N are used to cut the position of the cross section of three different slit width steel tubes As shown in Figure 17. As can be seen from the figure, as the cutting speed of the bead rope increases, the tangential cutting force acting on the bead rope also increases, and when the cutting section has a large slit width, the tangential cutting force is larger than the cutting section slit width and the tangential cutting force.

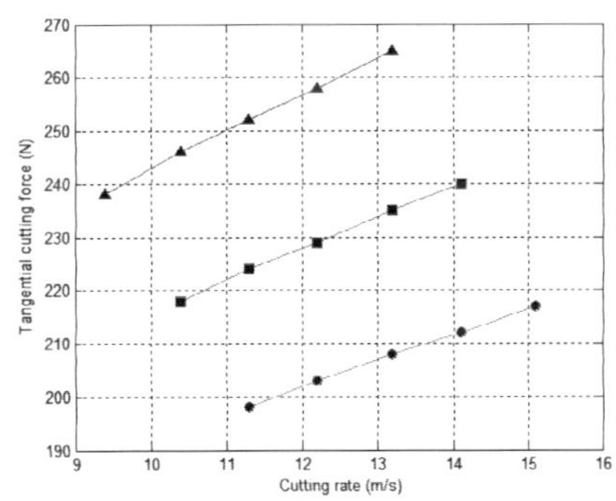

Figure 17. Relation between tangential cutting force and cutting rate

4.3. Macro Wear of Diamond Beaded Rope

4.3.1. Wear process of diamond beaded rope

During the cutting test, 10 beads were randomly selected as the research object to analyze the wear of the bead. The bead diameter was measured every 25 minutes. The curve of each bead diameter as a function of cutting time is shown in Figure 18, the average value of the curve is shown in Figure 19.

A powerful data fitting tool is provided in the MATLAB graphics window, which allows you to fit the plotted curve. Figure 20 plots the bead average wear data curve, the cubic polynomial fit curve, and the residual curve, showing the norm of the fitted polynomial and residual.

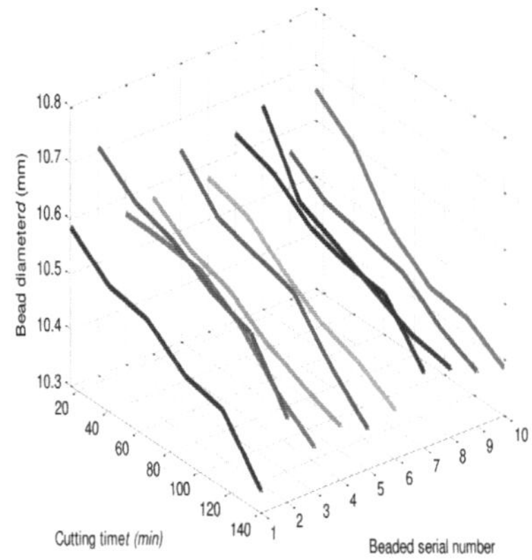

Figure 18. Graph of the wearing of diamond beads

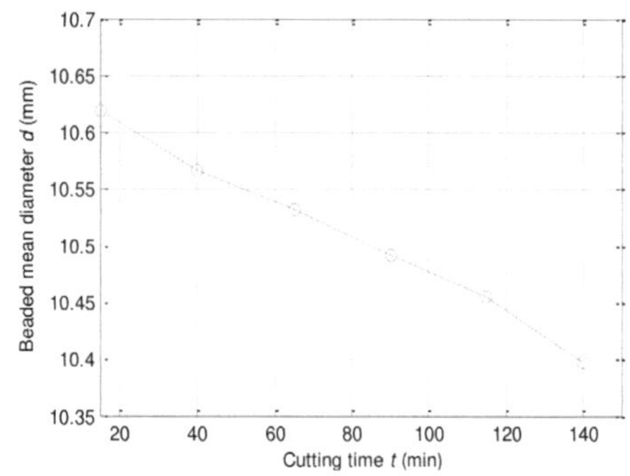

Figure 19. Graph of the average wearing of diamond beads

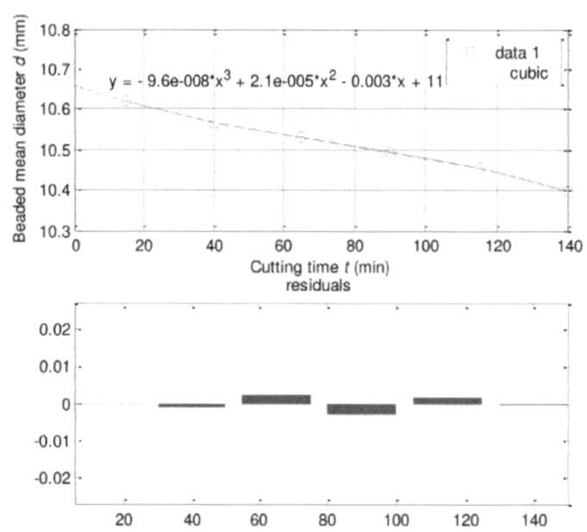

Figure 20. Data graph of the average wearing of diamond beads, thrice multinomial fitting curve and residual curve cutting force and feed rate

4.3.2. Relationship between wear resistance rate of beaded rope and feed rate

Cutting test using bead rope cutting speed v_q=13m/s and bead rope initial tension F=1400N, bead rope wear rate and feed the relationship between speeds is shown in Figure 21. It can be concluded from the figure that as the feed rate increases, the wear resistance rate of the beaded rope decreases; from the slope point of view, the slope of the cutting pipe slit width decreases faster, and the wear rate of the beaded rope decreases also faster.

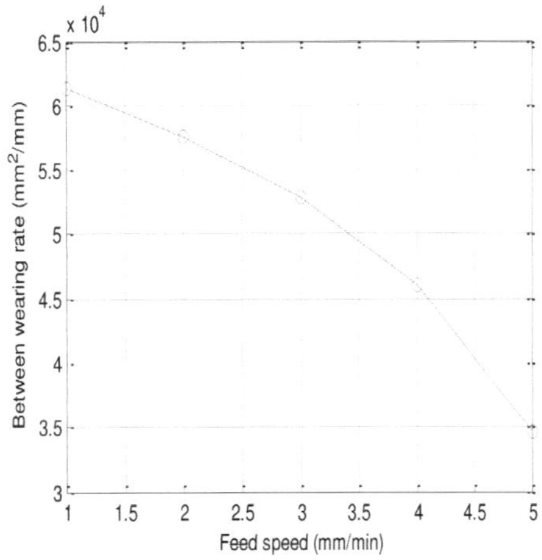

Figure 21. Relation between wearing rate and feed speed of the diamond beads

4.3.3. Relationship between wear resistance rate of beaded rope and cutting speed

The cutting test was carried out using the feed speed v_j=2.5 mm/min of the beaded rope and the initial tension of the beaded rope *F*=1400N. The relationship between the wear resistance rate of the beaded rope and the cutting speed is shown in Figure 22. It can be concluded from the figure that as the cutting speed increases, the wear resistance rate of the beaded rope decreases; from the slope point of view, the slope of the cutting pipe slit width decreases faster, and the wear rate of the beaded rope decreases faster.

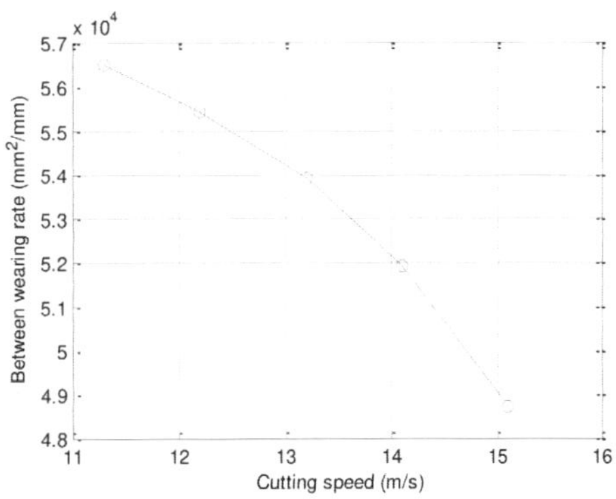

Figure22. Relation between wearing rate and cutting speed of the diamond beads

4.4. Oil Pipeline Cutting Test Results

The cross-section of the oil pipeline after cutting is shown in Figure 23.

Figure23. Section of cut-off oil pipeline

Through the analysis of the cutting test results of the oil pipeline, the following conclusions are drawn:

(1) The feed hydraulic motor can realize the work advance and retreat quickly; when the work is advanced, the motor can be continuously adjusted between the working speeds of 6~30r/min, and the low speed stable work can be realized without creeping and the wire saw is guaranteed, ensure that the feed is stable when the wire saw is cut;

(2) The motor can be continuously adjusted between the working speed of 500-1000r/min, and the driving device works smoothly when the wire sawing machine is cut;

(3) The hydraulic pump station works normally; the switch and the button work normally; the motor does not heat when working;

(4) The saw bow frame does not vibrate when cutting;

(5) The cut-off time of the cutting oil pipeline is 2.1 hours; when cutting the concrete at both ends of the pipeline, the feed motor speed is selected from 20 to 30r/min, and the driving motor speed is selected from 700 to 850r/min; the bead rope is cut to the steel pipe and the cut-through steel pipe position. When the feed motor speed is selected from 12 to 18r/min, the drive motor speed is selected from 500 to 750r/min; when cutting the middle part of the pipeline, the feed motor speed is selected from 20 to 26r/min, and the drive motor speed is selected from 700 to 800r/min;

(6) At present, a single-layer oil pipeline cutting is completed, and the beaded rope can only be used once; the bead rope wear is reduced from 11mm to 10.45mm; if the efficiency of use continues to decrease, it is recommended not to continue using it.

5. Conclusion

(1) Cutting the oil pipeline with the wire saw machine developed is an effective cutting method. It is difficult to process the cutting composite material and the irregularly shaped member by the general cutting method. Even if the processing can be performed, the processing technology is complicated; the process is simple and easy to operate by the wire sawing machine, and can meet the general processing technical requirements.

(2) The test of the oil pipeline was cut by a wire saw machine. The test verified the rationality of the design of the wire saw machine, and theoretically analyzed the test data, which laid a theoretical foundation for determining the optimal cutting process parameters.

(3) Due to the large amount of experimental work, only the specific specifications of the oil pipeline have been tested. The future research work is carried out on other composite materials and irregularly shaped components, so that the wire sawing machine cutting technology can be more Wide range of applications.

6. References

[1] Y. MA. Submarine pipeline maintenance technical equipment [J]. Oil and gas field ground engineering, 2001, 20 (3): 11-13.

[2] X. ZHU. Research on the disposal technology of offshore platform disposal [J]. Engineering Technology, 2017 (2): 284 – 285.

[3] H.WANG, L. ZHANG, Q. MENG, et al. Program Analysis and Research of Offshore Discarded Platform Piles Removal Wire Saw [J]. Machine Tool & Hydraulics, 2015, 43 (7): 39-42.

[4] Z. GAO. The application of the diamond wire saw cutting on engineering of concrete demolish [J]. Chinese Construction Message, 2015 (12): 68-69.

[5] L. YANG. Talking about the preparation and application of diamond wire saw in salt-soluble mining [J]. Super hard Materials Engineering, 2019, 31 (4): 45-48.

[6] L. CAO, Q. MENG, L. WANG, et al, Research on Underwater Diamond Wire Saw. Proceedings of the 2006 IEEE International Conference on Mechatronics and Automation, 2006: 1695-1700.

IOP Conf. Series: Materials Science and Engineering **746** (2020) 012037 doi:10.1088/1757-899X/746/1/012037

3D Motion Simulation of Engineering Equipment Working Mechanism Based on Reverse Drive Mode

Weijun Chu[1] and Zicheng Zhu[2]

[1]Army Engineering University of PLA, Nanjing, Jiangsu, 210007, China
[2]China Heavy Truck Group Co., Ltd, Jinan, Shandong, 250116, China
Email: chuwj2008@qq.com

Abstract. The working mechanism of engineering equipment has a relatively complicated structure which contains many kinematical pairs, its 3D motion simulation is a difficult problem. Based on the reverse drive mode linked by forward kinematics, 3D motion simulation model of working mechanism is analysed, and the kinematic chains are further established. General procedure of reverse drive mode is summed up, and the establishment strategy of reverse kinematic chain is expounded. This paper discussed these problems of adjusting motion axis and restraining DOF in reverse drive mode, and illustrated the controllers which control the form of 3D motion simulation. Through the above method, the 3d motion simulation of the working mechanism is realized. The analysis indicates that the 3D motion simulation of engineering equipment working mechanism based on reverse drive mode have more flexible use, convenient control and lifelike simulation result.

1. Introduction

The engineering equipment training can transform from real equipment training means to informatization means by constructing virtual training system, and there are great significance to the improvement of equipment support ability. The 3D motion simulation of engineering equipment working mechanism in virtual training system is generally more complex than others. It not only involves 3D geometric models, but also considers the dynamic changes of its position, motion, collision and deformation in the virtual scene. These problems are also the difficulties in the current virtual training system. The simulation of the interaction between model objects can be realized correctly and effectively by reasonably describing the pose of the objects, establishing the model of kinematic chain and the relationship between fit and constraint. The general 3D motion simulation approach is forward drive mode, which is to drive driven link in the kinematic chain to motion accordingly by the motion of driving link. In reverse drive mode based on destination oriented, the position and motion mode of driven link is determined first, and then the motion and position of other components including driving link in the kinematic chain are inversely calculated. Reverse drive mode is often easily used compared with forward drive mode, which can quickly create 3D movement of complex mechanism, improve design efficiency, and more intuitively verify the interaction effect of the designed mechanism members.

2. Establishment of 3D Motion Simulation Model of Working Mechanism

2.1. Mechanism Motion Analysis

Taking the motion of the working system of a certain engineering equipment as an example, the working system is composed of working mechanism (turning frame, arm, etc.), hydraulic system (hydraulic cylinder, control valve, etc.) and control system, which is used to complete the working movement such as turning, unfolding or folding and laying down or uplifting, as shown in figure 1. The turning frame is fixed on the front end of the roof deck of the equipment body by two pins, which is used to turn metal girders. The turning frame can achieve a maximum rotation of 80°. The arm which is connected to the turning frame by two pins is used for laying down and uplifting metal girders.

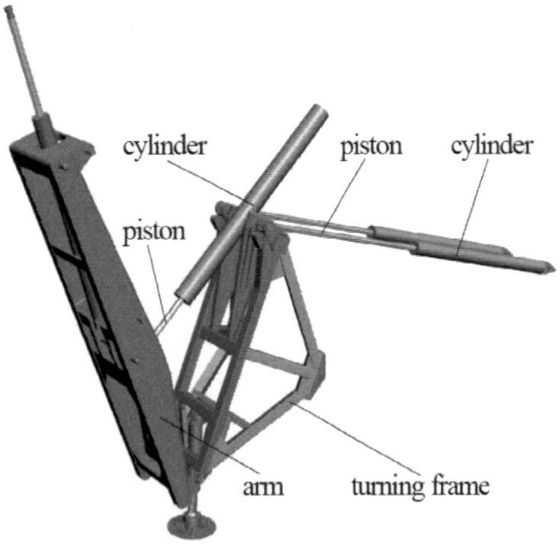

Figure 1. Working mechanism

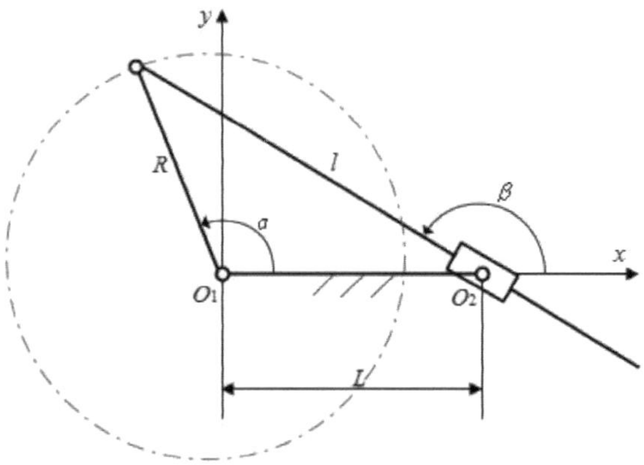

Figure 2. Movement diagram

The 3D motion simulation of erection system is built on the basis of its motion mathematical model, and the device can be simplified into two crank-shanking mechanism: the turning mechanism and the laying down mechanism, as shown in table 1.

Table 1. Form and member of the working mechanism

	Mechanism	Member	Form	DOF (Degree of Freedom)
1	turning mechanism	turning frame, cylinder, piston, frame	crank-shanking mechanism	1
2	laying down mechanism	arm, cylinder, turning frame	crank-shanking mechanism	1

The DOF of the whole working mechanism is 2, and the cylinders of the working mechanism are the driving links to drive and realize the given motion.

The movement diagram of a single crank-shanking mechanism can be simplified as shown in figure 2. A single crank-shanking mechanism is composed of crank, link and shanking. The connecting rod (piston rod) is the driving link, which moves in the shanking block (oil cylinder) and pushes the crank (turning frame) to rotate at a certain Angle.

As shown in figure 2, the crank radius is R, and the distance between the point O_1 and O_2 is L. When the working length of the connecting rod (the distance between the cylinder fulcrum and the fulcrum between the piston rod and the turning frame) is l at any moment, the position equation of a crank shanking mechanism is:

$$\left.\begin{aligned} R\cos\alpha - L = l\cos\beta \\ R\sin\alpha = l\sin\beta \end{aligned}\right\} \tag{1}$$

β and l can be obtained from formula (1):

$$\beta = \arctan 2[R\sin\alpha / (R\cos\alpha - L)] \tag{2}$$

$$l = \left(R^2 \sin^2\alpha + (R\cos\alpha - L)^2\right)^{1/2} \tag{3}$$

The variation of β and l can be set to design the motion simulation of working mechanism from the perspective of motion simulation. However the calculation of this method is relatively complex and the process is tedious. If the design is combined with the reverse driving mode, the 3D simulation process will be more flexible. Not affecting the motion simulation effect, the driving link is changed from original piston to crank, that is to say that forward drive mode is transformed into reverse drive mode. Reverse drive mode starts from child object rather than parent object. Change the drive to a crank here. The crank rotates as a driving link, so it (Angle α) can be easily controlled in motion design. Angle α can be obtained from formula (3):

$$\alpha = ar\cos\frac{R^2 + L^2 - l^2}{2R \times L} \tag{4}$$

2.2. Establishment of Kinematic Chain Based on Hierarchical Relationship
Because there are few members that move independently in mechanism, the movement of most members is restricted by other members or coordinated with other members to form various kinematic pairs (such as revolute joints, sliding joints, etc.).The whole mechanism is a kinematic chain system composed of several members connected by kinematic pairs and based on object hierarchical model.

When the mechanical device of engineering equipment realizes complex virtual movement simulation, its linkage relation is different from that of the real mechanism. This link relation is designed to simplify the process and facilitate the implementation of 3D motion simulation by linking objects together to form a virtual kinematic chain based on hierarchical relationship. Membership can be created by linking one object to another. In a linked chain, the movement of one chain may affect some or all of the chains, and the transformation applied to the parent object will also be passed to the child object. For the needs of simulation training, the mechanism motion model is transferred according to the hierarchical relationship of "driving link → driven link" or "parent object → child object", then kinematics chain is formed, as shown in figure 3.

The kinematics chain may consist of one or more sub-kinematics chains to form a complex linkage system. After establishing kinematic chain, it is convenient to determine the planning of object (member) action sequence, and it is also beneficial to control mechanism simulation movement. For example, it can affect the pose of child member by changing the pose of parent member, while the change of the pose of child member will not affect parent member.

3. Reverse Drive Mode

3.1. General Procedure of Reverse Drive Mode
Reverse drive mode using goal-oriented method reverses the direction of the kinematic chain based on hierarchical relationship, which can be used to locate the target object and set the position and direction of the end of the chain. It directly drives child member instead of parent member, and calculates the position and direction of the members in this kinematic chain through reverse calculation according to the kinematic link relation. In this way, the motion of the parent member is affected step by step, and finally parent member follows the motion of the child member. In this system, the whole kinematic chain system can move only by the motion of one child member in the kinematic chain. Its control program is shown in figure 4.

IOP Conf. Series: Materials Science and Engineering **746** (2020) 012037 doi:10.1088/1757-899X/746/1/012037

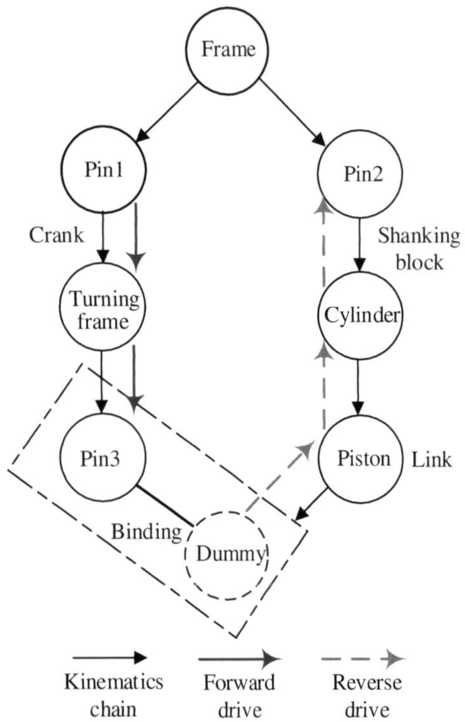

Figure 3. Kinematics chain

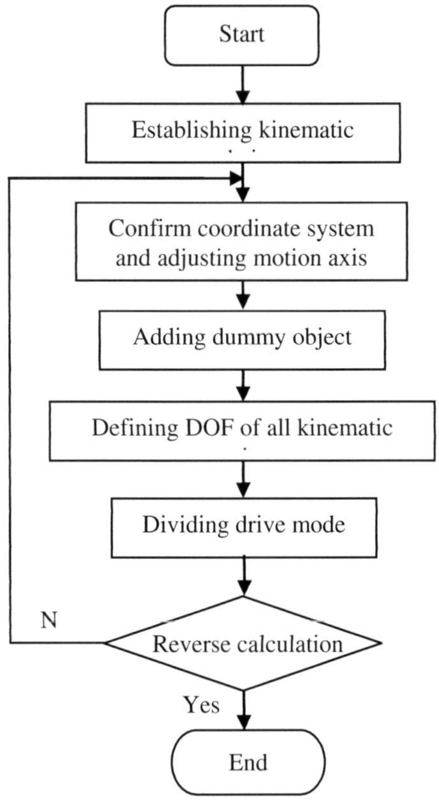

Figure 4. Flow chart

3.2. Establishment Strategy of Reverse Kinematic Chain

Reverse drive mode starts with and based on links and pivot positions, then two principles should be considered. Kinematic pairs are constrained by specific position and rotation properties, and the position and direction of parent member are determined by the position and direction of the child member.

Several kinematic chains can be created throughout the hierarchical instead of just one according to the mechanism motion complexity of engineering equipment. Of course, several separate hierarchy can also be created, rather than linking all the members together in one large hierarchy, as shown in figure 5.

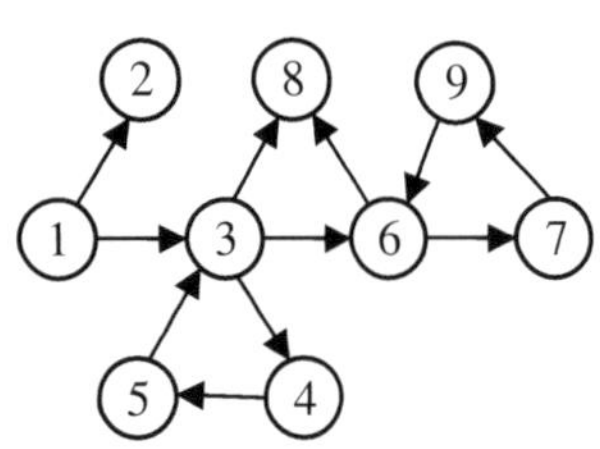

Figure 5. Hierarchy chain

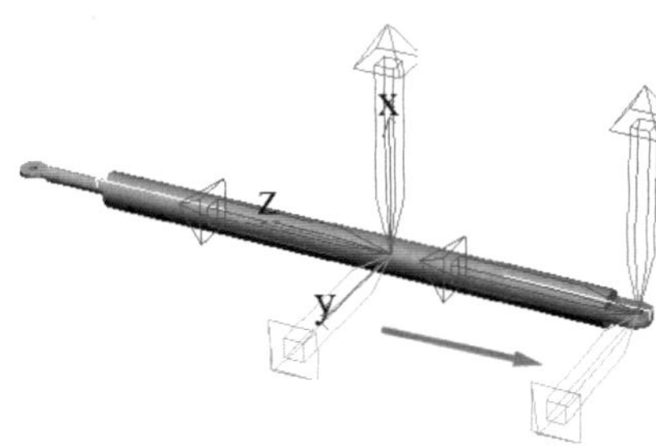

Figure 6. Adjusting pivot point

3.3. Adjusting the Motion Axis

An obvious advantage of reverse drive mode is that it is possible to define the working states of kinematic pairs. Kinematic pairs is actually at the pivot point of the member. The pivot point can be moved to change its actual position and direction relative to the member itself, as shown in figure 6. The working state control of kinematic pairs is actually to restrict the DOF of the member relative to the position and direction of the pivot point. Adjusting the pivot point of a member does not affect any child members that are linked to this member.

Axial, angular, or distance limits can be added to kinematic pair being defined. Moreover, the motion simulation can be made vivid by setting the priority order and resistance of kinematic pairs.

3.4. Restriction of DOF
Revolute joints, sliding joints have 3 DOF respectively. In reverse drive mode, the DOF direction of child member is set relative to the local coordinate axis of the parent member. The range of motion or rotation allowed on the active axis can be set a limit. For example, the piston can only slide within the length of the cylinder. Resistance can be used on the entire range of joint motion or rotation which simulating the natural effects of joint friction or inertia.

3.5. Determining Controller
Controller can store the values of moving key points, store the relevant settings of the program, and interpolate the values of moving key points. Track curve in x, y and z directions of controller is shown in figure 7. Controllers generally include floating point controller, Point3 controller, position controller, rotation controller, scaling controller, transformation controller, etc. As a position controller, Bezier controller is used to handle in most 3D motion simulation. Bezier controller can control acceleration, delay, and other types of motion by interpolating between key points of smooth curves. Euler XYZ is a common rotation controller whose rotation is divided into three separate "Bezier floating point" tracks. Angle α of the crank can be easily controlled by Euler XYZ in motion design.

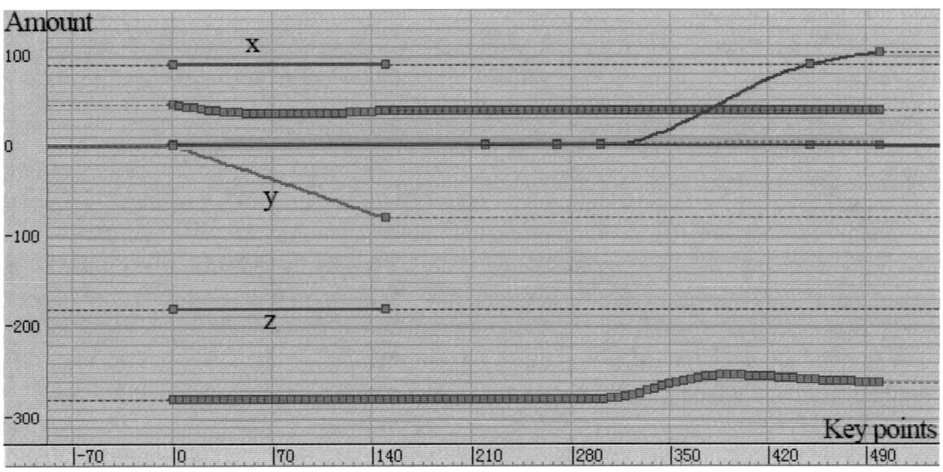

Figure 7. Track curve.

3.6. Calculation of Kinematic Chain in Reverse Drive Mode
In general, the object which rotate uniformly or move straight can be selected as the virtual driving member, so that its motion can be conveniently set and controlled, and then the motion process of other members in the kinematic chain can be determined through reverse calculation.

(a) (b)

Figure 8. The result of 3D motion simulation based on reverse drive mode

4. Conclusion

The forward drive mode uses a top-down approach, which starts from the position and direction of the parent member, and then goes to the position and direction of each child member hierarchy. So hierarchical linking should be in the order from parent hierarchy to child hierarchy. Axis points define joints between members. Child member inherits the transformation of the parent member.

Reverse drive mode is based on the concept of reversing kinematic chain, and the motion process of other members in the kinematic chain can be determined by one child member through reverse calculation. The solution is only to calculate the set key points, and all other motions are used by interpolation method. For more accurate result, all key points of the specified range can be also solved. Reverse drive mode is easier to use than forward one, which can quickly create complex motions. Of course, two modes should be combined to control the 3D motion simulation of engineering equipment complex mechanism more flexibly.

5. References

[1] Yun-wang LI, Hong-xin WANG and Su-mei DAI 2017 *Numerical Calculation and Simulation of Mechanical Principle* (Beijing: Science Press) p 70.

[2] Hai-lang HU, Chen-bo YIN, Jun-jing ZHOU, Wen-hua JIA 2018 Virtual Assembly and Motion Simulation of Jack Based on SolidWorks *Mechanical Research & Application (*No.2*)* (Nanjing) pp 117–126.

[3] Jia-hong ZHENG, Jie-lin LIU, Jun LI, Ya-xiong WANG, Jing CHEN, Ting-ting Mao 2018 Kinematics and error analysis of 3-PPR planar parallel mechanism *Journal of Shaanxi University of Science & Technology (*vol 36*)* ed Jia CHEN (Xian) pp 147–153.

[4] Qun-hai ZHOU, Zhao-hui WANG 2018 Kinematics Simulation Analysis of 6R Articulated Robot *Modular Machine Tool & Automatic Manufacturing Technique (*No.11*)* ed Xiu-min LI (Dalian) pp 62–66.

[5] Dan ZHAO, Wei-hua ZHANG 2018 Trajectory Planning and Kinematic Simulation of Hydraulic Excavator *Machine Building & Automation (*vol 31*)* (Lanzhou) pp 7–9.

IOP Conf. Series: Materials Science and Engineering **746** (2020) 012038 doi:10.1088/1757-899X/746/1/012038

Three-dimensional Simulation of Turbulent Rectangular Jet Spreading Technology of Carbon Fiber Bundle

Jinliang Chen[1], Xuejuan Niu[*, 1, 2] and Zhun Zhang[1]

[1]Mechanical Engineering Institute, Tiangong University, 300387 Tianjin, China
[2]Advanced Mechatronics Equipment Technology Tianjin Area Major Laboratory, Tiangong University, Tianjin 300387, China
Email: niuxuejuan@tjpu.edu.cn

Abstract. In order to smoothly spread large-sized carbon-fiber bundle, hot-jet blower is added to a turbulent spreading platform to conduct pre-spreading process. To achieve the best spreading performance, CFD model of the blower are simulated based on the Reynolds average Naiver-Stokes equation. The influence of the relevant geometric parameters and process parameters on the internal flow field are analyzed. The distributions of streamline, surface temperature and air pressure under different working condition are characterized and compared parametrically. The effective area of the nozzle directly contacts with fiber wall surface. Simulating model has been built and crossover experiments have been carried out on the spreading platform to investigate and optimize the influence of jet distance, nozzle air pressure and hot-jet temperature on the spreading performance of fiber bundle. According to the experimental data and simulating results, relevant geometric parameters and process parameters of the nozzle are optimized.

1. Introduction

As a kind of high performance conforming material with low density, high specific strength, high specific modulus, good heat resistance and chemical stability, carbon fibers are widely used in aerospace, automobile manufacturing, building reinforcement and other fields [1-3]. In order to meet the needs of different working conditions such as fiber fabrics, thin layer treatment of carbon fibers is very important. At present, there are many ways to thin carbon fibers, according to the relevant technical description given by Irfan [4].These methods can be divided into three types, one is the active method, the other is the passive method, and the third one is a mixture of the two. For the active methods, energy is used to spread the fiber bundle, such as fiber bundle through small gaps which depending on airflow (pressure or suction) or under (ultra-)sonic waves and vibration effects [5, 6]. The passive approach is applying tension and constant motion on different geometries to spreading the fiber bundle, such as using fiber rods, convex structures or geometric guiding elements to make the fiber bundle become the fiber tow [7-9]. Compared with other methods, jet spreading technology uses hot air as the working fluid. And hot air flow plays the following two important roles. On the one hand the perpendicular airflow make the force on the fiber bundle more concentrated. On the other hand impinging jet can produce high heat transfer rate around an impinging position on an impinging wall which can make the binder on the surface of fiber bundles can be softened sufficiently. Therefore, In order to make ensure the carbon fiber spreading smoothly and evenly in the spreader, it is necessary to pre-spreading.

Baucom [5] first used the airflow and completed the broadening of the fiber bundle with the help of air jet technology. The spread fiber bundle, here we call fiber tow, can be better fuse carbon fiber and resin powder. Chen [6] used the planar airflow fiber-spreading method to simulate the internal flow field of the spreader based on the Reynolds average Naiver-Stokes equation, and obtained the flow field characteristics inside the pneumatic spreader. It is concluded that the greater the air velocity, the more favorable the expansion of the fiber bundle. Sihn [7] used air flow negative pressure method, which fiber bundle was passed through the fiber guide equipped with a duct and a vacuum pump. The vacuum pump sucked air downward through the duct, and the fiber bundle can be easily diffused without damaging. El-Dessouky [8] used secondary airflow fiber-spreading technology. The fiber bundle was firstly spreading through a Type 1 fiber guide obtain the pre-spreading tow, and then passed through a Type 2 spreader to obtain the final fiber tow. By reducing the conventional $12k$ carbon fiber bundle by increasing the tow by $5mm$ to $25mm$, the weight per unit area is reduced by about 500%. Ren [9] used mechanical-pneumatic spreading method, combing the advantages of the pneumatic spreading and the traditional mechanical device by machining air groove on the spreading pin. It showed that when the airflow pressure is $0.2MPa$, the prepreg produced by carbon fiber after fiber spreading has good performance. Huang [10] spread the fiber bundle by multi-stage airflow spreading system, and simulated the internal flow field characteristics of the spreader by finite element method (FEM). Structural parameters of the facilitate fiber spreading are obtained, and the width of the tow can be stretched more than three times. However, most researchers analyzed the velocity of the air stream and the effect of the structure of the spreader on the fiber spreading, and most of them are nearly neglected the influence of the temperature of the air flow on the spreading of the fiber bundle.

In this work, a new pneumatic spreading, hot-jet spreading technology, are designed as a pre-spreading system which is very essential for large-sized fiber bundles to ensure the fully spread of carbon fiber bundles. By establishing the three-dimensional CFD model of hot-jet blower, the flow field in it is simulated and characterized. The influence of the relevant geometric parameters and process parameters, such as nozzle-to-fiber distance, nozzle pressure and temperature, on spreading performance of carbon fiber bundle have been studied. Based on the simulation data, hot-jet pre-spreading experiments have been carried out. Through comprehensive analysis, the optimum parameters for pre-spreading fibers are obtained.

2. Experimental Method

2.1. Materials

The raw materials used were carbon fiber (CF) bundle (purchased from Toray), and the size of the bundle is $24k$ which means the fiber bundle have 24000 filaments. The properties of carbon fiber (CF) tows as received is shown in Table 1.

Table 1. The properties of carbon fiber (CF) tows as received

Fiber type	No. of CF filaments/ tow	Tensile strength (GPa)	Tensile modulus (GPa)	Elongation (%)	Diameter of CF (μm)	Density (g/cm^3)	Size type	Size content (%)
CT24-5.0/270-E100	24,000	5.0	270	1.90	6.9	1.79	epoxy	1.0

2.2. Experimental Setup

The schematic experimental platform designed is shown in Fig.1. The jet pre-spreading is mainly consists of four parts, namely a sizing loosening system, a tension and speed controlling system, a fiber guiding system and a winding system. The components and operation parameters and functions are summarized in Table 2. The structure of the experimental platform is constructed of aluminum profiles. At the beginning of the experiment, a constant tension fiber is used, and two pairs of press rolls ensured that the tension of the fiber tow is constant. The tension sensor (JZHL-L1-5N) is used to determine the tension of the fiber tow between the two pairs of rolls. The hot air blower (FH6273) has the advantages of high power and convenient operation. The height of the hot air blower is adjustable to accommodate different fiber distances. The distance is adjusted according to the characteristics of the jet flow field.

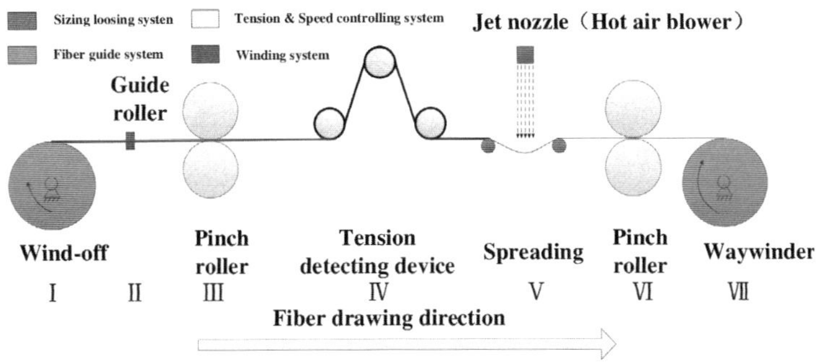

Figure 1. Schematic of Pre-spreading system experimental platform

Table 2. Summary of function, component and operating parameters of carbon fiber jet spreading system

System	Component	Operating parameter	Function
Sizing loosening system	V	Air flow rate and temperature of blower V	Hot air blowing make the fiber sizing softened, which is to prevent sizing on fibers from sticking the filaments together during the spreading process.
Tension and speed controlling system	Tension sensor III IV VI	Tension in fiber tow before blower V Speed of process	Low tension or tension-free state in the fiber bundle is the key for pre-spreading, which is beneficial to spread of filaments
Winding system	I VII	Radial winding under a given speed	Automatic winding to form UD tapes
Fiber guide system	II	Fiber bundle moving direction	Making the wind-off fiber bundle in the center of this system

When the tension sensor reaches the expected value, the two pairs of rollers are pressed to make sure that the tension of the fiber bundle will not change. The hot air blower is turned on and the airflow is directed from the inlet of the fan to the fiber bundle. After the airflow encounters the obstruction of the fiber bundle, forming a certain pressure on the surface of the fiber and fiber bundle will be in a tension-free state. With the help of pressure, the fiber bundle is spreading to be the fiber tow.

2.3. The Principle of Jet Flow Fiber Spreading
To better understand the principle of jet spreading, Fig.2 shows the distribution of fibers under jet airflow. Under the action of the air pressure, the fiber bundle moves rapidly to both sides, and the fiber bundle is easily widened and thinned. The initial state of the carbon fiber tow is shown in Fig. 2(a). Due to the blocking action of the carbon fiber tow, the air flow velocity v_1 should be less than the air flow velocity v_2 on both sides of the fiber bundle, therefore the pressure of P_1 is higher than the pressure of P_2. As shown in Fig. 2(b), under the action of pressure and the hot air flow, the cohesion of the binder between fibers is softened, and the external fibers break away from the fiber bundle one by one, and the carbon fiber bundle is unfolded. The same principle apply to Fig. 2(b), due to the blocking of the fiber tow, $v_1 < v_2 < v_3$, so $P_1 > P_2 > P_3$. As the airflow time increases, the width of the fiber expands to a stable level. As shown in Fig. 2(c), the fiber distribution finally reached a steady state.

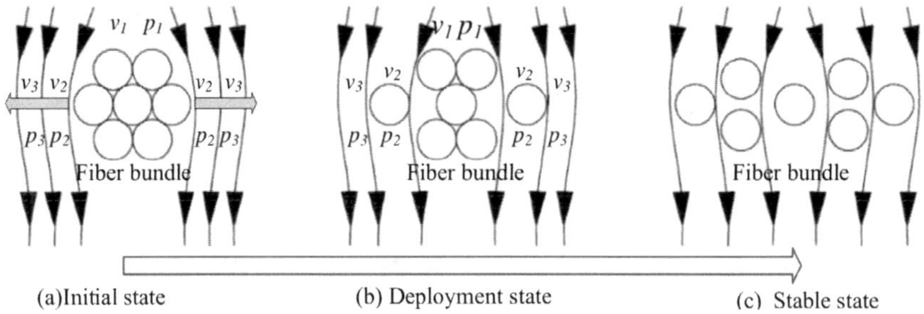

Figure 2. The principle of jet air flow fiber spreading

2.4. Collection of Experimental Data

For the convenience of experiments, the pressure drop and temperature of the fan are separately calibrated before the start of the experiment. During the experiment, the Mp115 differential pressure gauge is used to measure the pressure drop between the fan outlet and atmospheric pressure. Which range is $-500mbar$ to $500mbar$, and a resolution of $0.1mbar$. Change the gears of different speeds, measure the pressure drop three times for each gear position, and finally use the average as the pressure drop of the gear position. The TES1310 handheld thermometer, with a range of $223.15K$ to $1573.15K$, is used to measure the inlet temperature of the fan. Use the same way, by changing the different temperature gears, the inlet temperature of the fan is measured three times at each gear position. As shown in Table 3, the average value is calculated and used as the final value.

Table 3. Relationship of hot air blower and temperature

Temperature Gear position	Temperature(*K*)
Fist level	323
Second level	343
Third level	363

3. Numerical Study

3.1. Governing Equations

The continuity equation and the Naiver-Stokes equation used in establishing the jet flow field CFD model are as follows:

$$\begin{cases} \frac{\partial \rho}{\partial t} + div(\rho u) = 0 \\ \frac{\partial u_i}{\partial t} + \frac{\partial u_i u_j}{\partial x_j} + \frac{1}{\rho}\frac{\partial p}{\partial x_j} + \frac{\partial}{\partial x_j}\left(\overline{u_i' u_j'}\right) - \frac{\partial}{\partial x_j}\left[v\left(\frac{\partial u_i}{\partial x_j} + \frac{\partial u_j}{\partial x_i}\right)\right] = 0 \end{cases} \tag{1}$$

Where: ρ is the density; t is the time; $u_i(i = 1,2,3)$ is the velocity; p is the pressure; T is the temperature; v is the dynamic viscosity; u_i', u_j' is the velocity u_i and u_j of the pulsation value; $\overline{u_i' u_j'}$ is the Reynolds stress.

$$t_{ij} = -\rho\overline{u_i' u_j'} = \mu_t\left(\frac{\partial u_i}{\partial x_j} + \frac{\partial u_j}{\partial x_i}\right) - \frac{2}{3}\left(\rho k + \mu_t \frac{\partial u_i}{\partial x_i}\right)\delta_{ij} \tag{2}$$

Where k is kinetic energy, $k = \frac{\overline{u_i' u_j'}}{2} = \frac{1}{2}\left(\overline{u'^2} + \overline{v'^2} + \overline{w'^2}\right)$, μ_t is the turbulent viscosity; δ_{ij} is the "Kronecker delta" symbol ($\delta_{ij} = 1$ when $i = j$; $\delta_{ij} = 0$ when $i \neq j$).

The internal air velocity of the jet flow field is less than $100m/s$, which is an incompressible air, and the turbulence model adopts the Realizable k-ε model. The transport equation of the turbulence model is as follows:

$$\begin{cases} \frac{\partial(\rho k)}{\partial t}+\frac{\partial(\rho k u_i)}{\partial x_i}=\frac{\partial}{\partial x_j}\left[\left(\mu+\frac{\mu_t}{\sigma_t}\right)\frac{\partial k}{\partial x_j}\right]+G_k+Gb-\rho\varepsilon-Y_M \\ \frac{\partial(\rho\varepsilon)}{\partial t}+\frac{\partial(\rho\varepsilon u_i)}{\partial x_i}=\frac{\partial}{\partial x_j}\left[\left(\mu+\frac{\mu_t}{\sigma_\varepsilon}\right)\frac{\partial\varepsilon}{\partial x_j}\right]+\rho C_1 E\varepsilon-\rho C_2\frac{\varepsilon^2}{k+\sqrt{v\varepsilon}}+C_{1\varepsilon}\frac{\varepsilon}{k}C_{3\varepsilon}G_b \end{cases} \quad (3)$$

Where: $\mu_t=\rho C_\mu\frac{k^2}{\varepsilon}$, $C_{1\varepsilon}$, $C_{2\varepsilon}$, C_μ, C_2 are empirical constants; σ_k and σ_ε are the Prandtl numbers corresponding to the turbulent energy k and the dissipation rate ε, respectively; G_k is caused by the average velocity gradient; The term of the kinetic energy k, $G_k=\mu_t\left(\frac{\partial u_i}{\partial x_j}+\frac{\partial u_j}{\partial x_i}\right)\frac{\partial u_i}{\partial x_j}$. According to the recommended values of Launder [14], the turbulence model parameters in Fluent 15.0 are shown in Table 4.

Table 4. Realizable k-ε turbulence model parameters in Fluent15.0

$C_{1\varepsilon}$	$C_{2\varepsilon}$	$C_{3\varepsilon}$	C_2	σ_k	σ_ε
1.44	1.9	0.09	1.9	1.0	1.2

There are three states of flow: laminar, turbulent and in the transitional phase. The flow state can be specified by the value of Reynolds number (Re) which is defined as the ratio of inertial and viscous forces. Generally the Reynolds number $Re < 2100$ is a laminar flow state, $Re > 4000$ is a turbulent state, and $2100 < Re < 4000$ is a transition state. And Reynolds number can be defined as:

$$R_e=\frac{uD}{r} \quad (4)$$

$$\mathrm{D}=\frac{4A}{C}=\frac{4ab}{2(a+b)}=\frac{2ab}{a+b} \quad (5)$$

Where: u is average velocity; D is hydraulic diameter; r is kinematic viscosity; A is cross sectional area of the inlet; C is wetted perimeter of the jet; a is the length of rectangle; b is the width of rectangle.

3.2. Mathematical Model

The 3D model of the jet spreading device was carried out by the Cartesian orthogonal coordinate system. In order to facilitate the establishment of the CFD model of the jet flow field, the jet flow field is simplified and modeled. The whole model is simplified into a cuboid with a length of $400mm$, a width of $300mm$ and a height of $100mm$. According to the mouth shape of the blower, the jet nozzle is simplified to a rectangle of $60mm$ length and $18mm$ width.

The CFD model of the jet flow field is shown in Fig. 3. The $x-axis$ is the direction which the fiber bundles are spreading. The $y-axis$ is moving direction of the fiber bundle, and the $z-axis$ is the direction in which the jet air is incident. The jet nozzle is the inlet in the jet flow field, and uses the pressure inlet as a boundary condition. The wall, W_1, is the impacted wall by the jet air, and its boundary condition is a fixed and no-slip. According to the actual working condition, the periphery walls are connecting with the atmosphere. Therefore, it is most appropriate for them to set their BCs as Pressure outlet boundary conditions. The diameter of carbon fiber bundle is 6.9 um, the thickness of fiber bundle is $0.48mm$, and the original width is $7mm$. Compared with the flow filed of spreader, the size of carbon fiber bundle is too small and can be neglected. Therefore, the simulation is based on non-fiber simulation.

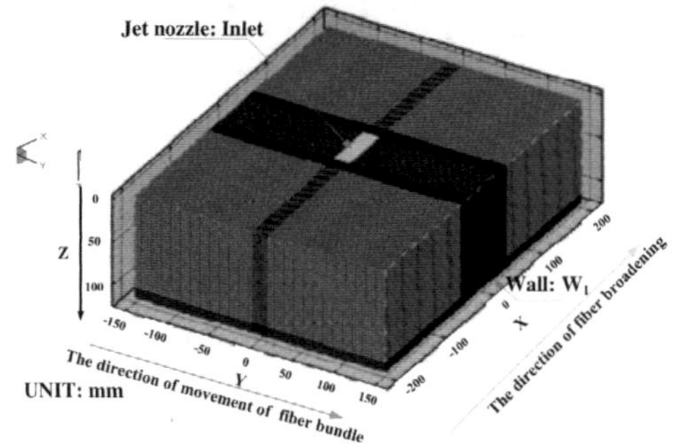

Figure 3. Establishment of CFD model for jet flow field

Since the simplified CFD model is a cuboid, it is divided into blocks by using multiple blocks technology in ICEM, and structured grid is used for meshing. To better observe the flow field distribution characteristics at the entrance and near the impact wall, mesh refinement technology is applied in these areas. The number of nodes and elements in the fluid domain are 288288 and 304271, respectively. Among them, there are 271825 with units type HEXA_8 and 32446 with units type QUAD_4. Convergence checks are made on the mesh to ensure that the number of meshes.

The SIMPLE algorithm was used to treat the pressure-based coupling. The convective terms and the diffusion terms of the governing equations were discretized with the upwind scheme both with second order accuracy. By solving with double precision, the convergence criteria for the continuity, momentum and the energy equations were defined to reach the scaled residues of 10^{-4}, $10^{-4}$and 10^{-6}, respectively. All simulations were performed on a computer with ten $2.4GHz$ processors and $64GB$ of main system RAM.

3.3. Mesh Sensitivity and Model Validation

In order to check the mesh dependence, three cases with 189973, 304271 and 640387 mesh elements, respectively, were calculated for the CFD model. The air velocity along the nozzle axis and the temperature distribution perpendicular to radial direction at $z = 80mm$ are shown in Fig. 4(a) and (b), respectively. With the pressure perpendicular to radial direction at $z = 80mm$ shown in Fig. 4(c). As shown in Fig.4, there are little differences in the results with the increase of grid number. Therefore, we can conclude that the grid dependence is acceptable, and the mesh with 304271 elements is used in the following simulations.

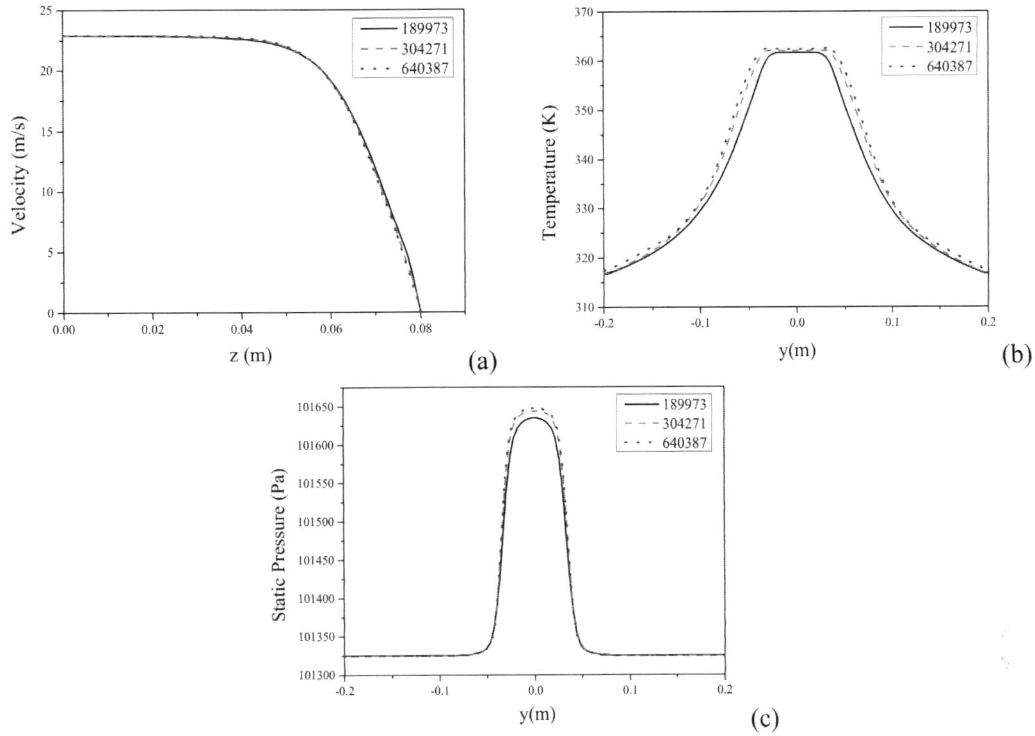

Figure 4. Grid dependence check ((a) air velocity along the nozzle axis; (b) air temperature along theradial direction at z = 80 m; (c) air pressure along the radial direction at z = 80 m)

In order to verify the correctness of the simulation model and reliable results, the velocity on the jet axis measured at different spray distances was compared with the simulation data. The experimental equipment in this study consists of a rectangular made of synthetic plastics with a hydraulic diameter of $28mm$. As shown in Fig. 5. It can be visually seen that the velocity on the central axis of the jet flow field is through a constant velocity zone, and then the velocity begins to decrease gradually. The velocity of this phase is in the free jet zone; when approaching the fiber wall surface, the velocity is drastically reduced, and finally the attenuation on the wall is zero. As we can see that the simulation results and the experimental data are match well which indicating that the simulation is reliable. However, the simulated results are slightly larger than the experimental data. This is due to the difference in temperature of the surrounding environment during the experiment test. While in the process of simulation, the ambient temperature is set to be constant and will not be affected by external factors.

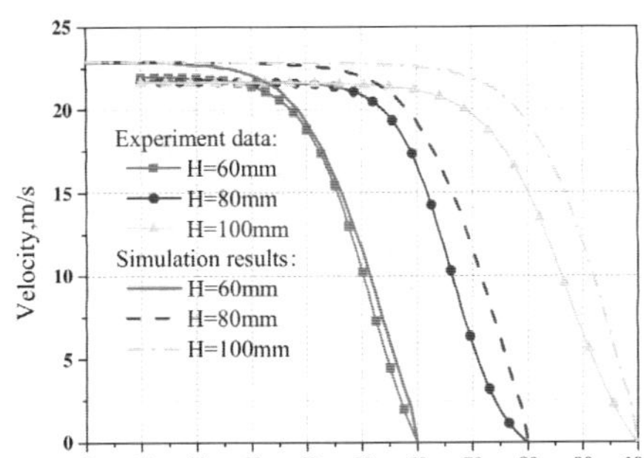

Figure 5. The velocity distribution of the jet central axis

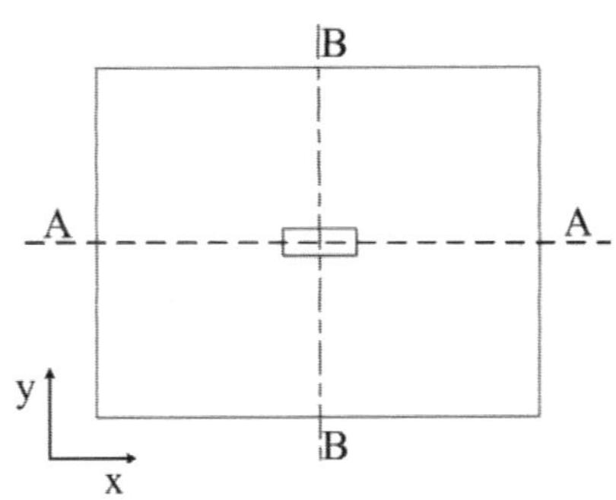

Figure 6. Lines selected for plotting results

3.4. Data Reduction

Numerical results are obtained after solving the governing equations. The results are presented in non-dimensional form. The non-dimensional pressure coefficient is drawn along lines A-A and B-B. Figure 6 shows the lines selected for plotting the results.

4. Results

4.1. Computational Results

4.1.1. *Analysis of jet flow field characteristics*

For the convenience of description, some parameters are defined. Such as the inlet temperature, T. The jet nozzle pressure, P_i. The distance, D, from the nozzle to the impact wall surface, the distance from the nozzle to the fiber bundle is called the fiber distance. The outlet pressure, P_o. And the pressure drop, P_d, is the difference between inlet and outlet pressure. Therefore, $P_d = P_i - P_o$. As shown in Fig.7, the jet flow field has a constant velocity flow area, a free jet area, a sluggish area and a near wall jet area. When the air flow from the nozzle is directed towards the wall surface, a constant velocity area and a free jet area are formed successively. Along with the airflow approaches the wall, the velocity of the airflow gradually decays to zero. In the meantime, the dynamic pressure is also drop down to zero, and a sluggish area is formed at the near wall surface. When the air flow is blocked by the wall, the airflow begins to spread to the surroundings, and the flow direction of the airflow changes, forming a near-wall jet region around the impact region.

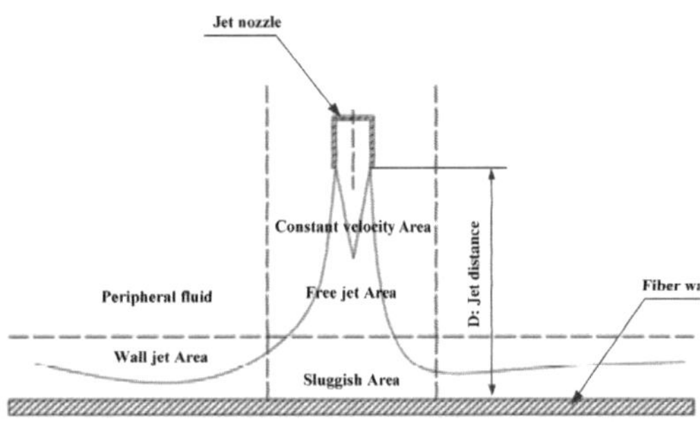

Figure 7. Characteristics of jet flow field

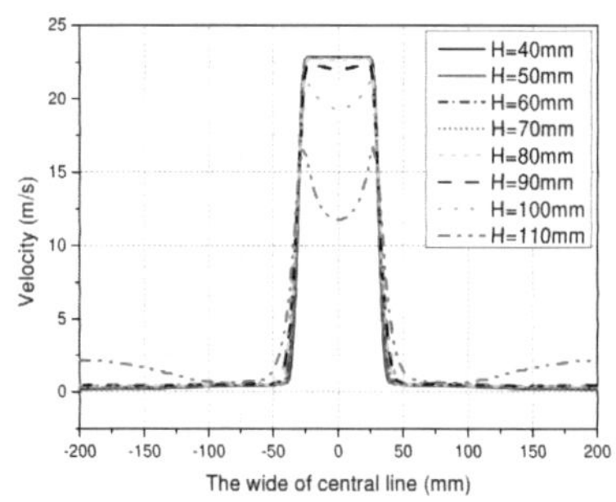

Figure 8. Velocity changes in the centerline of different sections

4.1.2. Discussion on the fiber distance of the jet flow field

For the sake of appropriate fiber distance under the jet flow field, a simulation of the jet with a spray distance of $120mm$ is carried out. When the boundary condition is set as: $P_i = 101645Pa$, $P_o = 101325Pa$, $T = 323K$, the jet flow field is simulated and analyzed.

The flow field adopts fiber-free simulation. To find a suitable fiber-optic distance, it is assumed that the distance between the nozzles on the $z - axis$ and the fiber bundle on the H-plane is evenly distributed. By analyzing the velocity on the $y - axis$ centerline of the H-plane, the force of the fiber in the flow field can be easily judged. Therefore, the cross section, H, in the $z - axis$ direction of the jet flow field is selected as the fiber surface, the $x - axis$ direction is the direction of the fiber bundle spreading, and the $y - axis$ is the moving direction of fiber bundle. Select the fiber cross section in the positive direction of the $z - axis$. The distance from jet nozzle to the selected fiber cross section are $40mm$, $50mm$, $60mm$, $70mm$, $80mm$, $90mm$, $100mm$, $110mm$. The velocity at the centerline of the section ($y = 0$) is analyzed as shown in Fig.8. When the fiber cross section changes from $40mm$ to $80mm$, the velocity on the centerline has hardly changed. The velocity between $-50mm$ and $50mm$ is a constant value, which is benefit for fiber bundle spreading. When the fiber cross-section is changed from $80mm$ to $110mm$, the air flow velocity is V-shaped in the range of $-50mm$ to $50mm$. The intermediate air flow velocity is small, and the air flow velocity on both sides is large. This phenomenon is not conducive to the uniform deployment of the fibers, which causes the fibers to alternately float and the spacing between the fibers becomes larger. Therefore, fiber spacing is most suitable for selection in the constant velocity area, that is to say a choice from $40mm$ to $80mm$ is advantageous for the fiber bundle spreading uniformly.

4.1.3. Analysis of the influence of pressure drop on jet spreading

In order to study the influence of pressure drop on the jet flow field, the jet flow field under the fixed fiber distance is simulated and analyzed. As can be seen from Fig.8, the suitable fiber spacing is $40-80mm$. In this section, the fiber wall distance of $80mm$ is selected as the impact wall surface of the jet flow field. The boundary conditions are set as: $P_i = 101645Pa$, $P_o = 101325Pa$, $T = 323K$. The flow field of the jet spreading flow is simulated under different pressure drops.

Fig.9 shows the detail distribution of the air velocities in the A-A cross section and B-B cross section (shown in Fig.6). Compared Fig.9a and Fig.7, it can be seen that the air velocity under the inlet of the flow field is constant. This area is called the undisturbed jet core. As approaching the fiber surface at different pressure drops, air flow goes through the free jet area and the stagnation zone, and its velocity decreases gradually. In the Stagnation Zone, the velocity rapidly decreases to zero and a negative pressure is formed. With the increase of the pressure drop (P_d), the initial velocity of the nozzle airflow increases from $16m/s$ to $25.6m/s$. There are obvious turbulence phenomenon in the near fiber area, which cause the vibration of the fibers and contribute to disperse of the fiber bundle greatly. Under the impact of the hot-air jet flow, the cohesive material in fiber bundle are soften and the fibers are dispersed rapidly toward the clapboards on each side.

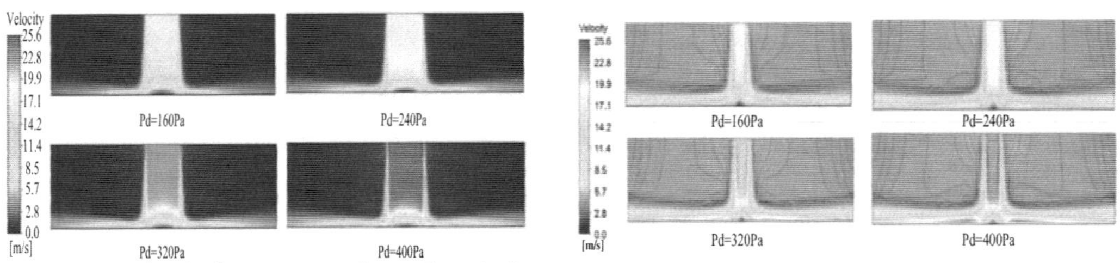

(a) Computed velocity contours alone plane A-A (b) Computed velocity contours alone plane B-B

Figure 9. Velocity distribution on the central cross section

From the velocity cloud chart of Fig.9, it can be seen that the velocity of the airflow is reduced to zero near the fiber wall surface, forming a negative pressure. The fiber bundles are forced to unfold under the action of pressure. In order to better describe the pressure on the fiber wall surface, the pressure distribution diagram of the fiber wall under different pressure drops are shown in Fig.10. The pressure distribution of the fiber wall surface is an elliptical shape, the pressure distribution ranges from $-50mm$ to $50mm$ in the x direction and from $-30mm$ to $30mm$ in the $y-axis$ direction. The range of pressure in the $x-axis$ direction is significantly larger than the $y-axis$ direction, which is closely related to the shape of the jet nozzle. The $x-axis$ direction is the direction which the fiber bundle spreading, the direction of fiber movement is the $y-axis$. Therefore, the faster the pressure gradient changes along the x-axis, the more advantageous the fiber bundle broadening. From Fig. 10(a) to Fig. 10(d), as the pressure increases, the pressure distribution in the $x-axis$ direction does not change significantly. This may be due to the effect of fiber spacing and nozzle shape. Therefore, a reasonable pressure needs to be verified according to the experiment.

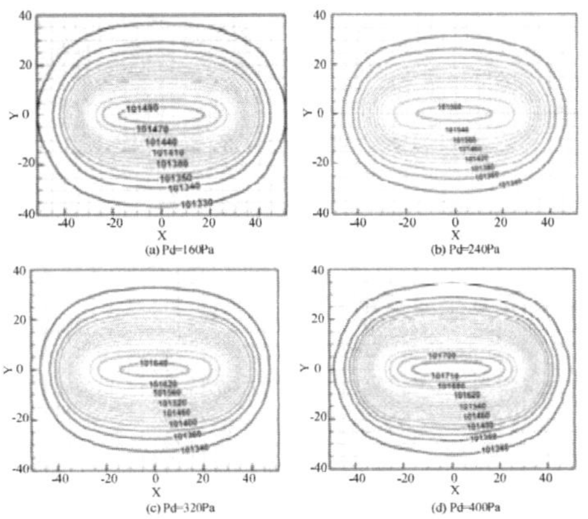

Figure 10. Pressure distribution of the fiber wall

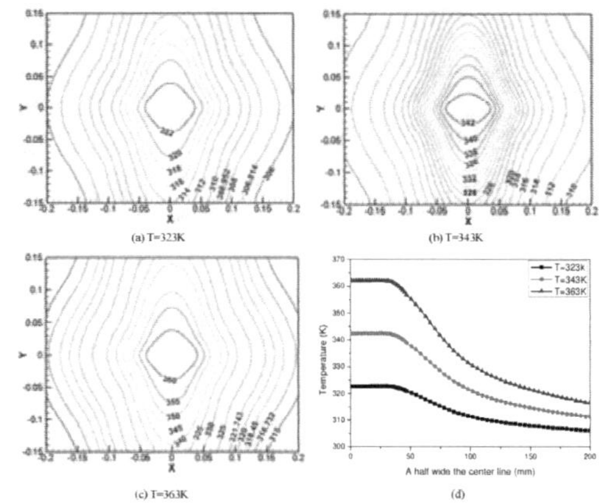

Figure 11. Temperature distribution of fiber wall

4.1.4. Analysis of the effect of temperature on jet spreading

In order to study the action mechanism of thermal jet impact on fiber spreading, the wall surface of different temperature airflows was analyzed. The fiber spacing is $80mm$, the boundary condition are $P_i = 101645Pa$, $P_O = 101325Pa$, $P_d = 320Pa$. The set temperatures are $323K$, $343K$, and $363K$, respectively. The results of the analysis are shown in Fig. 11.

Fig. 11(a) and Fig. 11(b) are temperature distribution diagrams of the fiber wall surface at temperatures of $323K$ and $343K$, respectively, and the temperature diffusion on the fiber wall surface is relatively uniform, and the temperature gradient is $2K$. The temperature of Fig.11(c) is $363K$, the change of temperature gradient is $5K$, and the speed of diffusion is relatively fast. It can be seen that the higher the nozzle temperature, the more obvious the exchange heat exchange with the surrounding gas. After encountering the gas with lower temperature, the temperature gradient drops faster. Fig. 11(d) is a graph showing the temperature diffusion in the positive direction of the x-axis of the fiber wall at different temperatures. Under the influence of the jet nozzle, the temperature change is relatively slow from $0mm$ to $50mm$, and the temperature drops sharply from $50mm$ to $200mm$.

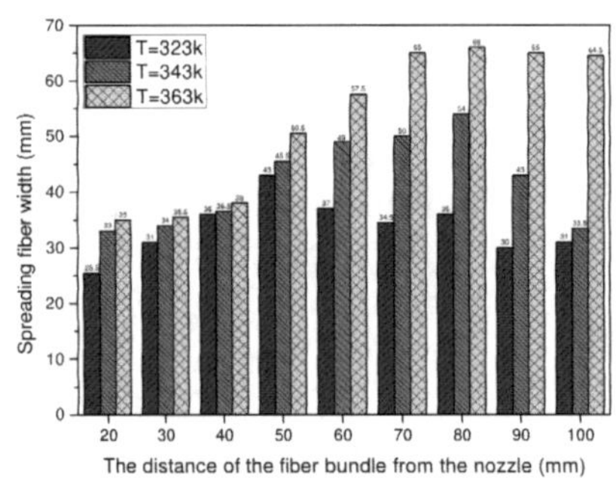

Figure 12. Comparison of fiber expansion width at different temperatures

4.2. Experiment Results

Combined with the results of the simulation, when the boundary conditions are set as, $D = 120mm$, $P_i = 101645Pa$, $P_O = 101325Pa$, $P_d = 320Pa$, the experimental results are shown in Fig.10. When the temperature is constant, the width of the fiber bundle broadening tends to increase first and then decrease as the fiber-to-nozzle distance increases. When the temperature is $323K$, the width of the fiber broadening reaches a maximum at a fiber-to-nozzle distance of $50mm$, and the maximum broadening width is $43mm$. When the temperature is $343K$ and $363K$, the width of the fiber broadening reaches a maximum at a fiber-to-nozzle distance of $80mm$, and their maximum values are $54mm$ and $66mm$, respectively. It can be seen that when the fiber is fixed from the nozzle, the width of the carbon fiber spreading increases remarkably as the temperature increases. When the fiber-to-nozzle changes from $80mm$ to $100mm$, as the temperature and the fiber distance increases, the width of the fiber bundle unfolds or even decreases. This is because the fiber bundle is in the free jet region, which the jet velocity changes relatively, the fiber bundle easy to overlap.

5. Conclusion

Through the three-dimensional simulation analysis of the hot-jet flow field, the data which is favorable for jet spreading. And the spreading under different boundary conditions are obtained. The simulation results are verified by experiments, and the optimal boundary conditions for jet spreading are obtained.

1) Through the simulation analysis of the hot-jet flow field, it can be seen that there is a constant velocity area, a free jet area, a sluggish area and a wall jet area in the flow field.

2) When the spray distance $D = 120mm$, the optimal fiber distance is $40 - 80mm$, and the constant velocity area of the jet in this area is beneficial to the homogenization of the fiber.

3) On the basis of the simulation, the hot-jet spreading test is carried out to obtain the pressure drop and temperature which are favorable for the jet spreading. When the temperature is $363K$, the pressure drop is $320Pa$, and the fiber-to-nozzle distance is $80mm$, the fiber bundle spreading effect is optimal, which can be spread to $65mm$.

6. Acknowledgment

Financial support from the Natural Science Foundation of Tianjin (No: 18JCYBJC89000) is acknowledged.

7. References

[1] Lu R, He X M 2012 The Application of Carbon Fiber Concrete in the Intelligent Bridge Reinforcement. *Advanced Materials Research.* 594-597: 1539-1542.
[2] Chen, Yan Y 2014 The Application of Carbon Fiber Materials in Sports Equipment. *Applied Mechanics and Materials.* 687-691: 4240-4243.
[3] Wang F F 2017 Study on the Application of Carbon Fiber Composite Materials in High-Speed Trains. *Materials Science Forum.* **893**: 31-34.
[4] Irfan M S, Machavaram V R, Mahendran R S, Shotton-Gale N, Wait C F, Paget M A, Hudson M, Fernando G F 2011 Lateral spreading of a fiber bundle via mechanical means. *J. Compos Mater.* **46** (3): 311-330.
[5] Sebaey T A, Mahdi E, Manufacturing 2016 Using thin-plies to improve the damage resistance and tolerance of aeronautical CFRP composites. *Compos. Pt. A-Appl. Sci. Manuf.* **86**: 31-38.
[6] Borg C 2015 An introduction to spread tow reinforcements: Part 1 – Manufacture and properties. *Reinforced Plastics.* **59** (4): 194-198.
[7] Park S M, Kim M S, Choi Y S, Lee E S, Yoo H W, Chon J S 2015 Carbon Fiber Tow Spreading Technology and Mechanical Properties of Laminate Composites. *Composites Research.* **28** (5): 249-253.
[8] Baucom R M, Marchello J M 1990 LaRC powder prepreg system. 21: 4.
[9] Chen J C, Chao C G 2005 Numerical simulation and experimental investigation for design of a carbon fiber tow pneumatic spreading system. *Carbon.* **43** (12): 2514-2529.
[10] Sihn S, Kim R, Kawabe K, Tsai S 2007 Experimental studies of thin-ply laminated composites. *Compos. Sci. Technol.* **67** (6): 996-1008.
[11] El-Dessouky H M, Lawrence C A 2013 Ultra-lightweight carbon fiber/thermoplastic composite material using spread tow technology. *Compos. Pt. B-Eng.* **50**: 91-97.
[12] Ren F, Yu Y, Cao M, Li Y, Xin C, He Y 2017 Effect of pneumatic spreading on impregnation and fiber fracture of continuous fiber-reinforced thermoplastic composites. *J. Reinf. Plast. Compos.* **36** (21): 1554-1563.
[13] Huang B, Yuan S, Xue J 2015 Spreading process simulation and experimental investigation for a carbon fiber pneumatic spreader. *Journal of Xian Jiaotong University.*
[14] Launder B E, Spalding D B 2014 The numerical computation of turbulent flows. *Comput. Meth. Appl. Mech. Eng.* **3** (2): 269-289.

IOP Conf. Series: Materials Science and Engineering **746** (2020) 012039 doi:10.1088/1757-899X/746/1/012039

Analysis of Man-machine-environment System in Industrial Design and Comprehensive Evaluation of Products Man-machine Relationship

Wei Chen

Jiangsu Open University, Nanjing, China
Email: chenwei@jstvu.edu.cn

Abstract: The study of safety, reliability and amenity in industrial design is to analyze the factors in the "man-machine-environment" system by means of ergonomics so as to provide reliable design parameters and test data for product design. According to ergonomics, how about the design of product man-machine relationship? Does safety, reliability and suitability meet the design requirements? This requires a more objective and fair evaluation standard. The evaluation of product man-machine system is not only the judgment of the value of the product designed, but also the judgment of the product design results. Based on the analysis of various factors in the "man-machine-environment" system, this paper presents a comprehensive evaluation method of man-machine relationship of products by using fuzzy theory.

1. Introduction

Industrial design is center on a man design, all services are for people, the core of which is product design. In addition to according the aesthetic principles of product design, but also attaches great importance to the use of products safety, reliability and pleasant, to improve the working environment of users; And ergonomics is a design object based on man in the "man-machine-environment" system, based on the physiological and psychological characteristics of human beings and the aim of improving the quality of human work.

This paper mainly studies the relationship between man and machine, man and environment, machine and environment, and takes human factor as an important condition and principle of system design. It provides a theoretical basis and method for making product design into a "man-machine-environment" system which is easy to operate, labor-saving, safe, reliable, efficient and comfortable.

2. Analysis of Man-machine-environment System

The "man-machine-environment" system is referred to as man-machine system. Man, machine and environment, which constitute the "three factors" of man-machine system, can be regarded as three relatively independent subsystems of man-machine system. Although each of them has its own factor attributes, as a systematic research, it does not mean the sum of the factors attributes of each subsystem, but depends on the organizational structure of the system and the degree of synergy within the system. Therefore, to study the factors in the "man-machine-environment" system, we should not only study the factors in each subsystem of man, machine and environment, but also study the overall structure and attributes of the system in order to make the best use of man, machine and environment so as to make the whole system safe, efficient and comfortable to people and life support function, the ultimate goal is to make the system comprehensive use efficiency highest.

From this we can see that in man-machine system, besides man, machine and environment factors, there are also comprehensive factors formed from this, that is, the factors in man-machine-environment are composed of man, machine, environment and comprehensive factors.

2.1. Man's Factors

1) The parameters of human body size mainly include the working posture and the space range of human activities under static and dynamic conditions.

2) Man's mechanical parameters mainly include maneuvering force, speed and frequency, accuracy and endurance limit.

3) The ability of man's information transmission mainly includes the ability of information acceptance, storage, memory, transmission and output, and the physiological limit ability of various sensory channels.

4) Human reliability and adaptability to work mainly include the ability of psychological adjustment in the process of labor, the mechanism of psychological reflex, and the possibility and cause of human error under normal conditions.

2.2. Machine Factor

1) Information display mainly refers to all kinds of display devices that give feedback information to people after the "machine" receives instructions from people. Such as a variety of displays (screen display, analog display, digital display, etc.), instrument display, sensory information devices (audio information transmission, tactile information transmission, olfactory information transmission, etc.).

2) Operational control mainly refers to the various devices that the "machine" recipient sends out instructions, such as buttons, knobs, handles, control rods, hand wheels, rockers, steering wheels, keyboards, etc.

3) Safety protection mainly refers to the safety protection facilities and devices when the "machine" has errors or human errors, such as overload protection (external force overload protection and electrical overload protection), rescue and escape devices.

2.3. Environmental Factors

Environmental factors include a wide range of contents, whether outdoors or indoors, people are faced with different environmental conditions, they directly or indirectly affect people's work, system operation, and even affect the safety of human life. Generally speaking, the environmental factors that affect people are:

1) The physical environment includes temperature, humidity, lighting, noise, vibration, radiation, air pressure, gravity, magnetic field, etc.

2) The chemical environment mainly refers to toxic gases and vapors, industrial dust and smoke, and water pollution.

3) The psychological environment mainly refers to the aesthetic factors (shape, color, texture, decoration and functional music, etc.) of the machine being used, and the working space (size, height of workshop, layout of the machine, road traffic, etc.).

4) The social environment mainly refers to social status (political, economic), employment status, interpersonal relationship and so on.

2.4. The Comprehensive Factors

The comprehensive factors can't be simply understood as the sum of man, machine and environment factors, but the comprehensive use efficiency formed by them after the integration of man-machine system, that is, rational division and cooperation of man-machine, man-machine information transmission and exchange. Man and the machine between the continuous transmission and exchange of information, constantly changing the state, constantly realizing people's wishes, so that the system keeps coordinating work.

2.4.1. Rational division and coordination of man and machine

The rational division and cooperation of man and machine should comprehensively consider the characteristics and functions of man and machine, so as to make them develop their advantages and avoid their disadvantages, and give full play to the comprehensive utilization efficiency of man-machine system.

For a rough comparison of man and machine in nine aspects, such as creativity, information processing, reliability, control ability, work efficiency, sense ability, learning ability, induction, durability, you can refer to literature [2]. All the heavy, fast, fine, regular, monotonous, high-speed operation, complex operation of the work, are suitable for the "machine" to undertake; all the "machine" system design, maintenance, monitoring, fault handling, as well as procedures and instructions for the arrangement, are suitable for people to undertake.

2.4.2. Man-machine information transfer and exchange

Man-machine information transmission and exchange refers to the person issue instructions to "machine" through the executive organs (hands, feet, mouth, body, etc.), and to accept the "machine" feedback information through the sensory organs (eyes, ears, nose, tongue, body, etc.) .The intermediary area of man-machine information transfer and exchange is called "man-machine interface" [3]. It includes the man-machine interface of operation control system, information display system and the environment.

In the modern production process, the original data of the production equipment is input into the "machine" as information, which is reflected by the display device on the "machine" and constitutes the man-machine interface of the information display system. The display device feeds back to the human auditory or visual organs through visual or auditory signals and transmits them to the human brain. After analysis and judgment, the human brain makes a decision, to add a control device to the "machine" by his hands or feet and to give the operation action, to form the human-machine interface of the operation control system. Due to operation action, the control signal generated by the control device is then transmitted to the "machine" so that the whole process of production is completed. In the whole production process, human and "machine" are affected by the environment, forming an environment man-machine interface.

3. Comprehensive Evaluation of Man-machine Relationship in Products

Based on the factors of man-machine-environment, the evaluation principle and system of man-machine relationship are put forward, and then the man-machine relationship is evaluated comprehensively.

3.1. Evaluation Principle of Man- machine Relationship

"Safety", "reliability" and "agreeableness" are the principles that should be followed in the evaluation of man-machine relationship of products. They are referred to as "three principles" for short. The "three principles" contain all the factors in the "man-machine-environment" as mentioned above.

3.1.1. Security

Safety includes objective, behavioral, physiological and psychological factors that cause accidents ("machines" and people). Objective factors mainly refer to comprehensive factors and machine factors, such as inappropriate distribution of human and "machine" functions; Design errors such as the use of tools and workplaces (such as inappropriate layout of displays and controllers, alarm devices in hard-to-see locations, etc.); Lack of necessary safety devices and protective measures. Behavioral factors mainly refer to human factors, including training and skills, memory ability, age and experience, life stress and so on. Physiological factors mainly refer to human factors and physiological and biological rhythms of human beings. Psychological factors mainly refer to the character of the person (introverted, extroverted or adventurous).

3.1.2. Reliability

Reliability mainly includes the use function of the machine, the service life of the machine and the working environment conditions of the machine. In addition, the manufacturing, assembly, management, maintenance, repair, transportation, packaging, inventory and so on of the machine are also factors to be considered for reliability.

IOP Conf. Series: Materials Science and Engineering **746** (2020) 012039 doi:10.1088/1757-899X/746/1/012039

3.1.3. Agreeableness

Agreeableness includes whether machine is easy to operate; Whether people are easy to fatigue when operating; Whether the design of machine conforms to man physiological characteristics; Whether the shape, color and texture of machine and its decoration meet human psychological requirements; And whether micro-climate (physical factors, such as temperature, humidity, lighting, noise, vibration, radiation, air pressure, etc. and chemical factors, such as toxic gases, industrial dust and smoke as well as water pollution) are suitable for people.

3.2. Evaluation System of Human Machine Relationship

According to the principle of product man-machine relationship evaluation, a good product man-machine relationship must be the combination of "safety", "reliability" and "agreeableness". The man-machine relationship evaluation system is shown in the following Fig.1

- Man-machine relationship evaluation system
  - Safety (0.55)
    - Objective factors (0.36)
    - Behavioral factors (0.30)
    - Physiological factors (0.14)
    - Psychological factors (0.20)
  - Reliability (0.25)
    - Usage function (0.56)
    - Service life (0.20)
    - Working environment conditions (0.24)
    - Operability (0.30)
  - Agreeableness (0.20)
    - Fatigue (0.20)
    - Satisfying physiology (0.10)
    - Satisfying psychology (0.24)
    - Micro-climate (0.16)

Figure 1. Man-machine relationship evaluation system

Explain: The values in brackets are the weights of a set of factors in the man-machine relationship that have been statistically processed for a CNC machine tool manufactured by a factory.

3.3. Basic Method of Fuzzy Synthetic Evaluation of Man-machine Relationship

Fuzzy evaluation is a method of evaluating things after fuzzy transformation according to the given evaluation system, and comprehensive evaluation is a general evaluation of things affected by multiple factors. The establishment of fuzzy comprehensive evaluation model can be summarized as follows:

1) We are given that a factor sets $u = \{u_1, u_2, \ldots\ldots\ldots u_n\}$;

A evaluation sets $v = \{v_1, v_2, \ldots\ldots\ldots, v_n\}$.

A factor sets is a common set of various factors that affect the evaluation object, and a evaluation sets is a set of evaluation results that the judge may make on the evaluation object.

2) Establishing evaluation matrix

A single factor constitutes a fuzzy evaluation vector: $r\ (r_{i1}\ r_{i2}, \ldots\ldots, r_{im})$, the fuzzy evaluation vector of all single factors is composed of fuzzy evaluation matrix R [4]:

$$R = \begin{bmatrix} r_{11} & r_{12} & \cdots & r_{1m} \\ r_{21} & r_{22} & \cdots & r_{2m} \\ \cdots & \cdots & \cdots & \cdots \\ r_{n1} & r_{n2} & \cdots & r_{nm} \end{bmatrix}$$

3) Determining the weight of each factor in evaluation

The degree of influence of different factors on evaluation objects are different. Therefore, when evaluating, the importance of each factor is different. In order to make the evaluation more scientific, each factor should be assigned different weights.

$$A = (a_1, a_2, \dots, a_n), \ (\sum_{i=1}^{n} a_i = 1, \ a_i \geq 0)$$

4) Carry out the compound operation of fuzzy matrix, get the result of fuzzy comprehensive evaluation, and normalize it.

$$B = A \cdot R = (b_1, b_2, \dots, b_m)$$

Here, $b_j = \bigvee_{i-1}^{n} (a_i \wedge r_{ij})$

It is normalized fuzzy comprehensive evaluation index.

The above is a first level fuzzy comprehensive evaluation model. When a complex system is synthetically evaluated, there are many factors involved, and there are different levels among them. If the first-level evaluation can not reach a conclusion, it is necessary to establish a mathematical model of multi-level fuzzy comprehensive evaluation, see reference [5].

3.4. Weight Distribution

The determination of the factors that affect the weight plays an important role in the fuzzy comprehensive evaluation. Whether the weight is reasonable or not will directly affect the result of the evaluation. There are many ways to determine weights. In this paper, fuzzy statistics is used to distribute weights.

In order to assign weights to the comprehensive evaluation weight of man-machine relationship, first of all should organizes the relevant experts, technical personnel and management cadres who engaged in ergonomics for a long time to carefully analysis the designed "machine", to consider the influence degree of various factors for man-machine relationship. Each one is put forward a weight distribution for each factor; and then respectively carries on statistics, and remove excessive dispersed numerical values. Finally, according to a maximum value and a minimum value in the data groups, calculates their class interval and frequency within the group, to determine the factor's membership degree value.

The values in the brackets in the front figure are the weights of a group of factors in the man-machine relationship that have been statistically processed for a CNC machine tool manufactured by a factory. The weight of different products is different. The judgment set of each factor should be given by a considerable number of experts.

3.5. Application of Fuzzy Comprehensive Evaluation Method

Taking a CNC machine tool as an example, fuzzy comprehensive evaluation is made for its man-machine relationship.

The safety evaluation matrix, the reliability evaluation matrix and the agreeableness evaluation matrix are respectively:

$$R_1 = \begin{bmatrix} 0.18 & 0.52 & 0.28 & 0.02 \\ 0.16 & 0.64 & 0.17 & 0.03 \\ 0.24 & 0.69 & 0.07 & 0.00 \\ 0.54 & 0.36 & 0.10 & 0.00 \end{bmatrix} \quad R_2 = \begin{bmatrix} 0.35 & 0.30 & 0.25 & 0.10 \\ 0.32 & 0.38 & 0.20 & 0.10 \\ 0.36 & 0.40 & 0.20 & 0.04 \end{bmatrix} \quad R_3 = \begin{bmatrix} 0.45 & 0.35 & 0.10 & 0.10 \\ 0.44 & 0.35 & 0.21 & 0.00 \\ 0.46 & 0.48 & 0.06 & 0.00 \\ 0.30 & 0.35 & 0.30 & 0.05 \\ 0.28 & 0.56 & 0.16 & 0.00 \end{bmatrix}$$

Weight distribution:

$$A_1 = \{0.36, \ 0.30, \ 0.14, \ 0.20\} \quad b_1 = A_1 \bullet R_1 = [0.18 \ \ 0.36 \ \ 0.28 \ \ 0.03]$$

Normalization treatment:

$$b_1 = [0.21 \ \ 0.42 \ \ 0.33 \ \ 0.04]$$

Weight distribution:

$$A_2 = \{0.56,\ \ 0.20,\ \ 0.24\} \qquad b_2 = A_2 \bullet R_2 = [0.35 \quad 0.30 \quad 0.25 \quad 0.10]$$

Normalization treatment:

$$b_2 = [0.35 \quad 0.30 \quad 0.25 \quad 0.10]$$

Weight distribution:

$$A_3 = \{0.30 \quad 0.20 \quad 0.10 \quad 0.24 \quad 0.16\} \quad b_3 = A_3 \bullet R_3 = [0.30 \quad 0.30 \quad 0.16 \quad 0.16]$$

Normalization treatment:

$$b_3 = [0.33 \quad 0.33 \quad 0.17 \quad 0.17]$$

According to the principle of maximum membership, the evaluation of safety is better; the evaluation of reliability is good; the evaluation of agreeableness is better.

Because the first grade judgement has a clear conclusion, there is no two level judgement.

4. Conclusion

Through the above analysis, we can draw the following conclusions:

1) The overall factors affecting the man-machine relationship of products are human factors, machine factors, environmental factors and comprehensive factors. Specifically, objective factors, behavioral factors, physiological factors, psychological factors, use function, service life, working environment conditions, operability, fatigue, satisfying physiological, satisfying psychological and microclimate.

2) The principle of evaluating man-machine relationship is safety, reliability and agreeableness.

3) There are many factors involved in the evaluation of man-machine relationship of products, and most of the factors are fuzzy information, which can be quantitatively evaluated by fuzzy theory.

4) Applying the fuzzy comprehensive evaluation method to evaluate the man-machine relationship of the products can obtain more objective and fair conclusions.

5) The analysis of various factors in "man-machine-environment" and the comprehensive evaluation of man-machine relationship of products provide a more scientific basis for product design.

5. Acknowledgments

This research is supported by the "2019 curriculum innovation and excellence program" of Jiangsu Open University (1027154ZL2019/001).

6. Reference

[1] Shi-yong LI, Engineering fuzzy mathematics and its application, Harbin Institute of Technology press, 2015. 2.

[2] Yu-hong ZHANG, Human - machine engineering and industrial design, China water conservancy and hydropower press, 2011. 4.

[3] Wei CHEN. Study on the functions and modes of human-machine systems. Human ergonomics, 1998, 11 (1).

[4] Wei CHEN. Fuzzy comprehensive evaluation of machine tool molding quality, Combined machine tool and automatic processing technology. 1998, 4.

[5] Wei CHEN. Research and Application of Fuzzy Synthetic Evaluation Method about the Design Quality of Industrial Products, Trans Tech Publication, 2014. 5.

IOP Conf. Series: Materials Science and Engineering **746** (2020) 012040 doi:10.1088/1757-899X/746/1/012040

The Research on Vibration Mode of a Ship Pump Group

Binbin Zhou[a] and Guoxiang Hou[*]

School of Naval Architecture & Ocean Engineering, Huazhong University of Science and Technology, East 2 room 306c, Luoyu Road, Hongshan District, Wuhan, China
[*]Corresponding author email: huoguoxiang@163.com, [a]zhoubinbin1996@live.com

Abstract. With the intention of reducing the vibration and noise of a ship pump group, rubber dampers were installed between the motor base and the foundation. The finite element analysis method was used to analysing the influence of the stiffness of the rubber damper on the damping effect. It was found that the rubber shock absorber significantly reduces the low-order modal frequency of the group. When the structure has good symmetry, the frequency of the translational mode is proportional to the square root of the stiffness of the rubber damper in the corresponding direction. And some impedance characteristics of the pump group were obtained by harmonic response analysis, which are beneficial to pick the right rubber damper to avoid the modal resonance frequency.

1. Introduction

Many of the auxiliary equipment in the ship are connected to the hull deck through the base, so the vibration of the auxiliary equipment is mainly transmitted to the hull through the base. Therefore, the impedance characteristics of the base are of great significance to the vibration and noise reduction of the whole group. In this paper, rubber dampers were installed between the motor cabinet and the foundation, which made the impedance of the system change abruptly. The vibration energy is absorbed and consumed by the rubber damper in the transmission path of vibration energy, thereby achieving the purpose of vibration and noise reduction. The research object of this paper is a ship water pump group, whose working speed is 300 RPM.

2. The Basic Theory

For the vibration of a multi-degree-of-freedom system, the complex-modal vibration differential equation [1] of a general viscous system can be described as

$$M\ddot{x} + C\dot{x} + Kx = f(\mathrm{t}) \tag{1}$$

M and *K* are the mass matrix and stiffness matrix of the system respectively, and *C* is the general viscous damping matrix. Let the exciting force as *f(t)* and the steady-state displacement response as *x(t)*.

$$f(t) = Fe^{j\omega t} \tag{2}$$

$$x = Xe^{j\omega t} \tag{3}$$

The velocity is the derivative of the displacement

$$v = \dot{x} = i\omega X^{j\omega t} \tag{4}$$

The acceleration is the derivative of the velocity

$$a = \dot{v} = -\omega^2 X e^{jwt} \tag{5}$$

Substitute (2) - (5) to (1), and the original differential equation can be rewritten as:

$$\mathrm{X} = H(\omega)F, \mathrm{H}(\omega) = (\mathrm{K} + j\omega C - \omega^2 \mathrm{M})^{-1} \tag{6}$$

$$\mathrm{Z}_D(\omega) = H(\omega)^{-1} \tag{7}$$

$H(\omega)$ represents the frequency response function matrix and $Z_D(\omega)$ for Displacement impedance [2] matrix.

3. Finite Element Analysis for Models

3.1. The Finite Element Models

The foundation was modelled in ANSYS [3] with solid 187. In this model, the opening of the pump is aligned with the X direction, the opening of the motor cabinet is aligned with the Z direction, and the vertical direction is aligned with the Y direction. In the actual case, the foundation is connected to a test platform with high stiffness by 26 M20 bolts. The inner face and end face of the bolt hole were fixed to simulate the bolt connection between the base and the test platform, and the vertical displacement of the bottom surface of the base was restrained to set precise boundary conditions [4]. The foundation and the motor cabinet are made of special material high manganese steel, the other structures are made of common structural steel. The rubber dampers are installed in the four corners of the foundation.

3.2. Modal Analysis

Vibration modes are inherent and integral characteristics of the elastic structure. Modal analysis can be used to find out the characteristics of the main modes of a structure in a range of susceptible frequencies [5]. Then, the actual vibration response of the structure in this frequency band, generated by external or internal vibration sources, can be predicted.

3.2.1. Results of foundation

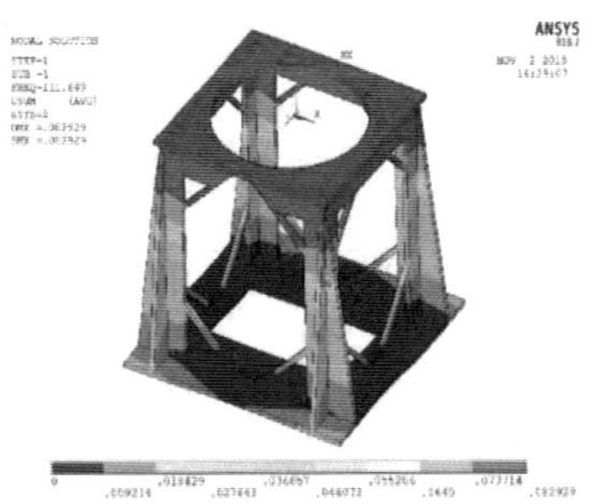

Figure 1. The first mode

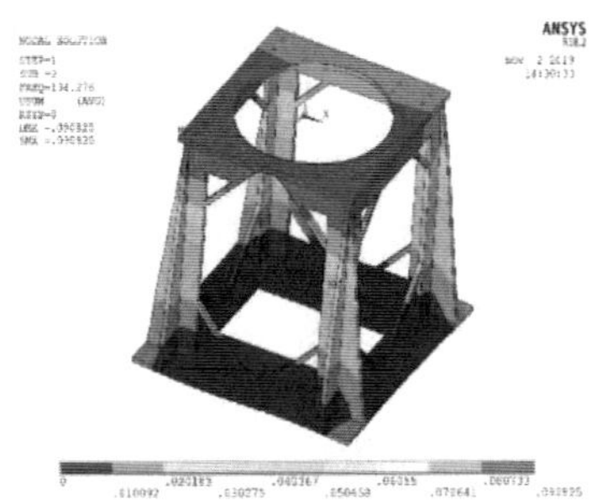

Figure 2. The second mode

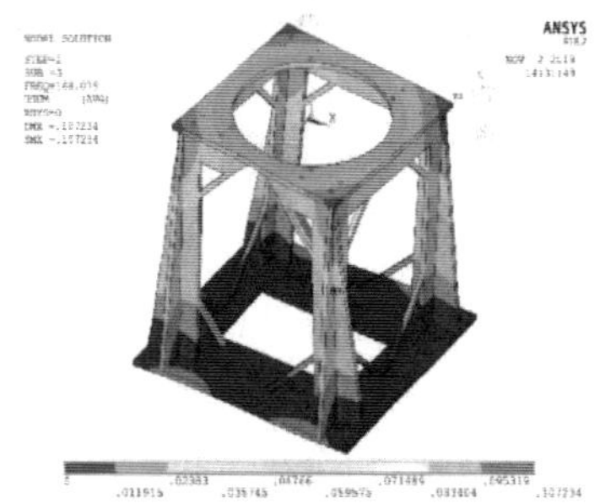

Figure 3. The third mode

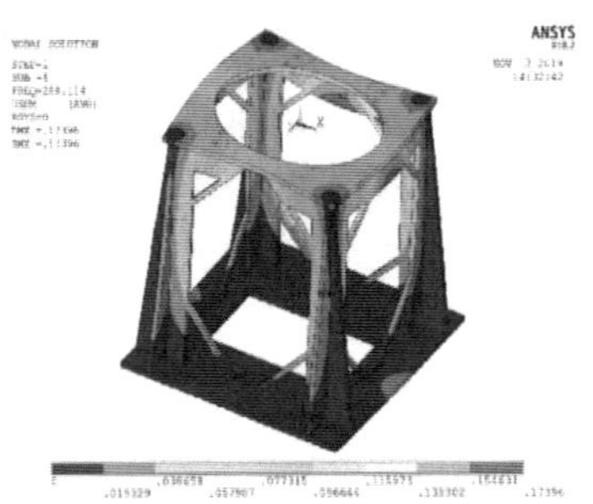

Figure 4. The fourth mode

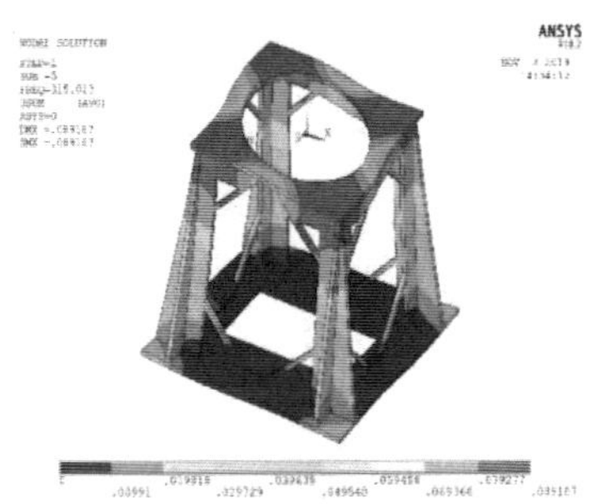

Figure 5. The fifth mode

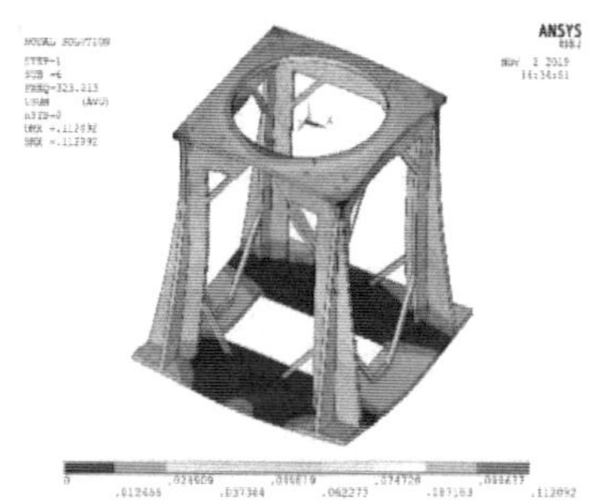

Figure 6. The sixth mode

IOP Conf. Series: Materials Science and Engineering **746** (2020) 012040 doi:10.1088/1757-899X/746/1/012040

The frequency of first-order mode is 111.67Hz, which shows the translation of the whole structure along the x direction. The frequency of the second-order mode is 138.24Hz, and its mode shape is the translation of the structure along the z direction. The third-order mode has a frequency of 168.08Hz, and its mode shape is that the structure rotates around the vertical axis. The fourth-order mode has a frequency of 288.11Hz, and its mode shape is the stretching and compression of the opposite side. The fifth-order mode has a frequency of 315.01Hz, and its mode shape is diagonal stretching and compression. The frequency of the fifth-order mode is 323.01Hz, which shows the vertical translation

3.2.2. Results of motor cabinet

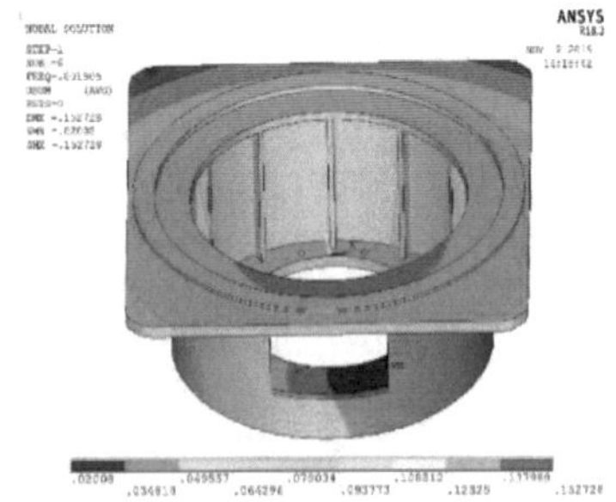

Figure 7. The seventh mode

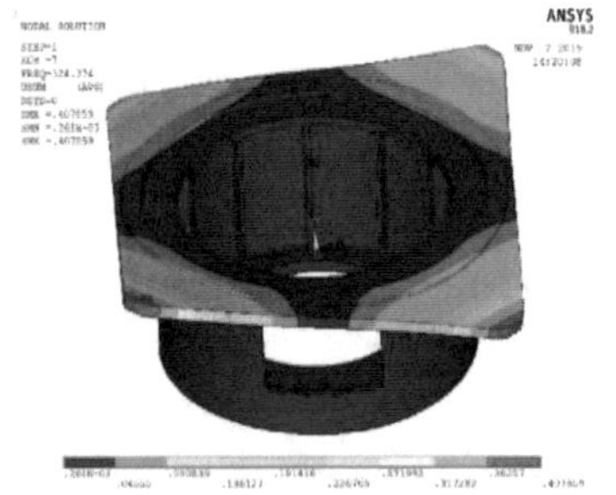

Figure 8. The eighth mode

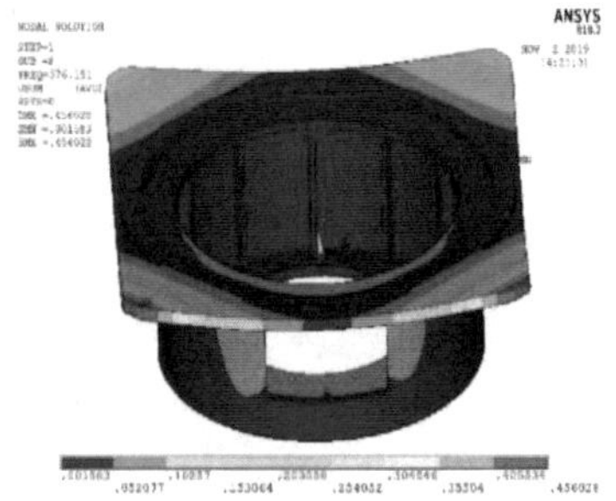

Figure 9. The nineteenth mode

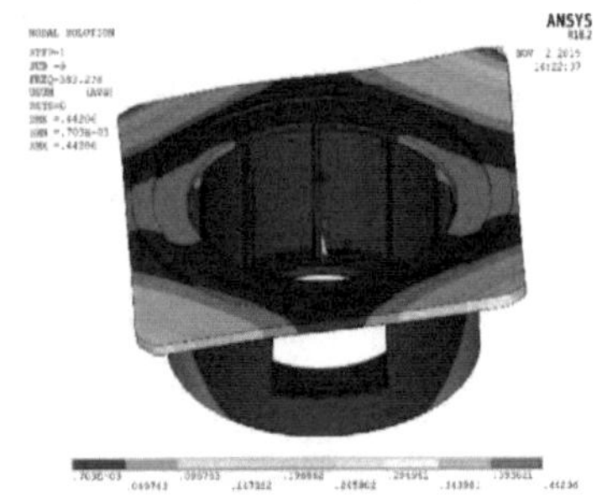

Figure 10. The eleventh mode

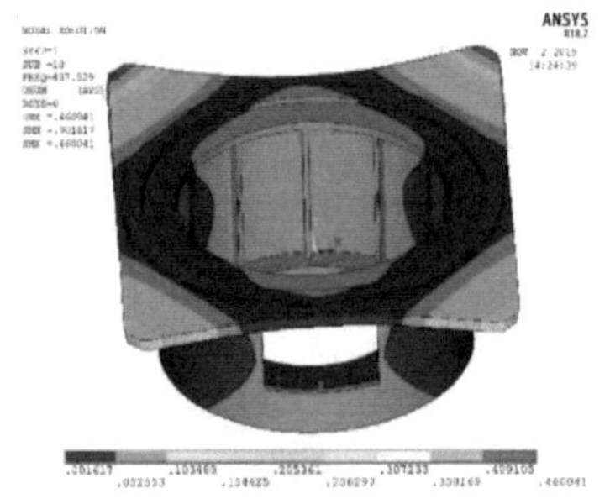

Figure 11. The twelfth mode

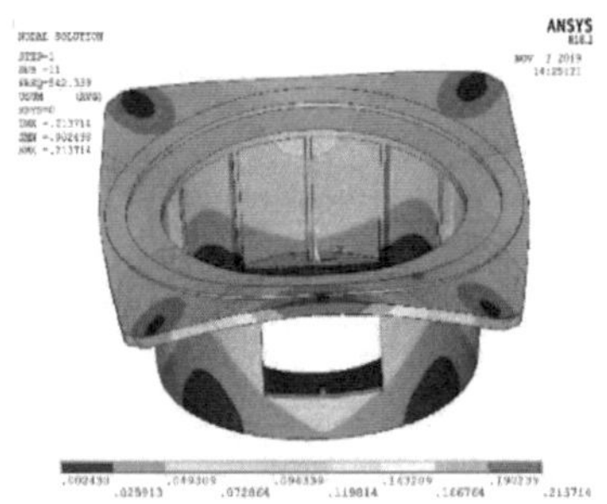

Figure 12. The thirteenth mode

The motor cabinet was modelled with shell 181 with the boundary condition of completely free. The above are the mode shapes of the first six modes whose frequencies are not 0. It can be seen the four corners of the upper plate deformation are obvious, are the most vulnerable part of the cabinet.

3.2.3. Results of the assembly of the motor cabinet and foundation

Springs with three directional stiffness were used to simulate the rubber dampers between the base and the motor base. The mode of 33.317 Hz shows the motor cabinet moves along the X direction. The mode of 50.081 Hz shows the translation of motor cabinet in the Z direction. The frequency of the third-order mode is 58.738 Hz, and its mode shape is the motor seat is rotated around the vertical axis. The mode of 64.051 Hz describe vertical translation of motor base. Modes of 71.489 Hz and 81.779 Hz describe rotations around the Z and X direction correspondingly.

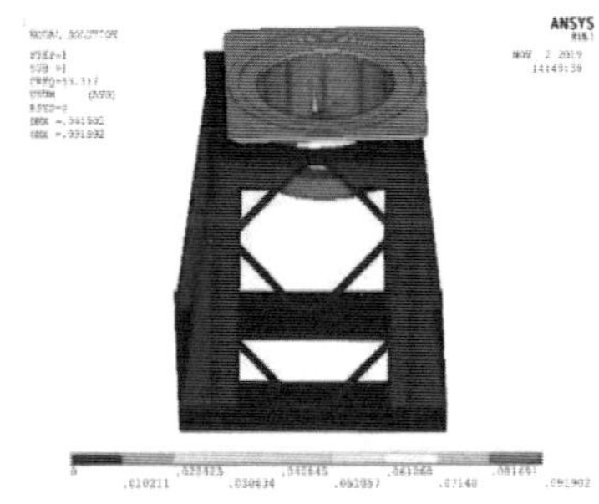

Figure 13. The first mode

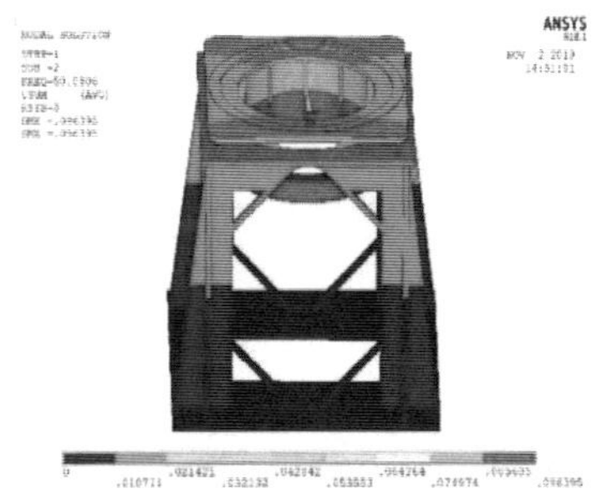

Figure 14. The second mode

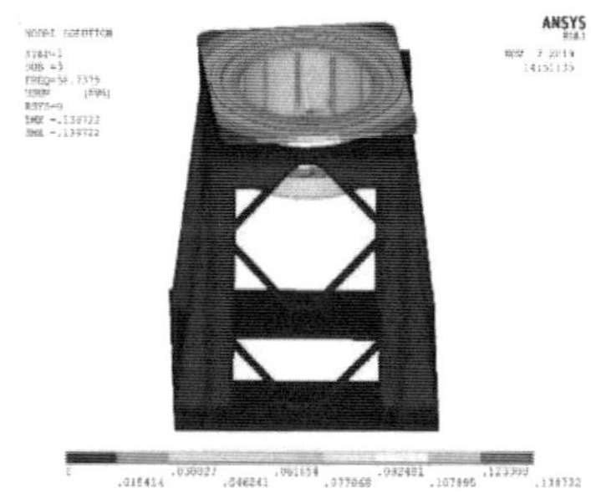

Figure 15. The third mode

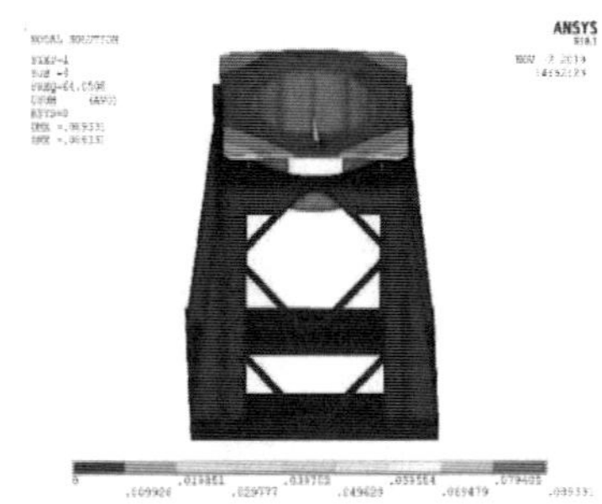

Figure 16. The fourth mode

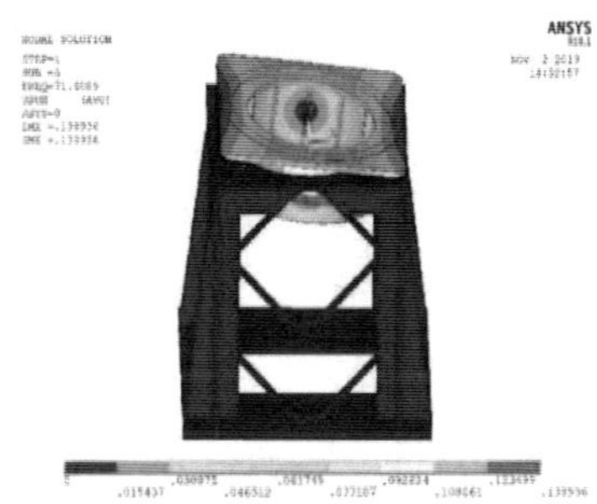

Figure 17. The fifth mode

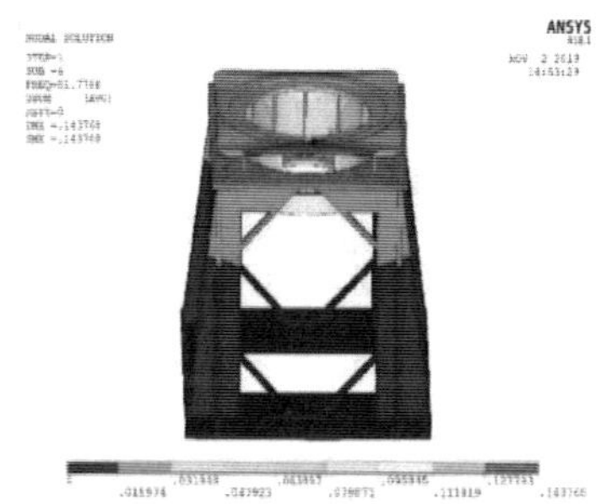

Figure 18. The sixth mode

The foundation and motor cabinet have high modal frequency respectively, but after assembly with a rubber damper, the modal frequency of the assembly drops significantly. It is obvious that foundation has almost no movement, because the stiffness of the spring is very small compared with the foundation and motor cabinet, and the spring connection is the weakest part, and the stiffness of the spring connection has a significant influence on the mode. The foundation and motor cabinet were modelled according to the real models. Because the internal structure of the rotor is very complicated and its stiffness is large, the pump and the motor have little effect on the modal analysis results. To simplify the calculation, only the approximate shape of the pump and the motor is retained [6]. By adjusting its density, the position of the centre of gravity of the model is consistent with the position of the centre of gravity of the pump group in the real situation. At the same time, when the pump group is working, the pump inlet and outlet will be connected to the pipeline which also has a restraining effect on the pump group. The spring with three directional stiffness was used to simulate the pipeline. After the motor and pump body were assembled, the model mass increased and the frequency of first six modes decreased significantly, but the mode shape did not change much.

3.3. Summary and Simplified Hypothesis

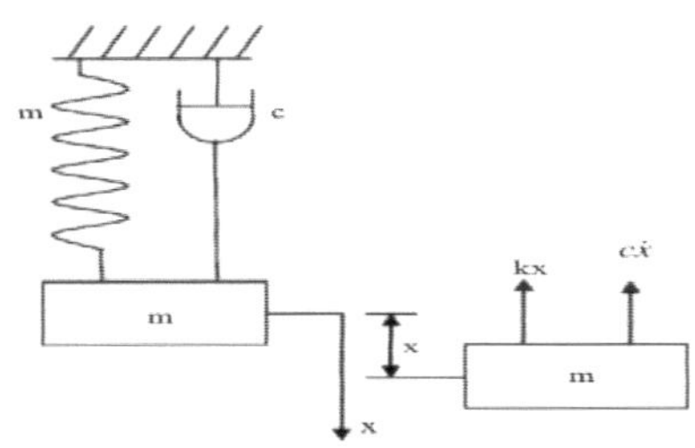

Figure 19. Vibration model of single degree of freedom spring

Most vibration structures can be discretized into multi-degree-of-freedom systems with finite degrees of freedom. Within the linear range, the vibration response is the linear superposition of the main vibration by degrees. If only the motion of a certain degree of freedom is considered, it can be simply treated as a single-degree-of freedom system.

The differential equation of vibration in regular form for a single degree of freedom vibration system [8] is shown in the figure above can be written as:

$$\ddot{x} + 2\sigma\dot{x} + \omega_0^2 x = 0 \tag{8}$$

Attenuation damping coefficient $\sigma = \dfrac{c}{2m}$, undamped natural frequency $\omega_0 = \sqrt{\dfrac{k}{m}}$.

Table 1. The first six mode frequencies

	Foundation	Motor cabinet	Assembly of foundation and motor cabinet	pump group
1	111.67Hz	0 Hz	33.317 Hz	10.694 Hz
2	134.28 Hz	0 Hz	50.081 Hz	14.383 Hz
3	168.08 Hz	0 Hz	58.738 Hz	15.087 Hz
4	288.11 Hz	$4.0559*10^{-4}$ Hz	64.051 Hz	18.162 Hz
5	315.01 Hz	$1.5602*10^{-3}$ Hz	71.489 Hz	21.208 Hz
6	323.01 Hz	$1.9086*10^{-3}$ Hz	81.779 Hz	21.946 Hz

It can be seen from equation (8) that the natural frequency is proportional to the square root of the stiffness and inversely proportional to the square root of the mass. Since the stiffness of the spring is smaller than that of the foundation and motor cabinet, the reduction of stiffness makes the modal frequency of the assembly decrease, which is in accordance with the law. After assembling the motor and pump, the increase of mass makes the modal frequency of the assembly further decrease, which also conforms to the law.

4. Verification and Analysis

With the attention to further study the effect of the spring stiffness on the modal frequency of the structure, three different sets of stiffness were set for the spring. The three sets of parameters are respectively from the rubber damper of the EA400, BE300 and E400. The translational motion in three directions were studied, and the motion in this direction was analogous to a single-degree-of-freedom spring-proton system. At the same time, assuming the motions in the three directions are orthogonal to each other, then the low-order natural frequency should be proportional to the square root of the stiffness of the spring.

Table 2. Spring stiffness setting

	EA400(kg/cm)	BE300(kg/cm)	E400(kg/cm)	(BE300/EA400)^0.5	(E400/EA400)^0.5
KX	1700	5500	2400	1.798692	1.188177
KY	6500	6000	13000	0.960769	1.414214
KZ	5000	17000	6200	1.843909	1.113553

The table below shows the 0-12th order modal frequencies. It can be seen the spring stiffness has a significant effect on the first six modal frequencies, but has little effect on the frequency of the seventh and higher modes. The energy of higher-order modes accounts for a lower proportion of the vibration energy of the structure and has little effect on the vibration of the entire structure [8]. Since the working speed of the pump is 300RPM that the exciting frequency of the structure is not high, it is enough to focus only on the first six modes.

IOP Conf. Series: Materials Science and Engineering **746** (2020) 012040 doi:10.1088/1757-899X/746/1/012040

Table 3. Frequencies of 0 to 12th mode

Order	Foundation (Hz)	Motor cabinet (Hz)	Assembly of foundation and motor cabinet (Hz)	Pump group with EA400(Hz)	Pump group with BE300(Hz)	Pump group with E400(Hz)
1	111.67	0	33.317	10.694	13.274	12.634
2	134.28	0	50.081	14.383	15.201	17.804
3	168.08	0	58.738	15.087	18.683	19.418
4	288.11	0.00041	64.051	18.162	20.432	22.745
5	315.01	0.00156	71.489	21.208	25.559	24.758
6	323.01	0.00191	81.779	21.946	36.854	28.901
7	346.66	324.37	131.32	114.844	114.8	115.54
8	379.58	376.15	139.42	119.653	119.66	119.68
9	429.65	393.28	174.51	126.337	148.387	129.68
10	480.38	407.53	289.77	138.905	153.521	140.72
11	489.34	542.34	319.03	173.846	177.736	175.45
12	489.9	598.58	325.57	175.332	180.091	175.86

The lowest natural frequencies of translational motion in three directions were found, and the corresponding proportional relationship was calculated with the data of EA400 as the reference. It can be seen that except for the Z direction, that is, the direction of the opening on the motor cabinet, the simulation values of the other two directions are not much different from the expected values.

There are two reasons for the error of the results in the Z direction: one is that the foundation and the motor cabinet have openings on only one side, which destroys the symmetry of the structure in the Z direction; the other is that the eccentric installation of the motor also destroys the symmetry of the structure. This result confirms that the stiffness at the joint can significantly and clearly affect the modal frequency of the pump group once again. Since the translational mode frequencies in the vertical and X direction are proportional to the square root of the spring stiffness in the corresponding direction, it is convenient to modify the spring stiffness to change the frequency of the mode to reduce the coincidence with the working frequency and its multiple frequency, so as to avoid resonance.

Table 4. Results of simulation and prediction

	Direction	Simulation value	Predictive value	Error (%)
BE300	X	1.181447	1.188177	0.566414
	Y	1.362755	1.414214	3.638700
	Z	0.980293	1.113553	11.96710
E400	X	1.747146	1.798692	2.865749
	Y	0.963434	0.960769	0.277382
	Z	1.407275	1.843909	23.67980

5. Harmonic Response Analysis

The harmonic response analysis is used to determine the steady-state response of a linear structure subjected to a sinusoidal load over time [9]. The purpose of harmonic response analysis is to calculate the curve of structure response with frequency. Thus, the designer can predict the continuous dynamic characteristics of the structure and verify whether the structure can overcome the harmful effects caused by resonance, fatigue and other forced vibration.

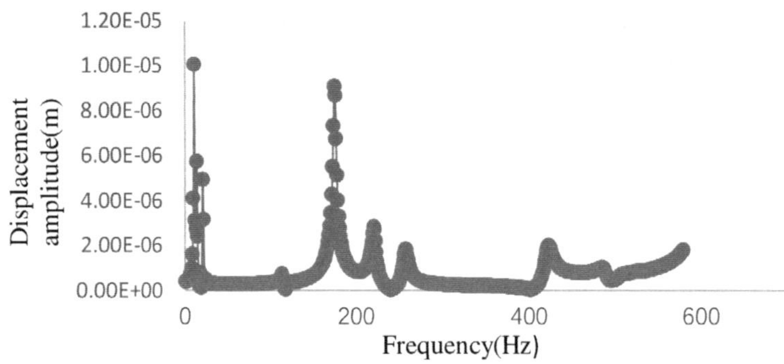

Figure 20. Response with EA400

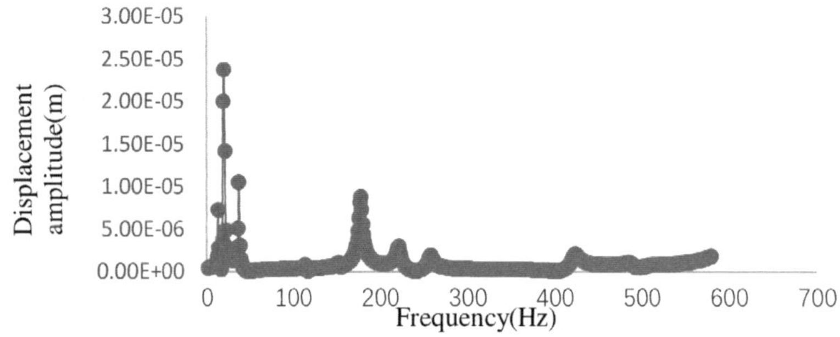

Figure 21. Response with BE300

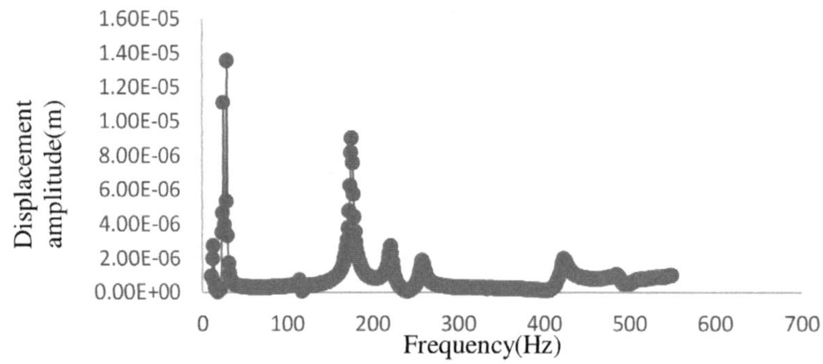

Figure 22. Response with E400

Calculating with the harmonic response module in ANSYS. A vertical downward force of 2000N was applied to the four corners of the upper panel of the motor base. According to the natural frequency range of the structure, the results of harmonic response in 0 Hz ~ 700 Hz are analyzed. The value calculated here is the displacement response of the centre of the upper plate of the motor cabinet.

Table 5. Results of harmonic response analysis

Magnitude sort	1	2	3	4	5
EA400(Hz)	11	175	14	21	221
BE300(Hz)	20	37	178	15	222
E400(Hz)	29	24	176	222	13

It can be seen from the results of the harmonic response analysis that the frequencies with large amplification factor are still concentrated in the low-order frequency band. From the displacement response in figures 20-22, the resonance frequencies are keeping the same with the natural frequencies in mode analysis. The rationality of the results of the harmonic response analysis is verified. The working speed of the pump is 300 RPM, so the mode frequency should avoid 5Hz and its multiplier to prevent resonance. From the results of modal analysis and harmonic response analysis, the E400 can be used to avoid resonance.

6. Conclusions

The least rigid components in the assembly have a significant impact on the low-order mode and low-frequency vibration characteristics of the structure. Therefore, it is possible to use a rubber damper to adjust the modal frequency of the structure. After the rigidity of the base is raised to a certain extent, it has almost no effect on the improvement of the low-order mode frequency of the structure, and can only affect the frequency of the high-order mode. When the structure has good symmetry, the lowest order mode frequency of the corresponding direction is mainly proportional to the square root of the stiffness in the corresponding direction. From the results of modal analysis and harmonic response analysis, the E400 can be used to avoid resonance.

7. Acknowledgments

We acknowledge the financial support from the National Natural Science Foundation of China (Grant No. 51679099 and No. 51979115), and the Fundamental Research Funds for the Central Universities HUST. 2019kfyXKJC041. The author gratefully acknowledges financial support from China Scholarship Council (No. 201906160110).

8. References

[1] Shuqian Cao 2014 *Modal Analysis of Vibration Structure--Theory Experiment and Application* (Tianjin: Tianjin University Press) p 27.

[2] Hixson E L 2001 MECHANICAL IMPEDANCE *J. Shock & Vibration Handbook* pp 194-196.

[3] Heylen W, Lammens S, Sas P 1997 Modal Analysis Theory and Testing *J. Communications Magazine IEEE* 22 (5) pp 64-70.

[4] Xiaodong Yang 2008 Modal analysis of cantilever beam based on ANSYS *J. Mechatronics* 14 (6) pp 58-60.

[5] Zhongjie Yu, Wenrong Ye, Xinpeng Yao 2019 Vibration characteristics analysis of intermediate support of a ship main engine *J. China Shiprepair* 32 (05) pp 21-23.

[6] Zhang yue 2014 Analysis and research on vibration characteristics of three-screw pump *D. Harbin engineering university*, p32.

[7] Danxia Wang, Jianwen Zhang 2017 *J. Boundary Value Problems* 2017 (1).

[8] Baoping Tang, Guowen Zhang, Libo Meng 2012 Elimination of false modes in random subspace based on mode energy *J. Journal of Huazhong University of science and Technology (NATURAL SCIENCE EDITION)*, 40 (03) pp 94-98.

[9] Hongcai Zhang 2012 *ANSYS 14.0 Theoretical Analysis and Engineering Application Examples* (Beijing: China Machine Press) p 383.

IOP Conf. Series: Materials Science and Engineering **746** (2020) 012041 doi:10.1088/1757-899X/746/1/012041

Research on Fuzzy Logic Algorithm and Analogue Experiment on Active Suspension Control of Vehicle

Chuanyin Tang[1], Jinghong Li[2,3], Di Zhang[1] and Hongtie Zhang[1]

[1]School of Mechanical Engineering&Automation, Northeastern University, Shenyang 110819, China
[2]Faculty of Mechanical, Shanghai Dianji University, Shanghai, 201306, China
[3]Faculty of Mechanical Engineering and Automation, Zhejiang Sci-Tech University, Hangzhou, 310018, China
Email: monsigor@aliyun.com

Abstract. A seven-degree of freedom whole body model with an active suspension system is described in the paper. First, construction of fuzzy control system is obtained. One fuzzy control discipline is designed. The fuzzy controller for vertical motion, fuzzy controller for pitch motion and roll motion is established. Second, controller for logical operation is designed, and the sketch of the composed fuzzy control system is presented. In addition, simulation analysis is obtained. The simulation results indicate that the ride comfort and the control stability is ameliorated by the proposed controller. Finally, simulation analogue of experiment is gained and the simulation environment of electronically controlled air suspension is constructed.

1. Introduction

The function of vehicle suspension is to bear the force between the car body and the wheel, complete the transfer of force and torque between the wheel and the car body, mitigate the impact load from the road to the car body, and reduce the vibration caused by impact load, to ensure the normal running of the vehicle. The ride comfort and handling stability of the vehicle is greatly determined by the suspension system of a vehicle, and the vibration reduction effect and other operational performance is affected by the suspension too. The contradiction between ride comfort and control stability is the problem of vehicle dynamics. The research on the control law of the suspension electronic control system has been widely concerned by scholars. The design and simulation analysis of the active and semi-active suspension system are conducted by many researchers.

One control method for the whole vehicle system model is presented in the paper. The fuzzy algorithm controller for vertical vibration, the fuzzy algorithm controller for pitch motion, and the fuzzy controller for vehicle body roll motion are designed respectively. The motion state of roll and pitch is different from that of vertical motion, and the fuzzy rules of roll and pitch are proposed according to the simulation model in this paper. At last, according to the motion of the vehicle, the logic operation controller of the three controllers was obtained, and the simulation software was used to conduct the modelling of the seven-degree-of-freedom vehicle system. Based on the simulation results, the fuzzy controller for the vehicle model is analysed and studied. At the same time, the simulation environment of electronically controlled air suspension is constructed.

IOP Conf. Series: Materials Science and Engineering **746** (2020) 012041 doi:10.1088/1757-899X/746/1/012041

2. Seven Degrees of Freedom Model for Vehicle System

The seven degrees of freedom model for vehicle system [2] is shown in figure 1. Differential equation of vehicle system is presented as:

$$m_i \ddot{z}_i = k_{ti}(y_i - z_i) + k_{si}(z_i' - z_i) + c_i(\dot{z}_i' - \dot{z}_i) + F_i, (i = 1,2,3,4)$$

$$z_1 = z - a\phi + d\theta;\ z_2 = z + b\phi + d\theta;\ z_3 = z - a\phi - c\theta;\ z_4 = z + b\phi + c\theta$$

$$m\ddot{z} = k_{s1}(z_1 - z_1') + k_{s2}(z_2 - z_2') + k_{s3}(z_3 - z_3') + k_{s4}(z_4 - z_4')$$

$$c_1(\dot{z}_1 - \dot{z}_1') + c_2(\dot{z}_2 - \dot{z}_2') + c_3(\dot{z}_3 - \dot{z}_3') + c_4(\dot{z}_4 - \dot{z}_4') - F_1 - F_2 - F_3 - F_4$$

$$J_y \ddot{\phi} = -[k_{s1}(z_1 - z_1') + c_1(\dot{z}_1 - \dot{z}_1') + k_{s3}(z_3 - z_3') + c_3(\dot{z}_3 - \dot{z}_3')]a + (F_1 + F_3)a$$

$$+[k_{s2}(z_2 - z_2') + c_2(\dot{z}_2 - \dot{z}_2') + k_{s4}(z_4 - z_4') + c_4(\dot{z}_4 - \dot{z}_4')]b - (F_2 + F_4)b$$

$$J_x \ddot{q} = -[k_{s3}(z_3 - z_3') + c_3(\dot{z}_3 - \dot{z}_3') + k_{s4}(z_4 - z_4') + c_4(\dot{z}_4 - \dot{z}_4')]c + (F_3 + F_4)c$$

$$+[k_{s1}(z_1 - z_1') + c_1(\dot{z}_1 - \dot{z}_1') + k_{s2}(z_2 - z_2') + c_2(\dot{z}_2 - \dot{z}_2')]d - (F_1 + F_2)d$$

Where: a- the distance from the centre of mass of the car body to the axis of the front wheel (labelled with 1 and 3); b- distance from the centre of mass of the body to the axis of the rear wheels (labelled with 2 and 4); c- distance from the centre of mass of the car body to the axis of the right wheel (labelled with 3 and 4); d- the distance from the centre of mass of the car body to the axis of the left wheel(labelled with 1 and 2) j_x -the inertia of the vehicle body with respect to the X-axis; j_y - the inertia of the vehicle body with respect to the Y-axis; $z_i (i = 1,2,3,4)$ - vertical displacement of the mass of unspring; $z_i' (i = 1,2,3,4)$ - vertical displacement at the four end points of the vehicle body; $y_i (i = 1,2,3,4)$- excitation of the road.

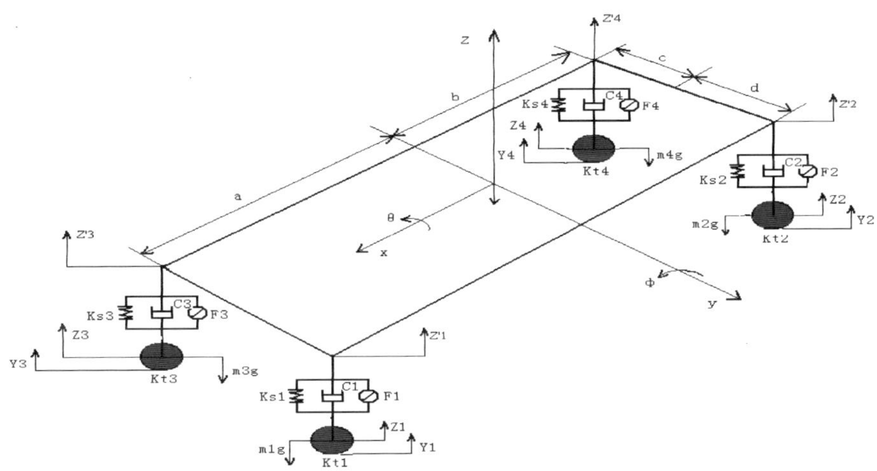

Figure 1. Model of vehicle with seven degrees of freedom

IOP Conf. Series: Materials Science and Engineering **746** (2020) 012041 doi:10.1088/1757-899X/746/1/012041

3. Construction of Fuzzy Control System

According to different control objects, different control variables are adopted to control the seven-degree-of-freedom vehicle model [3, 4]. Based on the simulation results, the effectiveness of the control method for the vehicle model is analysed in this paper. The fuzzy control system includes the fuzzy algorithm controller for vertical vibration, the fuzzy algorithm controller for pitch vibration, the fuzzy algorithm controller for control of roll motion, and the logic operation controller. In the vertical vibration controller, the speed and acceleration of the body are input variables. After the calculation of the vertical controller, the first intermediate output variable is obtained. In the pitch vibration controller, the angular velocity of the body pitch and the angular acceleration of the body pitch are taken as input variables. After the operation of the anti-nod controller, the second intermediate output variable is obtained. In the tilting vibration controller, the tilting angular velocity and tilting angular acceleration of the body are designed as input variables. After the calculation of the anti-tilting controller, the third intermediate output variable is obtained. Three intermediate output variables are taken as input variables for the logic operation controller and four output control forces are obtained after mathematical operation and logic operation.

3.1. Fuzzy Controller for Vertical Motion

The vehicle speed and acceleration are selected as input variables, and the first intermediate variable is designed as the output variable, in the fuzzy controller of vertical motion. In the design of the fuzzy system, according to the range of acceleration of the car body obtained by simulation, the fuzzy domain of the car body speed is set as [-2.5 2.5], and the fuzzy domain of the car body acceleration is set as [-2.5 2.5]. Seven fuzzy sets, NB, NM, NS, ZE, PS, PM, PB, are defined for vehicle speed. Seven fuzzy sets are defined for body acceleration: NB, NM, NS, ZE, PS, PM, PB; seven fuzzy sets are defined for intermediate variables: NB, NM, NS, ZE, PS, PM, PB. The membership function of gauss2mf is selected as the input and output membership function after repeated attempts. The fuzzy rules of vertical motion are shown in table 1.

Table 1. Rule for vertical fuzzy control system

U		EC						
		NB	NM	NS	ZE	PS	PM	PB
	NB	PB	PB	PB	PB	PM	ZO	ZO
	NM	PB	PB	PB	PB	PM	ZO	ZO
	NS	PM	PM	PM	PM	ZO	NS	NS
E	ZE	PM	PM	PS	ZO	NS	NM	NM
	PS	PS	PS	ZO	NM	NM	NM	NM
	PM	ZO	ZO	NM	NB	NB	NB	NB
	PB	ZO	ZO	NM	NB	NB	NB	NB

3.2. Fuzzy Controller for Pitch Motion and Fuzzy Controller for Roll Motion

In the design of fuzzy controller for pitch motion, the two variables of vehicle body pitch angle velocity and pitch angle acceleration are taken as the input variables, and the second intermediate variable is designed as the output variable. The fuzzy domain of pitch angle velocity is defined as [-2.5 2.5], and the fuzzy domain of pitch Angle acceleration is defined as [-2.5 2.5]. Seven fuzzy sets, NB, NM, NS, ZE, PS, PM, PB, are defined for pitch Angle velocity. Seven fuzzy sets of pitch angle acceleration are defined as: NB, NM, NS, ZE, PS, PM, PB. Seven fuzzy sets are defined for intermediate variables: NB, NM, NS, ZE, PS, PM, PB. The membership function of the input and output variables is selected gauss2mf function after repeated attempts. When pitching, set the car body right and left symmetry. After repeated experiments, the fuzzy rules of pitch motion are obtained. According to the same principle, the design of the controller for the roll motion takes the two variables of the body pitch Angle velocity and the body pitch Angle acceleration as the input and the third intermediate variable as the output variable. The fuzzy theory domain of roll velocity is defined as [-3 3], and the basic theory domain of roll acceleration is [-3 3]. Seven fuzzy sets are defined for the roll velocity: NB, NM, NS, ZE, PS, PM, PB; roll acceleration is defined as: NB, NM, NS, ZE, PS, PM, PB. Seven fuzzy sets are defined for intermediate variables: NB, NM, NS, ZE, PS, PM, PB. The membership function of gauss2mf is selected as the input and output membership function after repeated attempts. When tilting, it is assumed that the car body is symmetrical to before and after, and the motion state of roll motion is different from that of pitch motion and vertical motion. After repeated tests, the fuzzy rules of vertical motion are shown in table 2.

Table 2. Rule for roll fuzzy control system

U			EC						
			NB	NM	NS	ZE	PS	PM	PB
	NB	U_r	NB	NB	NM	NM	NS	NS	ZO
		U_l	PB	PM	PM	PS	PS	ZO	ZO
	NM	U_r	NB	NB	NM	NM	NS	NS	ZO
		U_l	PB	PM	PS	PS	ZO	ZO	ZO
	NS	U_r	NM	NM	NS	NS	ZO	ZO	PS
		U_l	PM	PS	PS	ZO	ZO	NS	NS
E	ZE	U_r	NM	NS	NS	ZO	ZO	PS	PS
		U_l	PM	PS	ZO	ZO	NS	NS	NM
	PS	U_r	NS	ZO	ZO	PS	PS	PM	PM
		U_l	PS	PS	ZO	ZO	NS	NS	NM
	PM	U_r	ZO	PS	PS	PM	PM	PB	PB
		U_l	ZO	ZO	ZO	NS	NS	NM	NM
	PB	U_r	ZO	PS	PS	PM	PM	PB	PB
		U_l	ZO	ZO	NS	NS	NM	NM	NB

3.3. Controller for Logical Operation

In the design of the logic operation controller, three intermediate output variables of the fuzzy controller are taken as the input variables of the logic operation controller, and the output variable of the logic operation controller is the control force of the system. Parameter a represents the distance from the front axis to the centre of mass, and parameter b represents the distance from the rear axis to the centre of mass. Parameter c represents the distance from the mass of the left wheel of the vehicle to the centre of mass of the axle, and parameter d represents the distance from the mass of the right wheel of the vehicle to the centre of mass of the axle.

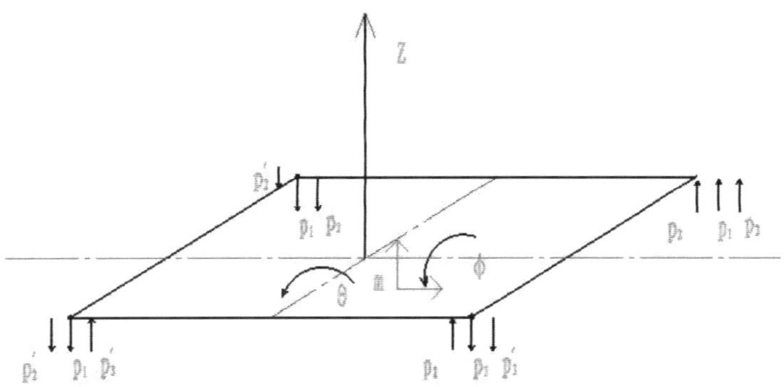

Figure 2. The sketch of the composed system

IOP Conf. Series: Materials Science and Engineering **746** (2020) 012041 doi:10.1088/1757-899X/746/1/012041

The overall vibration of the automobile body is regarded as the combined movement of the body in the vertical direction and the rotating motion around the body's centre of mass. The force exerted on the system after the integration is shown in figure 2. After the operation of the logic controller, the output is shown as

$$F_1 = P_1 + P_2^{'} - P_3^{'};\qquad F_2 = P_1 - P_2 - P_3$$
$$F_3 = P_1 + P_2^{'} + P_3;\qquad F_4 = P_1 + P_3^{'} - P_2$$
$$P_2.a = P_2^{'}.b;\qquad P_3^{'}c = P_3.d$$

4. Simulation Analysis

Simulation analysis shows that the road condition is b-class road and the speed is 30m/h. The results of the control strategy are shown in figure 3 to figure 6. The effects of the control strategy are shown in table 3. From the simulation results, the vehicle body vertical acceleration response of the body pitching angle acceleration response and body roll angle acceleration response are decreased, and tire dynamic deformation response are reduced, the effect of suspension stroke is slighted improved .The effect of the control method is shown in table 3.

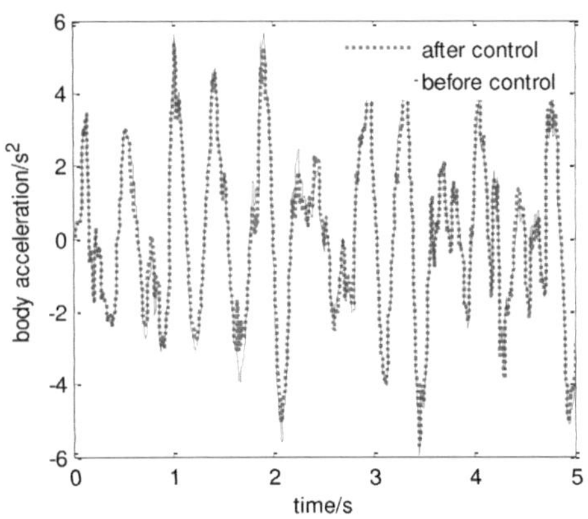

Figure 3. Response of vertical acceleration

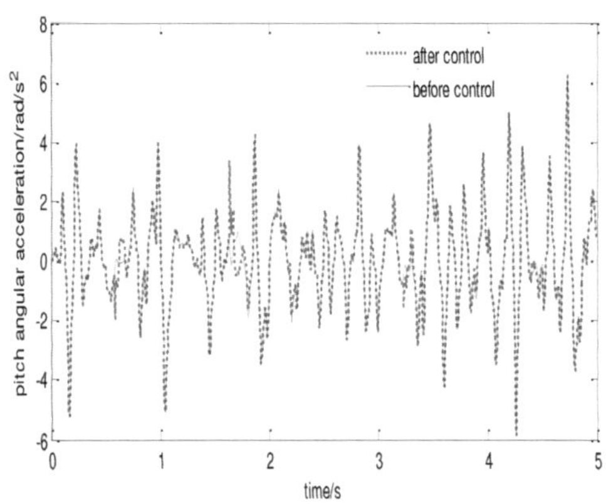

Figure 4. Response of pitch acceleration

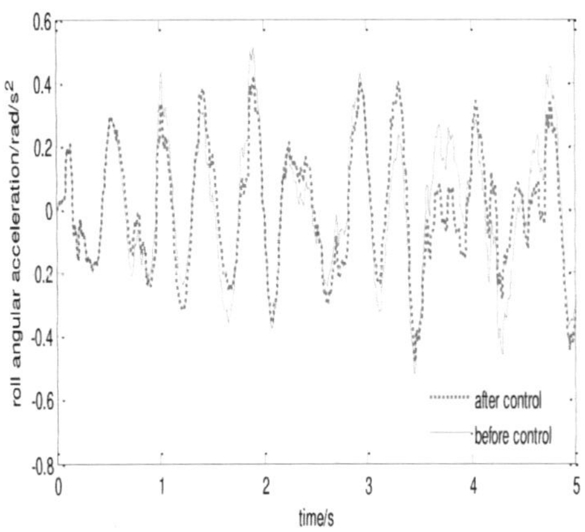

Figure 5. Response of roll acceleration

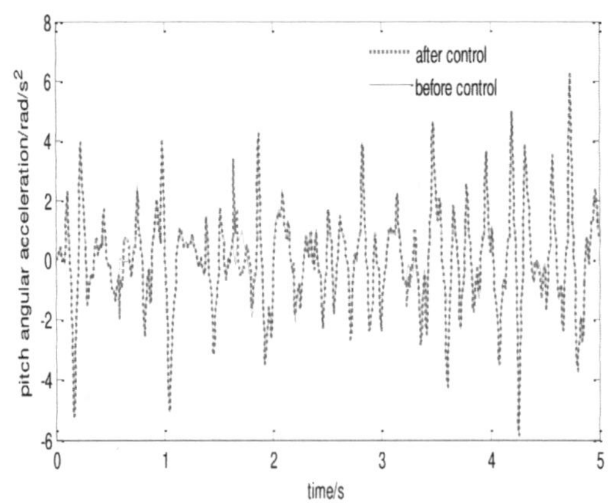

Figure 6. These two figures have been placed side-by-side to save space. Justify the caption

Table 3. Effect of the controlled method

	Performance improved of method two(%)
Body vertical acceleration	14.71
Body vertical pitch acceleration	16.1
Body vertical roll acceleration	26.2
Dynamic travel of suspension	3.4
Dynamic deformation of tire	4.8

5. Simulation Analogue of Experiment

The electronically controlled air suspension is simulated, and the control unit is shown in figure 7. The electronic control air suspension system is executed with ECU for overall control. ECU receives the electric signal detected by the sensor and sends instructions to the solenoid valve or stepper motor. Through the solenoid valve, the air spring is inflated and released, and the main and secondary air chamber opening and stiffness are controlled by the stepper motor. Due to the limited experimental conditions, the existing MC9SDP256MPVE and some peripheral equipment were used for simulation experiments. The experimental equipment is included as: MC9SDP256 development board was used to simulate and control solenoid valve through the control relay, and dc motor was used for the control of stepping motor. The overall structure of the simulation experiment principle is shown in figure 8.

Figure 7. Response of vertical acceleration

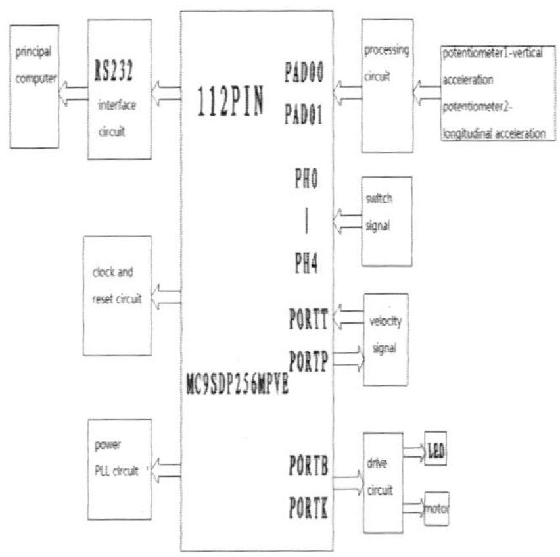

Figure 8. Response of pitch acceleration

6. Conclusion

The fuzzy controller for vertical movement, pitch movement and the roll movement is obtained respectively in the paper. The fuzzy controllers of different motions were established respectively, and the vertical acceleration of vehicle body was improved to 14.71% degree, the dynamic travel of suspension was improved to 3.4% degree, and the dynamic deformation of tire was improved to 4.8% degree. The comfort and control stability of the system were improved to some extent. Finally, the simulation environment of electronically controlled air suspension is constructed. The control program is programmed with language C and Codewarrior IDE is used to program, compile and download the programs online of the experiment.

7. Acknowledgments

Thanks for the found of NSFC (51505071) and the support of Northeastern University.

8. References

[1] Quanmin G, Haowen Z, Yan W 2019 Vehicle Coordination Strategy for Semi-active Suspension Vehicles, *Journal of System Simulation.*

[2] Eski I, Yildirim S. 2009 Vibration control of vehicle active suspension system using a new robust neural network control system. *Simulation Modelling Practice and Theory*, vol 17 pp 778-793.
[3] HUANG Z G 2005 Control system simulation and test research for air suspension. *Jinlin University*.
[4] GUCLU R. 2005 Fuzzy Logic Control of Seat Vibrations of a Non-Linear Full Vehicle Model. *Nonlinear Dynamics*, vol 40 pp 21-34.
[5] Xin G, Wenfeng C, Xin J, Lizeng Z 2017 Research on longitudinal control correction and switching method of intelligent vehicle, Automotive engineering, vol 39 pp 1047-1051.

IOP Conf. Series: Materials Science and Engineering **746** (2020) 012042 doi:10.1088/1757-899X/746/1/012042

Based on the Shipping Prosperity and Ship Renewal Decision

Jiale Xiang[1, a], Xuezeng Gan[2, b] and Honggang Zhang[3]

[1]Marine College of Zhejiang Ocean University, Zhoushan 316000, P. R. China
[2]Marine College of Zhejiang Ocean University, Zhoushan 316000, P. R. China
[3]Marine College of Zhejiang Ocean University, Zhoushan 316000, P. R. China
[a]Email:18868005626@163.com,[b]Email:gxz0310@sina.com,
[c]Email:332537068@qq.com

Abstract. As the carrier of the shipping industry, the carrying capacity of the ship is determined by the state of the ship in a large extent, and affects the amount of related costs incurred in the operation of the ship. The timely renewal of the ship can not only keep the ship in good operating condition, but also guarantee the minimum transportation capacity of the shipping enterprise. This paper will study and analyze the relationship between shipping prosperity and ship renewal, to find the best ship renewal opportunity, provide the best ship renewal scheme for ship operation enterprises, to ensure that ships operate in the optimal state, and at the same time, ensure the continuous renewal of shipping capacity and create the greatest economic benefits for shipping enterprises.

1. Introduction

The transportation capacity of the shipping company to a large extent is determined by the state of the ship which is one of the important decisive factors of the operation state of the shipping enterprise. At the same time, to a certain extent, the economic returns of a shipping enterprise is influenced by the state of the ship. As a means of transportation, there must be many common problems, such as the aging of equipment, the loss of hull, the backward function and so on. How to use it properly to analyze the shipping market trend of the whole shipping industry, so as to choose the right time to update the ship is particularly important. This paper will analyze and study the prosperity of the whole shipping market, and find the relationship between the prosperity of shipping market and the factors that affect the renewal of ships. At the same time, the trend of shipping prosperity and ship renewal are combined to determine the best ship renewal opportunity through shipping prosperity, and the decision of ship renewal is obtained, to create the greatest economic benefits for shipping enterprises.

2. Analysis and Research on Shipping Prosperity

The analysis of shipping prosperity mainly focuses on the analysis of the shipping market trend. Monitoring the current situation of the shipping market and forecasting the overall trend of the future shipping market are obtained through the analysis of the overall trend of the shipping market. Through the systematic analysis of the shipping market, it is realized by updating the fleet reasonably to maintain the strong competitiveness of shipping enterprises in the current and future shipping market.

2.1. Model Analysis

The fluctuation of shipping prosperity is mainly reflected in the influence of shipping market fluctuation on shipping prosperity. The analysis of shipping prosperity can be done by monitoring the volatility of shipping market. The dynamic model of monitoring shipping market is established. BDI, CCFI, CSPI, HRCI, Clarkson index and other shipping market related indexes are selected. The relevant data are analyzed by time difference analysis method and K-L information quantity method to find out the correlation and sensitivity. The fluctuation of China's shipping market can be reflected by CCFI and CSPI, but the monitoring effect of global shipping market is not ideal. Clarkson index is a response to the trend of shipbuilding market. HRCI is the container ship charter price index. Although the container transportation volume is on the rise in recent years, the market share of container transportation is still far lower than that of dry bulk transportation. Therefore, BDI index is selected as the barometer of shipping market as the benchmark index of this paper. By analyzing the trend of BDI index, shipping market is divided into three different states: cold, warm and hot. According to the boom interval theory, when the critical value of BDI index is less than 1000, the shipping market appears in the "cold" range; when the critical value of BDI index is 1000-3000, the shipping market appears in the "warm" range as a whole; when the critical value of BDI index is more than 3000, the shipping market appears in the "hot" range. The dynamic analysis model is as follows:

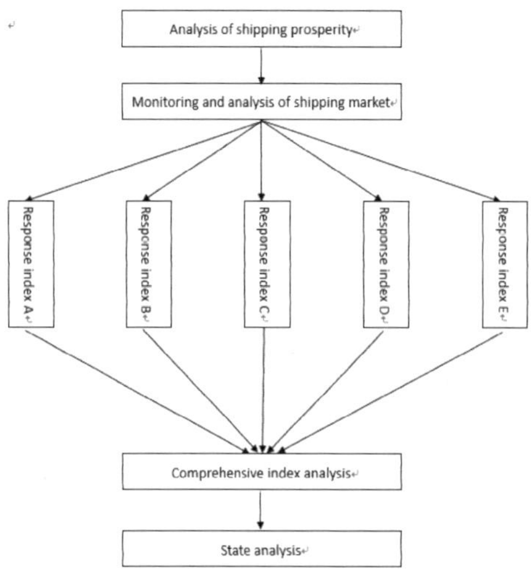

Figure 1. Dynamic analysis model of shipping prosperity

Through the monitoring and analysis of A, B, C, D, E and other reaction indexes in the model, the monitoring and analysis results are put into the model. After the analysis of the reaction index, each reaction index is combined into a new comprehensive index. The comprehensive index has the basic characteristics of all response indexes and is the comprehensive embodiment of all response indexes. Through the analysis of comprehensive index, we can monitor the status of shipping market.

In the correlation analysis, the paper analyzes the correlation by using the BDI index instead of the reaction index. The correlation analysis, the calculation of the correlation index should be mainly focused. The correlation index directly reflects the correlation between the response index and the fluctuation of the shipping market. The benchmark index of correlation is determined. The benchmark indicators for correlation are determined during the analysis process, and the benchmark indicators are used as the leading indicators to measure the correlation.

K-L information method is to calculate the distribution of the difference between the relevant indicators and the benchmark indicators to reflect the impact of the response indicators on the model. During the calculation, the difference between the correlation index data and the benchmark index data is constantly changed. When the K-L information is the smallest, the correlation index data and the final index data are obtained. Move the selected index data one month before and after the benchmark index, and observe the change of K-L information difference. When the amount of K-L information is the smallest, it shows that the response index can best reflect the real degree of the model.

After analyzing the BDI index, the shipping market is divided into three situations: cold, warm and hot. The index which reflects the state of shipping market is affected by various complex factors and has great uncertainty. X-12arma method is used to modify the index time and dynamic, and the time interval with large fluctuation is adjusted to stable time. Then the BDI index is integrated into the analysis of this time to determine the zero point of different interval of BDI index. The regional division and distribution map is as follows:

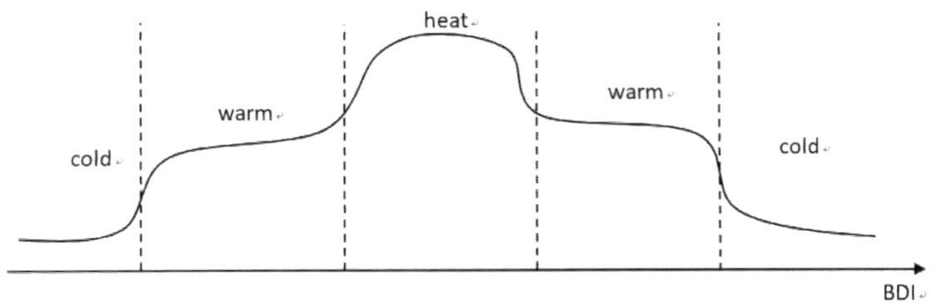

Figure 2. Regional distribution of shipping market

2.2. Systems Analysis
Shipping market is one of the most complex and changeable markets, which has many characteristics, such as wide range, large amount, difficult to determine property disputes and so on. It has complicated relations with financial market, national policies, international laws and regulations, total amount of market trade circulation, etc. Therefore, the shipping market is difficult to be monitored. The indexes which can reflect the shipping market state are analyzes in the paper,that is, the reaction indexes A, B, C, D and E in the above model. Through statistics related knowledge, taking into account the impact of related indexes, more than 100 indexes affecting the shipping industry were screened, and the most suitable correlation index was selected for monitoring and analysis. Among them, the indexes that have great influence on the shipping market are BDI, CCFI, CSPI, HRCI, Clarkson index, etc. In order to directly reflect the impact of the response index on the shipping market, BDI index, known as the barometer of the shipping market, is selected to represent the response index. Through the analysis of BDI index, the trend of the shipping market is monitored. BDI index in the first half of 2019 is shown in Figure 3:

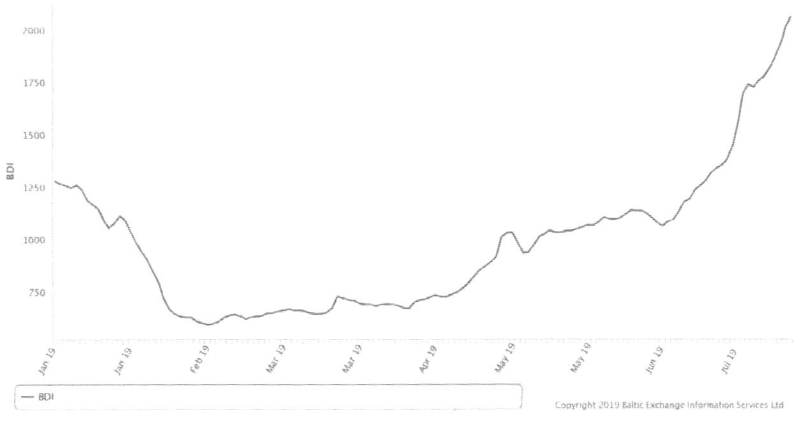

Figure 3. BDI index in the first half of 2019

One year in 2011, the BDI index experienced an unprecedented shock. In one year, the highest point of the BDI index was 11459, and the lowest point was 663. In such a short period of time, the highest point and lowest point successively appeared, which is unprecedented since the BDI index was released. And in the future, BDI index is still falling. Although it has picked up occasionally, there is no higher index point again. Between 2014 and 2018, the BDI index fluctuated back and forth between 200 and 1700, with the highest not exceeding 1700. It can be seen that the prosperity of the shipping market is relatively low, and the overall state of the shipping market is not good. But from April to

September 2019, the BDI index is in an upward trend, with only a small fall back in July. It can be seen that since 2019, the recovery of shipping prosperity is obvious. The overall status of the shipping market is good.

In this uncertain situation, the attention of enterprises to the BDI index of shipping market is self-evident. The trend of BDI affects the trust of shipping enterprises to the whole shipping market to a great extent. When the BDI index fluctuates greatly, corresponding countermeasure are likely to be made by the shipping enterprise according to the current shipping market. One of the biggest aspects is to update the ships of the enterprise fleet so that the company's transport capacity can better adapt to the current shipping market. There are three situations:

(1) When the BDI index continues to decline, it indicates that the current shipping market is not stable, and large fluctuations in the shipping market were led various uncertain factors at any time. Shipping enterprises do not have strong confidence in the market, and under this influence, various indexes will also fall significantly. At this time, the shipping market is the most complex.

(2) When the BDI index fluctuates in a small range in a period of time, it indicates that the current shipping market is relatively stable and will not fluctuate greatly in a short period of time. The shipping market will not fall or rise in a short period of time can be predict by shipping companies.

(3) If the shipping enterprises monitor that the BDI index rises continuously in a period of time, and there is a trend of continuous rise, it indicates that the current shipping prosperity is in an upward trend, the whole shipping market is in good condition, and the operating income of the shipping enterprises will have a significant growth trend.

3. Analysis of the Factors Affecting Ship Renewal

Ship renewal is mainly affected by the internal decision-making and external shipping market environment. For the internal decision-making factors, different decisions and different countermeasures can be made by shipping companies. However, for the changes of external shipping market environment, shipping companies need to strengthen monitoring and make timely countermeasures for the changes of different shipping markets.

3.1. The Influence of Ship's Economic Life on Ship Renewal

Determining the economic life of a ship indirectly determines the economic benefits that a ship from the end of its construction to its abandonment can be create by a ship. The competitiveness of shipping enterprises can be greatly increase by the good ship renewal plan and keep the fleet's capacity at a high level.

As a kind of transportation carrier, ship will inevitably cause wear and tear and consumption in the process of operation, and the value of ship itself will be reduced accordingly. This process is called depreciation of ships. Since the ship has wear and tear and consumption, it is necessary to maintain it. When the basic operation process can't be completed by the ship after maintenance although the operation can be complete by it, but the benefits it creates are lower than the cost needed for maintenance, the maintenance of the ship is completely unnecessary. Therefore, it is very important to determine the economic life of ships and then determine the timing of ship renewal.

3.2. The Influence of Shipping Enterprises' Development Focus on Ship Renewal

The investment of shipping enterprise is a kind of investment with high risk. The company's ship purchase and sale, ship leasing have a huge circulation of capital flow, and there is a huge investment risk in this process. Therefore, the enterprise recognizes the risks existing in the operation process of the enterprise itself, and makes correct preventive measures, so as to be able to respond stably when the risks occur. Among them, it is particularly important for shipping companies to find the right development focus.

The Great Depression of the global economy in 2008 was caused by the economic crisis of, and many industries were in great turbulence in this economic crisis. Shipping industry is no exception. BDI index fluctuates and is unstable before and after the economic crisis. The shipping industry is in the doldrums. At this time, in order to cope with the impact of the economic crisis, great adjustments are made to the company's internal by the shipping enterprises, and focus on one aspect of development, so as to have greater competitiveness in this aspect of the shipping market and reduce their own losses.

4. Shipping Prosperity and Ship Renewal Decision

It can be seen from the above discussion that the prosperity of shipping market can be divided into three trends, and the specific trend can be predicted through the trend analysis of BDI index.

(1) When the overall trend of BDI index is low, it can be predicted that the current shipping market is in a depressed state. At this time, due to the instability of the shipping market, the transportation capacity and market demand will be unbalanced. The transportation capacity invested by shipping enterprises is larger than the market demand, resulting in excess tonnage. It is necessary to consider whether the ship needs to be sealed up and stop the operation of the ship.

(2) When the BDI index fluctuates up and down in a small range, it indicates that the current shipping market is in a relatively stable state. If there is no serious impact factor, the shipping boom should not have a large fluctuation in the future, and there will be no rapid decline and rise. At this time, shipping enterprises should monitor the trend of shipping market, strengthen the maintenance of the ship, and make the ship in the best seaworthy condition. To deal with the fluctuation of shipping market at any time in the best condition.

(3) When the BDI index continues to rise for a period of time, and there is a trend of continuing to rise, shipping enterprises can appropriately increase the company's transportation capacity, expand the company's shipbuilding, or according to their own business conditions, when chartering, they can appropriately increase the ship charter period. We will appropriately increase the number of ships manufactured and expand the shipping capacity of enterprises.

5. Conclusion

In recent years, the global shipping market continues to decline, and many shipping companies have been holding a negative attitude towards the continuous decline of the shipping market, but in recent years, the shipping market has a slow recovery. In the 11 years since the financial storm in 2008, the shipping market is slowly recovering from the impact of the financial storm. The future shipping market is still very broad. Shipping companies may need to strengthen ship renewal when the shipping market is about to recover, so as to deal with various unstable factors smoothly in the future complex shipping market. At the same time, it can occupy its own advantages in the increasingly enhanced market competition, expand its own transport capacity, stand out in the competition, and create the maximum operating efficiency for the company.

6. References

[1] Wan Zhengtian. Research and application of ship renewal based on green concept. Master Thesis of Wuhan University of Technology 2011.
[2] Zhang Jiahui, He hongdi, Multifractal research on BDI index and CBFI index before and after financial crisis. Computer engineering and application 2019. 55.
[3] Yan Jiaming, Economic analysis and Countermeasures of large container ships in the economic downturn. Master Thesis of Shanghai Jiaotong University 2012.
[4] Qing Qian. Research on the prosperity index of global dry bulk shipping market. Master Thesis of Shanghai Jiaotong University 2012.
[5] Zhou Dequan, Zhen Hong. Development trend of China's shipping industry based on China's shipping prosperity index. Water transport management March 2010.
[6] Zhou Dequan, Zhen Hong, Yue Meilong. China Shipping Prosperity Monitoring and early warning system. System engineering July 2015.

IOP Conf. Series: Materials Science and Engineering **746** (2020) 012043 doi:10.1088/1757-899X/746/1/012043

Bi-directional Evolutionary Structural Optimization of Continuum Structures with Multiple Constraints

Yunkai Gao[1], Chao Ma[1], Bo Feng[1], Linli Tian[2, 3]

[1]School of Automotive Studies, Tongji University, Shanghai 201804, China
[2]Hubei Key Laboratory of Advanced Technology for Automotive Components, Wuhan University of Technology, Wuhan 430070, China
[3]Hubei Collaborative Innovation Center for Automotive Components Technology, Wuhan University of Technology, Wuhan 430070, China
Email: gaoyunkai@tongji.edu.cn, machaomit@163.com

Abstract. Most of the structural topology optimization problems involved in practical engineering applications are problems with many constraints. Thus, this paper presented an improved bi-directional evolutionary structural optimization (BESO) method for problems with many constraints. Slack variables were introduced to transform the inequality constraints into equality constraints. Then the Lagrange multiplier method converted the multi-constrained optimization problem to an unconstrained optimization problem and an advanced multiplier calculation was proposed. Elemental sensitivity numbers were derived according to the obtained multipliers. The design variables were updated by the BESO method. One numerical example that aimed to minimizing the structural mean compliance and involved volume, fundamental frequency and displacement constraints was used to validate the proposed method. The new method extends the BESO method to topology optimization that has a number of constraints, such as volume, displacement and frequency constraints.

1. Introduction

The evolutionary structural optimization (ESO) is a topology optimization method proposed by Xie and Steven et al. [1, 2]. ESO gradually removes inefficient materials from the structure and evolves the structure to the optimal material distribution based on certain criteria. ESO is easy to be implemented and computationally inexpensive. However, ESO has problems such as easily trapped in local minima, got a mesh-dependency and checkerboard design [3-6]. To overcome these shortcomings, ESO has been extended to bi-directional evolutionary structural optimization (BESO) by Querin et.al [7-9]. BESO not only removes inefficient material but also recovers the useful material to the solid with its material addition and removal scheme. Compared with conventional topology optimization methods such as homogenization approach and density approach, BESO has no intermediate density elements [10]. Therefore, a black-and-white topology can be obtained. Programs for BESO can be easily developed by the programming languages. Those programs can solve problems in a wide range of disciplines by interfacing the commercial finite element (FE) codes [11-13]. Therefore, BESO has a great value to explore.

Early BESO method can only deal with problems have a single constraint, and most often it is a volume constraint which is easier to control [14]. Huang et al. solved the compliance minimization problem has volume constraint and local displacement constraint using the Lagrange multiplier method [15]. Wang et.al solved the compliance minimization involving volume constraint and stress constraint using a global stress measure based on aggregation function [16]. However, to our best knowledge, few studies are focused on problems involving multiple types of constraints though most of the real engineering design problems involving multiple types of constraints besides volume constraints, such as displacement constraints, frequency constraints. In this study, an improved BESO method is proposed based on Huang's work [17]. Besides the volume constraint, the displacement and natural frequency constraints are also included. To construct the Lagrange function, first, slack variables are introduced to transform the inequality constraints to equality constraints, and then the Lagrange multipliers are introduced to transform the multi-constrained optimization problem into the unconstrained optimization problem. In order to simplify the derivation of Lagrange multipliers, an improved multiplier search scheme is proposed. This method enables the BESO method to handle multi-constrained optimization problems. Finally, the effectiveness of the proposed method is verified by one numerical example with several constraint combinations.

2. Lagrange Function

2.1. Lagrange Relaxation

Lagrange multipliers method is usually used for multi-constrained problems. Not lose generality, a multi-constrained problems can be formulated as:

$$\begin{aligned} &\text{Find} \quad \boldsymbol{X} \\ &\min \quad f(\boldsymbol{X}) \\ &s.t. \quad h_j(\boldsymbol{X}) = 0 \quad j = 1,2,\cdots,l \\ &\qquad\;\; g_k(\boldsymbol{X}) \le 0 \quad k = 1,2,\cdots,m \end{aligned} \tag{1}$$

Where $f(\boldsymbol{X})$ is the objective function, $h(\boldsymbol{X})$ is the equality constraint, $g(\boldsymbol{X})$ is the inequality constraint, $\boldsymbol{X}$ represents the design variable vector and can be expressed by $\boldsymbol{X}=\{x_1, x_2, \ldots, x_n\}^T$, l and m represents the number of equality and inequality constraints, respectively.

To formulate the Lagrange function, a slack variable S_k is introduced to transform an inequality constraint into an equality constraint:

$$g_k(\boldsymbol{X}) + S_k^2 = 0 \tag{2}$$

Thus, the Lagrange function of a multi-constrained optimization problem is obtained.

$$L(\boldsymbol{X}, \boldsymbol{\lambda}, \boldsymbol{\mu}, \boldsymbol{S}^2) = f(\boldsymbol{X}) + \sum_{j=1}^{l} \lambda_j h_j(\boldsymbol{X}) + \sum_{k=1}^{m} \mu_k (g_k(\boldsymbol{X}) + S_k^2) \tag{3}$$

Where λ_j, μ_k are Lagrange multipliers for the equality and inequality constraints. Thus, the multi-constrained optimization problem formulated in Eq. (1) is transformed into the minimization of $L(\boldsymbol{X}, \boldsymbol{\lambda}, \boldsymbol{\mu}, \boldsymbol{S}^2)$. The Eq. (3) is equivalent to $f(\boldsymbol{X})$ in Eq. (1) if all constraints are satisfied. $\boldsymbol{\lambda}$, $\boldsymbol{\mu}$ and $\boldsymbol{S}^2$ can be treated as additional variables. Therefore, the derivatives of Lagrange function with respect to these variables are determined as the following additional derivatives.

$$\begin{cases} \dfrac{\partial L}{\partial x_1} = 0, \quad \dfrac{\partial L}{\partial x_2} = 0, \quad \cdots \\ \vdots \\ \dfrac{\partial L}{\partial \lambda_j} = h_j(X) = 0 \\ \vdots \\ \dfrac{\partial L}{\partial \mu_m} = g_m(X) + S_m^2 = 0 \\ \dfrac{\partial L}{\partial S_m} = 2\mu_m S_m \end{cases} \tag{4}$$

A solution could be reached based on Eq. (4). However, it is difficult to demonstrate this solution is the global optimum without the Hessian matrix. It is impractical to calculate the Hessian matrix since large amounts of variables in the design domain.

Nevertheless, a heuristic estimation can be made based on Eq. (4) at least. For problems that minimize the objective functions. If the derivative of Lagrange function with respect to λ_j is greater than zero. λ_j should be reduced so as to reduce the Lagrange function in the next iteration, and vice versa. If the derivative of Lagrange function with respect to λ_j is equal to zero, the *j*-th constraint is satisfied and λ_j should remain unchanged.

For the inequality conditions, if the derivative of Lagrange function with respect to μ_k is greater than zero. That means $g_k(\boldsymbol{X}) + S_k^2$ is positive. The multiplier μ_k should be reduced to minimize the Lagrange function. If S_k is not equal to zero, μ_k needs to be zero to satisfy the *k*-th inequality constraint.

2.2. *Advanced Search Scheme for the Lagrange Multipliers*

Note that, the Lagrange multipliers are continuous in the infinite domain. It is infeasible to find a solution of these additional variables. Thus λ_j and μ_k are redefined by a scaling function of replacement factors γ_j and φ_k that belong to a finite domain [0, 1).

$$\lambda_j = \frac{\gamma_j}{1-\gamma_j} \tag{5}$$

$$\mu_k = \frac{\varphi_k}{1-\varphi_k} \tag{6}$$

If the corresponding constraint is satisfied, the Lagrange multiplier needs to be zero which can be realized by making replacement factor equal to zero. The Lagrange multipliers are increased as the replacement factors increased. As the replacement factors approach one, the Lagrange multipliers would be infinity. Therefore, this scaling function makes the determination of Lagrange multipliers easy to be implemented.

Sensitivity numbers are sensitive to the redistribution of material. To avoid a large fluctuation of the sensitivity numbers, the replacement factors are increased or reduced step by step. In addition, the increment of the replacement factor $\Delta\gamma$ or $\Delta\varphi$ (for simplicity, $\Delta\gamma$ and $\Delta\varphi$ are denoted by $\Delta\varphi$ in the rest part of this paper) is determined by following steps:

Step 1. If a constraint is inactive, the replacement factor should be reduced, and vice versa. To avoid the numerical problems, the replacement factor should not be greater than 0.999. If the response is less than 0.9 times that of the specific constraint value R^*, the replacement factor subtracts $\Delta\varphi$. If the response is greater than 1.1 times that of R^*, the replacement factor adds $\Delta\varphi$. $\Delta\varphi$ is selected in [0.01, 0.02].

Step 2. If the response is greater than 0.9 times that of R^* and is less than 1.1 times that of R^*. The increment $\Delta\varphi$ is determined by following function:

$$\Delta\varphi = sign(R - R^*)\cdot 100\tau_m \cdot \left|\frac{R-R^*}{R^*}\right|^2 \tag{7}$$

Step 3. Particular responses of the constraints could approach or exceed the imposed constraint values as the volume fraction decreased. To make sure these responses returned to the specific limit in the next several iterations, the corresponding replacement factors are initialized as a small value 0.1 when these responses are greater than 0.95 times that of R^* for the first time. This procedure is illustrated in Figure 1.

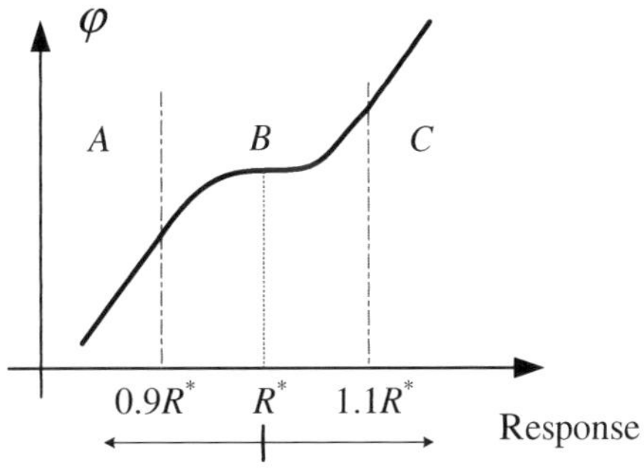

Figure 1. Ranges of the replacement factor

3. Topology Optimization Formulation

This study considers compliance minimization with the volume constraint, displacement constraint and natural frequency constraint. Thus, the topology optimization problem for BESO can be formulated as:

$$\begin{aligned} \min:\ & C \\ s.t.\ & \sum x_i v_i = V^* \\ & u \le u_* \\ & \omega \ge \omega_* \end{aligned} \tag{8}$$

Where C denotes the compliance of the structure, x_i is the topology design variable assigned with the finite element i. If x_i equals one, element i is presence in the FE model for next iteration. If x_i equals to a small value, e.g. 0.001, element i is absence in the FE model for next iteration. v_i represents the volume of element i. u and ω represent the displacement of the node and the natural frequency of the structure. V^*, u_*, ω_* represent the prescribed volume, displacement and natural frequency.

4. Sensitivity Analysis

4.1. Sensitivity of Lagrange Multipliers

On the basis of Eq. (3) and procedures in [17, 18], the total sensitivity of the Lagrange function can be expressed by:

$$\alpha_i = -\frac{\partial L}{\partial x_i} = \alpha_{i,C} + \sum_{j=1}^{l} \lambda_j \alpha_{i,u} + \sum_{k=1}^{m} \mu_k \alpha_{i,\omega} \tag{9}$$

Where $\alpha_{i,C}$ and$\alpha_{i,\omega}$ represents the sensitivity for compliance and the sensitivity for natural frequency, respectively. The derivation of sensitivity was proposed in [18, 19]. $\alpha_{i,u}$ represents the sensitivity for displacement and the derivation is proposed in Section 4.2.

4.2. Displacement Sensitivity

It is assumes that the arbitrary node j in the structure subject to a specific displacement u_j. The derivative of this displacement with respect to the design variable x_i is expressed as du_j/dx_i:

A virtual unit load f_j is introduced in which only the corresponding j-th component is equal to one. The other components are equal to zero. Thus, the specific displacement u_j can be expressed as:

$$u_j = \boldsymbol{f}_j^T \boldsymbol{u} \tag{10}$$

The equilibrium equation of the structure subjected to this load can be expressed as:

$$\boldsymbol{f}_j = \boldsymbol{K}\boldsymbol{u}_* \tag{11}$$

IOP Conf. Series: Materials Science and Engineering **746** (2020) 012043 doi:10.1088/1757-899X/746/1/012043

Where $\boldsymbol{K}$ is the global stiffness matrix of the structure. $\boldsymbol{u}^*$ is the nodal displacement vector of the structure. The equilibrium equation of the structure subjected to a real load $\boldsymbol{F}$ is expressed as:

$$\boldsymbol{F}=\boldsymbol{K}\boldsymbol{u} \tag{12}$$

The differentiation on both sides of Eq. (12) with respect to arbitrary variable x_i can be expressed as:

$$\frac{\partial \boldsymbol{K}}{\partial x_i}\boldsymbol{u}+\boldsymbol{K}\frac{\partial \boldsymbol{u}}{\partial x_i}-\frac{\partial \boldsymbol{F}}{\partial x_i}=0 \tag{13}$$

It is assumed that the real load is independent from the design variables, and the Eq. (13) can be simplified as:

$$\frac{\partial \boldsymbol{u}}{\partial x_i}=-\boldsymbol{K}^{-1}\frac{\partial \boldsymbol{K}}{\partial x_i}\boldsymbol{u} \tag{14}$$

The differentiation on both sides of Eq. (10) can be expressed as:

$$\frac{du_j}{dx_i}=\boldsymbol{f}_j^T\frac{\partial \boldsymbol{u}}{\partial x_i}=\boldsymbol{u}_*^T\boldsymbol{K}^T\frac{\partial \boldsymbol{u}}{\partial x_i} \tag{15}$$

Substituting Eq. (14) into Eq. (15), the derivative with respect to x_i can be expressed as:

$$\frac{du_j}{dx_i}=-\boldsymbol{u}_*^T\frac{\partial \boldsymbol{K}}{\partial x_i}\boldsymbol{u} \tag{16}$$

To avoid the re-meshing and singularity of the global stiffness matrix, a solid isotropic material with penalization (SIMP) like material interpolation scheme is introduced:

$$\rho_i=x_i\rho_0 \quad , \quad \boldsymbol{D}_i=x_i^p\boldsymbol{D}_0 \tag{17}$$

Where ρ_i represents the material density of the arbitrary element i, ρ_0 represents the material density of the solid material. $\boldsymbol{D}_i$ and $\boldsymbol{D}_0$ represent the constitutive matrix of arbitrary element i and the solid material. p represents the penalty exponent and equals to three in this study. The stiffness matrix of element i can be expressed as:

$$\boldsymbol{k}_i=\iiint_\Omega \boldsymbol{B}_i^T\boldsymbol{D}_i\boldsymbol{B}_i d\Omega=px_i^{p-1}\iiint_\Omega \boldsymbol{B}_i^T\boldsymbol{D}_0\boldsymbol{B}_i d\Omega=px_i^{p-1}\boldsymbol{k}_{i,(0)} \tag{18}$$

Where $\boldsymbol{k}_i$ denotes the elemental stiffness matrix, $\boldsymbol{k}_{i,(0)}$ is the stiffness matrix of solid element. $\boldsymbol{B}_i$ relates the displacement and strain. The derivative of global stiffness matrix with respect to x_i is only determined by the stiffness matrix of element i. Therefore, Eq. (16) is simplified as:

$$\frac{du_j}{dx_i}=-\boldsymbol{u}_*^T\boldsymbol{k}_i\boldsymbol{u}=-px_i^{p-1}\boldsymbol{u}_*^T\boldsymbol{k}_{i,(0)}\boldsymbol{u}_i \tag{19}$$

Thus, the sensitivity for the displacement is expressed as:

$$\alpha_{i,u}=-\frac{du_j}{dx_i}=px_i^{p-1}\boldsymbol{u}_*^T\boldsymbol{k}_{i,(0)}\boldsymbol{u}_i \tag{20}$$

5. Numerical Implementation

BESO is an easily implemented optimization method. This study proposes a program interfaces the commercial FE code ABAQUS. The implementation of BESO can be summed up as follows:

Step 1. Discretize the design domain with a FE model.

Step 2. Define optimization parameters, such as evolutionary ratio *ER*, sensitivity filter radius r_{min}, maximal admission ratio AR_{max} and the penalty exponent *p*.

Step 3. Perform FE analysis, calculate the sensitivity number based on Eq. (9).

Step 4. The procedures proposed in [10] are introduced to stabilize the optimization process.

Step 5. Determine the volume of the structure for next iteration first and update the design variables based on their sensitivity number. A 'soft-kill' technique is utilized in this study. It assigns the element with a soft material if the corresponding design variable is equal to x_{min}.

Step 6. Repeated the Steps 3~5 until the convergence criteria is satisfied. The convergence criteria is the same as that proposed in [17].

6. Numerical Example

The numerical example is illustrated in Figure 2. A plate simply supported at the lower corners and a vertical load is applied to the middle point of the top edge. The material Young's modulus is 1GPa, density is 8000kg/m^3 and Poisson's ratio is 0.3. The design domain is discretized with 200×100 four-node plane stress elements. The study aims at minimizing the mean compliance with the volume, displacement and natural frequency constraints. The final volume fraction is set to 0.3. The *x*-direction displacement of the lower right corner denoted as point d_A is set as displacement constraint. The first-order natural frequency of the structure is set as natural frequency constraint. The evolutionary volume ratio *ER* is 0.02, the maximal admission ratio AR_{max} is 0.01, the filter radius r_{min} is 1.5mm. The increment of replacement factor $\Delta\varphi$ is 0.015, the penalty exponent *p* is 3.

To demonstrate the effectiveness of the method proposed in this study. Four different optimization cases are design as shown in Table 1.

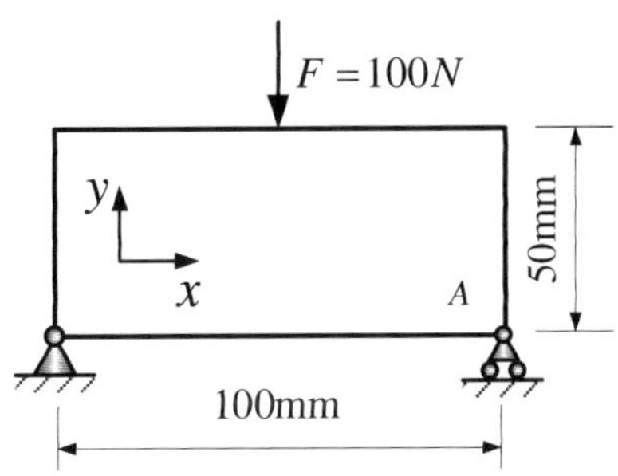

Figure 2. Dimensions and load conditions of example 1

Table 1. Constraint combinations of example 1

Cases	Constraints
Case 1	No constraints on d_A andω_1
Case 2	d_A<1.4mm, ω_1>240Hz
Case 3	d_A<1.3mm, ω_1>240Hz
Case 4	d_A<1.4mm, ω_1<160Hz

The optimal designs of the four cases are shown in Figure 3. The performances of the four cases are listed in Table 2. They are almost the same as those obtained in [17]. The optimal design of Case 1 is shown in Figure 3 (a). The distribution of the material is a symmetric configuration. The optimal designs of Case 2 and Case 3 have similar material distribution for they have the same natural frequency constraint and similar displacement constraint. It is noted that the optimal design of Case 2 is stiffer than that of Case 3 because of the bigger imposed displacement constraint value. The optimal design of Case 4 approximates a symmetric configuration.

Case 2 is solved by the existing method proposed in [17] and the corresponding results are list in the last row of Table 2. It is found that the errors between results of optimal designs in this study and the imposed constraint values are smaller than that obtained in [17] though it needs more iterations. Compliance of optimal design due to Case 2 in this study is lower than that obtained in [17]. The advanced Lagrange multipliers search scheme proposed in this study is robust than that proposed in [17] for the increment of the replacement factor is smaller when the response approaching the imposed constraint value. This robustness also stabilizes the optimization process.

(a) Optimal design of Case 1 (b) Optimal design of Case 2

(c) Optimal design of Case 3 (d) Optimal design of Case 4

Figure 3. Optimal designs of example 1

Table 2. Comparison of different Cases

Cases	Iteration number	C (Nmm)	d_A (mm)	ω_1 (Hz)
Case 1	54	102.76	1.61	236.6
Case 2	145	106.37	1.399	248.53
Case 3	98	111.23	1.296	251.23
Case 4	96	110.98	1.386	152.86
Case 2 [17]	78	113.13	1.346	257.4

7. Conclusions

This paper presented an improved topology optimization method for the multi-constrained topology optimization of continuum structures. The general inequality constraints were transformed into equality constraints using the slack variables. A Lagrange function was established by taking all constraints into the objective function. Thus, the multi-constrained optimization problem was converted to an unconstrained optimization problem. An advanced Lagrange multipliers search scheme was proposed to find a solution of the additional Lagrange multipliers. The design variables were updated by the BESO method.

A numerical example that minimizing mean compliance of the structure involving displacement and natural frequency constraints besides the volume constraint was solved by the proposed topology optimization method. The result shows that different constraint combinations result in different material distributions. Compared with the existing method, though the improved multiplier search scheme makes the optimization process converged with more iterative steps, the mean compliance is converged to a smaller value and all of the constraints are satisfied. The responses of optimal designs and the imposed constraint values are closer. The advanced Lagrange multipliers search scheme is robust which stabilize the optimization process.

8. Acknowledgments

This work was supported by the National Key Research and Development Program of China [grant number 2016YFB0101602].

IOP Conf. Series: Materials Science and Engineering **746** (2020) 012043 doi:10.1088/1757-899X/746/1/012043

9. References

[1] Y. Xie, G. P. Steven.A simple evolutionary procedure for structural optimization [J]. Computers& Structures, 1993, 49 (5): 885-896.

[2] Y. Xie, G. Steven. Evolutionary structural optimization for dynamic problems [J]. Computers & Structures, 1996, 58 (6): 1067-1073.

[3] O. Sigmund, J. Petersson.Numerical instabilities in topology optimization: A survey on procedures dealing with checkerboards, mesh-dependencies and local minima [J]. Structural optimization, 1998, 16 (1): 68-75.

[4] Liang Xia, Li Zhang, et al. Stress-based topology optimization using bi-directional evolutionary structural optimization method [J]. Computer Methods in Applied Mechanics and Engineering, 2018, 333: 356-370.

[5] T. E. Bruns, D.A. Tortorelli. An element removal and reintroduction strategy for the topology optimization of structures and compliant mechanisms [J]. Int. J. Numer. Methods Eng. 2003, 57: 1413–1430.

[6] Zhou M, Rozvany GIN. On the validity of ESO type methods in topology optimization [J]. Struct Multidisc Optim 2001, 21: 80–83.

[7] Querin OM, Steven GP, Xie YM. Evolutionary structural optimization (ESO) using a bi-directional algorithm [J]. Eng Comput, 1998, 15: 1031–1048.

[8] O.M. Querin, G.P. Steven, Y.M. Xie. Evolutionary structural optimisation using an additive algorithm [J]. Finite Elements in Analysis and Design, 2000, 34 (3): 291- 308.

[9] X. Y. Yang, Y.M. Xie, G.P. Steven, et al. Bidirectional evolutionary method for stiffness optimization [J]. AIAA J. 1999, 37 (11):1483–1488.

[10] X. Huang, Y.M. Xie. Convergent and mesh-independent solutions for the bi-directional evolutionary structural optimization method [J]. Finite Elements in Analysis and Design, 2007, 43 (14): 1039-1049.

[11] H. Hofmeyer, J.M. Davila.Automated design studies: Topology versus One-Step Evolutionary Structural Optimisation [J]. Advanced Engineering Informatics, 2013, 27 (4): 427-443.

[12] Hélio Luiz S., Valério S. et al.A smooth evolutionary structural optimization procedure applied to plane stress problem [J].Engineering Structures, 2014, 75: 248-258.

[13] Yunlong T., Aidan K., et al.Bidirectional Evolutionary Structural Optimization based design method for lattice structure to be fabricated by additive manufacturing [J]. Computer-Aided Design, 2015, 69:91-101.

[14] J. H Rong, Y.M Xie, X.Y Yang.An improved method for evolutionary structural optimisation against buckling [J]. Computers & Structures, 2001, 79 (3): 253-263.

[15] X. Huang, Y. M. Xie. Evolutionary topology optimization of continuum structures with an additional displacement constraint [J]. Structural and multidisciplinary optimization. 2010, 40: 409 -416.

[16] X. Wang, H. Liu, K. Long, et al. Stress-constrained topology optimization based on improved bi-directional evolutionary optimization method [J]. Chinese Journal of Theoretical and Applied Mechanics, 2018, 50 (02): 385-394.

[17] Z. H. Zuo, Y. M. Xie, X. Huang. Evolutionary topology optimization of structures with multiple displacement and frequency constraints [J]. Advances in Structural Engineering, 2012, 15 (2):359-372.

[18] X. Huang, Y. Xie. Bi-directional evolutionary topology optimization of continuum structures with one or multiple materials [J].Computational Mechanics, 2009, 43:393-401.

[19] X. Huang, Z. H. Zuo, Y. M. Xie. Evolutionary topological optimization of vibrating continuum structures for natural frequencies [J].Computers & Structures, 2010, 88 (5-6): 357- 364.

IOP Conf. Series: Materials Science and Engineering **746** (2020) 012044 doi:10.1088/1757-899X/746/1/012044

Design and Optimization of Symmetrical Pulsating Pressure Generator Based on Gas Medium

Jing Nie, Yinfeng Wu, Shiqi Xu, Ning Li, Lulu Zhang and Xiaofeng Meng

Science and Technology on Inertial Laboratory, Beihang University, Beijing 100191, China

Email: niejing@buaa.edu.cn

Abstract. In order to realize pulsating pressure in wide frequency and pressure range, this paper designs a symmetrical pulsating pressure generating device structure, which solves the problem of offset movement of the piston movement due to offset pressure and the problem of bias pressure retention. In addition, the main factors that limited the performance of the pulsating pressure generating device are analysed in detail, and the pulsating pressure generating demand in a wide frequency range is realized by segmentation processing.

1. Introduction

In recent years, the requirements of supersonic wind tunnel and other gas media, high frequency and high pressure dynamic pressure test with offset pressure have been increasing. Before measuring dynamic pressure, the pressure sensor first needs dynamic pressure calibration. Therefore, the research on dynamic calibration technology of such pressure sensors is of great significance [1-3].For dynamic pressure calibration, the key lies in how to generate a pressure signal with stability, good repeatability, accurate amplitude and time index of signal, and convenient traceability to the international standard system of units (SI).

From the signal form, the pressure source of dynamic pressure calibration can be divided into three types: pulse pressure signal, step pressure signal and continuous period pressure signal. Step pressure and pulse pressure are adjustable frequency range is relatively narrow, and continuous in the form of a periodic signal is more close to the actual measurement of dynamic pressure signal, and its energy concentration, output stability, deterministic signal is good, can carry on the frequency sweep to the pressure sensor calibration and other characteristics, can get more accurate frequency response of the sensor or system, is a kind of static calibration methods [4-6], become the focus of the dynamic pressure calibration direction. Sinusoidal excitation signals are commonly used in continuous period pressure calibration. Corresponding to pressure sensors, sinusoidal pressure signals containing offset pressure are pulsating pressure signals. The conduction medium of pulsating pressure can be divided into liquid medium and gas medium. Generally speaking, since the compressibility of liquid medium is far less than that of gas medium, the pressure generation mode of liquid medium has great advantages in generating high frequency and high pressure pulsating pressure signals. However, pressure sensors used in wind tunnel testing, engine performance testing and other occasions should not be calibrated with liquid medium. Therefore, it is very important to study the generating device that can generate high frequency and high pressure pulsating pressure in the gas medium.

2. Structural Design of Pulsating Signal Generator

As shown in Fig. 1(a), the whole device consists of three main parts: a cylinder, a vibrating table and a biasing pressure applying structure. The piston rod is rigidly connected with the table of the shaking table through the flange, so that the linear reciprocating vibration generated by the shaking table can directly drive the piston movement, making the volume of the air chamber on both sides of the piston in the cylinder change and generating pulsating pressure. The pressure sensor is sealed on the cylinder head by a specially designed mounting base to ensure that the additional volume brought by the installation of the sensor is as small as possible. A volume-matching unit is designed in the symmetrical position of the cylinder head to keep the pressure distribution on the end cover symmetrical. Similarly, the open valve of the offset pressure-applied design needs to be as close to the cylinder as possible to prevent the introduction of unnecessary initial cavity volume. In order to improve the piston movement origin deviation and gas leakage caused by offset pressure, a symmetrical gas pulsation pressure generator is designed and implemented in this paper. When the piston is at the original position of movement, both sides of the piston have the same offset pressure. Therefore, the force on the piston is balanced, which solves the problem of balance origin deviation. At the same time, due to the same pressure on both sides of the piston, the gas leakage problem has been improved. Finally, the schematic model of symmetrical gas pulsation pressure generator is obtained as shown in Fig. 1(b).

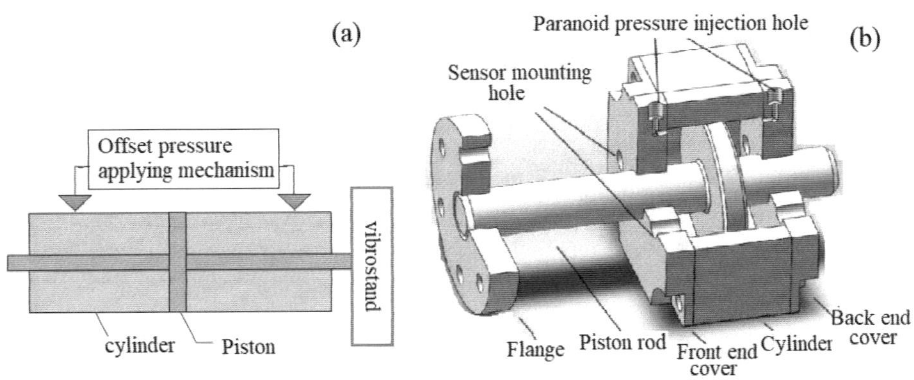

Figure 1. Device structure: (a) is structure block diagram; (b) is mechanical diagram

3. Optimization of Structural Parameters

In the symmetrical gas pulsation pressure generator, the initial offset pressure Ps, the initial volume V_{dz0} of the air chamber and the volume change V_{dz} caused by reciprocating vibration of the piston are the three key parameters that determine the peak-to-peak (P-P) value of the pulsation pressure. In this paper, according to the demand of pulsating pressure (Table 1), the vibration displacement of the vibration table will be analysed by referring to the selection of the vibration table, and the main parameters of the cylinder will be designed accordingly. Vibration table is the power source of piston vibration and one of the core components of pulsating pressure generator. The main technical indexes that can be achieved, such as maximum thrust, maximum acceleration and maximum displacement, directly affect the design of pulsating pressure generating device in all aspects, and are also the main factors restricting the amplitude and frequency of pulsating pressure. According to the requirements of pulsating pressure, the vibration table with the technical parameters as shown in table 2 is selected in this work.

Table 1. Gas pulsation pressure generation requirements

Pulsating pressure frequency: 5 Hz ~ 100 Hz		Pulsating pressure frequency: 100 Hz ~ 500 Hz	
Offset pressure (KPa)	P-P pressure (KPa)	Offset pressure (KPa)	P-P pressure (KPa)
30	5	30	0.5
100	10	100	1
450	20	450	2

Table 2. Main technical parameters of vibration table

Technical index	Parameters	Unit
Frequency range	5~5000	Hz
Rated sinusoidal thrust	2.94	kN
Maximum acceleration	980	m/s^2
Maximum speed	2	m/s
Maximum displacement	40	mm
Mass of moving parts	3	kg

The peak range of $L_{dZ}(f)$ of unilateral displacement of cylinder piston vibration at a certain frequency can be expressed as:

$$L_{dZ,\ min}(f) \le L_{dZ}(f) \le L_{dZ,\ max}(f) \tag{1}$$

Where, $L_{dZ,\ min}(f)$ is the minimum unilateral displacement of cylinder piston at a certain frequency under the condition of guaranteeing the quality of vibration waveform. It is limited by the nonlinear characteristics of the vibration table, the accuracy of the vibration measuring sensor and the ambient noise level. $L_{dZ,\ max}(f)$ is the maximum unilateral displacement of the cylinder piston at a certain frequency, which is limited by the maximum thrust of the vibration table.

$$l_{dZ}(t) = L_{dZ} \cdot \sin(2\pi \cdot f \cdot t) \tag{2}$$

$$v_{dZ}(t) = 2\pi \cdot f \cdot L_{dZ} \cdot \cos(2\pi \cdot f \cdot t) \tag{3}$$

$$a_{dZ}(t) = -(2\pi \cdot f)^2 \cdot L_{dZ} \cdot \sin(2\pi \cdot f \cdot t) \tag{4}$$

$$L_{dZ,\ max}(f) = \frac{F_{max}}{(2\cdot\pi\cdot f)^2\cdot M} \tag{5}$$

Where, F_{max} is the maximum thrust of the shaking table, M is the mass of moving parts (including load mass) of vibration table, f is the vibration frequency. The vibration velocity and acceleration are bounded due to the restriction of vibration table index, so the range of vibration amplitude at the specified frequency is limited. If the maximum thrust is 3000 N and the load is 3 kg, the mass of the moving part is 6 kg and the maximum acceleration is about 50 g. Within 1 Hz to 100 Hz, there is the maximum amplitude-frequency curve as shown in Figure 2(a), and within 100 Hz to 500 Hz, there is the maximum amplitude-frequency curve as shown in Figure 2(b). The maximum amplitude of the vibration table from 5 Hz to 500 Hz is constrained as shown in table 3. It can be seen that with the increase of frequency, the vibration table will work under the limitation of maximum displacement, maximum velocity and maximum acceleration successively.

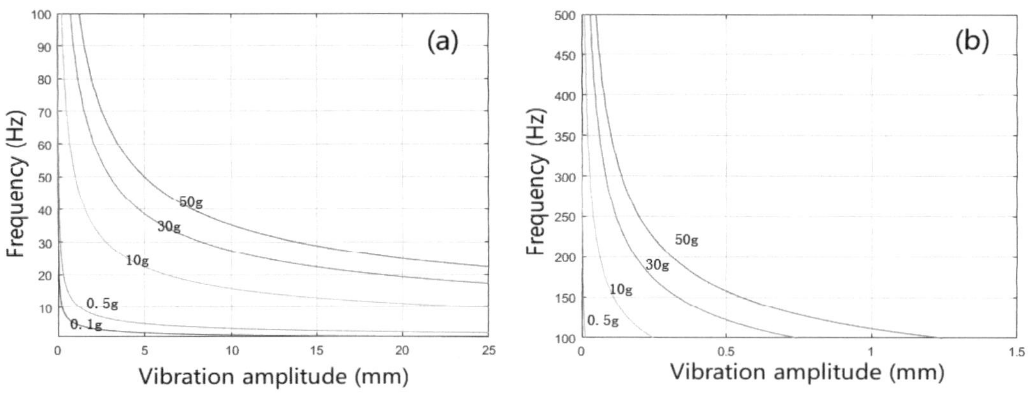

Figure 2. Maximum amplitude-frequency curve: (a) within 1 Hz to 100 Hz, (b) within 100 Hz to 500 Hz.

Table 3. Vibration table maximum amplitude constraints

Frequency (Hz)	Piston maximum unilateral amplitude (mm)	Maximum speed (m/s)	Maximum acceleration (g)
500	0.05	0.157	50 (upper limit)
100	1.25	0.785	50 (upper limit)
25	12.8	2 (upper limit)	32
5	20 (upper limit)	0.628	2

When the vibration frequency f=5 Hz, if the peak-to-peak pulsating pressure with a peak-to-peak value of 5 kPa is to be generated under the bias pressure 30 kPa, the static-static pressure ratio should be 0.083, and the unilateral equivalent length of the cylinder should not be shorter than 5/0.083≈60 mm. At 100 Hz, even when the vibration amplitude reaches the maximum, the equivalent length of the cylinder side shall not be longer than 1.5/0.083 ≈15 mm. Therefore, the symmetrical cylinder of a single length cannot meet the requirements of pulsating pressure from 5 Hz to 100 Hz. It is necessary to further segment the frequency range and design the size of the symmetrical pulsating pressure generating device respectively. Among them, the cylinder with 195 mm length on one side is required to complete the pulsating pressure from 5 Hz to 30 Hz. The cylinder with 15 mm length on one side is used to complete the pulsating pressure generation requirements from 20 Hz to 100 Hz. Similarly, at 500 Hz, in order to generate the peak-to-peak pulsating pressure of 500 Pa at 30 kPa, the ratio of station-static pressure should reach 0.83%. Then, the equivalent length of one side of the cylinder should not be longer than 0.05/ (0.83%) ≈6 mm. A cylinder with a unilateral length of 6 mm is selected to complete the pulsating pressure generation requirements from 100 Hz to 500 Hz.

The theoretical calculation results are respectively simulated and verified. Taking a cylinder with a unilateral length of 15 mm as an example, the pulsating pressure that can be generated within the frequency range of 20 Hz to 100 Hz is shown in Figure 3(a). Where, the black line box represents the demand for pulsating pressure at different frequencies. The red curve (the innermost part) represents the amplitude-frequency relationship of pulsating pressure generated by vibration of the vibration table when the maximum acceleration is 0.5 g. The blue curve (outermost) represents the amplitude-frequency relationship of pulsating pressure generated by vibration of the shaking table at a maximum acceleration of 50 g. It can be seen that under the bias pressure of 30 kPa and 100 Hz, the maximum output acceleration of the vibration table is 50 g, corresponding to the maximum unilateral amplitude of 1.25 mm. At this time, the positive peak value of the pressure is 2.7 kpa, the negative peak value is 2.3kpa, and the negative peak value is less than 5 kPa/2= 2.5 kpa. Therefore, it is necessary to shorten the effective length of the cylinder and increase the negative peak pressure.The unilateral length of the cylinder was corrected to be 13mm, and the simulation was conducted again. At this time, under the bias pressure of 30 kPa and 100 Hz, the maximum output acceleration of the vibration table is 50 g, corresponding to the maximum unilateral amplitude of 1.25 mm. At this time, the positive peak value of pressure is 3.2 kpa, the negative peak value is 2.6 kpa, and the negative peak value is greater than 5 kPa/2 = 2.5 kpa, which meets the preliminary requirements.

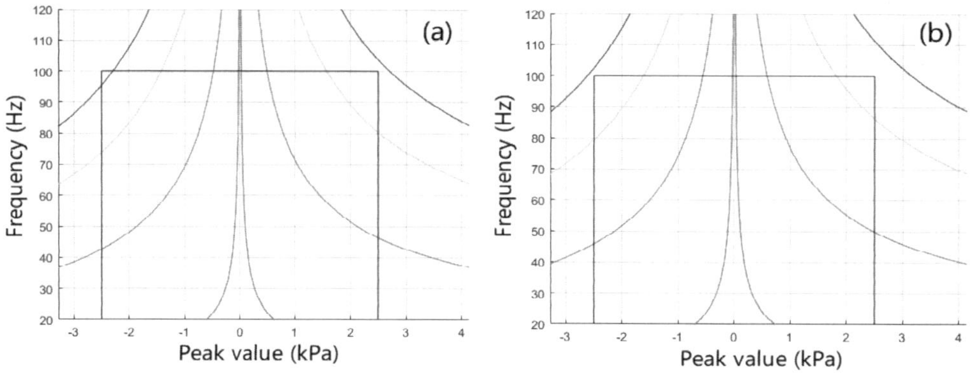

Figure 3. Amplitude-frequency curve under offset pressure of 30kPa. (a) The initial length of air cavity is 15 mm, (b) The initial length of air cavity is 13 mm

As shown in Figure 4, at the 450 kPa offset pressure and 25 Hz, the minimum acceleration of sinusoidal vibration of the shaking table is guaranteed to be 0.5 g, corresponding to the minimum unilateral amplitude of 0.2 mm, corresponding to the positive peak value of 7.03 kpa and the negative peak value of 6.82 kpa, which also meet the pulsating pressure of the peak value of 20 kPa.

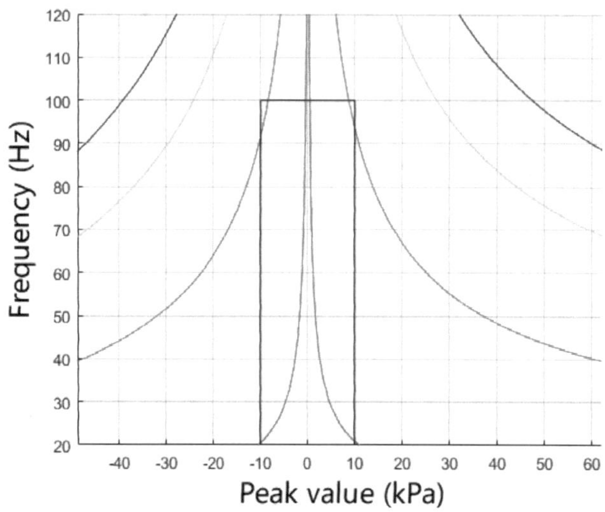

Figure 4. Amplitude-frequency curve under offset pressure of 450 kPa within the 13 mm air cavity

Similarly, three symmetrical cylinders with effective lengths, as shown in Table 4, were finally determined to meet the requirements of pulsating pressure from 5 Hz to 500 Hz.

Table 4. Cylinder unilateral length and applicable scope

Effective unilateral length of cylinder	Working pressure	Frequency range	Maximum speed
6 mm		20~500 Hz	
13 mm	10 kPa ~ 750 kPa	10~300 Hz	2 m/s
195 mm		5~50 Hz	

4. Conclusion

In this work, the constraint conditions of various parameters of the pulsating pressure generator are analysed in detail, and the requirements for pulsating pressure generation in a wide frequency range are realized through piecewise design, and the feasibility of the structure is determined through simulation analysis, which provides a theoretical basis for the realization of the pulsating pressure generation device.

5. Acknowledgments
This work is supported by the National Defence Technology Basic Research Projects (Grant: JSJL2017601B001).

6. References
[1] Jan H. Dynamic Measurement of pressure - A Literature Survey. BORAS: SP Swedish National Testing and Research Institute, 2002, 34: 10-13, 44-53.
[2] Bean V. E. Dynamic pressure metrology. Metrologia, 1993/94, 30: 737-741.
[3] Alberto C. G. C. Diniz, Alessandro B. S. Oliveira, João Nildo de Souza Vianna, et al. Dynamic calibration methods for pressure sensors and development of standard devices for dynamic pressure. Rio de Janeiro. Metrology for a Sustainable Development. Brazil: IMEKO, September, 17 – 22, 2006.
[4] Wang S. H. et al. Hydraulic square-wave pressure generator with a specific rotating valve. Measurement, 2009, 43: 672-677.
[5] Wang S. H. et al. Method of generating a hydraulic step wave with a shot rise time. Measurement, 2009, 43:935-940.
[6] Kobata T, Ooiwa A. Method of evaluating frequency characteristics of pressure transducers using newly developed dynamic pressure generator. Sensors and Actuators, 2000, 79:97-101.

IOP Conf. Series: Materials Science and Engineering **746** (2020) 012045 doi:10.1088/1757-899X/746/1/012045

Thermal-structural Coupling Researches on Transmission Shafting of the Intermediate Gearbox of a Certain Helicopter

Ao Wang, Xiaozhou Hu, Minggui Wu and Cheng Luo

College of Mechanical and Electrical Engineering, State Key Laboratory of High Performance Complex Manufacturing, Central South University, Changsha 410083, China

Email: csujxwa@163.com

Abstract. To study the lubrication performance and structural strength of transmission shafting of the intermediate gearbox of a certain helicopter, applying theories of tribology and heat transfer, thermal-structural coupling simulations of the entire transmission shafting, which mainly consists of a spiral bevel gear pair, two bearings, are performed based on the software platform of ANSYS Workbench. Both steady-state temperature field and stress field are obtained, and influences of rotational speed, torque on the thermal-structural coupling effect are discussed.

1. Introduction

Spiral bevel gear pair and bearings which are the main components of the intermediate gearbox of a certain helicopter, require good working and lubrication conditions. Under the conditions of high speed and heavy load, both the meshing of the bevel gear pair and the relative motion of different parts of the bearings cause friction, which lead the temperature of the transmission system to rise. Excessive temperature will cause severe thermal stress and thermal deformation on gears and bearings, which would easily lead to pitting and gluing of gears and bearings, endangering the stability of the entire transmission system. Therefore, in order to facilitate the analysis and design of the intermediate gearbox and its lubrication system and improve the working performance, it is very important to study the thermo-mechanical coupling effect of the spiral bevel gear transmission shaft system of the intermediate gearbox in the helicopter.

As early as 1995, Handschuh and Kicher [1] investigated the thermal behavior of spiral bevel gears by a finite element method, and verified the accuracy of the method through experiments. Since then, many scholars have conducted researches on the temperature distribution or stress of the geared transmission. Among them, Li and Tian [2] studied the temperature characteristics of a geared transmission, and obtained the influence of friction heat generation on gear bearing capacity and transmission sensitivity. Argyris et al. [3]applied a finite element method to conduct stress analysis; Mehrabi et al. [4]established a finite element model for thermal analysis of spiral bevel gears in helicopter's main gearboxes, and studied the effects of initial temperature, heat flux and lubricant on system temperature; Sun et al. [5] established geometric models to analyze the heating of bearings under oil air lubrication and spray lubrication conditions respectively, and obtained the influence of different lubrication conditions on the bearing temperature field.

The thermal-stress coupling analysis is also the research focus. Linjamaa et al. [6] developed a multi-physical model to analyze the elastic deformation and thermal deformation of hybrid journal bearings, then evaluated the performance of the bearings. Zhang et al. [7] presented a thermo-elastic-hydrodynamic coupling model to study the thermal characteristics of the spiral bevel gear and obtain the temperature rise of the gear surface. Mei et al. [8] studied the thermal characteristics of the gearbox, and obtained the relationship between deformation, structural load and thermal load. Liu et al. [9] established a thermo-elastic coupled finite element model to study the thermal deformation of the spiral bevel gear at different rotational speeds; Lu et al. [10] studied the thermal characteristics of the main shaft, and obtained the thermal deformation law under different rotational speeds and lubrication conditions.

In this work, the spiral bevel gear drive shaft system of the intermediate gearbox of a helicopter is taken as the research object, and following investigations are performed: firstly, the thermal-structural coupling simulation of the drive shaft system is carried out based on the finite element analysis method, then the steady temperature field and stress field distribution are obtained. Secondly, the influence of speed, torque and lubricating oil input temperature on the thermo-coupling effect is studied. Finally, the simulation results of steady-state temperature field is verified by experiments.

2. Determination of Boundary Conditions for Thermal Analysis

The heat generated by the transmission shaft of the intermediate gearbox mainly comes from load loss and non-load loss. The load loss mainly includes the friction power loss of the spiral bevel gear and the bearing. The transfer of heat in the transmission shaft system and its surroundings is mainly through three modes: heat conduction, heat convection and heat radiation [11]. Due to the heat transfer effect of heat radiation is the smallest, this paper ignores.

2.1. Transmission Shaft Heat Generation Analysis

2.1.1. Friction power loss of spiral bevel gear

Sliding friction is the main source of heat generated by spiral bevel gear transmission. According to the literature, the average heat flux density Q generated by the sliding friction of the tooth surface is

$$Q = f_0 \sigma_0 v_s \tag{1}$$

Among this, f_0 is tooth surface sliding friction coefficient; σ_0 is the maximum contact stress of meshing zone;v_s is the relative sliding speed of the driving and driven wheels at the meshing point.

The average heat flux density at the meshing point into the main and driven wheels q_1, q_2 is

$$\begin{cases} q_1 = \dfrac{b\beta_0 Q\omega_1}{2\pi v_1} \\ q_2 = \dfrac{b(1-\beta_0)Q\omega_2}{2\pi v_2} \end{cases} \tag{2}$$

Among this, b is the half width of contact at the meshing point; β_0 is the heat partition coefficient; ω_1, ω_2 is the angular velocity of the driving and driven wheels; v_1, v_2 is the tangential velocity at the meshing point of the driving and driven wheels.

2.1.2. Friction power loss of the bear

This paper uses the Palmgren calculation model to calculate the frictional heat of the bearing. The heat generated by the friction of the bearing H is

$$H = \frac{\pi n M}{30} \tag{3}$$

Among this, n is the bearing speed; M is the bearing friction torque, M=M_l+M_v, M_l is friction torque associated with bearing load; M_v is torque associated with the properties of the lubricating oil.

2.2. Analysis of Convective Heat Transfer of Transmission Shaft

Convective heat transfer at the end face of a spiral bevel gear can be simplified into a convective heat transfer model of a rotating cone. The calculation formula h_s of the end face heat transfer coefficient in this state is

$$h_s = N_u \lambda \sqrt{\frac{\omega}{v_f \sin \alpha_b}} \tag{4}$$

Among this, N_u is Nusslet number; λ,v_f is the heat transfer coefficient and kinematic viscosity of the fluid on the end face;α_b is half cone angle of spiral bevel gear.

According to the literature [12], the convective heat transfer coefficient of spiral bevel gear meshing surface can be expressed as:

$$h_t = 1.418 \left(\frac{v_f c_f \rho_f}{\lambda z} \right)^{\frac{1}{4}} \frac{\sqrt{\lambda c_f \rho_f \omega}}{2\pi} \tag{5}$$

Among this, c_f,ρ_f is specific heat and density of lubricating oil on the mating surface; z is the number of gear teeth.

The convective heat transfer coefficient of the root surface, the tooth top surface and the non-meshing tooth surface can be approximated within the range of 1/3 to 1/2 of the end surface convective heat transfer coefficient.

2.3. Convective Heat Transfer of Bearing

This paper uses Harris calculation model to calculate the convective heat transfer of tapered roller bearings. The convection heat transfer coefficient h_a of bearing is

$$h_a = 0.332 \frac{k}{x} P_r^{\frac{1}{3}} R_e^{\frac{1}{2}} \tag{6}$$

Among this: k, P_r is the thermal conductivity and Prandtl number of lubricating oil; R_e is Reynolds number.

3. Simulation Analysis

3.1. Establishment of Finite Element Model of Transmission Shafting

The transmission shafting of the intermediate gearbox of a certain helicopter includes a input spiral bevel gear shaft, an output spiral bevel gear and a pair of reverse mounted tapered roller bearings of the same type, as shown in Fig. 1. The basic parameters of the spiral bevel gears and he tapered roller bearings are shown in Table 1 and 2. The physical parameters of the drive shaft system are as follows: The rotating speed of driving wheel is 4455 r/min, the torque is 200 N·m, and the lubricating oil's input temperature and ambient temperature are both 50 °C. According to the analysis and calculation of the force of the tapered roller bearings, the bearing 1 is "relaxed", the radial force is 20342 N, the axial force is 5983 N; the bearing 2 is "compressed" and the radial force is 20342 N, the axial force is 8429N.

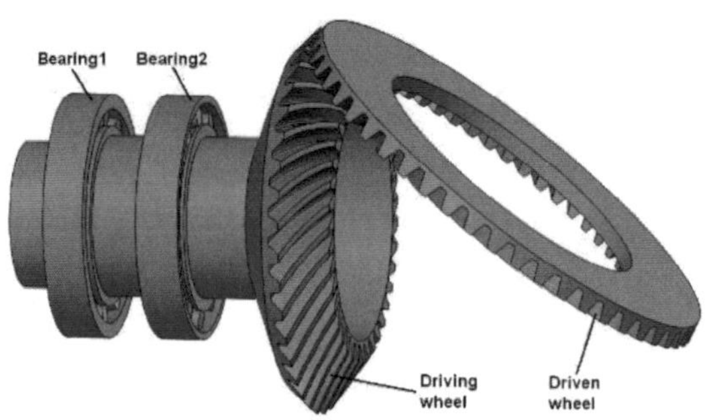

Figure 1. Three-dimensional model of a transmission shaft system

Table 1. Basic parameters of spiral bevel gear

	Number of teeth Z	*Modulus* m/(mm)	Tooth width B/(mm)	Helix angle β/(°)	pressure angle α/(°)	Axis angle Σ/(°)	Rotation
Driving wheel	39	3.75	26	35	20	128	Left-handed
Driven wheel	51	3.75	26	35	20	128	Right-handed

Table 2. Basic parameters of tapered roller bearings

Inside diameter d/(mm)	Outer diameter D/(mm)	Width T/(mm)	Number of rolling elements	Dynamic load rating C/(kN)	Static load rating Co/(kN)
60	95	17	12	57.57	76.12

3.2. Finite Element Analysis of Steady-state Temperature Field

In the steady state thermal analysis module of ANSYS Workbench, divide all entities of the imported driveshaft 3D model for free meshing. In order to improve the accuracy of simulation calculation, the mesh of the gear meshing area and the bearing roller and the inner and outer raceway contact areas are refined. The total number of nodes divided is 1080122, and the number of units is 656359. The divided transmission shaft system mesh model is shown in Fig. 2.

Figure 2. Drive shaft system mesh model.

Apply the thermal analysis boundary conditions to the corresponding surfaces of the drive shaft system model, and the steady state temperature field of the drive shaft system is obtained by simulation, as shown in Figure 3.

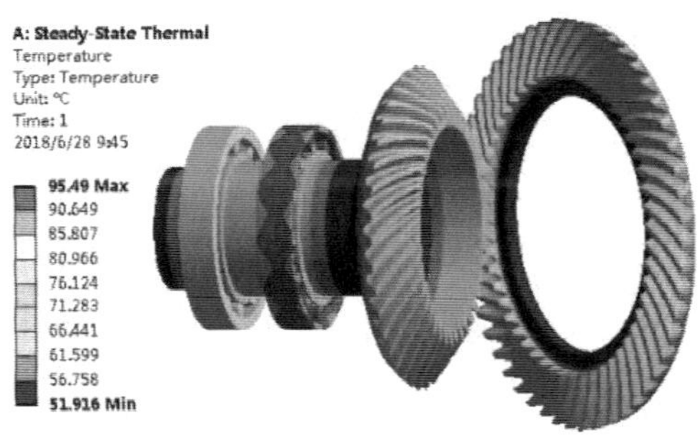

Figure 3. Steering axis steady state temperature field

As can be seen from Fig.3:

(1) In the entire spiral bevel gear transmission shaft system, the temperature of the outer ring raceway and the roller contact area of the bearing 2 is the highest, reaching 95.49 ° C. The main reason is that under high-speed and heavy-load conditions, the bearing 2 is in a state of being "pressed". The frictional torque is large, resulting in a large frictional heat generation, while the small size results in a weak convective heat transfer capability.

(2) For the helical bevel gear pair, the gear meshing zone has the highest temperature, reaching 78.84 °C. The main reason is that the friction of the gear meshing region is the main source of heat, and the convective heat transfer coefficient of this region is small.

3.3. Thermal-structural Coupling Analysis

This paper uses the sequential coupling method to analyze the thermal-structural coupling of spiral bevel gear transmission shafting. The steady-state temperature field of the drive shaft system calculated in the thermal analysis module is used as a boundary condition and imported into the structural analysis module for simulation calculation.

The simulation results show the stress field of the drive shaft under the thermal-structural coupling effect, as shown in Fig. 4. In order to facilitate the comparative analysis, the stress field of the drive shaft system is calculated without considering the temperature, as shown in Fig. 5.

Figure 4. Stress field under thermal-structural coupling.

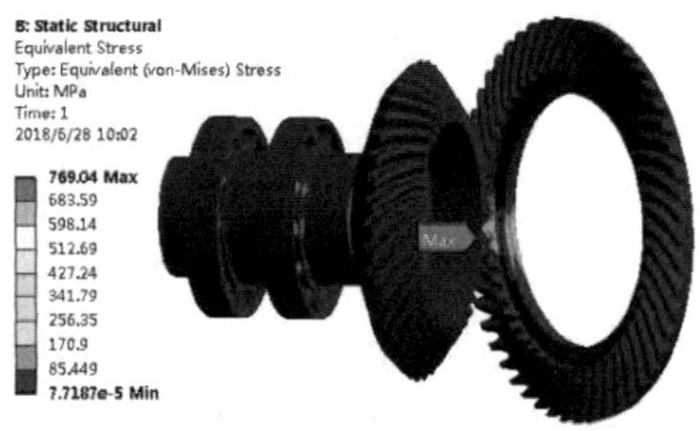

Figure 5. Stress field without considering temperature.

From Fig. 4 and 5, it can be seen that it is in accordance with the actual force of the transmission shaft of the intermediate gearbox under high speed and heavy load conditions, while the maximum equivalent stress of the drive shaft system also appears at the gear meshing position. After the application of the thermal load, the maximum stress at the gear meshing position increases by 10.68 MPa, an increase of 1.4%; the maximum stress of the bearing pair increases by 239.60 MPa, an increase of 52.5%. The main reason is that the thermal stress caused by temperature and the stress generated by the load overlap each other, so that the stress increases, the larger temperature rises, and the greater the temperature rise, the greater the thermal stress increase. Therefore, for bearings, the thermal stress can not be ignored.

4. Analysis of Influencing Factors

Speed, torque are the main factors affecting the thermal-structural coupling effect of the spiral bevel gear drive shaft. This paper mainly analyzes and discusses the impact of these three factors.

4.1. Analysis of the Influence of Speed

In the fixed working condition where the driving wheel torque is 200 N·m and the lubricating oil input temperature is 50 °C, the gear shaft rotation speed is set to be 4455 r/min, 5455 r/min, 6455 r/min,

7455 r/min, 8455 r/min. The maximum temperature and maximum stress of the shafting system are shown in Figs.6 and 7.

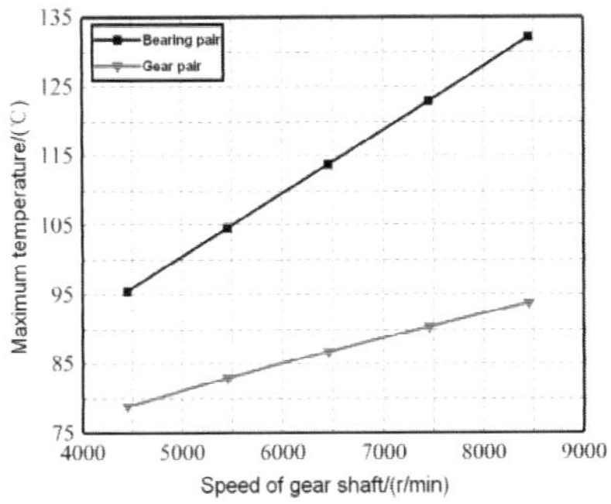

Figure 6. Maximum temperature at different speeds

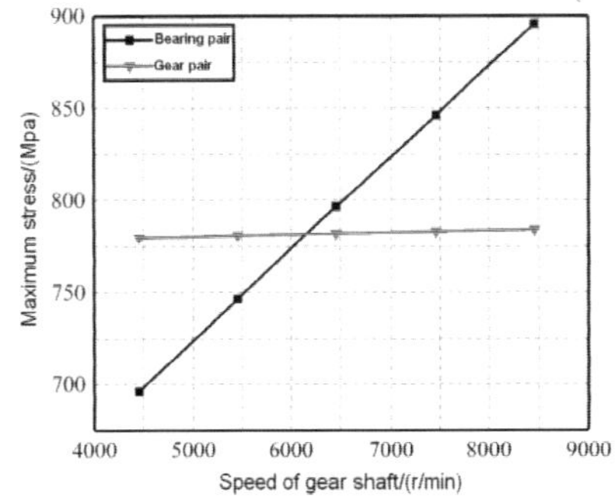

Figure 7. Maximum stress at different speeds.

It can be seen from the variation law of Fig. 6:

(1) When the speed of the gear shaft is in the range of 4455 ~ 8455 r/min, the maximum temperature of the bearing pair is always higher than the gear pair.

(2) The maximum temperature of the bearing pair and gear pair increases with the increase of the rotational speed, because the increase of the rotational speed causes the frictional heat generation and the convective heat transfer coefficient of the bearings and gears increase.

(3) When the rotational speed increased by 89.8%, the maximum temperature of the bearing pair and the gear pair increased by 38.3% and 18.9% respectively, indicating that the influence of the rotational speed on the bearing temperature field is more significant than that on the gear temperature field.

It can be seen from the variation law of Fig. 7:

(1) The maximum stress of the bearing pair increases significantly with the increase of the rotational speed. The main reason is that as the rotational speed increases, the maximum temperature of the bearing pair increases significantly, causing the thermal stress generation also to increase significantly. Besides, the maximum stress of the gear pair gradually increases with the increase of the rotational speed.

(2) When the rotational speed increases by 89.8%, the maximum stress of the bearing pair and the gear pair increases by 28.6% and 0.6% respectively, indicating that the influence of the rotational speed on the bearing stress field is more significant than that on the gear stress field.

4.2. Analysis of the Influence of Torque

Under the fixed working conditions of the driving wheel speed of 4455 r/min and the lubricating oil input temperature of 50 °C, the transmission is calculated when the driving wheel torque is 150 N·m, 200 N·m, 250 N·m, 300 N·m, 350 N·m. The maximum temperature and maximum stress of the shafting system are shown in Figs. 8 and 9.

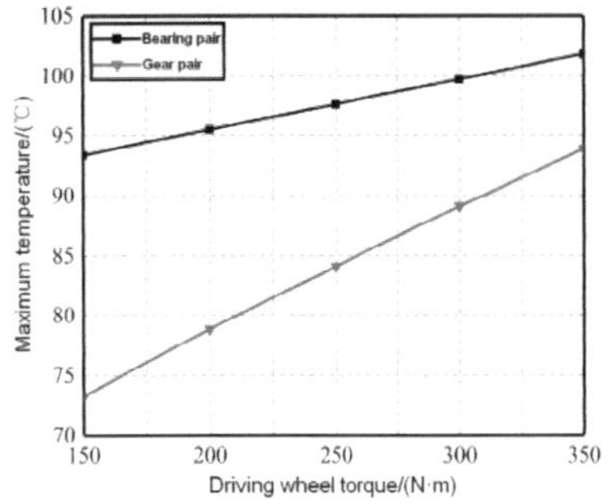

Figure 8. Maximum temperature at different torques.

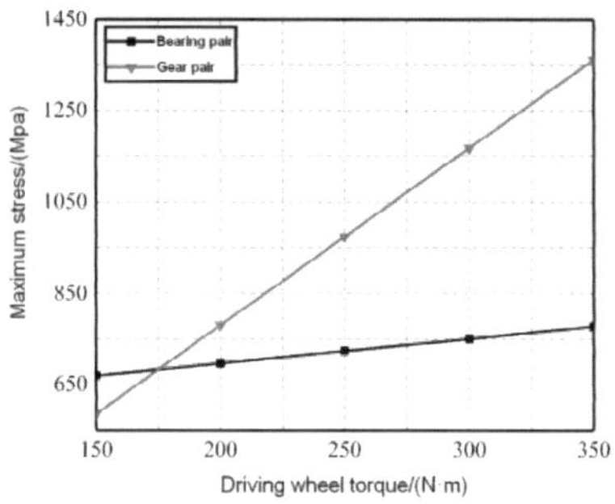

Figure 9. Maximum stress at different torques.

It can be seen from the variation law of Fig. 8:

(1) When the driving wheel torque is in the range of 150 ~ 350 N·m, the maximum temperature of the bearing pair is higher than that of the gear pair.

(2) The maximum temperature of the bearing pair increases with the increase of the driving wheel torque. The main reason is that as the driving wheel torque increases, the bearing friction torque increases gradually, so that the friction heat generation increases, causing the maximum temperature of the bearing to increase. The maximum temperature of the gear pair also increases with the increase of the driving wheel torque. The main reason is that the increased driving wheel torque increases gear contact pressure, which in turn leads to an increase in the heat flow density and cause the maximum temperature of the gear to increase.

It can be seen from the variation law of Fig. 9:

(1) When the torque is in the range of 150 to 175 N·m, the maximum stress of the bearing pair is larger than that of the gear pair. When the torque is in the range of 175 to 350 N·m, the maximum stress of the gear pair is larger than that of the bearing pair.

(2) The maximum stress of the bearing pair gradually increases with the increase of the driving wheel torque. The main reason is that as the driving wheel torque increases, the maximum temperature of the bearing pair gradually increases, so that the generated thermal stress gradually increases. At the same time, the radial and axial forces of the bearing increases gradually, so that the load stress is also gradually increased. The maximum stress of the gear pair increases significantly with the increase of the driving wheel torque. The main reason is that as the driving wheel torque increases, the maximum temperature of the gear pair increases significantly, so that the generated thermal stress increases significantly. At the same time, the contact pressure of the gear pair is significantly increased, so that the load stress is also significantly increased.

5. Conclusion

(1) Under the condition of high speed and heavy load, the temperature of the contact area between the outer ring raceway and the roller of the bearing 2 in the drive shaft system is the highest, thus the lubrication here needs to be strengthened.

(2) Under the combined influence of load stress and thermal stress, the stress of the drive shaft system is mainly distributed at meshing zone of the spiral bevel gears, the bearing roller support and gear shaft journal. The stress at the gear meshing is the largest, and the deformation is most likely to occur here, thus structural optimization can be appropriately performed to improve stress concentration.

(3) As the rotational speed, driving wheel torque increase, the maximum temperature and maximum stress of the bearing and gear will also increase. The influence of the rotational speed on the bearing temperature field and the stress field is more significant. The influence of driving wheel torque on the gear temperature field and stress field is more significant. The effect of temperature rise on bearing stress field is more significant than that on gear stress field.

6. References

[1] Handschuh R F, Kicher T P. Experimental and analytical assessment of the thermal behavior of spiral bevel gears [J]. Handschuh R F, Kicher T P. Experimental and analytical assessment of the thermal behavior of spiral bevel gears [J]. 1995.

[2] LI W, Tian J Y. Unsteady-state temperature field and sensitivity analysis of gear transmission [J]. Tribology International, 2017, 116: 229-243.

[3] Argyris J, Fuentes A, Litvin F L. Computerized integrated approach for design and stress analysis of spiral bevel gears [J]. Computer Methods in Applied Mechanics & Engineering, 2002, 191 (11): 1057-1095.

[4] Mehrabi M, Mohammadimehr M, Fatehi M R, et al. Thermal analysis, effect of lubricant and fatigue life of spiral bevel gears of main gearbox [J]. Modares Mechanical Engineering, 2017, 17(8): 38-48.

[5] Sun Q G, Wang Y F, Wang Y, et al. Comparing of Temperatures of Rolling Bearing under the Oil-air Lubrication to the Spray Lubrication[C]//Applied Mechanics and Materials. Trans Tech Publications, 2013, 395: 763-768.

[6] Linjamaa A, Lehtovaara A, Larsson R, et al. Modelling and analysis of elastic and thermal deformations of a hybrid journal bearing [J]. Tribology International, 2018, 118: 451-457.

[7] Zhang J, Liu S, Fang T. Determination of surface temperature rise with the coupled thermo-elasto-hydrodynamic analysis of spiral bevel gears [J]. Applied Thermal Engineering, 2017, 124: 494-503.

[8] Mei Y, Wang F P, Liu Q Y, et al. Finite Element Analysis on Thermal Structures of Heavy Machinery Gearbox[C]//Advanced Materials Research. Trans Tech Publications, 2011, 228: 651-655.

[9] Liu S, Liu Z, Lyu Y. Investigation Into the Temperature Field and Thermal Deformation of a Spiral Bevel Gear Pair at Different Rotational Speeds [C]//ASME Turbo Expo 2018: Turbomachinery Technical Conference and Exposition. American Society of Mechanical Engineers Digital Collection, 2018.

[10] Lu Y, Gao S D, Hao Z P. Study on steady-state temperature field and thermal deformation for permanent magnet synchronous motorized spindle [C]//Advanced Materials Research. Trans Tech Publications, 2011, 305: 340-343.

[11] Hu X Z, Liu S J, Peng J. Steady state thermal analysis of transmission shaft system of main reducer of helicopter [J]. Journal of Central South University (Science and Technology), 2015, 46 (12): 4469-4475 (in Chinese).

[12] Patir N, Cheng H S. Prediction of the Bulk Temperature in Spur Gears Based on Finite Element Temperature Analysis [J]. ASLE Transactions, 1979, 22 (1): 25-36.

IOP Conf. Series: Materials Science and Engineering **746** (2020) 012046 doi:10.1088/1757-899X/746/1/012046

Application of USB for Communication Verification Device in Theodolite

Xinghong Li, Lingling Zhang and Yongfeng Lei

Department of Automation Engineering, The Engineering & Technical College of Chengdu University of Technology, Leshan, Sichuan, China
Email: 545178279@qq.com

Abstract. It is primary that difficult to verify the data for a long time in the data communication system of present theodolite. So take advantage of CY7C68013 to Designing the data communication verification device and composing the program for the test of the data communication verification. It is profitable to realize the data communication of theodolite and the realtime or later verificate data, and making the debugging more convenient. Meantime realize the high speed transmission.

1. Introduction

The photoelectric theodolite is the high precision photoelectric detection instrument of space mobile target tracking and positioning for the modernization in range, while the new weapons and equipment must achieve real-time, high precision, dynamic tracking and realize real-time image reproduction performance, which requires the communication system has good real-time and reliability.

2. System Structure

2.1. Data Communication System of Theodolite

The data communication system is the data exchange hub of the large photoelectric theodolite. According to different requirements, the theodolite subsystem adopts different MCU, so as to realize the communication function. If the real time, the transmission distance is not high and the transmission speed is high, the RS-422 interface can be discarded and the USB2.0 interface is used for data transmission.

2.2. USB Data Communication Verification Device of Hardware Development

The block diagram of the whole system is shown in figure 1. The system selects CY7C68013 as USB2.0 controller, which is responsible for USB transaction processing, and also as the USB external chip CPU. In order to give full play to the advantages of fast transmission speed of USB interface chip, the hardware design adopts the high speed 32 bit fixed point TMS320F2812.

2.2.1. Design of communication interface circuit

Asynchronous serial transmission mode is the most theodolite data communication methods, namely uses RS-422 transmission mode mostly, then before the serial data received by the DSP, the level requires interface conversion, while DSP is the standard TTL 3.3V, therefore should be level conversion for the serial data received.

2.2.2. Slave FIFOs hardware connection and working principle

In the whole communication system, the subsystem of SCM and DSP as the controller and the check device is connected, select Slave FIFOs mode as the work mode. The hardware connection is shown as figure 2.

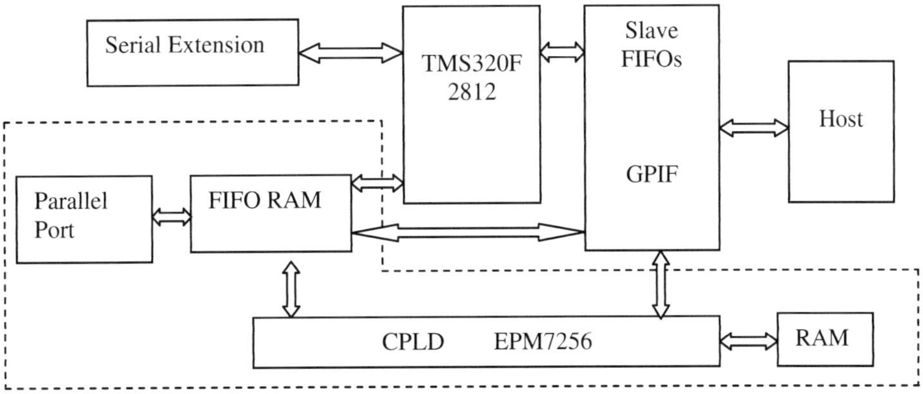

Figure 1. USB data communication Verification Device

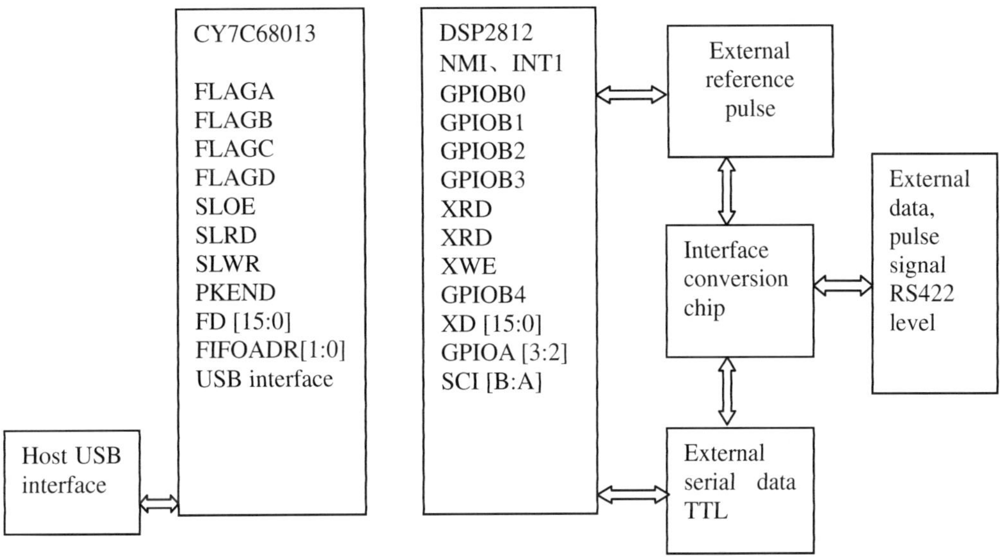

Figure 2. The hardware connection of USB Slave FIFOs

3. Software of Data Communication Verification Device

All based on the MCU and its peripheral circuit of the normal work of functional devices are inseparable from the participation of software, USB devices are no exception.

In the process of data communication verification device, DSP is responsible for receiving serial data and sending the data to the host. In order to ensure the normal communication between DSP and USB controller, it is necessary to establish a standard communication protocol between them. DSP is responsible for continuously receiving external serial data and sends the data to the USB controller receiving end point of operation, it not only improves the transmission efficiency of USB, but also the DSP concentration to send data, thus increasing the transmission efficiency of DSP.

4. Test of the Data Communication Verification Device

The maximum transmission speed of up to 42.9Mbps USB interface between the data communication verification device and the main CPU, from the theoretical analysis, data check device can be extended ten serial experiments, and each serial communication baud rate can reach 4.1Mbps.

4.1. Functional Test of Data Communication Verification Device

The design of the system is as follows: the function of the communication system.The data communication verification device through the USB interface is connected with the notebook, data source is provided by

IOP Conf. Series: Materials Science and Engineering **746** (2020) 012046 doi:10.1088/1757-899X/746/1/012046

another TMS320F2812 port. This system chooses the highest baud rate of 4.1Mbps data source, the frequency is 100HZ, the contents of the packet is known, the experiment data are shown in table 1.

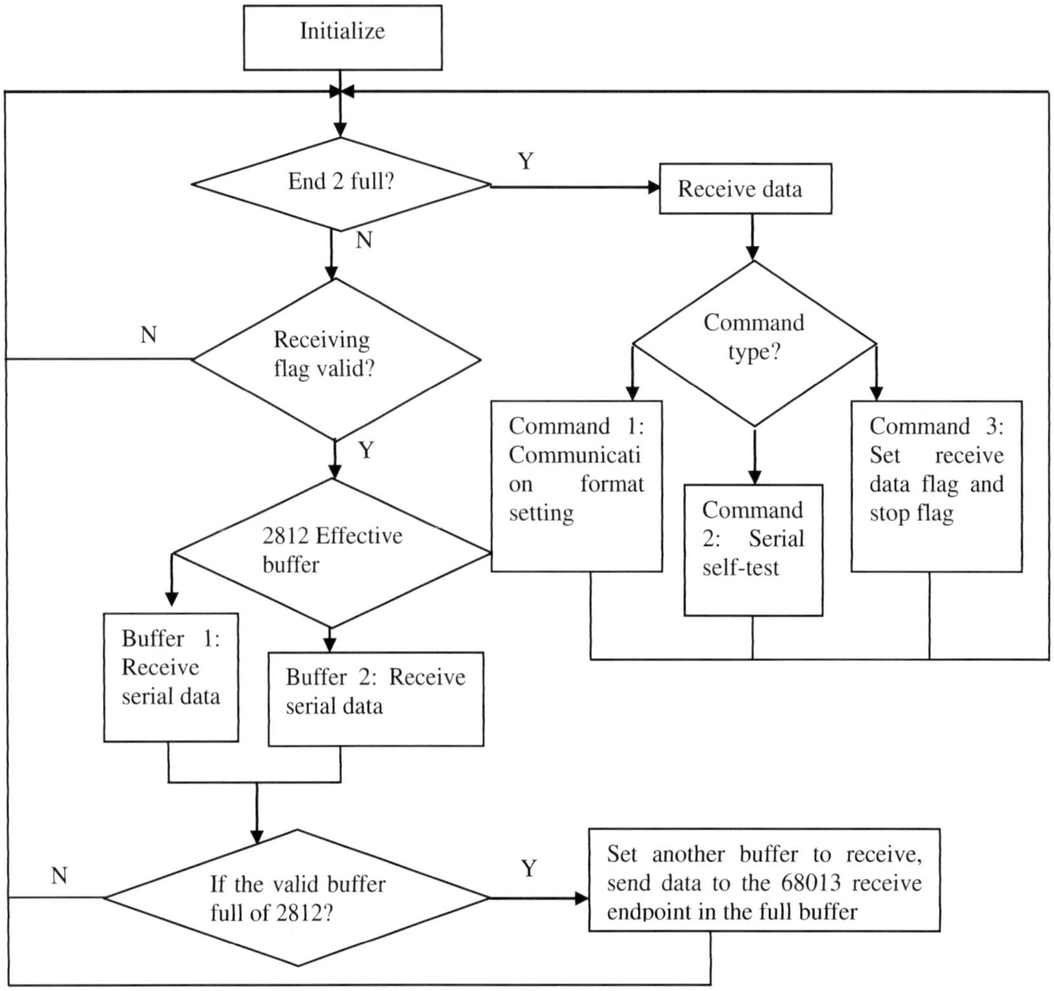

Figure 3. The flow chart of DSP program

Table 1. The communication data of communication Verification Device and TMS320F2812

baud rate	Stop bit	Number of significant bits	Number of packets	Error percentage
481.6 Kbps	2 bits	8	4159	0%
963.2 Kbps	2 bits	8	5050	0%
4.252 Mbps	1 bit	8	5332	0%

From the table 1, in the communication verification device and PC communication process, not only can flexibly set the baud rate, and transmission rate gradually increased, but also the right to transmit data.

4.2. Transmission Speed Test of the Data Communication Verification Device
TMS320F2812 as a data source, CY7C68013 provides read and write signal, the host selects various types of files for senting to DSP, DSP opens array reception, when the array is full and then returns the data to the USB controller, the host saves the receive data at the same time, finally compares the received files and the original file of the host, the test results are shown in table 2.

Table 2. The actual data of transmission velocity

file name	file size	transmission time	speed	bit error
video.avi	30.64 MB	11.68 s	38.12 Mbps	0
text.txt	5.26 MB	1.98 s	38.64 Mbps	0
audio.mp3	4.52 MB	1.69 s	38.38 Mbps	0

From the table 2, when the transmission bandwidth is 8, transmission speed is about 38Mbps, because TMS320F2812 external I/O read and write speed is slow, DSP read and write frequency is 4.46MHz in the working environment of the system, if using faster operating frequency, transmission speed will be increased.

5. Summary and Innovation
The data communication verification device used in the theodolite data communication verification, can receive and store data long time; real time check and post check combination, can accurately determine the time and number of errors occurred in the process of data communication, which greatly facilitates the system debugging, to ensure the stability of data communication in the process; can also be used for receiving video and image transmission speed higher requirements on the occasion.

6. Acknowledgement
This project is supported by the National Natural Science Foundation of China: 11375055.

7. Reference
[1] Li Yingwei, Wang Chengru, USB2.0 principles and engineering development [M], National Defense Industry Press, 2007.3-368.
[2] ZLG, PDIUSBD12 USB firmware and driver development [M], Beihang University press, 2012.11-1-105.
[3] Wuan River, Tai Ming, Yu Hongtao, Windows 2000/XP WDM device driver development of [M], electronic industry university press, 2005. 5-353.
[4] Zhao Zhiyong, Hou Jiguang, [J] design, USB2.0 communication interface automation, 2015-03-62-63.
[5] Wang Hong, millet, Liu Xinguo, based on the CY7C68013 high-speed multi-channel synchronous data acquisition [J], electronic technology, 2014-08-48-5.

IOP Conf. Series: Materials Science and Engineering **746** (2020) 012047 doi:10.1088/1757-899X/746/1/012047

Software Implementation Analysis of Fingerprint Identification System

Ruijie Liu

School of software, Qufu Normal University, 57 Jingxuan West Road, Qufu City
Email: njnulmh@163.com

Abstract. With the fast development of E-time, the scope of Human-Computer Interaction Techniques' application has become broader and broader. At the same time, the information security and identity verification have drawn increasing attention. Fingerprint recognition technology, which has played an important role in finance, internet, medicine and security fields, develops first, and is relatively more mature in these three. This paper mainly discusses the software part of the fingerprint recognition system. The extract of fingerprint's image is directly from existing fingerprint database. Then a series of preprocessing, such as the grid, segment, smooth and converge of images, and so on, will be done. By doing these, image features could be extracted, so as to match the images of fingerprint. This paper using an implementation method, based on VC6.0 environment, in the sections of image's preprocessing and feature matching, which contains 2 easy-operation function modules.

1. Introduction

1.1. Current status of application field development
The information society is developing more and more rapidly, and mobile terminal devices have become the "protagonist" of people's lives. Some manufacturers have added fingerprint recognition technology to their mobile devices, and the wearable boom is also unfolding, and more direct man-machines. The way of interaction will have greater development prospects in the future. Fingerprint pattern recognition as an entry point will take me closer to such an area.

1.2. The Composition and Process of the Fingerprint Identification System

1.2.1. Overview of the module of the fingerprint identification system
The entire fingerprint identification system is divided into the following modules, which are relatively independent but functionally related to each other, as shown in Figure 1.1.

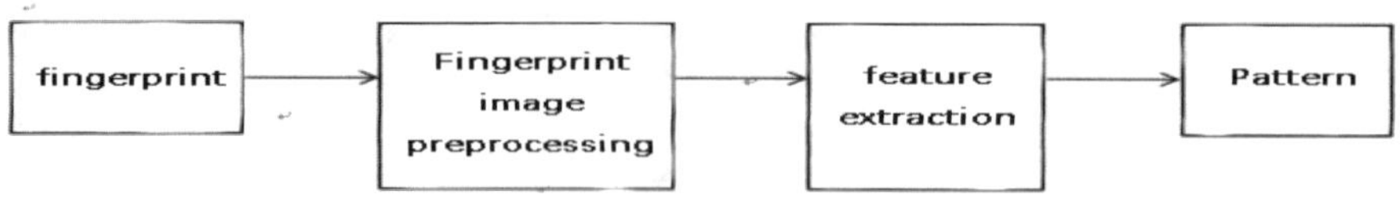

Figure 1.1.

(1) Fingerprint acquisition module

This module takes a static sampling and directly uses the fingerprint image in the fingerprint database for testing.

IOP Conf. Series: Materials Science and Engineering **746** (2020) 012047 doi:10.1088/1757-899X/746/1/012047

(2) Preprocessing module

The original fingerprint image is subjected to distortion correction, equalization, smoothing, etc. to complete the image realization, and a series of enhancement and refinement processes such as segmentation, denoising, enhancement, and binarization of the obtained real image are extracted, and the original fingerprint image is extracted. The feature points will eventually be compared to the data information needed to be delivered to the next module, the fingerprint pattern matching module. The processing effects of some pre-processing sub-modules are shown in Figure 1.2, 1.3, 1.4, and 1.5.

Figure 1.2.
Image after equalizationequalization

Figure 1.3.
Smooth processed image

Figure 1.4.
Binary image processing

Figure 1.5.
Image after enhancement

(3) Feature extraction module

The module mainly extracts feature values from image information. Figures 1.6 and 1.7 list two common feature points.

Figura 1.6.
End of ridge feature

Figura 1.7.
Line feature fork

1.2.2. Fingerprint identification system for process planning

(1) Process planning of fingerprint identification system preprocessing module

The preprocessing stage mainly completes the process of authenticating and digitizing the original image, and provides registration and comparison parameters for the matching module. Figure 1.8 shows the specific links and processes contained in the preprocessing module.

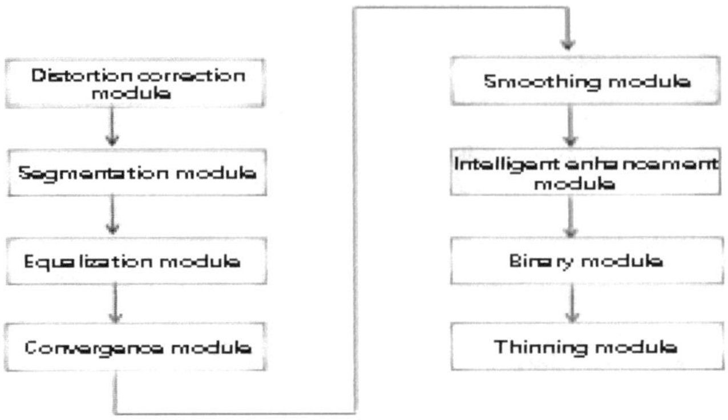

Figure 1.8.
Flow chart of fingerprint image preprocessing module

(2) Matching module process planning

By comparing the center point, the triangle point, the direction field registration point and some other feature points, the result of the comprehensive comparison finally finds a match. The process is shown in Figure 1.9.

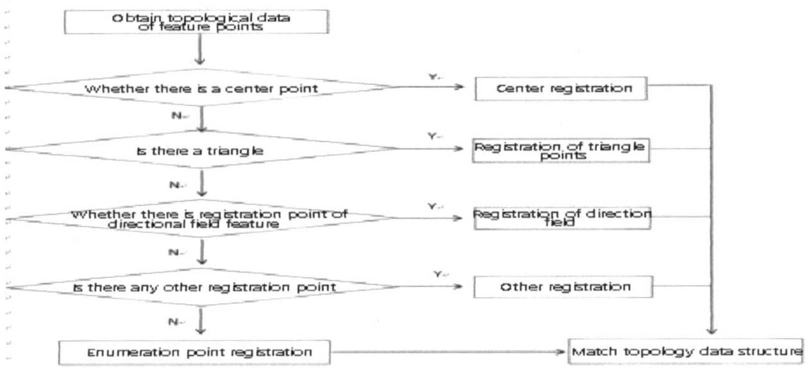

Figure 1.9.
Flow chart of fingerprint feature matching module

2. Fingerprint Image Preprocessing

2.1. Image distortion correction

The fingerprint image of this thesis is directly derived from the test fingerprint database, which skips the process of fingerprint collection, and mainly explains the distortion correction of the simple mathematical model.

2.1.1. Geometric correction

We subdivide the fingerprint image on a large grid of many small grids, and take the necessary linear stretching or filling for irregularities and distortions.

2.1.2. Grayscale enhancement

After geometric correction, the photon will appear brighter or darker on a certain area when imaging, which requires grayscale enhancement of the image.

2.1.3. Mathematical fitting modeling

Abstract a simple mathematical model that linearizes complex images while preserving certain error values so that we can correct and integrate the images that have been distorted.

2.2. Fingerprint Image Field and Its Calculation

2.2.1. Intensity field of fingerprint image field

The concept of image field and intensity field is introduced to visualize such physical quantities.

Set function f(x, y), let the function characterize the fingerprint image field, and the intensity field is expressed as V, the gamma field amount is represented by $|V(x,y)| = f(x,y)$ to describe the gray value of the pixel in the dot area.

2.2.2. The gradient field of the fingerprint image field.

In the study of fingerprint images, the fingerprint lines are classified, and the ridges and steep drops of the line trend can be recorded and characterized, and the direction of the fingerprint lines can also be recorded, so that they can be mapped to On the mathematical model, the gradient field can be calculated by performing partial differential processing on the direction vector.

Let ∇(x, y, z) be a gradient, which can be defined as $(\frac{\partial f}{\partial x}, \frac{\partial f}{\partial y}, \frac{\partial f}{\partial z})$, among them:

(1) Gradient operator: $\nabla=\frac{\partial}{\partial x}i+\frac{\partial}{\partial y}j+\frac{\partial}{\partial z}k$.

(2) Two-dimensional image gradient field definition.

The gradient vector of the gray function f(x, y) is $\nabla f = (\frac{\partial f}{\partial x}, \frac{\partial f}{\partial y})$. Let the gradient field of the image be T(x, y), then T(x, y) =∇f(x, y).

2.3. Segmentation of Fingerprint Images

From the intensity field of the original image, a histogram of the difference between the reaction image and the background intensity can be obtained. This histogram is visually regarded as a map of the reaction gradient due to the thinning of the horizontal axis, and the part of the high gradient value is dug. Come out, this is the split.

The following formula describes a simple way to approximate the gradient magnitude: $|T|=\sqrt{(\frac{\partial f}{\partial x})^2 + (\frac{\partial f}{\partial y})^2} \approx |\frac{\partial f}{\partial x}|\,|\frac{\partial f}{\partial y}|$.

2.4. Balance of Fingerprint Images

2.4.1. Processing image gray balance

When the fingerprint image is processed, the original image with a larger color is equalized.

2.4.2. processing image grayscale contrast

When the fingerprint image is processed, the contrast of the image is adjusted.

Figure 2.1.
Image before equalization

Figure 2.2.
Image after equalization

2.5. Convergence of Fingerprint Images

If the pixel is too distorted or "dark", the computer cannot recognize and associate the pixel points that should be associated with it, causing the computer to "blindly recognize" the image, classifying it as a distorted image, thereby causing system error. Convergence operations on such images are required. Figure 2.3 and Figure 2.4 show the image effects before and after the convergence operation.

Figure 2.3.
Before convergence

Figure 2.4.
After convergence treatment

2.6. *Smoothing of Fingerprint Images*

The input of the fingerprint image sometimes brings the external light and some stains on the fingerprint into the original image. In such a case, the impression or the spot is often left on the original image, or the fingerprint line is blocked and the line appears. The faults are reflected in the intuitive sense that some places appear too sharp and rough. In order to achieve the denoising operation, a convolution operation definition is given here to mathematically describe the smoothing operation of the image.

$$g(x,y) = T * f(x,y) = \sum_{i=0}^{m-1}\sum_{j=0}^{m-1} T(i,j)\, f(x - j + \frac{m-1}{2}, y - j + \frac{m-1}{2}$$

When m=3:

$$g(x,y)=T(0,0)f(x+1,y+1)+T(0,1)f(x+1,y)+T(0,2)f(x+1,y-1)+ \\ T(1,0)f(x,y+1)+T(1,1)f(x,y)+T(1,2)f(x,y-1)+ \\ T(2,0)f(x-1,y+1)+T(2,1)f(x-1,y)+T(2,2)f(x-1,y-1)$$

Applying the above formula to the matrix is a smoothing operation. The following describes a matrix template commonly used for smooth denoising such as edge detection, and the rectangular neighborhood of the domain average smoothing template operator (3*3 matrix).

3*3 large template: $T^3 = \frac{1}{9}\begin{pmatrix}1 & 1 & 1\\ 1 & 1 & 1\\ 1 & 1 & 1\end{pmatrix}$, $T_C^3 = \frac{1}{4}\begin{pmatrix}0 & 1 & 0\\ 1 & 1 & 1\\ 0 & 1 & 0\end{pmatrix}$

Figure 2.5.
original image

Figure 2.6.
After smoothing

2.7. *Enhancement of Fingerprint Images*

2.7.1. *Gabor wavelet definition*

The Gabor wavelet function consists of a Gaussian function and a trigonometric function, forming a function of periodic oscillation.

The one-dimensional expression of the Gabor wavelet function is:

$$H(x) = \frac{1}{\sqrt{2\pi}\sigma}\exp\{-\frac{x^2}{2\sigma^2}\}\cos(2\pi fx)$$

The two-dimensional expression of the Gabor wavelet function is:

$$H(x,y) = G(x',y')\exp(2\pi fx'i)$$

The formula is specified as follows.

(1) Gaussian component:

$$G(x,y) = \frac{1}{2\pi\gamma\sigma^2}\exp\{-\frac{\left(\frac{x}{\gamma}\right)^2+y^2}{2\sigma^2}\}$$

(2) The rotation formula of the coordinate axis:

$$\begin{bmatrix} x' \\ y' \end{bmatrix} = \begin{bmatrix} \cos(90^\circ-\theta) & \sin(90^\circ-\theta) \\ -\sin(90^\circ-\theta) & \cos(90^\circ-\theta) \end{bmatrix} \begin{bmatrix} x \\ y \end{bmatrix} = \begin{bmatrix} \sin\theta & \cos\theta \\ -\cos\theta & \sin\theta \end{bmatrix}$$

(3) Parameter Description.
f is the frequency.
σ is the standard deviation.
γ refers to the ratio of the long and short axes of the ellipse below the image.
θ refers to the angle formed by the positive direction of the x-axis.

3. Feature Matching of Fingerprint Images

3.1. Fingerprint Feature Extraction

The extraction of singular points of fingerprint features mainly takes two steps.

1. Poincare calculations.
2. Calculate the final singularity by averaging.

We get a few of the surrounding singularities we get by averaging. False singularities may occur due to noise or other effects in some fingerprint images.

The multi-directional matrix parameter table implemented according to this algorithm is given below, with i, j as the center, and the direction vector parameter values of the surrounding directions are as shown in Table 3.1.

Table 3.1. Mean value of direction field of singular point 5*5

	3	2	1	
4	3	2	1	12
5	4	(i, j)	8	11
6	5	6	7	10
	7	8	9	

3.2. Registration and Matching of Fingerprint Images

3.2.1. Registration method of fingerprint image

(1) Special point method.

Look for the special position of the fingerprint image. The most common ones are the center point and the triangle point. Compare the multiple pairs of special points one by one to determine whether the registration is successful.

3.2.2. Matching of fingerprint images

The matching of the fingerprint image has been gradually transformed into the implementation of the algorithm, which involves the realization of the algorithm.

The first thing to consider is that the algorithm implementation flow chart is shown in Figure 3.1.

IOP Conf. Series: Materials Science and Engineering **746** (2020) 012047 doi:10.1088/1757-899X/746/1/012047

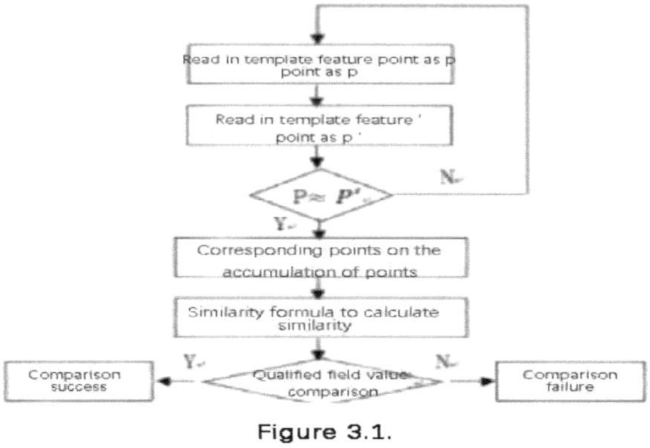

Figure 3.1.

Comparing P with P', if the prior feature template is satisfied, that is, P is approximately equal to P', and within the error value, then the match can be determined, so that the match of the array P and P' can determine that the match is successful.

3.3. Test Results and Demonstration

3.3.1. Preprocessing module

By selecting a fingerprint image, it performs various detail processing and feature extraction in the preprocessing stage, thus providing reference data for the matching module. The test interface is shown in Figure 3.2.

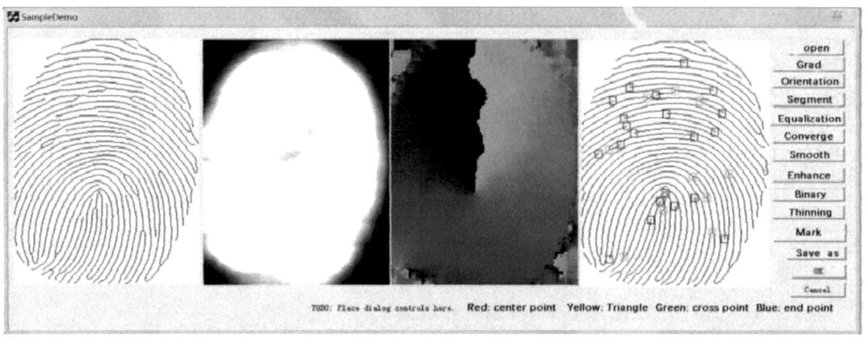

Figure 3.2. Pretreatment interface

3.3.2. Matching module

The data obtained from the preprocessing module implements matching of two fingerprint images in this function module, and the operation interface is as shown in Figure 3.3.

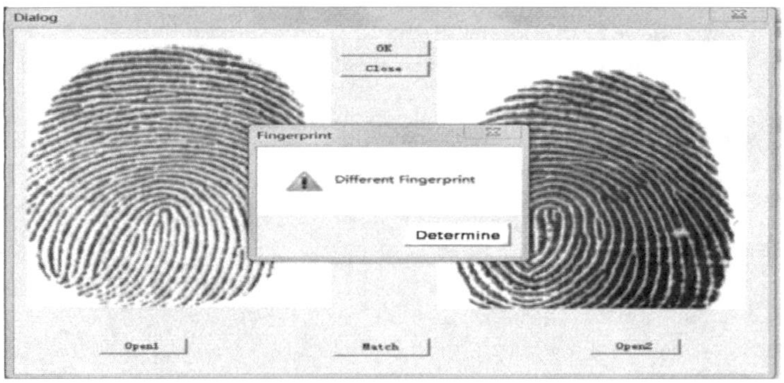

Figure 3.3. Fingerprint matching interface

IOP Conf. Series: Materials Science and Engineering **746** (2020) 012047 doi:10.1088/1757-899X/746/1/012047

4. References

[1] Zhao Xiangxin et al. Fingerprint. Beijing: Mass Publishing House, 1983.

[2] Zhang Honglin. Visual C++ digital image pattern recognition technology and engineering practice. Beijing: People's Posts and Telecommunications Press, 2003.

[3] Yang Zhiling, Wang Kai Visual C++ Digital Image Acquisition, Processing and Practical Application. Beijing: People's Posts and Telecommunications Press, 2003.

[4] Lu Fengjun. Introduction to Digital Image Processing Programming. Beijing: Tsinghua University Press, 1999.

[5] Shi Jiming, Application of Wavelet Transform in Image Compression [J]. Journal of Zhejiang Business and Technology College, 2002.

[6] Li Wei, Fu Wei. Internet of Things Fingerprint Identification System Algorithm and Implementation. Beijing: People's Posts and Telecommunications Press, 2011.

[7] Richard Spillman. Classical Cryptography and Modern Cryptography. Beijing: Tsinghua University Press, 2005.

[8] Sun Shuling. Applied Cryptography. Beijing: Tsinghua University Press, 2004.

[9] Gong Jianwei, Xiong Guangming. Visual C++/Turbo C Serial Communication Programming Practice. Beijing: Publishing House of Electronics Industry, 2005.

[10] Research and Development Department of NTT COMWARE Co., Ltd. Status and development trend of RFID. Beijing: People's Posts and Telecommunications Press, 2007.

[11] Chen Guoliang et al. Parallel algorithm practice. Beijing: Higher Education Press, 2003.

IOP Conf. Series: Materials Science and Engineering **746** (2020) 012048 doi:10.1088/1757-899X/746/1/012048

A Fractal Prediction Model for Evaluating Normal Contact Stiffness of Micro-pitting Gear

Xiaopeng Wang and Shijun Liu

Zhengzhou Research Institute of Mechanical Engineering, China Academy of Machinery Science and Technology, Zhengzhou, Henan, China, 450052.
Email: wangxiaopeng4109@hotmail.com

Abstract. Estimating normal contact stiffness of micro-pitting is one important subject to study the dynamic fault characteristics of gear system. Combined with the fractal parameters of surface morphology and the distribution area of contact points on the microscopic scale, a fractal model of joint surface with normal contact stiffness was established by investigated the elastic-plastic contact mechanical properties of micro-convex. The estimated fractal parameters for micro-pitting characteristic was established with the macroscopic physical dimension of involute gear and microscopic physical cause of micro-pitting, to investigate normal contact stiffness of involute gear under different machining technology of tooth surface and different levels of micro-pitting. The result shows that the surface microscopic elements have different effects on normal load and normal contact stiffness. The surface contact coefficients are different along with the engagement point of the tooth profile, and improving the accuracy of the tooth surface (reducing the roughness) can improve the normal contact stiffness of meshing tooth surface. The normal contact stiffness of tooth is lower significantly with micro-pitting. The simulation results are consistent with the actual engineering. The proposed mathematical model of micro-pitting can be used as a basis for an analysis of dynamic response and expansion of micro-pitting.

1. Introduction

When cyclical contact stress of the tooth surface exceeds the fatigue limit of materials, micro-cracks are generated at the surface or subsurface of the tooth with the cyclical contact stress, and the pits are formed which the minute metal piece falls off from the tooth surface with developing of crack. This phenomenon is called fatigue pitting [1]. The position of pitting is more common on the lower side of the pitch line, as shown in Figure 1.

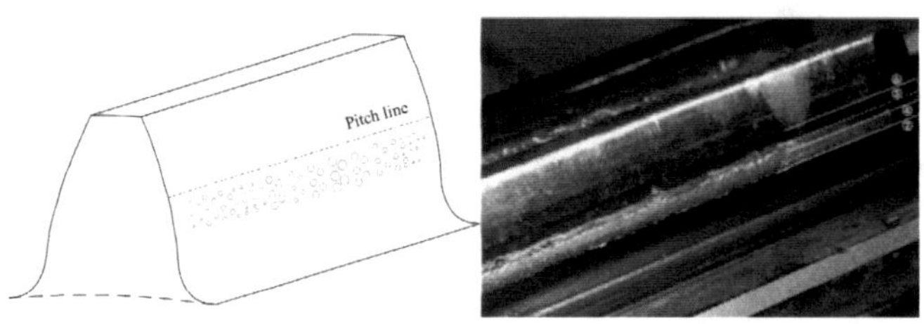

Figure 1. Schematics of fatigue pitting failure of tooth surface

Pitting, as one important feature of gear fatigue failure, not only affects contact strength of the tooth surface but also affects contact stiffness of the contact interface. The formation mechanism of pitting features and the strength properties and dynamic response characteristics after pitting have been the subject of continuous research by scholars [2-4]. Micro-pitting is an initial failure feature of the tooth surface in the fatigue pitting. But relative research is still lacking with the different effects of spatial scale. With the research on contact characteristics of contact surface micro-morphology in the micro spatial scale, contact characteristics of contact interface are closely related to an external load, surface micro-morphology and material properties [5-7]. Tian [8, 9] investigated normal contact stiffness of the fixed interface by a mathematical and mechanical model of asperity surface with microscopic geometry. Goerke [10] established a mathematical model and experiment system of the normal fractal contact surface to analyze the influence of different parameters (morphology, hardness, etc.) on the load capacity of the contact surface. The effect of fractal properties of asperity surface on the normal and tangential load capacity of the contact surface and normal contact stiffness were investigated in the references [11, 12]. Yuan [13] proposed an improved M-B model for the asperity contact surface by combination with elastic-plastic theory. Contact strength and contact stiffness of the tooth surface were analyzed qualitatively with fractal theory by Zhao [14] and Chen [15].

Based on the above research theory and engineering application of the fractal model for contact surface, the load capacity and normal contact stiffness of the tooth surface is future investigated combine with the material parameters and load effect of operating surface in a real meshing process of gear pair. A mathematical fractal prediction model of surface morphology of micro-pitting is established to investigate quantitatively the influence of micro-pitting feature on normal contact stiffness combine with the forming physical phenomenon of micro-pitting. The calculation mathematical /mechanical model of contact stiffness for gear system dynamics and response characteristics of tooth surface fatigue pitting are further improved and perfected.

2. Micro-pitting Formation

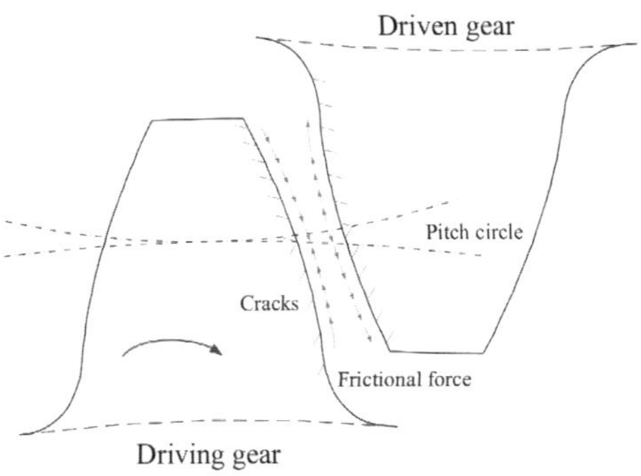

Figure 2. Crack initiation direction of tooth surface

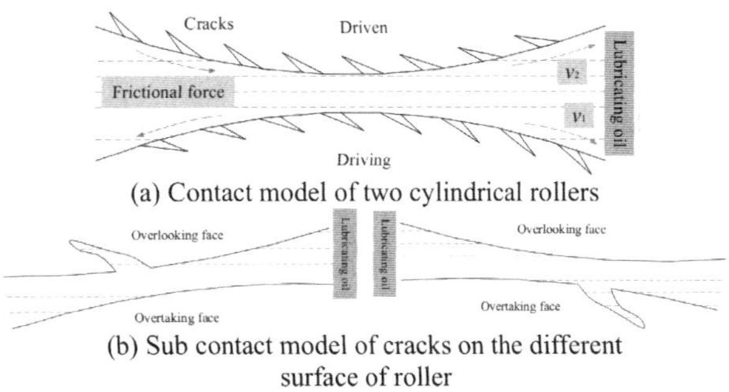

Figure 3. The formation mechanism of pitting

Under the action of cyclical contact stress, cracks always appear along the direction of frictional force by the sliding of tooth surface in the meshing process of a pair of involute spur gears, as shown in Figure 2. The pitch line is as a boundary, the friction direction and cracks direction are alternately presented on the top and down tooth surface. The contact model of any engagement position on tooth surface can be approximately simulated as the contact of two cylindrical rollers. The contact model of two cylindrical rollers (driving and driven) is shown in Figure 3, the physical symbols of v_1 and v_2 are the speed of driving and driven roller, and v_1 is greater than v_2. The driving roller is defined as an overtaking face, and the driven roller is an overlooking face. While the clearance between two rollers is filled with lubricating oil, the oil also flows into the cracks on the surface. The friction direction and crack direction are generated according to the movement direction which is shown in Figure 3a. And the surface makes contact from right to left. In the sub contact model of cracks on the overlooking face (left of Figure3b), the crack notch is first contacted and sealed subsequently with the contact points moves. And the lubricating oil pressure is increased in the crack, stress concentration exists in the crack tip in the meantime so that causing the crack growth. But in the sub contact model of cracks on the overtaking face (right of Figure3b), the crack flank is contacted and the lubricating oil in the crack is squeezed out. And the lubricating oil pressure is decreased in the crack which slows down the crack growth. According to the contact characteristic of tooth surface in the meshing process, the overtaking face is along with tooth profile from pitch line to the tooth root, and the overlooking face is along with tooth profile from pitch line to the tooth tip. Therefore, cracks are expanded for pitting easily below the pitch line, which is consistent with the schematics of pitting failure feature in Figure 1.

3. Fractal Contact Model of Asperity Surface

3.1. Contact Model of Micro-convex

Based on the fractal function of W-M and dimension independence of fractal parameters, the elastic-plastic contact is equivalently simplified to the contact between asperity surface and rigid ideal surface in the fractal contact M-B model [16] of asperity surface, as shown in Figure 4. In the schematic diagram, z is the height of a given asperity measured from the mean of original asperity heights. d is the separation between the rigid flat surface and mean of original asperity heights. δ is the total deformation of asperity surface. Therefore, the deformation of the given asperity surface under the load is written as $\delta = z - d$. And from the Weierstrass-Mandelbrot function, the roughness is a cosine wave for which the micro-convex before deformation is given as equation (1), the schematic diagram is given as Figure 5.

$$z(x) = G^{D-1} l^{2-D} \cos\left(\pi x / l\right), \quad -l/2 < x < l/2 \tag{1}$$

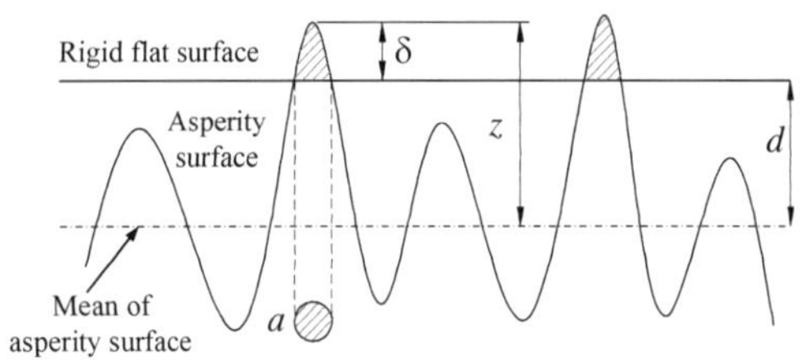

Figure 4. Contact diagram between asperity surface and rigid surface

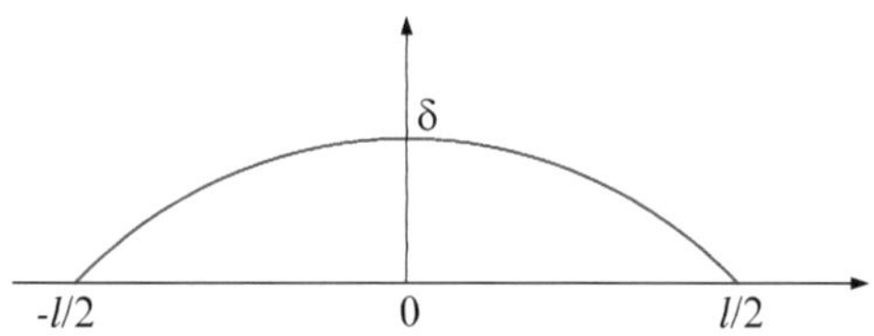

Figure 5. Geometry of contact point with the length l

Which D is the fractal dimension, G is the fractal roughness parameter of surface and l denotes the length scale. Then combine with the cosine function of micro-convex and calculation method of any curvature radius, the curvature radius of micro-convex and max deformation of the peak can be derived as equation (2) and equation (3).

$$R = \lim_{x \to 0} \left| \left(1 + \frac{dz(x)}{dx} \right)^{3/2} / \frac{d^2 z(x)}{dx^2} \right| = \frac{l^D}{\pi^2 G^{D-1}} = \frac{a^{D/2}}{\pi^2 G^{D-1}} \tag{2}$$

$$\delta_{\max} = G^{D-1} a^{(2-D)/2} \tag{3}$$

Substitute curvature radius R into equation (4), the critical deformation δ_c which exchanges from elastic to plastic is then derived. The critical contact area a_c of elastic-plastic transformation is derived by equation (3) and equation (4), which is shown in equation (5). The mathematical relationship of elastic contact area, curvature radius, and elastic deformation can be obtained as equation (6) by the contact geometric deformation of micro-convex.

$$\delta_c = \left[\pi K \sigma_y / (2E) \right]^2 R = \left(K\phi / 2 \right)^2 \cdot a^{D/2} / G^{D-1} \tag{4}$$

$$a_c = \left(K\phi / 2 \right)^{2/(1-D)} G^2 \tag{5}$$

$$a = \pi R \delta \tag{6}$$

Where, a is the elastic contact area, H is the surface hardness, σ_y is the material yield strength, and E is the synthetic Yang's modulus. K is a ratio of surface hardness to the material yield strength, which is expressed as $K=H/\sigma_y$, and the material parameter ϕ is expressed as $\phi=\sigma_y/E$.

$$\begin{cases} p_e=\frac{4}{3}ER^{1/2}\delta^{3/2} \\ p_p=\mathrm{H}a \end{cases} \Rightarrow \begin{cases} p_e(a)=\frac{4}{3}\sqrt{\pi}EG^{D-1}a^{(3-D)/2} & \text{Elastic} \\ p_p(a)=K\sigma_y a & \text{Plastic} \end{cases} \tag{7}$$

From the Hertz elastic contact theory, the normal elastic load of a micro-convex is expressed in equation (7), and when the plastic deformation of micro-convex is produced, the plastic load related to hardness and contact area can be expressed in equation (7).

3.2. Model of Normal Load and Normal Contact Stiffness

To defined a_l as the max contact area of a micro-convex, any contact point of the tooth surface is simplified as two cylindrical contacts. The contact coefficient of cylindrical surface λ_c is introduced since the number of micro-convex is different between two cylindrical contacts and one cylinder and one flat contact. So that the area distribution function of tooth surface contact points $n(a)$ is given as equation (8). Which B is the tooth width, R_1 is the curvature radius of contact point for pinion and R_2 is the curvature radius for gear.

$$n(a)=\left|\frac{dN(A>a)}{da}\right|=\frac{1}{2}\lambda_C D\frac{a_l^{D/2}}{a^{D/2+1}} \quad and \quad \lambda_c=\left[(\frac{4B}{\pi E}\frac{R_1R_2}{R_1+R_2})^{1/2}/(\pi R_1+\pi R_2)\right]^{\frac{1}{R_2}+\frac{1}{R_1}} \tag{8}$$

When the a_l is greater than the a_c, the contact state of asperity surface includes two contact formation, one is elastic contact, and the other is plastic. The total normal load can be derived as equation (9) by combine with elastic load and plastic load.

$$P=\int_{a_c}^{a_l}p_e(a)n(a)da+\int_{a_s}^{a_c}p_p(a)n(a)da=\frac{4}{3}\lambda_c\sqrt{\pi}EG^{D-1}\frac{D}{3-2D}a_l^{D/2}\left(a_l^{(3-2D)/2}-a_c^{(3-2D)/2}\right)+\frac{D}{2-D}\lambda_c K\sigma_y a_l^{D/2}a_c^{(2-D)/2} \tag{9}$$

When the a_l is smaller than the a_c, the asperity surface exhibits a plastic contact state, and its total normal load only contains a plastic load, which is derived as equation (10).

$$P=\int_{a_s}^{a_l}p_p(a)n(a)da=D\lambda_c K\sigma_y a_l/(2-D) \tag{10}$$

According to the relation between elastic load and deformation in equation (7), the normal contact stiffness is obtained by differential calculation of the contact deformation amount by the elastic load, as shown in equation (11).

$$k_n=dp_e/d\delta=2ER^{1/2}\delta^{1/2} \tag{11}$$

Substitute equation (6) into equation (11), the total normal contact stiffness related to the contact area is derived as equation (12) by combine with the area distribution function of contact points $n(a)$ (equation (8)).

$$K_n=\sum k_n=\int_{a_c}^{a_l}k_n n(a)da=\frac{2E\lambda_c}{\sqrt{\pi}}\frac{D}{1-D}a_l^{D/2}(a_l^{(1-D)/2}-a_c^{(1-D)/2}) \tag{12}$$

3.3. Nondimensionalization of the Proposed Model

While the a_l is greater than the a_c, the tooth contact is in the form of incomplete plasticity. In order to further investigate the relationship between total load and contact area and reduce the input of parameters in numerical simulation, a fractal contact model considered the asperity surface in the dimensionless form is given as equation (13), and the dimensionless parameters and simplified expressions in equation (13) satisfy the following formulas in equation (14).

$$\begin{cases} \text{For } D \neq 1.5, \; P^* = \dfrac{4\lambda_c\sqrt{\pi}}{3} G^{*D-1} g_1(D) A_r^{*D/2} \left[\left(\dfrac{2-D}{D\lambda_c} A_r^* \right)^{(3-2D)/2} - a_c^{*(3-2D)/2} \right] + \lambda_c K \phi g_2(D) A_r^{*D/2} a_c^{*(2-D)/2} \\ \text{For } D = 1.5, \; P^* = \sqrt{\pi}\lambda_c G^{*1/2} \left(\dfrac{A_r^*}{3\lambda_c} \right)^{3/4} \ln\left(\dfrac{A_r^*}{3\lambda_c a_c^*} \right) + 3\lambda_c K\phi \left(\dfrac{A_r^*}{3\lambda_c} \right)^{3/4} a_c^{*1/4} \end{cases} \tag{13}$$

$$g_1(D) = \frac{D}{3-2D}\left(\frac{2-D}{D\lambda_c}\right)^{D/2}, g_2(D) = \left(\frac{D}{2-D}\right)\left(\frac{2-D}{D\lambda_c}\right)^{D/2} \quad and \quad P^* = \frac{P}{A_a E}, G^* = \frac{G}{\sqrt{A_a}}, A_r^* = \frac{A_r}{A_a}, a_c^* = \frac{a_c}{A_a} \tag{14}$$

Where, P^* is the dimensionless normal load, G^* is dimensionless fractal roughness parameter, A_r^* is the dimensionless real contact area, a_c^* is the dimensionless critical contact area, A_r is the real contact area, A_a is the nominal contact area, and $g_1(D)$ and $g_2(D)$ are the expressions of fractal dimension and contact coefficient of cylindrical surface. The normal contact stiffness K_n^*(equation (12)) considered the asperity surface in a dimensionless form is given as equation (15). $g_3(D)$ is the expression of fractal dimension and contact coefficient of cylindrical surface.

$$K_n^* = \frac{2\lambda_C}{\sqrt{\pi}} g_3(D) A_r^{*D/2} \left[\left(\frac{2-D}{D\lambda_C} A_r^* \right)^{(1-D)/2} - a_c^{*(1-D)/2} \right] \quad and \quad g_3(D) = \frac{D}{1-D}\left(\frac{2-D}{D\lambda_c} \right)^{D/2} \tag{15}$$

From equation (13) and equation (15), the dimensionless normal load P^* and dimensionless normal contact stiffness K_n^* have a certain mathematical expression correlation with the dimensionless real contact area A_r^* and the dimensionless critical contact area a_c^* respectively. the mathematical relationship of the dimensionless normal load P^* and dimensionless normal contact stiffness K_n^* is established by the above relationship, which is given as equation (16).

$$P^* \sim (A_r^*, a_c^*), K_n^* \sim (A_r^*, a_c^*) \Rightarrow (K_n^* \sim P^*) \tag{16}$$

4. Simulation and Discussion of the Proposed Model

In order to visualize the simulation data display, the drawing method with a single logarithmic coordinate is used to draw the simulation results. Figure 6 (λ_c=0.7, G^*=10^{-10}) shows that the normal load increases with the increase of real contact area, and decreases first and increases subsequently with the increase of fractal dimension. It means that the normal load has a minimum extreme value about the fractal dimension D. Therefore, fractal dimension D from 1.50 to 1.65 is selected for the detailed simulation analysis of normal load, the result is shown in Figure7. The three-dimensional topological diagram of normal load present concave with the variable of fractal dimension and real contact area in Figure7a. And the curved surface concave is more obvious with the increase of the real contact area. It can be seen in the contour diagram (Figure 7b) that the minimum extreme value of normal load approximately appears when the fractal dimension is equal to 1.57. Figure8 shows that the normal contact stiffness increases with the increase of fractal dimension, and increases with the increase of real contact area.

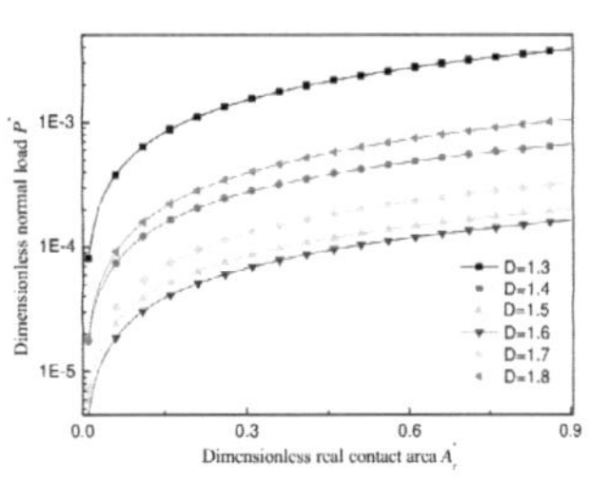

Figure 6. Relationship of D and P^*

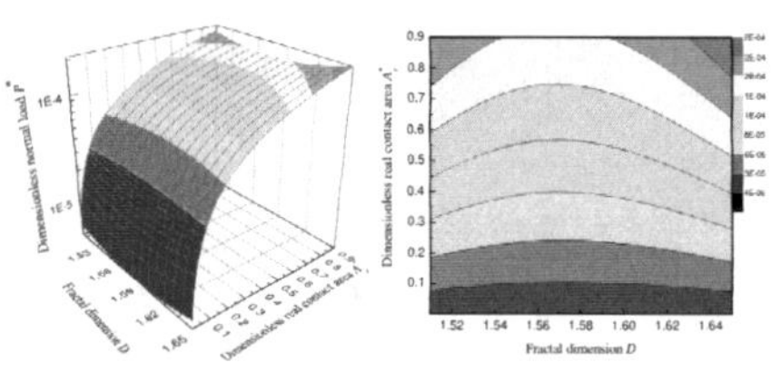

(a) D ~ P^* (b) The contour map

Figure 7. Relationship of D and P^*

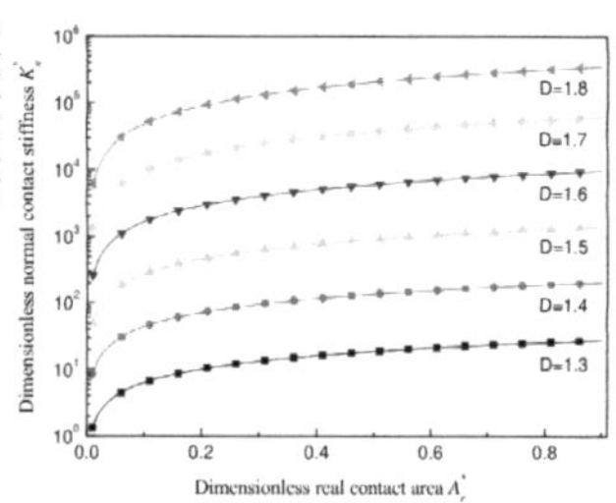

Figure 8. Relationship of D and K_n^*

The influence of roughness parameter on normal load and normal contact stiffness are analyzed with the fractal dimension and contact coefficient of cylindrical surface (D=1.6, λ_c=0.7). In Figure 9, the normal load increases with the increase of roughness parameter, but the normal contact stiffness decreases with the increase of roughness parameter. It can be obtained from equation (5) that critical contact area which exchanges from elastic to plastic decreases with the decrease of roughness parameter under the same conditions of hardness coefficient, material parameter, and nominal contact area, then the surface smoothness and normal contact stiffness are improved. Considering the result of Figure6, Figure8, and Figure9, there is a certain coordination relation between the fractal dimension and the roughness parameter, which makes the optimal effect of normal load and normal contact stiffness.

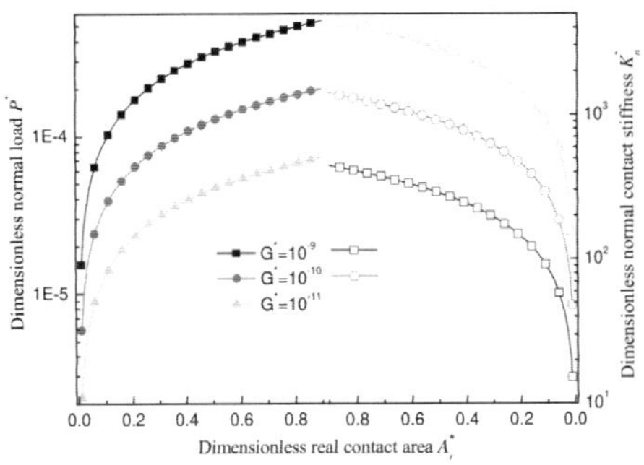

Figure 9. Effect of roughness parameter G^* on normal load P^* and normal contact stiffness K_n^*

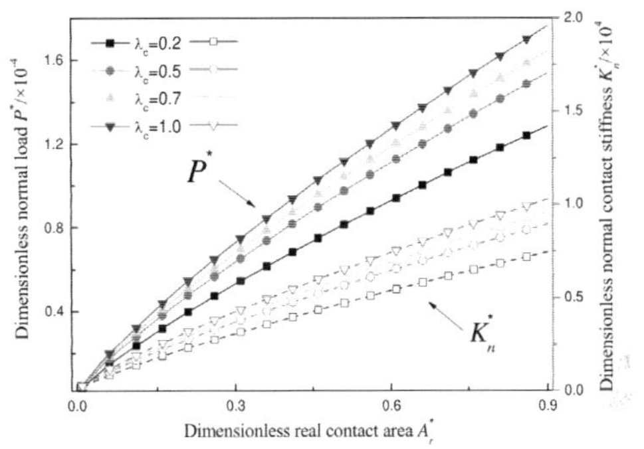

Figure 10. Effect of cylindrical surface contact coefficient λ_c on normal load P^* and normal contact stiffness K_n^*

Figure10 shows that the normal load increases with the increase of contact coefficient of cylindrical surface, and the increased curve presents nonlinearly, the load difference decrease gradually in the same real contact area. The effect of real contact area and contact coefficient of cylindrical surface on normal contact stiffness has the same variational tendency as the effect on normal load.

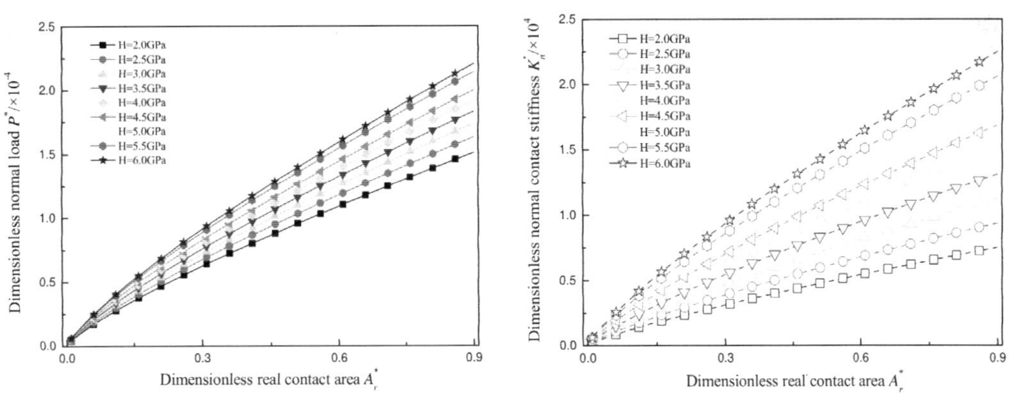

Figure 11. Effect of surface hardness on normal load P^* and normal contact stiffness K_n^*

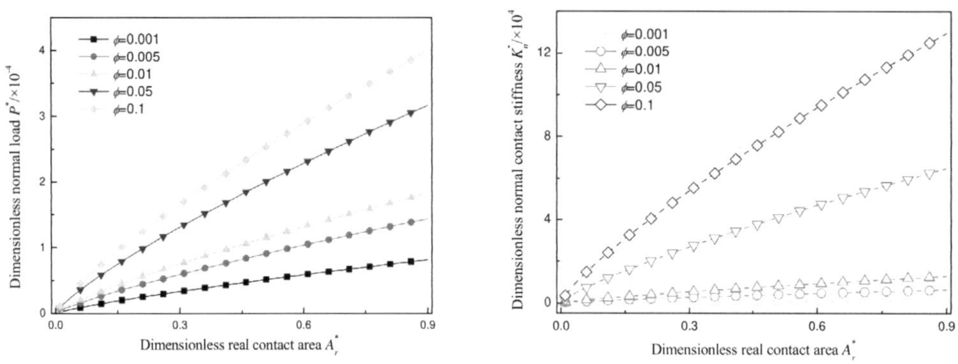

Figure 12. Effect of material parameter on normal load P^* and normal contact stiffness K_n^*

The effect of surface hardness and material parameter on normal load and normal contact stiffness is shown in Figure11 and Figure12, the normal load and normal contact stiffness both increase with the increase of the variable H and ϕ. And the effect of two variable on the above on the critical contact area are further analyzed. Figure13 shows that the critical contact area decreases nonlinearly with the increase of surface hardness and material parameter, but increases with the increase of roughness parameter. It means that under the same cold machining technology, the critical contact surface which exchanges from elastic to plastic is improved by the application of different material and heat treatment technology, so as to improve its load capacity and deformation resistance.

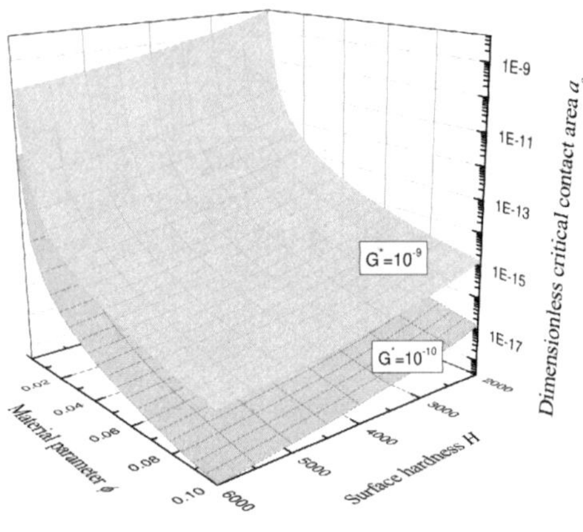

Figure13. Effect of material parameter and surface hardness H on critical contact surface a_c^*

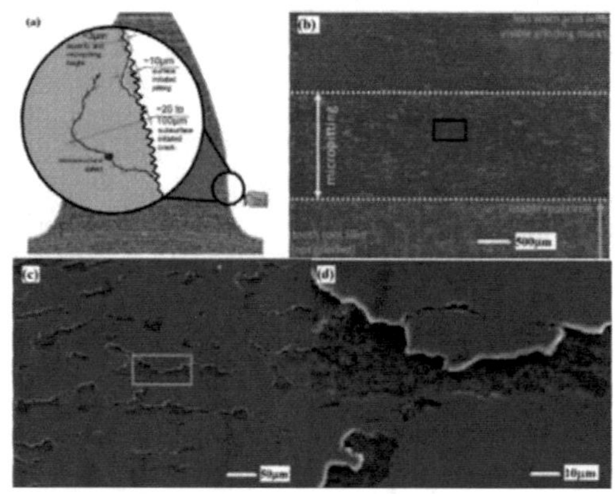

Figure14. Scanning electron microscope diagrams of a micro-pitting area [17]

Through the discussion of the influence of various parameters on the above, the parameters include fractal dimension, roughness parameter, contact coefficient of cylindrical surface, etc. do not affect the normal load and normal contact stiffness alone, but act together and to determine the surface micro-topography feature. Therefore, the influence of asperity surface factors on normal load and normal contact stiffness should be considered comprehensively for the engineering application of fractal theory.

5. Normal Contact Stiffness Model of Micro-pitting of Tooth Surface

The formation mechanism of micro-pitting and the relevant research in reference [18-20] shows that micro-pitting is produced firstly on the tooth surface of pinion, and the micro-pitting area mainly occurs from the pitch line to the dedendum (Figure1 and Figure14). The feature size of pitting ranges from 0.01mm (micro-pitting) to 0.8mm (extended pitting), and the failure feature of spalling occurs on the tooth surface in which the pitting expands under the load further. The cracks include surface crack and sub-surface crack are produced on the tooth surface as shown in Figure14a. Under the joint action of engagement force and lubricating oil press, the maximum feature size of a micro-pitting caused by surface crack is about 10 μm, while the feature size of micro-pitting caused by sub-surface crack ranges from 20 μm to 100 μm. Because the dimension feature of micro-pitting is very small, the mathematical analysis model of micro-pitting is established according to the simulation method of the roughness of asperity surface [21, 22], and to simulate the different size of micro-pitting by the change of tooth surface roughness [23].

According to the measured data and theoretical calculation of the asperity surface with different machining technology in the references [24, 25], the approximate relationship between roughness, fractal dimension and roughness parameter is obtained which presents power-exponential relation and is given as equation (17).

$$Ra^{0.042} = 1.528 / D = -5.26 / \lg G \tag{17}$$

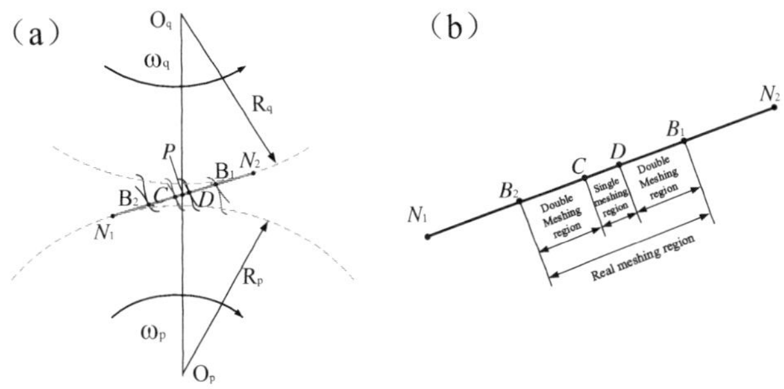

Figure 15. Meshing process of a spur gear pair

Table 1. Calculation parameters of involute spur gear pair

Physical parameters	Value
Number of teeth	25, 33
Module /mm	4
Face width /mm	28
Pressure angle /deg	20
Addendum coefficient	1
Tip clearance coefficient	0.25
Yang's modulus /N/mm2	2.06×10^5
Poisson's ratio	0.3
Yield strength of material/ N/mm2	835
Power /kW	30
Speed of revolution /r/min	1000

Figure15a shows that a meshing process of a spur gear pair in which the contact ratio ranges from 1 to 2, and the tooth surface along the meshing line is divided for single meshing region and double meshing region in Figure15b. The segment *CP* in the single meshing region, which the main region of micro-pitting occurs on the tooth surface, is selected to investigate the effect of micro-pitting on normal contact stiffness. The main calculation parameters of spur gear pair are shown in Table 1

With the drawn method of Bobillier and the basic parameter of tooth profile of involute gear, the relationship between curvature radius and engagement angle at any engagement point in segment *CP* of the single meshing region is derived as equation (18). And the contact half-width related to the nominal contact area and engagement angle is derived as equation (19) according to the Hertz elastic contact theory of two-cylinder.

$$\rho_{pm} = mz_p \cos\alpha \tan\alpha_{pm} / 2 \ and \ \rho_{qm} = mz_q \cos\alpha \tan\alpha_{qm} / 2 \tag{18}$$

$$F = 9549 \cdot \frac{P}{nR_{bp} \cos \alpha_{pm}}, a = \sqrt{\frac{8F}{\pi B} \frac{\rho_{pm} \rho_{qm}}{\rho_{pm} + \rho_{qm}} \left(\frac{1-\upsilon_1^2}{E_1} + \frac{1-\upsilon_2^2}{E_2} \right)}, A_a = 2aB \tag{19}$$

The derived parameter relation with engagement angle is obtained by bring the basic parameter of involute spur gear pair into the equation (18), equation (19) and equation (8). Figure16 shows that the comprehensive curvature radius and surface contact coefficient increase slightly from engagement point C to pitch point P. The engagement force and nominal contact area are also increased in Figure17.

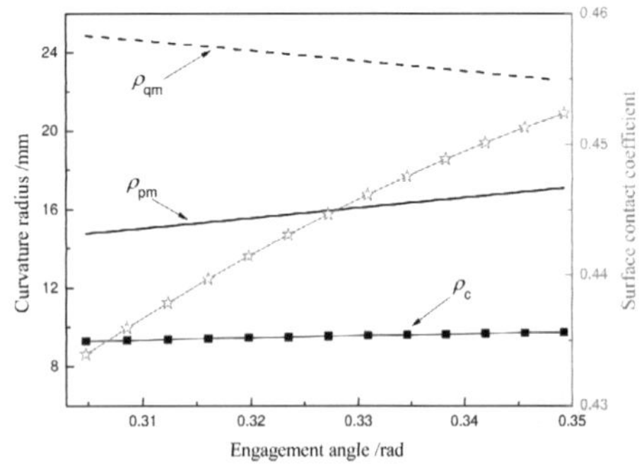

Figure 16. Diagram of curvature radius and surface contact coefficient with engagement angle

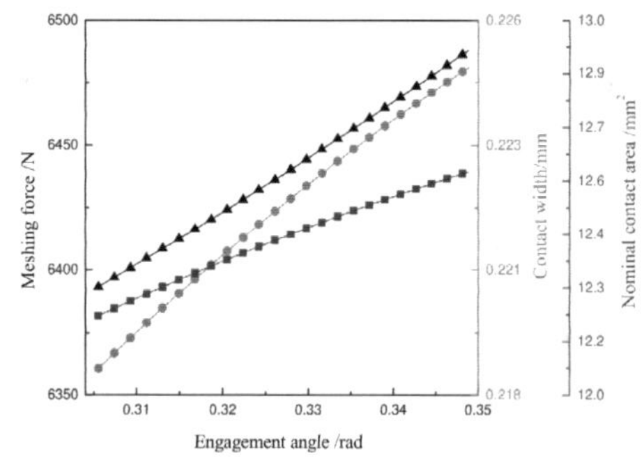

Figure 17. Diagram of load, semi-width of contact region and normal contact surface with engagement angle

The seven different roughness are selected to calculate the normal contact stiffness of tooth surface, which simulates the existed machining technology of gear (hobbling, grinding, finishing, etc.). the four first kinds are to simulate engineering application of roughness of tooth surface accuracy. And the latter three kinds are to simulate the different levels of micro-pitting of tooth surface topography. The fractal parameter of the proposed model is obtained by substitute the simulated value to equation (17). The result in Table 2 shows that the fractal dimension decreases with the increase of roughness, which means the fractal is rougher, but the roughness parameter increases with a roughness, which means the surface smoothness decrease. The variational tendency is consistent with the real surface topography of asperity surface.

Table 2. Fractal dimension for different roughness

	Roughness	Fractal dimension D	Roughness parameter G
Different machining technology	0.1	1.68	$10^{-5.79}$
	0.8	1.54	$10^{-5.31}$
	1.6	1.50	$10^{-5.16}$
	3.2	1.46	$10^{-5.01}$
Simulated micro-pitting features	12.5	1.37	$10^{-4.73}$
	20	1.35	$10^{-4.64}$
	50	1.30	$10^{-4.46}$

The design and calculation parameters are brought into the proposed micro-pitting model of normal contact stiffness. Figure18 shows that normal contact stiffness increases slightly along the segment CP with the different machining technology, which is different from the linear model of Yang-Sun [26]. The decrease of roughness, which means the accuracy of tooth surface is higher, gives rise to an increase of the normal contact stiffness. The level of normal contact stiffness at the Ra=0.1 is obviously higher than others, which has the same effect as improving the surface accuracy to enhance the contact fatigue strength of gear [27]. Comparison with the Yang-Sun model and semi-empirical formula which is proposed in the NASA technical report [28], the level of normal contact stiffness is 10^9 N/m under the ordinary machining technology between the proposed model and the two models. The curve of Yang-Sun model is between the resulting curve of Ra=0.8 and Ra=1.6. Since the

Yang-Sun model is independent of load and deformation, it means that the contact stiffness is a constant when the geometric parameters are determined, which is contrary to the actual condition. The semi-empirical formula is between the resulting curve of Ra=1.6 and Ra=3.2. The different result in the Figure19 is produced as the morphology of asperity surface is not taken into account in the two models. The normal contact stiffness has a different response with the different machining technology in the results. When high quality, high parameter and high performance of gear are needed for engineering, in addition to considering the macroscopic geometric size, the contact feature of gear needs to be analyzed qualitatively and quantitatively under different machining technology. And the Yang-Sun model and semi-empirical formula can be used as the initial stiffness prediction of gears with low-quality requirements or the initial stiffness prediction of new product design.

When the micro-pitting is produced on the tooth surface, the result of normal contact stiffness in Figure20 decreases obviously. Comparison with the contact stiffness of the non-pitting surface, the two results differ 1 or 2 order of magnitude. The load capacity of tooth is also reduced. The simulated result is consistent with the engineering application, and the proposed mathematical model of micro-pitting can be used as a basis for an analysis of dynamic response and expansion of micro-pitting.

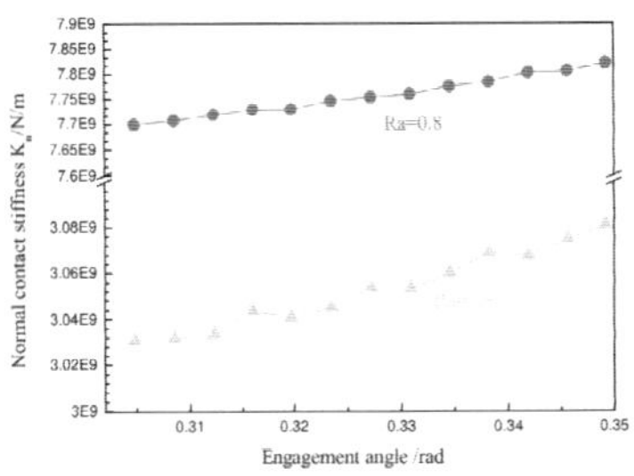

Figure18. Diagram of normal contact stiffness (Ra=0.8,1.6) with engagement angle

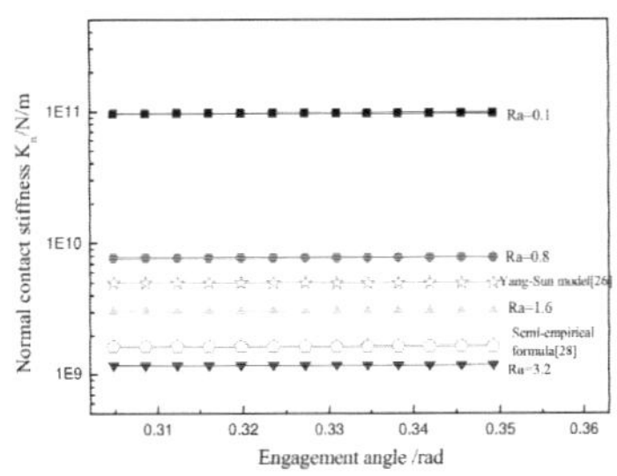

Figure 19. Normal contact stiffness with different machining technology

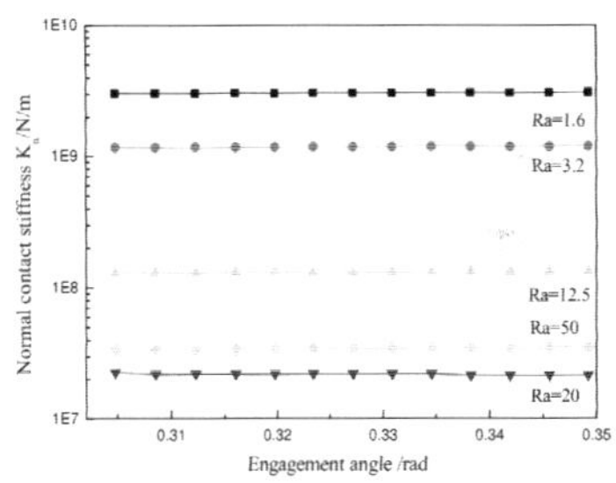

Figure 20. Normal contact stiffness of micro-pitting characteristics

6. Results

Based on the fractal theory of asperity surface, the mathematical fractal prediction model of surface morphology of micro-pitting is established to investigate quantitatively the influence of micro-pitting feature on normal contact stiffness combine with the forming physical phenomenon of micro-pitting. The results are shown in the below:

1) The normal load decreases first and increase subsequently with the increase of fractal dimension. It means that the normal load has a minimum extreme value about the fractal dimension. And the normal load also increases with the increase of roughness parameter, surface contact coefficient, surface hardness, and material parameter. the normal contact stiffness increases with the increase of the fractal parameters except the roughness parameter.

2) The normal contact stiffness of tooth surface is improved by improving the surface accuracy, which has the same effect as improving the surface accuracy to enhance the contact fatigue strength of gear. Comparison with the Yang-Sun model and semi-empirical formula, the proposed fractal model of normal contact stiffness can roundly predict the normal contact stiffness under different machining technology.

3) When the micro-pitting is produced on the tooth surface, the result of normal contact stiffness decreases obviously. The load capacity of tooth is also reduced. The simulated result is consistent with the engineering application, and the proposed mathematical model of micro-pitting can be used as a basis for an analysis of dynamic response and expansion of micro-pitting.

7. Acknowledgement

The authors gratefully acknowledge the support by the National Key Research and Development Plan of China through Grants No. 2018YFB2001700.

IOP Conf. Series: Materials Science and Engineering **746** (2020) 012048 doi:10.1088/1757-899X/746/1/012048

8. References

[1] Alban L E. *Systematic analysis of gear failures* (Ohio: American Society for Metals).
[2] Luo Y, Baddour N and Liang M *Mechanical Systems and Signal Processing* **119** 155-81.
[3] Liang XH, Zhang HS, Liu LB and Zuo MJ *Mechanism and Machine Theory* **106** 1-15.
[4] Sanchez MB, Pleguezuelos M and Pedrero JI *Mechanism and Machine Theory* **109** 231-49.
[5] Ge SR, Chen G. *Wear* **231** 249-55.
[6] Huang K, Zhao H, Chen Q *Tribology* **28** 529-33.
[7] Wang R, Zhu L and Zhu C *International Journal of Mechanical Sciences* **134** 357-69.
[8] Tian HL, Zhu DL and Qin HL *Journal of China Three Gorges University* **24** 83-88.
[9] Tian HL, Zhong XY, Qin HL, Zhao CH, Fang ZF, Zhu DL, Chen BJ and Zhang FJ. *Journal of Mechanical Engineering* **49** 108-22.
[10] Goerke D, Willner K. *Wear* **264** 589-98.
[11] Kucharski S, Starzyński G. Wear **440-441** 1-14.
[12] Pan W, Li X, Wang LL, Guo N and Mu JX. *European Journal of Mechanics - A/Solids* **66** 94-102.
[13] Yuan Y, Cheng Y, Liu K and Gan L. *Applied Surface Science* **425**1138-57.
[14] Zhao LL, Huang XP and Zhou ZW. *Journal of Chongqing University of Technology* **08** 78-83.
[15] Chen Q, Zhao H, Huang K and Xu S. *China Mechanical Engineering* **21** 1014-17.
[16] Majumdar A, Bhushan B. *ASME Journal of Tribology* **113** 1-11.
[17] Weibring M, Gondecki L and Tenberge P. *Tribology International* **131** 299-307.
[18] Li S, Kahraman A. *International Journal of Fatigue* **59** 224-33.
[19] Morales-Espejel G, Rycerz P and Kadiric A, *Wear* **38** 99-115.
[20] Shipley EE, Machine Design **39** 152–162.
[21] Chen H, Hu YZ, Wang H and Wang WZ. *Lubrication Engineering* **10** 52-55.
[22] Jiang YJ, Huang WQ, Sun ZY and Sun QC. *Machinery Design and Manufacture* **08** 8-10.
[23] Miao X, Huang X. *Wear* **309** 146-51.
[24] Chen Q. Hefei University of Technology. 2010.
[25] Ge SR, Zhu H *Fractal tribology* (China Machine Press).
[26] Yang DCH, Sun Z. *Journal of Mechanisms, Transmissions & Automation in Design* **107** 529-35.
[27] Zhao GH, Liu SJ, Li JQ and Wang XP. *Journal of Mechanical Strength* **395** 40-44.
[28] Lin HH, Townsend DP and Oswald, FB. *Profile Modification to Minimize Spur Gear Dynamic Loading* (NASA Technical Memorandum 89901) pp 1-22.

IOP Conf. Series: Materials Science and Engineering **746** (2020) 012049 doi:10.1088/1757-899X/746/1/012049

Parametric Design and Numerical Control Simulation of Cylindrical cam Based on Ug8.5

Xuanlin Ye

College of mechanical and electrical engineering, Yunnan Open University, Kunming 650500, China
E-mail: 190516046@qq.com

Abstract. This paper introduces the method and process of parametric modeling and numerical control programming simulation processing of cylindrical cam by using ug8.5 software, improves the design, manufacturing efficiency and precision of cylindrical cam, and realizes the serial design of cylindrical cam. It has been proved by practice that the performance of cylindrical cam is stable and efficient, which greatly shortens the machining cycle of cam and improves the efficiency.

1. Introduction

The key of cam design is to design the contour of cam working part. The traditional cam design and processing method adopts the drawing method to design the cam contour and the marking method to process the cam. The design efficiency is low and the design and processing accuracy is not high, which restricts the quality of mechanical transmission. With the development of science and technology, mechanical equipment is developing towards high-speed precision and automation, which puts forward higher requirements for the accuracy of cam mechanism [1]. Using ug8.5 software to carry on the parametric design and the numerical control processing programming to the cylinder cam, using the numerical control processing center to carry on the processing, has improved the design, the manufacture efficiency and the precision. With the development of computer technology, CAD / CAM is widely used in cam design and machining, which is of great significance to small and medium-sized machining enterprises in China [2, 3].

2. Parametric Design of Cylindrical cam

2.1. Design Thinking

According to the motion equation of the cam and the parameters in the parameter group, use the expression function of ug8.5 to draw the contour of the cam, substitute the set variable names in the expression, use the function of regular curve to draw the contour of the cam, then use the function of surface stretching and thickening to generate the cam entity, use the function of difference and cylinder difference to generate the cam workpiece, add the parameter variables, real Now it is parameterized.

2.2. Design Method and Steps Based on UG

2.2.1. Parameter analysis of cylindrical cam

Now the cylindrical cam is designed, the radius of the cylinder is r_0 = 40 mm, the height of the cylinder is H = 300 mm, the stroke of the follower is h = 20 mm, the width of the curve groove is b =

12 mm, the depth of the groove is 6 mm, the angle of motion of the thrust is 120°, the angle of far repose is 60°, the angle of motion of the return is 120°, the angle of near repose is 60°, and the thrust and return of the follower are in simple harmonic motion. (1) when the cam turns 120°, the follower will push h according to the cosine acceleration movement law; (2) when the cam turns 60°, the follower will stop moving; (3) when the cam turns 120°, the follower will return h according to the cosine acceleration movement law; (4) when the cam turns 60°, the follower will stop moving again. According to the above conditions, the actual contour curve of the cam is designed and the three-dimensional entity of the cam is generated.

2.2.2. *Parameter equation of cylindrical cam*

According to the knowledge of mechanical principle, the parameter equation of theoretical contour curve of cylindrical cam [4] is

$$\begin{cases} x = r_0 \cos j \\ y = r_0 \sin j \quad (0° \le j \le 360°) \\ z = s \end{cases} \tag{1}$$

Where x, y and z represent the coordinates of any point on the curve, s represents the motion law of the follower, r_o represents the radius of the base circle, and j represents the angle of the cam.

2.2.3. *UG expression of cam*

According to the above conditions, the working contour curve of the cam is designed. According to formula (1), the UG expression of each segment can be obtained as shown in Table 1. It should be noted that t is the UG system variable with a value of $0 \le t \le 1$.

Table 1. Ug expression of each segment of working curve of cylindrical cam

Motion process	Name	Formula	Notes
	X_{t1}	r_o *cos(j_1)	push X coordinate value variable
push stroke	Y_{t1}	r_o *sin(j_1)	Push Y coordinate value variable
	Z_{t1}	(h/2)*(1-cos(180*t))	Push Z coordinate value variable
	X_{t2}	r_o *cos(j_2)	Far rest Xcoordinate value variable
Far from rest	Y_{t2}	r_o *sin(j_2)	Far rest Y coordinate value variable
	Z_{t2}	h	Far rest Z coordinate value variable
	X_{t3}	r_o *cos(j_3)	Return X coordinate value variable
return stroke	Y_{t3}	r_o *sin(j_3)	Return Y coordinate value variable
	Z_{t3}	(h/2)*(1+cos(180*t))	Return Z coordinate value variable
	X_{t4}	r_o *cos(j_4)	Near rest X-coordinate value variable
Near rest	Y_{t4}	r_o *sin(j_4)	Near rest Y coordinate value variable
	Z_{t4}	0	Near rest Z coordinate value variable

2.2.4. *Parametric modeling*

Substituting $r_o = 40$ mm, h = 20 mm, $a_1 = 120°$, $a_2 = 60°$, $a_3 = 120°$, $a_4 = 60°$, $j_1 = a_1 * t$, $j_2 = a_1 + a_2 * t$, $j_3 = a_1 + a_2 + a_3 * t$, $j_4 = a_1 + a_2 + a_3 + a_4 * t$, into the above formula (where a1 is the thrust angle, a2 is the far angle of repose, a3 is the return angle, a4 is the near angle of repose), using the regular curve, each curve corresponds to each group of object parameters (X_t, Y_t, Z_t) in the table, Then use the derived curve function of ug8.5 to connect the four curves as a whole, as shown in Figure 1. Select ug8.5 surface lifting command, select the curve just connected to the section curve with a distance of 12 mm, as shown in Figure 2, select the thickening command of the offset / zoom function of ug8.5 with a direction inward and a distance of 6 mm, as shown in Figure 3, select the difference command of the combined function of ug8.5, select the cylinder as the target body, and select the body as shown in Figure 3 as the tool selection body, with the result as shown in Figure 4, and then Select the whole column feature command, and the result is shown in Figure 5.

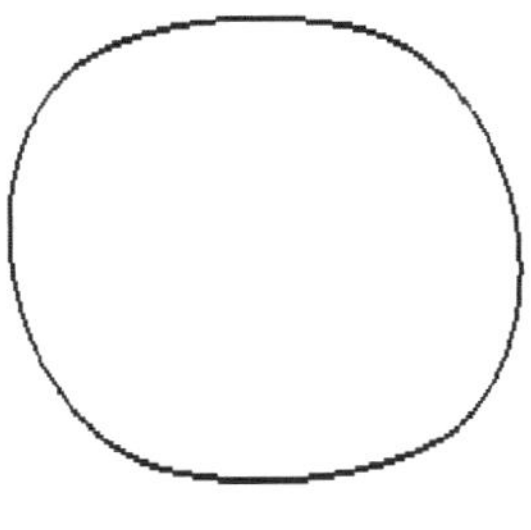

Figure 1. Cam curve

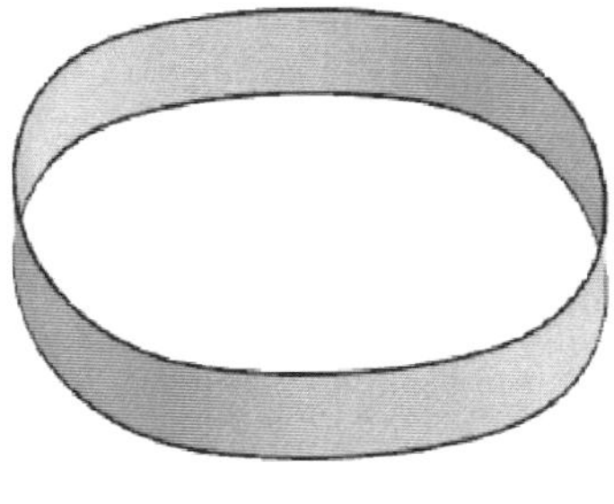

Figure 2. Extrude surface

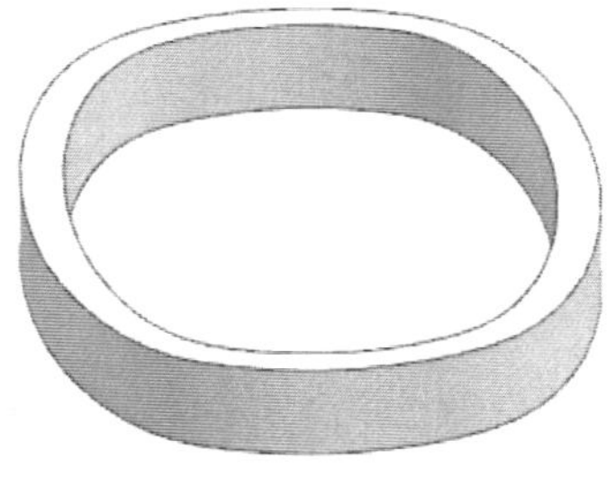

Figure 3. Thickened body

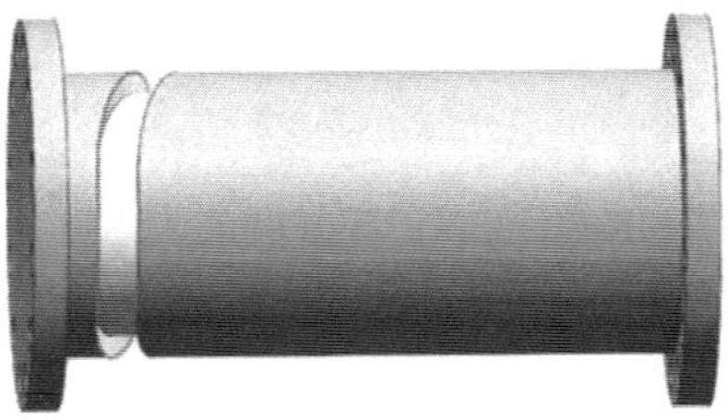

Figure 4. Cylindrical cam

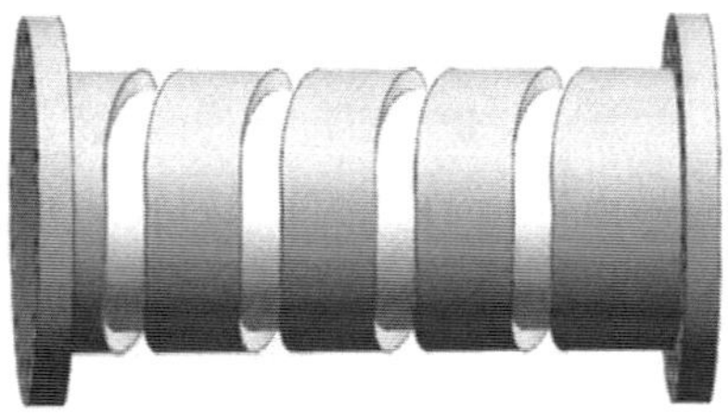

Figure 5. Whole row of rear cylindrical cams

3. Numerical Control Programming of Cylindrical Cam

3.1. Rough Machining

Using the derivative curve function of ug8.5, the on surface offset command offsets the contour curve of the bottom of the groove on the cylindrical surface of the bottom of the groove, selects the curve whose offset distance from the contour on the left side of the cam groove wheel to the middle of the groove is 6 mm as the driving curve, enters the machining environment to create the Φ 6 tool, creates the work order, selects the cylinder at the bottom of the groove as the part geometry, the driving method is curve driving, and the projection vector is The cutter axis and cutter axis shall be far away from the straight line, and the straight line shall be selected as the cam axis. In order to prevent over cutting at the turning point of the driving curve, the advance / retreat cutter mode shall be set as interpolation in the non cutting moving parameters, and multiple cutter paths shall be set in the cutting parameters, with the component offset of 6 mm, each cutter feeding of 1 mm, the component margin of 0.2 mm, and the tolerance of 0.03. The generated cutter path is shown in Figure 6, and the simulation machining is shown in Figure 7.

3.2. Semi Finish Machining

Using the derivative curve function of ug8.5, the on surface offset command offsets the contour curve of the cam at the bottom of the groove on the cylindrical surface of the cam groove. Select the curve whose offset distance from the left and right contour of the cam groove wheel to the middle of the groove is 2.05 mm as the driving curve, enter the processing environment to create the Φ 4 cutter, create the process, select the cylinder at the bottom of the groove as the part geometry, and the driving method is curve driving, The projection vector is the cutter axis, the cutter axis is far away from the straight line, and the straight line is the cam axis. In order to prevent over cutting at the turning point of the driving curve, the advance / retreat cutter mode is set as interpolation [5] in the non cutting moving parameters, and multiple cutter paths are set in the cutting parameters, with the component offset of 6 mm, each cutter feeding of 1 mm, the component allowance of 0.05 mm, and the tolerance of 0.01. The generated cutter path is shown in Figure 8, and the simulation processing is shown in Figure 9.

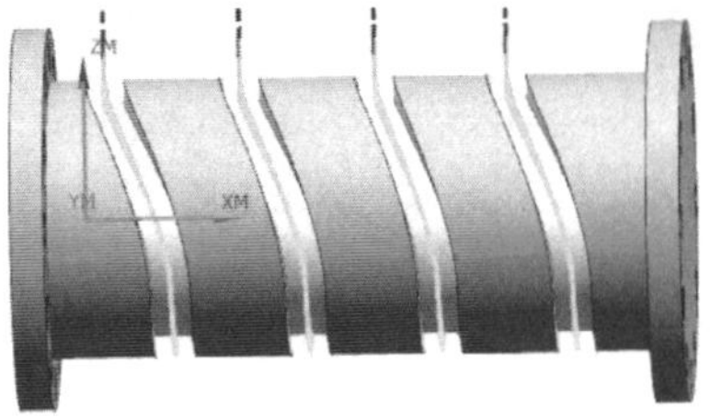

Figure 6. Path of rough machining tool

Figure 7. Rough machining simulation drawing

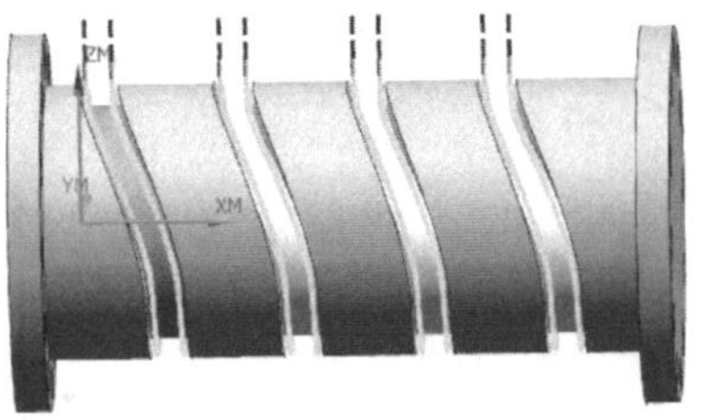

Figure 8. Path of semi finishing tool

Figure 9. Semi finishing simulation

3.3. Finish Machining
The processing of the bottom of the cam groove only needs to modify the corresponding parameters on the basis of the tool path of machining and semi finishing. For the processing of cam groove side, the driving method needs to be changed. The side of cam groove is selected as the driving surface and components, and the projection vector is selected to face the driving body [6]. The generated tool path is shown in Figure 10, and the simulation processing is shown in Figure 11.

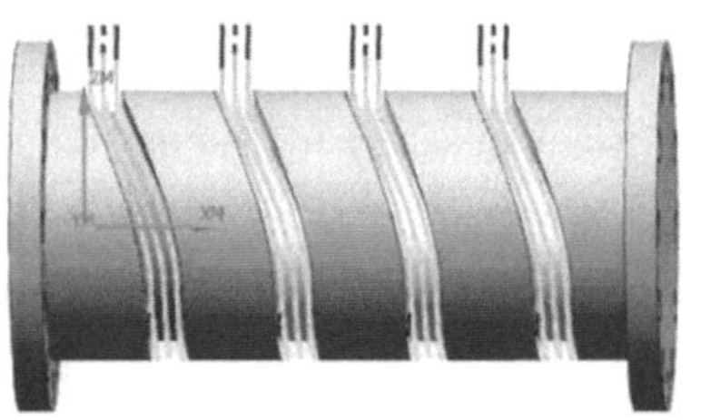

Figure 10. Path of finish machining tool

Figure 11. Finishing simulation drawing

4. Conclusion
The traditional design method, because of the complexity of the design process, the low precision, and the calculation results can not be directly used in NC machining (the NC machining of the cam needs CAD 3D modeling), has become more and more unsuitable for the current requirements of rapid and accurate cam design and NC machining UG software ensures the accuracy of part modeling, and provides a variety of simulation processing strategies, which has good operability, and generates NC program code quickly and accurately. Using CAD technology to replace the traditional cam design method can greatly shorten the design cycle, improve the design quality, and meet the objective requirements of NC machining.

5. Acknowledgements
The corresponding author of this paper is Xuan Lin Ye. This paper is supported by basic application research of Yunnan Science and Technology Department, Project number: 2013FD062, the People in charge of the project: Xuan Lin Ye.

IOP Conf. Series: Materials Science and Engineering **746** (2020) 012049 doi:10.1088/1757-899X/746/1/012049

6. Reference

[1] Y. L. Han: Manufacturing Technology & Machine Tool, Vol. 56 (2007) No. 11, p44-45.
[2] Q. Yu, R. C. Liu, C. M. Chen, S. Y. Ma, K. Zhang: Mechanical Research & Application, Vol. 23 (2006) No. 04, p78-79.
[3] D. Cheng: Mechanical Design Manual Single bank organization (Chemical Industry Publishing, China, 2004), p128-129.
[4] Y. H. Su: Coal Mine Machinery, Vol. 31 (2011) No. 02, p115-116.
[5] Beijing Zhaodi Technology Co., Ltd: UG NX 8.5 NC machining course (Mechanical Industry Press, China, 2013), p171-176.
[6] J. Yu: Modern Manufacturing Technology and Equipment, Vol. 52 (2017) No. 06, p34-35+37.

IOP Conf. Series: Materials Science and Engineering **746** (2020) 012050 doi:10.1088/1757-899X/746/1/012050

Research on Automatic Spray Equipment of Thin Spray Support Technology for New Material in Coal Mine Roadway

Huihui Dong[1], Andong Lou[2,*], Jianguo Zhu[2] and Fusong Min[1]

[1]Nanjing Coal Science & Technology Research Co., Ltd., Nanjing, Jiangsu 210018, China

[2]Luoyang Mining Machinery Engineering Design and Research Institute Co., Ltd., Luoyang, Henan 471039, China

[*]Corresponding author and e-mail: Andong Lou, citichma_louad@163.com

Abstract: Based on a new type of non-reactive polymer powder material for thin spray-on support technology in coal mine roadway, an automatic spray equipment for coal mine roadway thin spray-on support is designed and developed. It has Kinds of movements and motion trajectories. This paper introduce The technological basis and working principle of the equipment, design basis, and overall system composition of the equipment, which including: mechanical system, hydraulic system, and control system. Compared with the concrete spraying technology, the thin spray-on technology can reduce the material transportation volume by more than 90%, significantly reduce the labor intensity of workers, improve the construction efficiency and the construction environment. It is an ideal choice for closure and supporting for the future underground roadways.

1. Introduction

The development of clean, safe and fast support technology has become an important part of efficient and safe production in coal mines. With the development of thin spray material and thin spray support technology, more and more coal mines accept it. In the construction of large-scale mine roadways and large cross-sections tunnels, there are big problems in the existing construction technology and equipment. With the requirement of large amount of construction and high efficiency of construction, the length of the roadway for thin spray construction in some coal mines is up to km, part of the height is large, making it difficult for workers to implement, and the quality of spraying cannot be guaranteed. Ordinary thin spray construction technology and equipment have been unable to meet the construction requirements. The existing equipment is mechanized and has a low automation level, which is difficult to meet the requirements of high-speed tunneling. It requires thin spray equipment with a higher level of technology and automation.

Based on the thin-spray material roadway thin spray support technology, the research and development of automatic spray equipment for thin spray support in the roadway is developed. The corresponding products and results can also be applied to the support construction sites such as tunnels, which has great market and social significance.

2. The Process of Thin Spray Support Technology and Working Principle of Equipment

In order to ensure the cleanness of the spray support environment, the process is based on wet spraying. The thin spray automatic equipment is developed on the premise of basically realizing dust-free spray support.

Based on the research of the basic properties of thin spray materials, the optimal design of thin spray parameter matching was carried out. The spray arm adopts a the swinging spray structure of the extension to improve the flexibility of spray construction; The supporting automatic control technology mainly includes automatic extension and contraction of the spay arm, distance detection and identification, pressure and flow control of spraying materials, liquid and gas, so as to improve the spray automation, Control of the positioning accuracy and uniformity of the spray layer. On the premise of meeting the functional requirements, the structure of the spray arm is optimized, and the overall lightweight design is performed to improve the flexibility of the spray arm.

Referring to the working principle, control scheme and related software of industrial spraying robot , there are mainly technologies based on dynamics control, trajectory planning, vibration suppression, electromechanical coupling and other technologies for the spray equipment, especially its related control algorithm and theory, software design to control the spraying trajectory and path of spray gun scientifically and reasonably and a comprehensive and scientific optimization is carried out from both soft and hard aspects to control the spraying pressure, flow, time, spraying distance and other related technologies In order to reduce the waste of spraying materials and improve the uniformity of spray thickness and film-forming quality.

3. Design Basis

According to the analysis of the thin material spray process at the construction site of the roadway, the overall mechanical structure of the thin spray automatic equipment is determined, obtaining the size of the equipment and basic movement form of the spray arm for construction, and realizing the boom pitching (range 90°) Swing (range 240°), arm extension and spray gun attitude adjustment. The spray gun is equipped with two measurement sensors, which can sense the relative distance between the spray gun head and the wall of the roadway when the equipment is working, so as to maintain the stability of the spray distance and ensure the spray continuity and uniformity.

In order to adapt to the complex roadway conditions in coal mine, the walking mechanism of the thin spray equipment adopts a small crawler chassis to ensure good passability and flexibility of the equipment; the chassis is equipped with sensors to collect data and information of the surrounding environment of the walking mechanism, which has the function of automatically sensing and keeping straight-line walking when the direction of advance is unobstructed; Through the optimization of software and hardware system, it can ensure that the automatic construction operation of the spray arm is compatible with the control strategy, and scientifically control the movement trajectory of the spray gun and the spraying pressure and flow rate , etc., so as to ensure the film-forming quality and construction efficiency of thin spraying materials. The specific automatic spraying operation scenario is shown in Figure 1.

4. The Operation Process of the Thin Spray Automatic Equipment

1. The equipment enters the working area of the roadway through manual remote control; the equipment should be arranged in the middle of the roadway as far as possible, parallel to the roadway, and followed by the equipment of the thin material mixing system;

2. The power device is connected to take power for power supply, the thin material mixing and pumping system is connected to the water supply (air supply) source; each device is started, the spray arm is rotated and extensioned at the same time, and the relative distance between the nozzle and the wall of the roadway through the sensor to prepare for operation;

3. The equipment of the thin material mixing and pumping system is started, filling the thin material, adding water according to the process, after the mixing is completed, pump the mix material to the spray pipe to the nozzle, and the thin spray equipment begins to swing the arm to perform the spraying operation;

4. According to the requirement of spray-flim thickness, the spray arm sets the swinging speed or the spray period and times. The spray arm automatically adjusts the extension amount according to the distance between the nozzle and the roadway wall, and carries out spraying in the arch roadway;

5. After the spray of one section of the arched roadway is completed, the spray is stopped and the spray arm is shortened; the equipment automatically moves straight to the next operation position and

continues the next sprayoperation process;

6. After the spray operation of the entire roadway is completed, the spray arm is contracted, rotated, and folded to the initial position for easy transportation and parking.

The working process is shown in Figure 1.

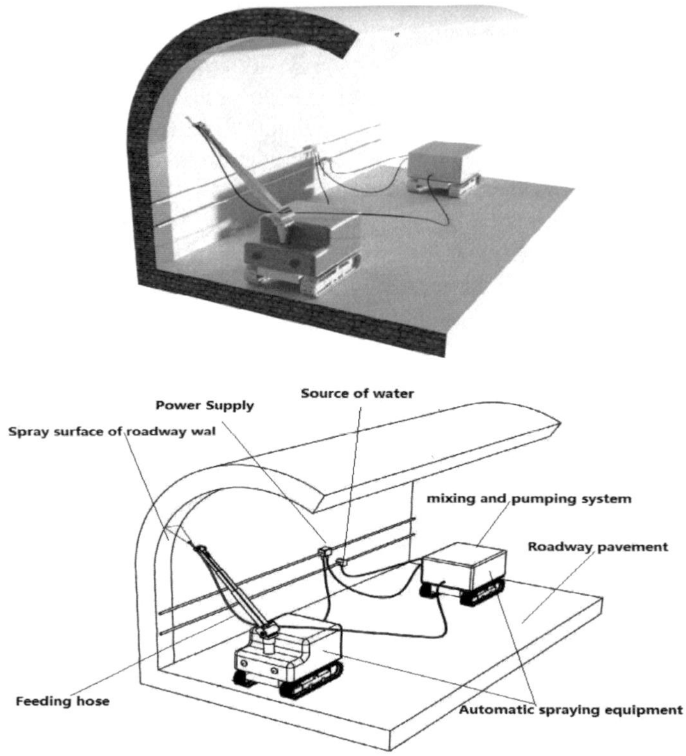

Figure 1. The automatic spraying operation scenario and working process

5. The Composition of Equipment System

The thin spray automatic equipment is composed of multiple systems including mechanical structure system (chassis+arm), spraying system, power system, hydraulic system, sensing and electro-hydraulic proportional control system. Each system is based on the coal safety explosion-proof design principle and constitutes a block diagram as shown in Figure 2.

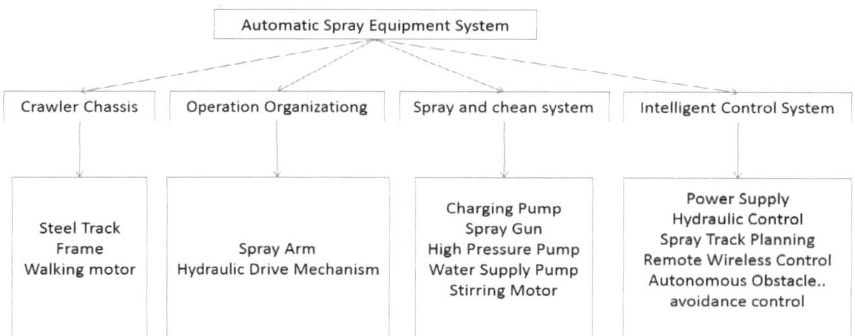

Figure 2. Constitutes a block diagram

5.1. Mechanical System and Working Space Range

The main body of the machine consists of the arm assembly, the lower body assembly, the upper body assembly, and the hydraulic drive mechanism. The upper body assembly is equipped with power system (electric motor), electrical control system, wireless communication system, hydraulic system and lighting system.the mechanical structure and working space range shown in Figure 3 .

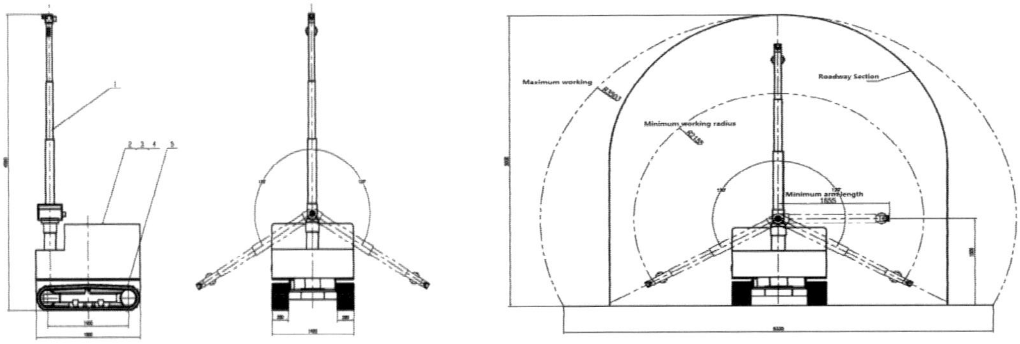

(a) Mechanical system (b) working space range

Figure 3. Mechanical system and working space range

In order to improve the road passability and walking stability of the equipment under various complex road conditions, the walking mechanism adopts the crawler structure, which is driven by two hydraulic rotary motors with brakes. When the speeds of the two tracks are the same, the equipment can move forward or backward; when the speeds of the two tracks are different, the equipment achieves steering movement.

5.2. *Power System*

The Power system is the power source of equipment to walk and work. Considering the environment and technical requirements of roadway operation site, the scheme adopts MA certified K3 explosion-proof motor drive.

The power source is taken from the underground power grid. The available voltage and frequency are 1140 / 660V, 50Hz. The motor starts directly, and the starting switch is equipped with a phase sequence protection device, so that the motor has the correct direction of rotation and is not affected by the phase sequence at the power output end of the power supply. The power system is mainly composed of motor installation, cooling system, protection system and partition assembly.

5.3. Hydraulic System

The hydraulic system is the driving source for each actuator, such as the rotation and extension of the spray arm, chassis travel. It is mainly composed of load sensitive variable piston pump, load-sensitive electro-hydraulic proportional directional valve, double -acting hydraulic cylinder, hydraulic rotary motor, walking motor, center rotary joint, safety valve, balance valve, solenoid valve and other hydraulic components. The whole hydraulic system can be divided into two parts: the main hydraulic circuit and the load feedback hydraulic circuit. The main pump equipped with the hydraulic system adopts the load sensitive variable pump, the control technology combining the load sensing and the independent flow distribution system independent of the load pressure is used. The high pressure load sensing proportional multi-way valve and LUDV proportional multi-way valve are selected as The control core components of the system make the output flow of the pump always meet the needs of the load, and have the advantages of good start-up performance and operating performance, energy saving and small system heating.

IOP Conf. Series: Materials Science and Engineering **746** (2020) 012050 doi:10.1088/1757-899X/746/1/012050

5.4. Electrical Control System

The electrical control system is mainly composed of motor start stop + phase sequence protection circuit, remote control system, controller, rectifier power supply, alarm circuit, lighting circuit and explosion-proof electrical cabinet. The composition diagram of the control system is shown in Figure 4.

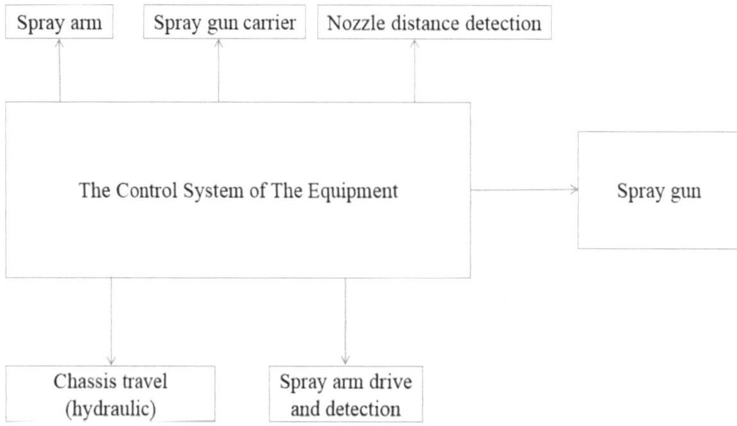

Figure 4. The composition diagram of the control system

All cable outlets of explosion-proof electrical cabinets use cable waterproof explosion-proof connectors and explosion-proof heavy-duty connectors. The entire explosion-proof control cabinet is fully enclosed, which can be effectively explosion-proof, waterproof, and dust-proof. Explosion-proof remote controllers and controllers are reliable components, with a wide ambient operating temperature and an IP65 protection level, fully adaptable to harsh working conditions. The electrical components in explosion-proof electrical cabinets are mainly imported brands with stable performance. The control loop power supply uses DC24V, which can effectively prevent electric shock accidents. The equipment is equipped with high-power LED work lights to ensure good illumination at the work site.

The hydraulic system is controlled by the electro-hydraulic proportional system based on the pulse width modulation. The basic components of the electro-hydraulic proportional control system include the control handle, controller, pulse width modulation (PWM) drive circuit, electro-hydraulic proportional valve and actuator. The control signal is generated by the remote control handle to control the output pressure of the electro-hydraulic proportional pilot valve, so as to control the position of the valve core of the main reversing valve, realize the control of the working speed of the working hydraulic cylinder.

The electro-hydraulic proportional valve is controlled by PWM, and the average current flowing through the proportional solenoid coil is adjusted by changing the duty ratio of PWM signal, so as to reduce friction, hysteresis and dead zone of the solenoid, and improve the response speed of the electro-hydraulic proportional valve.

In order to ensure that the control system of the equipment has good response characteristics and module expandability, the communication system of the equipment adopts the CAN bus communication mode, and the entire communication system integrating the equipment is established by using infrared remote control technology.

6. Conclusion

Carrying out research on thin-spray automatic spraying technology and equipment can solve the need for large workloads, closed construction demand of large roadway, and improve construction efficiency; improve the working environment of spray support, protect the health of workers; and improve the quality of spray-spray support , Reduce support costs; give full play to the performance and advantages of thin spray material, greatly reduce the labor intensity of workers and human factors on the spray quality to achieve standardization and consistency of spray. Automated thin spray

IOP Conf. Series: Materials Science and Engineering **746** (2020) 012050 doi:10.1088/1757-899X/746/1/012050

technology and equipment can further promote the application of thin spray materials in coal mines, and at the same time promote the rapid development of coal mines and the advancement of efficient and safe green mining technology.

7. Acknowledgement

Support from Special Key Projects of Science and Technology Innovation and Venture Capital of China Coal Technology & Engineering Group Corp (Grant No. 2018-2-ZD002) is acknowledged.

8. References

[1] L. Lacerda, M. Rispin. Current ground support membrane applications in north American underground mines [C], 2002 SME Annual Meeting, Feb. 25-27, Phonenix,Arizona.
[2] Lacerda, L. "GSM internal test report," Bullfrogmine, Lacminerals, beatty, NV, 1994.
[3] Thibodeau, D. "Masterseal Trial"Barrick gold strike underground division report, elko, NV, 2000.
[4] Atkar P N, Greenfield A, Conner D C, et al. Uniform coverage of automotive surface patches [J]. International Journal of Robotics Research, 2005, 24 (11): 883 ~ 898.
[5] Li Yunjiang et al. Mechanical structure design and research of shotcreting robot. Mining machinery, 1999 (8).
[6] Li shican, Xun Qiqiang, Ma Chunhong. Study on field application and improvement of push chain concrete shotcreting machine [J]. Shandong coal technology, 2015 (9): 95-96.

AUTHOR INDEX